W9-CNK-141

VITAMINS IN ANIMAL NUTRITION

ANIMAL FEEDING AND NUTRITION

A Series of Monographs and Treatises

Tony J. Cunha, Editor

Distinguished Service Professor Emeritus
University of Florida
Gainesville, Florida

and

Dean Emeritus, School of Agriculture
California State Polytechnic University
Pomona, California

Tony J. Cunha, SWINE FEEDING AND NUTRITION, 1977

W. J. Miller, DAIRY CATTLE FEEDING AND NUTRITION, 1979

Tilden Wayne Perry, BEEF CATTLE FEEDING AND NUTRITION, 1980

Tony J. Cunha, HORSE FEEDING AND NUTRITION, 1980

Charles T. Robbins, WILDLIFE FEEDING AND NUTRITION, 1983

Tilden Wayne Perry, ANIMAL LIFE-CYCLE FEEDING AND NUTRITION, 1984

Lee Russell McDowell, NUTRITION OF GRAZING RUMINANTS IN WARM CLIMATES, 1985

Ray L. Shirley, NITROGEN AND ENERGY NUTRITION OF RUMINANTS, 1986

Peter R. Cheeke, RABBIT FEEDING AND NUTRITION, 1987

Lee Russell McDowell, VITAMINS IN ANIMAL NUTRITION, 1989

D. J. Minson, FORAGE AND RUMINANT NUTRITION, 1989

Tony J. Cunha, HORSE FEEDING AND NUTRITION, SECOND EDITION, 1991

VITAMINS IN ANIMAL NUTRITION

Comparative Aspects to Human Nutrition

Lee Russell McDowell
Animal Science Department
University of Florida
Gainesville, Florida

Academic Press
San Diego New York Boston
London Sydney Tokyo Toronto

COPYRIGHT © 1989 BY ACADEMIC PRESS, INC.
ALL RIGHTS RESERVED.
NO PART OF THIS PUBLICATION MAY BE REPRODUCED OR
TRANSMITTED IN ANY FORM OR BY ANY MEANS, ELECTRONIC
OR MECHANICAL, INCLUDING PHOTOCOPY, RECORDING, OR
ANY INFORMATION STORAGE AND RETRIEVAL SYSTEM, WITHOUT
PERMISSION IN WRITING FROM THE PUBLISHER.

Academic Press, Inc.
A Division of Harcourt Brace & Company
525 B Street, Suite 1900
San Diego, California 92101-4495

United Kingdom Edition published by
ACADEMIC PRESS LIMITED
24-28 Oval Road, London NW1 7DX

Library of Congress Cataloging-in-Publication Data

McDowell, L.R., Date
Vitamins in animal nutrition.
(Animal feeding and nutrition)
Includes index.
1. Vitamins in animal nutrition. 2. Vitamins in
human nutrition. I. Title. II. Series.
SF98.V5M32 1989 636.08′52 88-16722
ISBN 0-12-483372-1 (alk. paper)

PRINTED IN THE UNITED STATES OF AMERICA
96 97 EB 9 8 7 6 5 4 3

This book is dedicated with appreciation
to my parents,
to my wife, Lorraine,
and to my daughters,
Suzannah, Jody, and Teresa

Contents

Foreword

This is the tenth in a series of books about animal feeding and nutrition. The books in this series are designed to keep the reader abreast of the rapid developments that have occurred in this field in recent years. As the volume of scientific literature expands, interpretation becomes more complex, and a continuing need exists for summation and for up-to-date books.

This book on vitamins is written by Dr. Lee R. McDowell, a distinguished scientist in animal nutrition, who was also editor of *Nutrition of Grazing Ruminants in Warm Climates* (Academic Press, 1985). For the past 23 years, he has been working on animal nutrition studies at the University of Florida and with numerous collaborating animal scientists in Latin America, Africa, Asia, and other areas. He has lectured throughout the world and has collected research data, photographs, and other materials of value for this book. His expertise and knowledge have been greatly enriched by contact with many of the world's leading nutrition scientists.

For many years, there has been a great need for a single book on vitamins covering both animals and humans. Dr. McDowell has done a magnificent job of reviewing the vitamin literature and condensing it into one textbook. He covers the basic chemical, metabolic, and functional role of vitamins. Moreover, he devotes proper consideration to vitamin supplementation. The book is well illustrated with a wealth of valuable photographs depicting vitamin deficiencies in livestock, laboratory animals, and humans.

In addition to worldwide use as a textbook, this book should also be of considerable value to research and extension specialists, teachers, students, feed manufacturers, farmers, and others dealing with the livestock industry. Animal industry personnel, veterinarians, and those interested in comparative nutrition will find this book very useful. It should be valuable for anyone concerned with vitamin nutrition.

New feed and food crops, improved methods of production and processing, increased productivity of animals and crops, changes in animal products including more lean and less fat in meat and less fat in milk, longer shelf life requirements of food products, and a myriad of new technological developments have resulted in a need to reevaluate vitamin supplementation. This book is timely and valuable in bringing the vitamin field up-to-date and in discussing supplementation needs.

Tony J. Cunha

Preface

Vitamins in Animal Nutrition contains 18 chapters of concise, up-to-date information on vitamin nutrition for both animals and humans. The first chapter deals with the definition of vitamins, general considerations, and the fascinating history of these nutrients. Chapters 2 through 15 discuss the 14 established vitamins in relation to history; chemical structure, properties, and antagonists; analytical procedures; metabolism; functions; requirements; sources; deficiency; supplementation; and toxicity. Chapter 16 deals with other vitamin-like substances, and Chapter 17 reviews the importance of essential fatty acids. The final chapter discusses vitamin supplementation considerations.

It is hoped that this book will be of worldwide use as a textbook and as an authoritative reference book for use by research and extension specialists, feed manufacturers, teachers, students, and others. An attempt has been made to provide a balance between animal nutrition and clinical human nutrition. Likewise, a comparison between the balance of chemical, metabolic, and functional aspects of vitamins and their practical and applied considerations has been made.

A unique feature is the description of the practical implications of vitamin deficiencies and excesses and the conditions that might occur with various animal species and humans. A large number of photographs illustrate vitamin deficiencies in farm livestock, laboratory animals, and humans. Unlike other textbooks, this one places strong emphasis on vitamin supplementation in each chapter and devotes the last chapter to this subject.

In preparing this book, I have obtained numerous suggestions from eminent scientists both in the United States and in other countries of the world. I wish to express my sincere appreciation to them and to those who supplied photographs and other material used. I am especially grateful to the following: C. B. Ammerman, L. B. Bailey, R. B. Becker, D. K. Beede, B. J. Bock, H. L. Chapman, J. H. Conrad, G. L. Ellis, R. H. Harms, J. F. Hentges, J. K. Loosli, R. M. Mason, R. Miles, R. L. Shirley, R. R. Streiff, and W. B. Weaver (Florida); R. T. Lovell and H. E. Sauberlich (Alabama); O. Balbuena, B. J. Carrillo, and B. Ruksan (Argentina); H. Heitman (California); J. M. Bell, M. Hidiroglou, and N. Hidiroglou (Canada); N. Ruiz (Colombia); N. Comben (England); M. Sandholm (Finland); L. S. Jensen (Georgia); T. B. Keith (deceased) (Idaho); A. H. Jensen (Illinois); V. Ramadas Murthy (India); A. Prabowo (Indonesia); V. Catron (deceased), and V. C. Speer (Iowa); G. L. Cromwell (Kentucky); G. F. Combs (Maryland); F. J. Stare (Massachusetts); R. W. Luecke, E. R. Miller, R. C. Piper, J. W. Thomas, and D. E. Ullrey (Michigan); R. T. Holman and T. W. Sullivan (Minnesota); V. Herbert, L. E. Kook, M. C. Latham, and M. L. Scott (New York); A. Helgebostad and H. Rimeslatten (Norway); D. E. Becker (Ohio); P. R. Cheeke, D. C. Church, O. H. Muth, and J. E. Oldfield (Oregon); D. S. McLaren (Scotland); J. R. Couch and T. M. Ferguson (deceased) (Texas); D. C. Dobson (Utah); J. P. Fontenot, M. D. Lindemann, and L. M. Potter (Virginia); J. R. Carlson, J. A. Froseth, and L. L. Madsen (deceased) (Washington); and M. L. Sunde (Wisconsin). Appreciation is expressed to company representatives, including G. Patterson (Chas. Pfizer Co); J. C. Bauernfeind, T. M. Fry, E. L. MacDonald, L. A. Peterson, and S. N. Williams (Hoffmann–LaRoche, Inc.); C. H. McGinnis (Rhône–Poulenc, Inc.); and A. T. Forrester (The Upjohn Co.). Special thanks go to J. P. Fontenot for the preliminary planning of this book, and to P. R. Cheeke and J. E. Oldfield for editing and providing useful suggestions for the entire publication.

I am particularly grateful to Nancy Wilkinson and Lorraine M. McDowell for their useful suggestions and assistance in the editing of the entire book. Likewise, I wish to acknowledge with thanks and appreciation the skill and care of Patricia French for overseeing the typing and proofreading of chapters, and also thank Vanessa Carbia for her valuable assistance. Also, I am indebted to the Animal Science Department of the University of Florida for providing the opportunity and support for this undertaking. Finally, I thank Tony J. Cunha for encouraging me to undertake the responsibility of writing this book, and for the experience of his expertise on this subject.

Lee Russell McDowell

1

Introduction and Historical Considerations

I. DEFINITION OF VITAMINS

Vitamins are defined as a group of complex organic compounds present in minute amounts in natural foodstuffs that are essential to normal metabolism and lack of which in the diet causes deficiency diseases. Vitamins consist of a mixed group of chemical compounds and are not related to each other as are proteins, carbohydrates, and fats. Their classification together depends not on chemical characteristics but on function. Vitamins are differentiated from the trace elements, also present in the diet in small quantities, by their organic nature.

Vitamins are required in trace amounts (micrograms to milligrams per day) in the diet for health, growth, and reproduction. Omission of a single vitamin from the diet of a species that requires it will produce deficiency signs and symptoms. Many of the vitamins function as coenzymes (metabolic catalysts); others have no such role, but perform certain essential functions.

Some vitamins deviate from the preceding definition in that they do not always need to be constituents of food. Certain substances that are considered to be vitamins are synthesized by intestinal tract bacteria in quantities that are often adequate for body needs. However, a clear distinction is made between vitamins and substances that are synthesized in tissues of the body. Ascorbic acid, for example, can be synthesized by most species of animals, except when they are young or under stress conditions. Likewise, niacin can be synthesized from the amino acid tryptophan and vitamin D from action of ultraviolet light on precursor compounds in the skin. Thus, under certain conditions and for specific species, vitamin C, niacin, and vitamin D would not always fit the classic definition of a vitamin.

II. CLASSIFICATION OF VITAMINS

Classically, vitamins have been divided into two groups based on their solubilities in fat solvents or in water. Thus, fat-soluble vitamins include A, D, E, and K, while vitamins of the B complex, C, and others are classified as water soluble. Fat-soluble vitamins are found in feedstuffs in association with lipids.

The fat-soluble vitamins are absorbed along with dietary fats, apparently by mechanisms similar to those involved in fat absorption. Conditions favorable to fat absorption, such as adequate bile flow and good micelle formation, also favor absorption of the fat-soluble vitamins (Scott *et al.*, 1982). Conversely, their absorption is impaired when conditions are unfavorable for normal fat absorption. Water-soluble vitamins are not associated with fats and alterations in fat absorption do not affect their absorption. Fat-soluble vitamins are stored in appreciable amounts in the animal body. Except for vitamin B_{12}, water-soluble vitamins are not stored and excesses are rapidly excreted. A continual dietary supply of the water-soluble vitamins is needed to avoid deficiencies. Fat-soluble vitamins are primarily excreted in the feces via the bile, whereas water-soluble vitamins are mainly excreted in urine. Water-soluble vitamins are relatively nontoxic but excesses of fat-soluble vitamins A and D can cause serious problems. Fat-soluble vitamins consist only of carbon, hydrogen, and oxygen, whereas some of the water-soluble vitamins also contain nitrogen, sulfur, or cobalt.

Table 1.1 lists 14 vitamins classified as either fat or water soluble. The number of compounds justifiably classified as vitamins is controversial. The term vitamin has been applied to many substances that do not meet the definition or criteria for vitamin status. Of the 14 vitamins listed, choline is only tentatively classified

TABLE 1.1

Fat- and Water-Soluble Vitamins with Synonym Names

Vitamin	Synonym
Fat soluble	
Vitamin A_1	Retinol
Vitamin A_2	Dehydroretinol
Vitamin D_2	Ergocalciferol
Vitamin D_3	Cholecalciferol
Vitamin E	Tocopherol
Vitamin K_1	Phylloquinone
Vitamin K_2	Menaquinone
Vitamin K_3	Menadione
Water soluble	
Thiamin	Vitamin B_1
Riboflavin	Vitamin B_2
Niacin	Vitamin pp, vitamin B_3
Vitamin B_6	Pyridoxol, pyridoxal, pyridoxamine
Pantothenic acid	Vitamin B_5
Biotin	Vitamin H
Folacin	Vitamin M, vitamin B_c
Vitamin B_{12}	Cobalamin
Choline	Gossypine
Vitamin C	Ascorbic acid

as one of the B-complex vitamins. Unlike other B vitamins, choline can be synthesized in the body, is required in larger amounts, and apparently functions as a structural constituent rather than as a coenzyme. *Myo*-inositol and carnitine are not listed in Table 1.1 even though they could fit the vitamin category but apparently for only several species. Chapters 2–15 in this book concern the 14 vitamins listed in Table 1.1, while Chapter 16 concerns vitamin-like substances and Chapter 17 considers essential fatty acids. The essential fatty acids are not vitamins, but a deficiency disease does result that is similar to vitamin deficiency. The final chapter deals with vitamin supplementation considerations.

III. VITAMIN NOMENCLATURE

When the vitamins were originally discovered, they were isolated from certain foods. During these early years, the chemical composition of the essential factors was unknown, therefore, these factors were assigned letters of the alphabet. The system of alphabetizing became complicated when it was discovered that activity attributed to a single vitamin was instead the result of several of the essential factors. In this way, the designation of groups of vitamins appeared (e.g., the vitamin ''B'' group). Additional chemical studies showed that variations in chemical structure occurred within compounds having the same vitamin activity but in different species. To overcome this, a system of suffixes was adopted (e.g., vitamin D_2 and D_3). The original letter system of designation thus became excessively complicated.

With the determination of the chemical structure of the individual vitamins, letter designations were sometimes replaced with chemical-structure names (e.g., thiamin, riboflavin, and niacin). Vitamins have also been identified by describing a function or its source. The term vitamin H (biotin) was used because the factor protected the ''Haut,'' the German word for ''skin.'' Likewise, vitamin K was derived from the Danish word ''koagulation'' (coagulation). The vitamin pantothenic acid refers to its source, as it is derived from the Greek work ''pantos,'' meaning ''found everywhere.''

The Committee on Nomenclature of the American Institute of Nutrition (CNAIN, 1981) has provided definite rules for the nomenclature of the vitamins. This nomenclature is used in this book. The official and major synonym names of vitamins are given in Table 1.1 and also in the respective vitamin chapters.

IV. VITAMIN REQUIREMENTS

Vitamin requirements for animals and humans are listed in the Appendix tables at the end of this book and in the appropriate chapter. While metabolic needs are similar, dietary needs for vitamins differ widely among species. Some vitamins are metabolic essentials, but not dietary essentials, for certain species, because they can be synthesized readily from other food or metabolic constituents.

Poultry, swine, and other monogastric animals are dependent on their diet for vitamins to a much greater degree than are ruminants. Tradition has it that ruminants in which the rumen is fully functioning cannot suffer from a deficiency of B vitamins. It is generally assumed that ruminants can always satisfy their needs from the B vitamins naturally present in their feed, plus that synthesized by symbiotic microorganisms. However, under specific conditions relating to stress and high productivity, ruminants have more recently been shown to have requirements, particularly for the B vitamins thiamin (see Chapter 6) and niacin (see Chapter 8).

The rumen does not become functional with respect to vitamin synthesis for some time after birth. For the first few days of life the young ruminant resembles a nonruminant in requiring dietary sources of the B vitamins. Beginning as early as 8 days and certainly by 2 months of age, ruminal flora have developed to the point of contributing significant amounts of the B vitamins (Smith, 1970). Production of these vitamins at the proximal end of the gastrointestinal tract is indeed fortunate for they become available to the host as they pass down the tract through areas of efficient digestion and absorption.

In monogastric animals, including humans, intestinal synthesis of many B vitamins is considerable (Mickelsen, 1956) though not as extensive or as efficiently utilized as in ruminants. Low efficiency of utilization is probably related to several factors. Intestinal synthesis in nonruminants occurs in the lower intestinal tract, an area of poor absorption. The horse, with a large production of B vitamins in the large intestine, is apparently able to meet most of its requirements for these vitamins in spite of the poor absorption from this area. Intestinally produced vitamins are more available to those animals (rabbit, rat, and others) that habitually practice coprophagy and thus recycle products of the lower gut. This behavior yields significant amounts of B vitamins to the host animal.

V. VITAMIN OCCURRENCE

Vitamins originate primarily in plant tissues and are present in animal tissue only as a consequence of consumption of plants, or because the animal harbors microorganisms that synthesize them. Vitamin B_{12} is unique in that it occurs in plant tissues as a result of microbial synthesis. Two of the four fat-soluble vitamins, vitamins A and D, differ from the water-soluble B vitamins in that they occur in plant tissue in the form of a provitamin (a precursor of the vitamin), which can be converted into a vitamin in the animal body. No provitamins are known for any water-soluble vitamin. Tryptophan, which can be converted into niacin, is usually known as an amino acid. In addition, fat- and water-soluble vitamins differ in that water-soluble B vitamins are universally distributed in all living tissues, whereas fat-soluble vitamins are completely absent from some tissues.

VI. HISTORY OF THE VITAMINS

The history of the discovery of the vitamins is an inspirational and exciting reflection of the ingenuity, dedication, and self-sacrifice of many individuals. Excellent reviews of vitamin history with appropriate references include Wagner and Folkers (1962), Marks (1975), Maynard *et al.* (1979), Scott *et al.* (1982), Widdowson (1986), and Loosli (1988). A brief sketch of important events emphasizing early history of vitamins is outlined in Table 1.2.

The existence of nutritive factors, such as vitamins, was not recognized until about the start of the twentieth century. The word "vitamin" had not been coined yet. However, what were to be later known as vitamin deficiency diseases, such as scurvy, beriberi, night blindness, xerophthalmia, and pellagra, had plagued the world at least since the existence of written records. Records of medical science from antiquity attesting to human association of certain foods with either the cause or prevention of disease and infirmity are considered the nebulous beginnings of the concept of essential nutrients (Wagner and Folkers, 1962). Even so, at the beginning of the twentieth century the value of food in human nutrition was expressed solely in terms of its ability to provide energy and basic building units necessary for life.

In the late 1800s and early 1900s some scientists believed that life could be supported with chemically defined diets. In 1860 Louis Pasteur reported that yeast could grow on a medium of sugar, ammonium salts, and ash of yeast. Justus von Liebig observed that certain yeasts were unable to grow at all under these conditions, while others grew only very slowly. The ensuing arguments between Liebig and Pasteur did not solve the question. Pasteur's (1822–1895) research showing that bacteria caused disease led scientists trained in medicine to be reluctant to believe the "vitamin theory" that certain diseases resulted from a shortage of specific nutrients in foods (Loosli, 1988).

The first phase leading to the "vitamin hypothesis" began with gradual recognition that cause of diseases such as night blindness, scurvy, beriberi, and rickets could be related to diet. Although the true cause, nutritional deficiency, was not suspected, these results marked the first uncertainty in the germ and infection theories of origin for these diseases. Finally, in the early 1900s many scientists in the field of nutrition almost simultaneously began to realize that a diet could not be adequately defined in terms of carbohydrate, fat, protein, and salts. At that time it became evident that other organic compounds had to be present in the diet if health was to be maintained.

TABLE 1.2

Brief History of Vitamins (Ancient History—1951)

2600 B.C.	Beriberi was recognized in China and is probably the earliest documented deficiency disorder.
1500 B.C.	Scurvy, night blindness, and xerophthalmia were described in ancient Egypt. Liver consumption was found to be curative for night blindness and xerophthalmia.
A.D. 130–200	Soranus Ephesius provided classical descriptions of rickets.
1492–1600	World exploration threatened by scurvy: Magellan lost four-fifths of his crew Vasco da Gama lost 100 of his 160 men
1747	Lind performed controlled shipboard experiments on the preventive effect of oranges and lemons on scurvy. Also developed method of preserving citrus juice by evaporation and conserving in acid form.
1768–1771	Captain Cook demonstrated that prolonged sea voyages were possible without ravages of scurvy.
1816	Magendie described xerophthalmia in dogs fed carbohydrate and olive oil.
1824	Combe described a fatal anemia, pernicious anemia, and suggested that it could be related to a disorder of the digestive tract.
1849	Choline was isolated by Streker from the bile of pigs.
1881	Lunin reported that animals did not survive on diets composed solely of purified fat, protein, carbohydrate, salts, and water.
1880s	Japanese physician Takaki prevented beriberi in Japanese Navy by substituting other foods for polished rice.
1897	Eijkman showed that beriberi (thiamin deficiency) from polished rice consumption could be cured by adding rice polishings back into the diet.
1906	Hopkins suggested substances in natural foods, termed "accessory food factors," were indispensable and did not fall into the categories of carbohydrate, fat, protein, and mineral.
1907	Holst and Frolich produced experimental scurvy in guinea pigs by feeding a deficient diet, with pathological changes resembling those in humans.
1909	Hopkins reported a rat growth factor in some fats.
1912	The term "vitamine" was first used by the Polish biochemist Funk to describe an accessory food factor.
1913	McCollum and Davis discovered fat-soluble A in butter that was associated with growth.
1919	Steenbock reported that the yellow color (carotene) of vegetables was vitamin A.
1919	Mellanby produced rickets in dogs, which responded to a fat-soluble vitamin in cod liver oil.
1920	Goldberger reported that pellagra was not caused by bacterial infection, but rather was an ill-balanced diet high in corn.
1922	McCollum established vitamin D as independent of vitamin A by preventing rickets after destroying vitamin A activity when bubbling oxygen through cod liver oil.
1923	Evans and Bishop discovered vitamin E. The deficiency caused female rats to abort, while male rats became sterile.
1926	Jansen and Donath isolated thiamin in crystalline form from rice bran.
1926	Minot and Murphy showed that large amounts of raw liver given by mouth daily would alleviate pernicious anemia.

(*continued*)

TABLE 1.2 (*Continued*)

1926	Steenbock showed that irradiation of foods as well as animals produced vitamin D_2.
1926	Goldberger and Lillie described a rat syndrome, later shown to be riboflavin deficiency.
1928	Bechtel and co-workers established that rumen bacteria of cattle synthesized B vitamins.
1928	Szent-Györgyi isolated hexuronic acid (ascorbic acid, vitamin C) from orange juice, cabbage juice, and cattle adrenal glands.
1929	Moore proved that the animal body converts carotene to vitamin A.
1929	Norris and co-workers reported a curled toe paralysis (riboflavin deficiency) in chicks.
1929	Castle showed that pernicious anemia resulted from the interaction of a dietary (extrinsic) factor and an intrinsic factor produced by the stomach.
1930	Norris and Ringrose described a pellagra-like dermatitis in the chick, later established as a pantothenic acid deficiency.
1931	Pappenheimer and Goettsch showed that vitamin E is required for prevention of encephalomalacia of chicks and nutritional muscular dystrophy in rabbits and guinea pigs.
1931	Willis demonstrated that a factor from yeast was active in treating a tropical macrocytic anemia seen in women of India.
1932	Choline was discovered to be the active component of pure lecithin previously shown to prevent fatty livers in rats.
1933	R. J. Williams and associates fractionated a growth factor from yeast and named it pantothenic acid.
1934	Szent-Györgyi named a factor that would cure dermatitis in young rats, vitamin B_6.
1934	Dam and Schönheyder described a nutritional disease of chickens characterized by bleeding, thus a new fat-soluble vitamin was discovered.
1935	Wald demonstrated the relation of vitamin A to night blindness and vision.
1935	Kuhn in Germany and Karrer in Switzerland synthesized riboflavin.
1935	Warburg and co-workers first demonstrated a biochemical function for nicotinic acid when they isolated it from an enzyme (NADP).
1935–1937	Cobalt, the central ion in vitamin B_{12}, was shown to be a dietary essential for cattle and sheep by Underwood and co-workers in Australia and in Florida by Becker and associates.
1936	Biotin was the name given to a substance isolated from egg yolk by Kogl and Tonnis that was necessary for the growth of yeast.
1936	R. R. Williams and colleagues determined the structure of thiamin and synthesized the vitamin.
1937	Elvehjem and associates found that nicotinic acid cured black tongue in dogs. It was quickly shown to be effective for pellagra in humans.
1939	Vitamin K was isolated by Dam and Karrer of Europe and a few months later in the U.S. from three different laboratories.
1940	Harris and associates completed the first synthesis of biotin.
1942	Baxter and Robeson crystallized vitamin A.
1943–1946	Chemists from the Lederle group crystallized and later synthesized folacin.
1948	Rickes and co-workers in the U.S. and Smith in England isolated vitamin B_{12}.
1951	Smith and co-workers showed that cobalt deficiency in sheep could be prevented by vitamin B_{12} injection.

Beriberi was probably the earliest documented deficiency disorder, being recognized in China as early as 2600 B.C. Scurvy, night blindness, and xerophthalmia were described in the ancient Egyptian literature around 1500 B.C. Substances rich in vitamin A as remedies for night blindness were used very early by the Chinese, and livers were recommended as curative agents for night blindness and xerophthalmia by Hippocrates around 400 B.C. In 1536 Canadian Indians cured Jacques Cartier's men of scurvy with a broth of evergreen needles. In 1747 James Lind, a British naval surgeon, showed that the juice of citrus fruits was a cure for scurvy, but its routine use was not started in the British Navy until 1795. Cod liver oil was used as a specific treatment for rickets long before anything was known about the cause of this disease, and was fed to farm animals as early as 1824. In the 1880s, the Japanese physician Takaki recognized the cause of beriberi in the Japanese Navy as stemming from an unbalanced white rice diet, and virtually eliminated this condition by increasing the consumption of vegetables, fish, and meat and by substituting barley for rice.

The period before the close of the nineteenth century was characterized by the discovery of diseases of nutritional origin in animals, which opened the way for controlled experimental studies of nutritional causes and cures for such diseases that were common to both humans and the lower animals. The rat undoubtedly contributed most to the discovery of vitamins from 1900 through the 1920s, although chickens, pigeons, guinea pigs, mice, and dogs also played their part (Widdowson, 1986). In 1890 Christiaan Eijkman, a Dutch physician working in a military hospital in Java, found that chickens fed almost exclusively on polished rice developed polyneuritic signs bearing a marked resemblance to those of beriberi in humans. A new head cook at the hospital discontinued the supply of "military" rice (polished) and thereafter the birds were fed on whole grain "civilian" rice with the result that they recovered. He also noted that beriberi in prisoners eating polished rice tended to disappear when a less highly milled product was fed. Many great advances in science have started from such chance observations pursued by men and women of inspiration.

In 1881 the Swiss biochemist N. Lunin reported that animals did not survive on synthetic diets composed solely of purified fat, protein, carbohydrate, salts, and water. Lunin proposed that natural foods such as milk contain small quantities of as yet unknown substances essential to life. In 1906 Frederick Hopkins in England suggested that unknown nutrients were essential for animal life and used the term "accessory growth factors." Hopkins was responsible for opening up a new field of discovery that largely depended on the use of the rat. When Hopkins later discovered that he was not the first to suggest that unknown nutrients were essential, or to conduct animal experiments, he was anxious to share his Nobel prize with Eijkman in 1929.

In 1912 Casimir Funk proposed the "vitamine theory." He had reviewed the literature and made the important conclusion that beriberi could be prevented or cured by a protective factor present in natural food. Funk named the distinct factor that prevented beriberi as a ''vitamine.'' This word was derived from ''vital amine.'' Later, when it became evident that not all ''vitamines'' contained nitrogen (amine), the term became vitamin.

After reviewing the literature between 1873 and 1906, in which small animals had been fed restricted diets of isolated proteins, fats, and carbohydrates, E. V. McCollum of the United States noted that the animals rapidly failed in health and concluded that the most important problem in nutrition was to discover what was lacking in such diets. By 1915, McCollum and M. Davis at Wisconsin discovered that the rat required at least two essential growth factors, a ''fat-soluble A'' factor and a ''water-soluble B'' factor. In addition to being required as factors for normal growth, the ''fat-soluble A'' factor was found to cure xerophthalmia and the ''water-soluble B'' factor cured beriberi. At the same time as their work in Wisconsin, T. B. Osborne and L. B. Mendel of Connecticut also established the importance of vitamin A.

With the pioneer work of Eijkman, Hopkins, Funk, McCollum, and others, scientists began to seriously consider the new class of essential nutrients. The brilliant research of scientists in the first half of this century led to the isolation of more than a dozen vitamins as pure chemical substances. The golden age of vitamin research was mainly in the 1930s and 1940s. For vitamin discovery, the general procedure employed was first to study the effects of a deficient diet on a laboratory animal and then to find a food that would prevent the deficiency. Using a variety of chemical manipulations, the particular nutrient involved was gradually concentrated from the food, and its potency was tested at each stage of concentration on further groups of animals (Wagner and Folkers, 1962). This laborious procedure has been simplified in recent years by the discovery that several vitamins are also growth factors for microorganisms that can therefore replace animals for potency testing. By such methods it is now possible to isolate vitamins and subsequently to identify them chemically. A remarkable achievement has been the direct synthesis by chemists of at least ten vitamins identified in this way. The last vitamin to be discovered was vitamin B_{12} in 1948, which brought the period of vitamin discovery to a close. On the other hand, the possibility that there are still undiscovered vitamins must be recognized (see Chapter 16). More detailed historical considerations for each vitamin are presented in the respective chapters (Chapters 2–15).

2

Vitamin A

I. INTRODUCTION

Although all vitamins are equally important in supporting animal life, vitamin A may be considered the most important vitamin from a practical standpoint. It is important as a dietary supplement for all animals, including ruminants. Vitamin A itself does not occur in plants, however, its precursors (carotenoids) are found in plants, and these can be converted to true vitamin A by a specific enzyme located in the intestinal walls of animals. Prior to the discovery of vitamin A, farmers complained that hogs in dry lot or barns did poorly when fed a ration consisting largely of white corn instead of yellow corn. Agricultural chemists would disagree and explain to farmers that chemical analysis showed that white corn and yellow corn were the same with the exception of color. Then came the vitamin era, which explained what the farmers already knew, that white corn has no carotene, the precursor of vitamin A (Ensminger and Olentine, 1978).

In human nutrition vitamin A is one of the few vitamins of which both deficiency and excess constitute a serious health hazard. Deficiency occurs in endemic proportions in many developing countries and is considered to be the most common cause of blindness in young children throughout the world. McLaren (1986) lists 73 countries and territories that are considered to have potentially serious vitamin A deficiency problems. Vitamin A toxicity usually arises from abuse of vitamin supplementation.

II. HISTORY

For thousands of years humans and animals have suffered from vitamin A deficiency, typified by night blindness and xerophthalmia. The cause was unknown but it was recognized that consumption of animal and fish livers had curative powers according to records and folklore from early civilizations. One of the earliest known reports was from Eber's papyrus, an ancient Egyptian medical treatise of about 1500 B.C., which recommended the livers of cattle or poultry as curative agents (Arykroyd, 1958). An early reference to vitamin A deficiency in livestock is the Bible (Jeremiah 14 : 6): "and the asses did stand in high places, their eyes did fail, because there was no grass."

It was not until early in the twentieth century that vitamin A was discovered. Its history has been reviewed by a number of authorities (Loosli, 1988; Sebrell and Harris, 1967). In 1909, Hopkins and Stepp found that certain fat-soluble substances were necessary for growth of mice and rats. In the years 1913–1915, McCollum and Davis described "fat-soluble A," a factor isolated from animal fats (unsaponifiable fraction of milk fat) or fish oils, which they associated with a growth-promoting activity. In their experiments, the growth of rats ceased prematurely when lard was used as the source of fat in the diet, whereas adequate growth was obtained when the dietary fat was either butter or fat extracted from egg yolk. At the same time, Osborne and Mendel also reported that something in butter appeared to be essential for life and growth in rats. Later Drummond suggested that the "fat-soluble factor A" should be named vitamin A. In 1919, Steenbock called attention to the fact that among vegetable foods vitamin A potency was associated with yellow color. He suggested that carotene was the source of the vitamin, but later recognized that the vitamin was not carotene itself because certain potent sources of the vitamin were colorless. Ten years later, Von Euler and associates in Stockholm obtained a definite growth response when carotene was added to vitamin A-deficient diets. In 1929, Moore produced proof that the animal body transformed carotene into vitamin A. Animals fed carotene had vitamin A in livers, whereas controls did not.

Research in the 1920s and 1930s demonstrated that most animal species need dietary vitamin A. The simultaneous use of chemical methods and experimental rats to test metabolic products resulted in the successful demonstration of vitamin A activity, making it the first confirmed vitamin rather than vitamin B or C, which had received earlier attention. Similar testing methods were used to identify most of the other vitamins.

Only a few years after vitamin A was discovered, it was thought that rickets was also a vitamin A deficiency. Proof that rickets was not caused by vitamin A deficiency was provided by McCollum and associates in 1922. This proof was obtained by oxidizing cod liver oil until vitamin A was destroyed, as shown by the inability of the oil to cure xerophthalmia, and then by demonstrating that the oxidized oil was still effective in curing rickets.

Vitamin A deficiency was shown to be responsible for xerophthalmia and certain forms of night blindness. A link between vitamin A and the visual process was demonstrated in 1935 when Wald, in a series of experiments, obtained a specific form of vitamin A (retinal) from bleached retinas. In 1944, Morton suggested that retinal from bleached visual purple (rhodopsin) might be identical with vitamin A aldehyde; he was able to prove this by synthesis.

The isolation of pure vitamin A became possible when a relationship was found between its growth-promoting activity and the intensity of the Carr–Price antimony trichloride color at 620 nm or the light absorption at 328 nm. Karrer and his group were thus able to obtain a pure oily retinol from vitamin A-rich concentrates. In 1930–1931, Karrer and co-workers proposed the exact structural formulas for vitamin A and β-carotene. Six years later the first crystals of vitamin A were obtained and still another growth-promoting factor—vitamin A_2—was isolated from freshwater fish liver oils.

In 1942, Baxter and Robeson crystallized pure vitamin A and several of its esters; five years later they also succeeded in isolation and crystallization of the 13-cis vitamin A isomer. Isler and co-workers synthesized the first pure vitamin A in 1947. In 1950, Karrer and Inhoffen reported the synthesis of β-carotene.

III. CHEMICAL STRUCTURE AND PROPERTIES

Vitamin A itself does not occur in plant products but its precursor, carotene (Fig. 2.1) does occur in several forms. These compounds are commonly referred to as provitamin A because the body can transform them into the active vitamin.

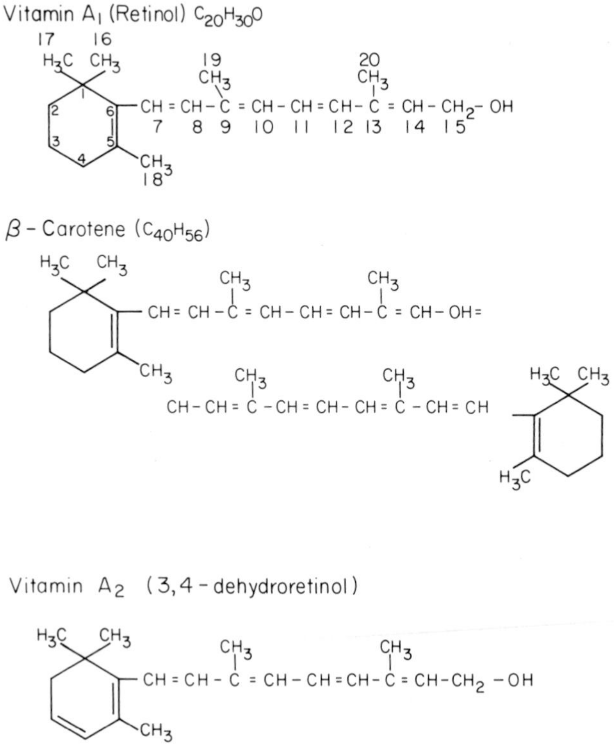

Vitamin A_1 (Retinol) $C_{20}H_{30}O$

17 16 H_3C CH_3 19 CH_3 20 CH_3

CH = CH – C = CH – CH = CH – C = CH – CH_2 – OH

7 8 9 10 11 12 13 14 15

1 2 3 4 5 6 CH_3 18

β - Carotene ($C_{40}H_{56}$)

H_3C CH_3 CH_3 CH_3

CH = CH – C = CH – CH = CH – C = CH – OH =

CH_3 CH_3 CH_3 H_3C CH_3

CH – CH = C – CH = CH – CH = C – CH = CH

H_3C

Vitamin A_2 (3,4 - dehydroretinol)

H_3C CH_3 CH_3 CH_3

CH = CH – C = CH – CH = CH – C = CH – CH_2 – OH

CH_3

Fig. 2.1 Chemical structure of vitamin A_1, β-carotene, and vitamin A_2.

This is how the vitamin A needs of farm animals are met, for the most part, because their rations consist mainly or entirely of foods of plant origin. The combined potency of a feed, represented by its vitamin A and carotene content, is referred to as its vitamin A value. Retinol is the alcohol form of vitamin A (Fig. 2.1). Replacement of the alcohol group by an aldehyde group gives retinal, and replacement by an acid group gives retinoic acid.

In addition to retinol, there is another form that is isolated from freshwater fish. It was originally distinguished on the basis of a different maximum spectral absorption and named A_2 to differentiate it from the previously isolated form. Vitamin A_2 is closely related to vitamin A_1 but contains an additional double bond in the β-ionone ring (Fig. 2.1). Liver oils of marine fish origin usually average less than 10% vitamin A_2 of the total vitamin A content. The relative biological activity of vitamin A_2 is 40–50% that of A_1.

Vitamin A is a nearly colorless, fat-soluble, long-chain, unsaturated alcohol with five double bonds. The vitamin is made up of isoprene units with alternate double bonds, starting with one in the β-ionone ring that is in conjugation with those in the side chain (Fig. 2.1). Since it contains double bonds, vitamin A can exist in different isomeric forms. More common isomeric forms of vitamin A and their relative biological activities are presented in Fig. 2.2.

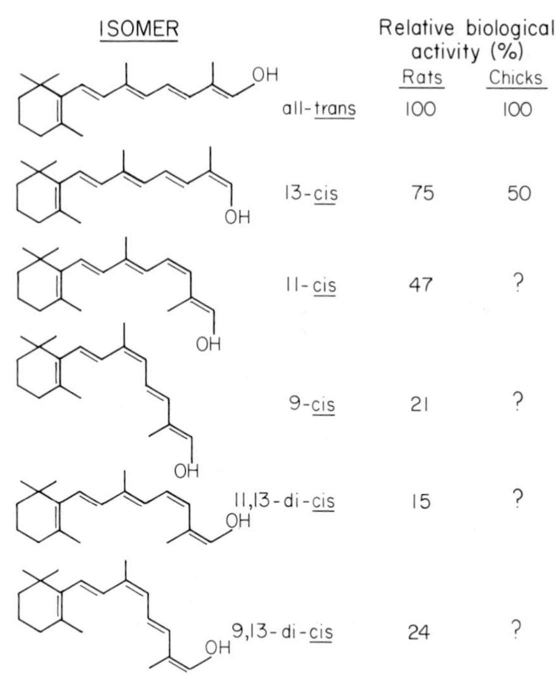

Fig. 2.2 Isomers of vitamin A (retinol). (Adapted from Ullrey, 1972.)

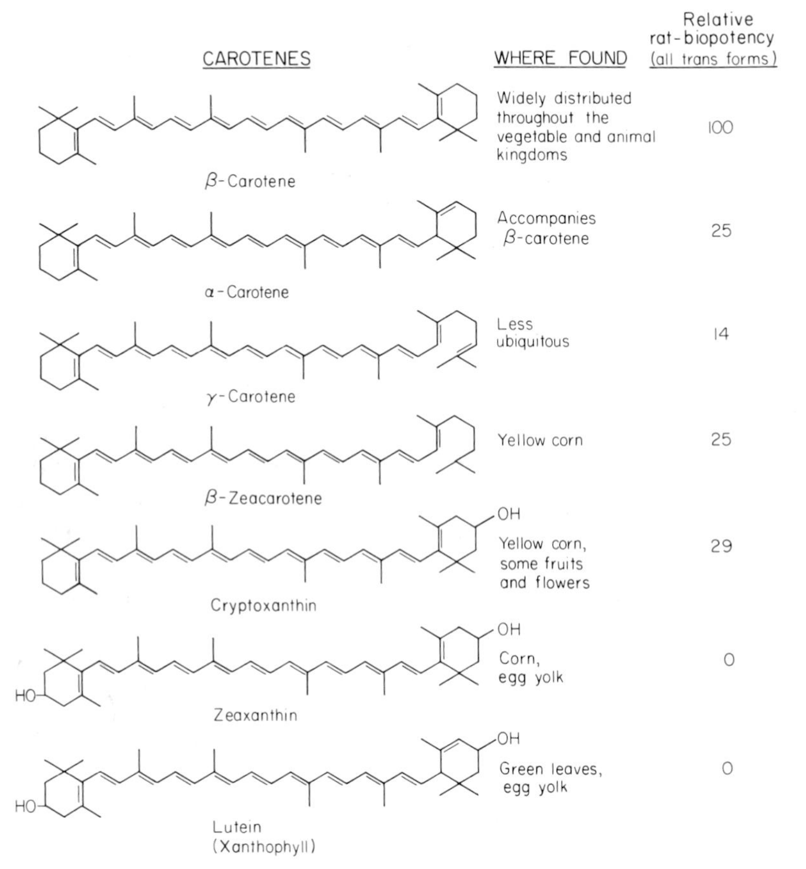

Fig. 2.3 The yellow carotenoids. (Adapted from Ullrey, 1972.)

The most active vitamin A form and that most usually found in mammalian tissues is the all-trans vitamin A. Cis forms can arise from the all-trans forms and a marked loss of vitamin A potency results. These structural changes in the molecule are promoted by moisture, heat, light, and catalysts. Therefore, conditions present during hay making and ensiling, dehydrating, and storage of crops are detrimental to the biological activity of any carotenoids present.

Precursors of vitamin A, the carotenes, occur as orange-yellow pigments mainly in green leaves and to a lesser extent in corn. Structures of some of the important carotenoid pigments and their distribution and relative biological activity are presented in Fig. 2.3. Four of these carotenoids—α-carotene, β-carotene, γ-carotene, and cryptoxanthine (the main carotenoid of corn)—are of particular importance because of their provitamin A activity. Vitamin A activity of β-carotene is substantially greater than that of other carotenoids. However, biological tests have consistently shown that pure vitamin A has twice the potency of β-carotene on a weight-to-weight basis. Thus, only one molecule of vitamin A is formed from one molecule of β-carotene.

IV. ANALYTICAL PROCEDURES

A number of methods are available for carotene and vitamin A determination (Pit, 1985). Biological methods include growth responses of rats or chicks, the storage test (liver), and quantitative evaluations of cell changes in vaginal smears (rats). Physicochemical methods include color reactions with antimony trichloride (Carr–Price method), gas chromatography, thin-layer chromatography, and spectrophotometric procedures. A number of recent reports (Grace and Bernhard, 1984; Hidriglou *et al.*, 1986) indicate excellent results and high recovery rates from a relatively new procedure, high-pressure liquid chromatography (HPLC).

Vitamin A activity is expressed in international units (IU) or, less frequently, in United States Pharmacopeia Units (USP), both of which are of equal value. An IU is defined as the biological activity of 0.300 μg of vitamin A alcohol (retinol) or 0.550 μg of vitamin A palmitate. One IU of provitamin A activity is equal in activity to 0.6 μg of β-carotene, the reference compound. Vitamin A may be expressed as retinol equivalents (RE) instead of IU. By definition, 1 retinol equivalent is equal to 1 μg of retinol, or 6 μg of β-carotene, or 12 μg of other provitamin A carotenoids. In terms of international units, 1 RE is equal to 3.33 IU of retinol or 10 IU of β-carotene.

V. METABOLISM

A. Digestion

A number of factors influence digestibility of carotene and vitamin A. Working with lambs, Donoghue *et al.* (1983) reported that dietary levels of vitamin A ranging from mildly deficient to toxic levels affect digestion and uptake. Percentage transfer from the digestive tract from supplemental dietary levels of 0, 100, and 12,000 μg retinol per kilogram were 91, 58, and 14%, respectively. Wing (1969) reported that the apparent digestibility of carotene in various forages fed to dairy cattle averaged about 78%. Variables that influenced carotene digestibility included month of forage harvest, type of forage (hay, silage, greenchop, or pasture), species of plant, and plant dry matter. In general, carotene digestibility was higher than average during warmer months and lower than average during winter.

A number of papers indicate that appreciable amounts of carotene or vitamin

A may be degraded in the rumen. Various trials with different diets have resulted in preintestinal vitamin disappearance values ranging from 40 to 70% (Ullrey, 1972). These notations of preintestinal disappearance are presumed by most workers to indicate destruction in the rumen, but the absorption of these nutrients from the rumen, while unlikely, has not been ruled out.

B. Absorption and Transport

Much of the conversion of β-carotene to vitamin A takes place in the intestinal mucosa. Provitamin A carotenoids must contain one unsubstituted β-ionine ring to be active. This conversion of β-carotene into vitamin A involves two enzymes. β-Carotene-15,15′-dioxygenase catalyzes the cleavage of β-carotene at the central double bond to yield two molecules of retinaldehyde. The second enzyme, retinaldehyde reductase, reduces the retinaldehyde to retinol. The cleavage enzyme has been found in many vertebrates but is not present in the cat or mink. Therefore these species cannot utilize carotene as a source of vitamin A.

In most mammals the product ultimately absorbed from the intestinal tract as a result of feeding carotenoids is mainly vitamin A itself. There is considerable species specificity regarding the ability to absorb dietary carotenoids. In some species, such as the rat, pig, goat, sheep, rabbit, buffalo, and dog, almost all the carotene is cleaved in the intestine. In humans, cattle, horses, and carp, significant amounts of carotene can be absorbed. Absorbed carotene can be stored in the liver and fatty tissues. Hence, these latter animals have yellow body and milk fat, whereas animals that do not absorb carotene have white fat.

In the case of cattle there is a strong breed difference in absorption of carotene. The Holstein is an efficient converter, having white adipose tissue and milk fat. The Guernsey and Jersey breeds, however, readily absorb carotene, resulting in yellow fat. The chick, on the other hand, absorbs only hydroxy carotenoids in the unchanged form and stores them in tissues. Hydrocarbon carotenoids with provitamin A activity are converted by the chick intestine and absorbed as vitamin A. Species specificity in vitamin A conversion may be due to presence or absence of suitable receptor proteins or to ability to form suitable micellar solutions in the intestinal lumen.

A number of factors affect absorption of carotenoids (Sebrell and Harris, 1967). Cis–trans isomerism of the carotenoids is important in determining their absorbability, with the trans forms being more efficiently absorbed. Dietary fat is important as illustrated by Roels *et al.* (1958). When small supplements of fat were given to vitamin A-deficient boys in a region of Central Africa, the absorption of dietary carotenoids increased remarkably (from less than 5% to about 50%). Dietary antioxidants (i.e., vitamin E) also appear to have an important effect on the utilization and perhaps absorption of carotenoids. It is uncertain whether the antioxidants contribute directly to efficient absorption or whether they protect both carotene and vitamin A from oxidative breakdown. Protein deficiency reduces absorption of carotene from the intestine.

Several reviews have summarized absorption and transport of vitamin A and carotenes, with most information based on rat data (Sebrell and Harris, 1967; Ullrey, 1972). Almost no absorption of vitamin A occurs in the stomach. The main site of lipid (including vitamin A and carotenoids) absorption is the mucosa of the proximal jejunum. Lipid micelles in the intestinal lumen serve as carriers by taking up vitamin A and carotene from emulsified dietary lipid and bringing these lipids into contact with the mucosal cell, where they diffuse from the micelle through the lipid portion of the microvillar membrane.

Vitamin A occurs in food primarily as the palmitate ester. This is hydrolyzed in the small intestine by retinyl ester hydrolase, which is secreted by the pancreas. Bile salts are required both for the activation of this enzyme and for the formation of the lipid micelle, which carries vitamin A from the emulsified dietary lipid to the microvillus. Normally, vitamin A is absorbed almost exclusively as the free alcohol, retinol. Within the mucosal cells, retinol is reesterified mostly to palmitate, is incorporated into the chylomicra of the mucosa, and is secreted into the lymph. A small amount of retinol may be oxidized first to retinal and then to retinoic acid, which may form a glucuronide and pass into the portal blood. Vitamin A is transported through the lymphatic system with a low-density lipoprotein in lymph acting as a carrier to the liver, where it is deposited mainly in hepatocytes and stellate and parenchymal cells.

Some vitamin A derivatives are reexcreted into the intestinal lumen via the bile. This is true for much of the retinoic acid and some retinol. The major vitamin A components of the bile are vitamin A glucuronides. An appreciable portion of these glucuronides are reabsorbed, thus creating an enterohepatic circulation for vitamin A derivatives and providing an opportunity for vitamin A conservation. It appears that this mechanism is not particularly important quantitatively however, because of the small amounts of vitamin A involved (Ullrey, 1972).

When vitamin A is mobilized from the liver, stored vitamin A ester is hydrolyzed prior to its release into the bloodstream, and vitamin A alcohol (retinol) then travels via the bloodstream to the tissues, where a metabolic requirement for the vitamin exists. Retinol is transported by a specific transport protein, retinol-binding protein (RBP). The RBP is synthesized and secreted by hepatic parenchymal cells (Goodman, 1980). Human RBP has a molecular weight of about 20,000 and contains one binding site for one molecule of retinol.

According to Peterson *et al.* (1974), 90% of plasma RBP is complexed to thyroxine-binding prealbumin. The retinol–prealbumin complex is transported to target tissues, where the complex binds to a cell-surface receptor and the retinol is transported into cells of target tissue. Metabolism, storage, and release of vitamin A by the liver are under several forms of homeostatic control with circulating RBP maintained over a wide range of total liver reserves (Hicks *et al.*, 1984). One factor that specifically regulates RBP secretion from liver is the nutritional vitamin A status of the animal. Thus, retinol deficiency specifically blocks secretion of RBP from the liver, so that plasma RBP levels fall and liver RBP levels rise.

Specific binding proteins known as cellular retinol-binding protein (CRBP) and cellular retinoic acid-binding protein (CRABP) have been detected in tissues and allow vitamin A access to the cell nucleus (Goodman, 1980). These proteins have very different ligand specificities and a different pattern of tissue distribution. In the rat, CRBP is found in all organs except heart, skeletal muscle, and serum. On the other hand, CRABP is detected only in eye, brain, testes, ovary, and uterus. In vitamin A deficiency, CRBP loses its ligand but the apoprotein concentration remains unaltered.

All fetal tissues appear to contain both CRBP and CRABP. CRBP continued to be present in adult life in all tissues except heart and skeletal muscle, whereas CRABP disappeared from all tissues except brain, eye, and skin. The observed changes in the concentration of the two binding proteins during development may indicate variable needs for retinol and retinoic acid during development of different organ types and a higher requirement for retinoic acid during embryogenesis than in later life (Anonymous, 1977).

C. Storage

Liver normally contains about 90% of total body vitamin A. The remainder is stored in the kidneys, lungs, adrenals, and blood, with small amounts also found in other organs and tissues. Vitamin A is highly concentrated in stomach oils of certain seabirds and in the intestinal wall of some fish. The entire vitamin A reserve of certain shrimp is in the eyes. Carotenoids are more evenly distributed in species that have the ability to absorb and store these precursors. Grass-fed cattle have large stores of carotene in their body fat, which is evidenced by a deep yellow color.

The liver can store large amounts of vitamin A in hepatocytes and stellate cells, with 10–20% of the total vitamin present in the parenchymal cells (Batres and Olson, 1987). The vitamin is usually stored as the retinyl ester, primarily palmitate. Several studies have shown that liver can store enough vitamin A to protect the animal from long periods of dietary scarcity. This large storage capacity must be considered in studies of vitamin requirements to ensure that intakes that appear adequate for a given function are not being supplemented by reserves stored prior to the period of observation. Measurement of the liver store of vitamin A at slaughter or in samples obtained from a biopsy is a useful technique in studies of vitamin A status and requirements. Figure 2.4 illustrates a liver biopsy precedure that was adapted from human medicine.

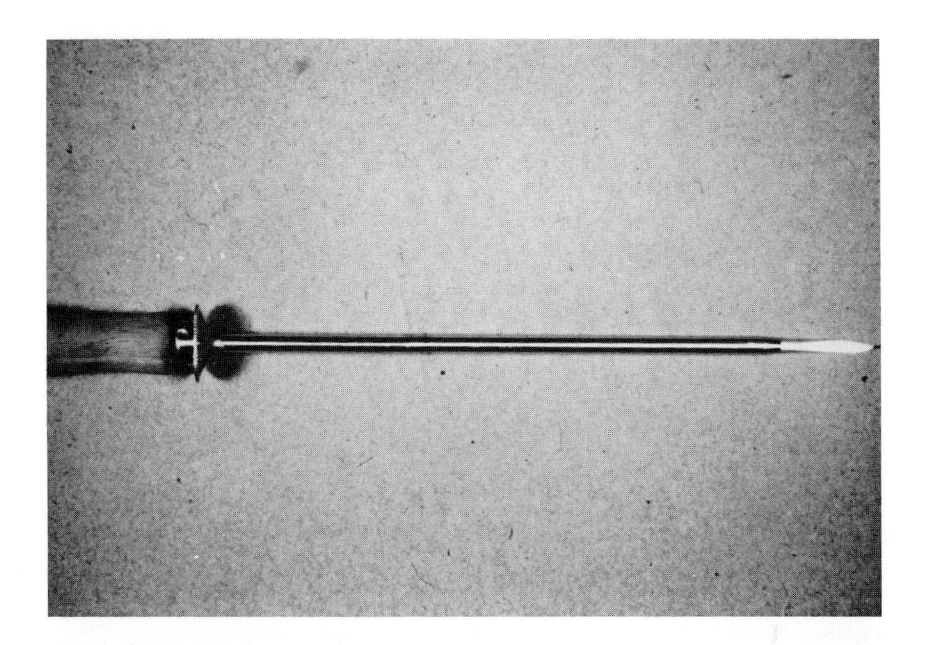

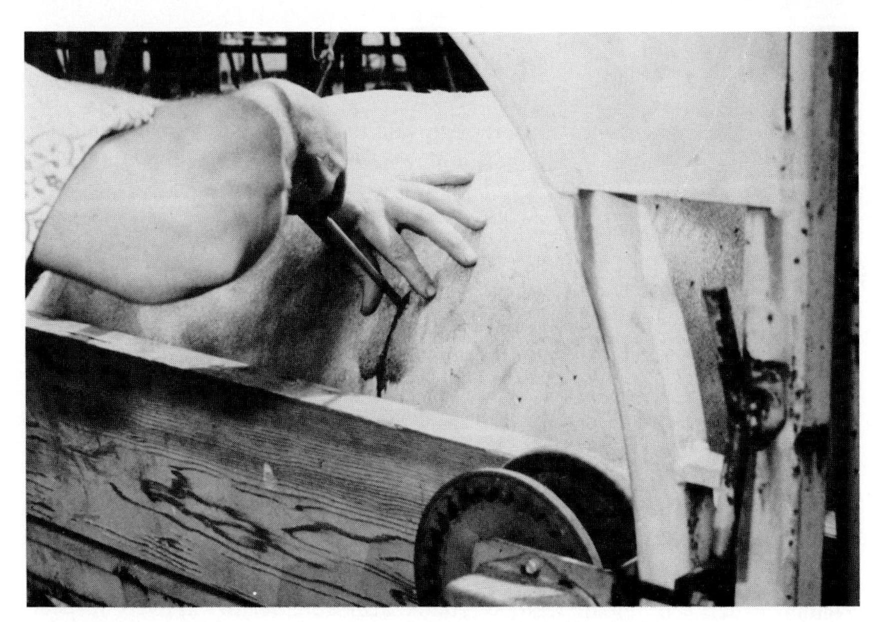

Fig. 2.4 Illustration of liver biopsy sample taken for both vitamin A and mineral analysis. Sample can be taken quickly with a trocar and cannula. (Courtesy of H. L. Chapman and L. R. McDowell, University of Florida.)

Fig 2.4 (*Continued*)

VI. FUNCTIONS

Vitamin A is necessary for support of growth, health, and life of higher animals. In the absence of vitamin A, animals will cease to grow and eventually die. The classic biological assay method is based on measurement of growth responses of weanling rats to graded doses of vitamin A. It is of primary importance in development of young, growing animals.

The metabolic function of vitamin A, explained in biochemical terms, is still incompletely known. Vitamin A deficiency causes at least four different and probably physiologically distinct lesions: loss of vision due to a failure of rhodopsin formation in the retina; defects in bone growth; defects in reproduction (i.e., failure of spermatogenesis in the male and resorption of the fetus in female animals); and defects in growth and differentiation of epithelial tissues, frequently resulting in keratinization.

The acid form of vitamin A, retinoic acid, has several interesting properties. Retinoic acid is unable to fulfill the function of vitamin A in vision (Fig. 2.5) or reproduction, although it maintains normal growth and a general state of health. Vitamin A-deficient female rats fed retinoic acid were healthy in every respect, with normal estrus and conception, but failed to give birth and resorbed their fetuses. When retinol was given even at a late stage in pregnancy, fetuses were

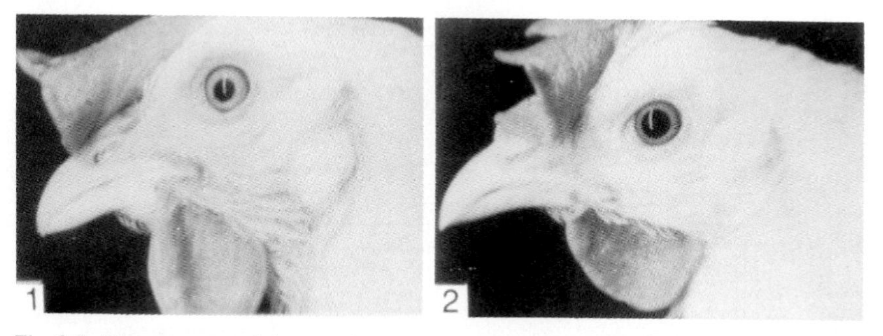

Fig. 2.5 The appearance of the eye of a blind hen (1) fed retinoic acid compared with the normal eye (2) of a hen fed retinol. (Courtesy of M. L. Scott, Cornell University.)

saved. Male rats on retinoic acid were healthy but produced no sperm, and without vitamin A both sexes were blind (Anonymous, 1977).

Retinoic acid cannot fully replace retinol in systems where very rapid division and differentiation of cells take place. Unlike retinol and its esters, the liver cannot store retinoic acid. Similarly, unlike retinol, retinoic acid has no special carrier for transport in blood and, like the long-chain fatty acids, it is transported in the blood by serum albumin; it is also less effectively transported across the plasma membrane.

A. Vision

The physiological function of vitamin A that has been most clearly defined on a biochemical basis is its role in vision (Wald, 1968). Vitamin A is an essential component of vision. When 11-*cis*-retinal (aldehyde form of vitamin A) is combined with the protein opsin, rhodopsin (visual purple) is produced. Chemical reactions involved in vision and the roles that *trans*-retinal and 11-*cis*-retinal play in this function are presented in Fig. 2.6. Rhodopsin breaks down in the physiological process of sight as a result of photochemical reaction. The all-trans retinaldehyde cannot form a stable complex with the opsin. Opsin opens through a series of changes that expose reactive groups. Finally, retinaldehyde is hydrolyzed off the opsin. The energy derived from this reaction is transported to the brain via the optic nerve and recorded in various intensities depending on the amount of light entering the eye.

Vitamin A deficiency, in terms of the needs for the resynthesis of rhodopsin, results in night blindness (nyctalopia), which is a clinical sign in both animals and humans. During the reactions in the retina some of the vitamin A is lost and is replaced by vitamin A from blood. If vitamin A blood level is too low, a functional night blindness will result. The deficiency first manifests itself as a

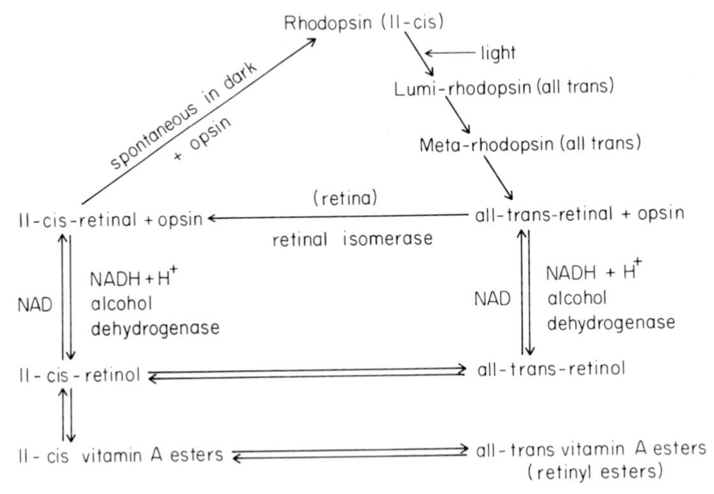

Fig. 2.6 The role of vitamin A in vision. (Adapted from Wald, 1968.)

Fig. 2.7 Vitamin A deficiency: a blind steer walking into a fence. (Courtesy of T. B. Keith, University of Idaho.)

slow, dark adaptation and progresses to total night blindness. At dusk or in moonlight, livestock with night blindness will bump into obstacles (Fig. 2.7) put intentionally in their path or into logs or stumps when driven at night.

In vitamin A deficiency, the outer segments of the rods lose their opsin, leading to their eventual degeneration. The entire structure becomes filled with tubules and vesicles. Even at a late stage, it is possible to regenerate rods, but continued deficiency results in disintegration of cones and total blindness. Vitamin A is needed for integrity of the visual cells as well as their normal regeneration.

Other eye clinical signs vary markedly among species, some of which represent secondary infections. Vision also can be impaired in a condition called xerophthalmia (from the Latin words for dry eye), a manifestation of vitamin A deficiency in which the conjunctiva (covering of the eye) dries out, the cornea becomes inflamed, and the eye becomes ulcerated.

Xerophthalmia is an advanced stage of vitamin A deficiency seen in all species. In children, dogs, fox, and rats, xerophthalmia is characterized by a dry condition of the cornea and conjunctiva, cloudiness, and ulceration. Copious lacrimation is a more prominent eye sign in cows (Fig. 2.8) and horses. In the case of

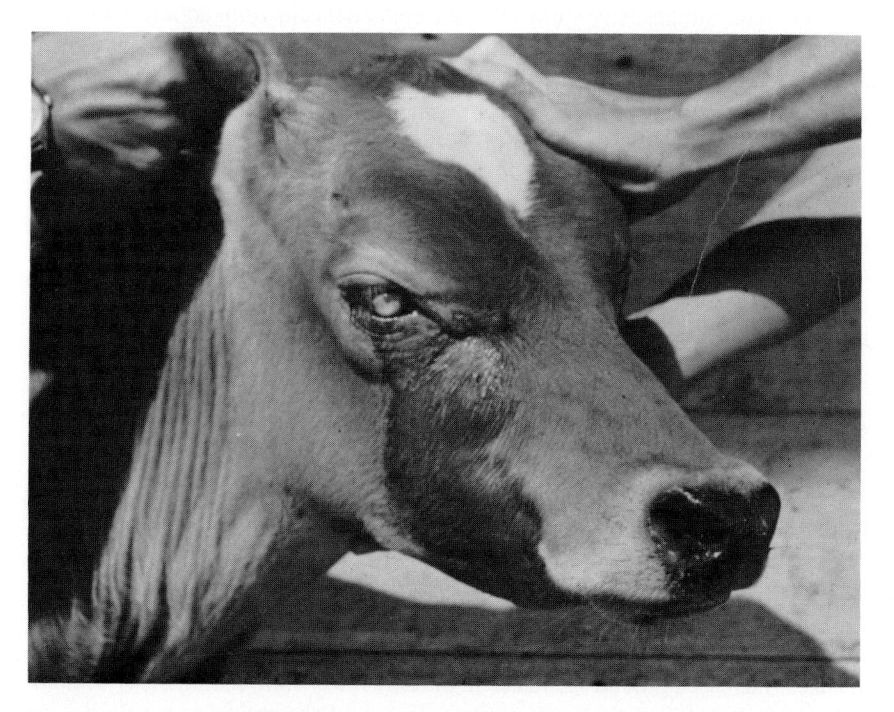

Fig. 2.8 Calf in the Philippines showing a vitamin A deficiency characterized by copious lacrimation and blindness. The 6-month-old animal had been fed reconstituted skim milk powder and poor quality bleached hay (practically devoid of carotene). (Courtesy of J. K. Loosli, University of Florida.)

chickens, on the other hand, the secretions of the tear glands dry up, and an infection may then occur, resulting in a discharge that causes the lids to stick together. Some of these conditions develop as a result of basic epithelial changes caused by a deficiency of the vitamin (Maynard *et al.,* 1979).

B. Maintenance of Normal Epithelium

Vitamin A is required for maintenance of epithelial cells, which form protective linings on many of the body's organs. The respiratory, gastrointestinal, and urogenital tracts, as well as the eye, are protected from environmental influences by mucous membranes. But if there is a deficiency of vitamin A, epithelial cells that make up the membrane will change their characteristic structure. It is postulated that vitamin A plays an important role in altering permeability of lipoprotein membranes of cells and of intracellular particles. Vitamin A penetrates lipoprotein membranes and, at optimum levels, may act as a cross-linkage agent between the lipid and protein, thus stabilizing the membrane (Scott *et al.,* 1982).

The normal mucus-secreting cells of epithelium in various locations throughout the body become replaced by a stratified, keratinized epithelium when vitamin A is deficient. Studies have shown that the epithelial cells from deficient animals fail to differentiate to mucus-secreting cells and mesenchymal cells fail to differentiate beyond the blast stage. This occurs in the alimentary, genital, reproductive, respiratory, and urinary tracts. Such altered characteristics make affected tissues more susceptible to infection. Thus, colds and pneumonia are typical secondary effects of a vitamin A deficiency.

Adequate dietary vitamin A is necessary to help maintain normal resistance to stress and disease. However, greater than optimal intakes of vitamin A will not aid in preventing infections. Aside from its curative effect on xerophthalmia, which may be secondary to a bacterial invasion, there is no evidence that administration of the vitamin after an infection has become established will shorten its course or lessen its severity.

There are many noninfective problems due to keratinization of epithelium, such as diarrhea. The formation of kidney and bladder stones is favored when damaged epithelium interferes with normal secretion and elimination of urine and sloughed keratinized cells may form foci for the formation of stones. There is a specific interference with reproduction caused by altered epithelium that is of great importance. Squamous metaplasia in the parotid gland is an early change in vitamin A-deficient calves and proves useful in diagnosing a deficiency. Elevated cerebrospinal fluid pressure observed in vitamin A-deficient animals, a very sensitive measure of the onset of vitamin A deficiency, is the result of cell changes. Increased ground substance in the dura mater surrounding the arachnoid villus and altered epithelial cells cause a decreased absorption of the fluid.

Part of the unsolved issue of vitamin A function centers around its role in maintaining cell differentiation. Two possible mechanisms have been considered

(Anonymous, 1982): (1) action of the cell nucleus, by altering expression of genetic information, or (2) extranuclear process, for example, the synthesis of cell surface glycoproteins. The case for a nuclear role of retinol is strong. Retinol bound to CRBP becomes attached to rat liver nuclear receptor sites. Isolated chromatin also binds retinol from retinol–CRBP but does not bind CRBP. The presence of vitamin A governs the level of mRNA for specific keratins.

There is evidence that vitamin A is necessary for the formation of large molecules containing glucosamine (Goodman, 1980). These are the mucopolysaccharides occurring in almost all tissues of mammalian organisms but principally in the mucus-secreting epithelia and in the extracellular matrix of cartilage, mainly as chondroitin sulfate. The intimate involvement of vitamin A in the biosynthesis of glycoproteins, which are constituents of membrane systems in cells, helps explain many biological effects of this vitamin.

In severe vitamin A deficiency, abnormalities in both RNA metabolism and protein synthesis have been reported. These changes in nucleic acid metabolism and protein synthesis may, however, reflect secondary efects of deficiency rather than the primary function of the vitamin.

C. Reproduction

In most livestock, the absence of vitamin A in the ration will dramatically reduce reproductive ability. Hatchability is significantly reduced when hens are fed a vitamin A-deficient ration. One of the first signs of vitamin A deficiencies in rabbits is a reduction in fertility and an increased incidence of abortion in pregnant does.

For a number of species, vitamin A deficiency in the male results in a decline in sexual activity and failure of spermatogenesis, and in the female the resorption of the fetus, abortion, or birth of dead offspring. Retained placenta may be a characteristic of a vitamin A deficiency in some species. Often the reproductive problems associated with vitamin A deficiency are actually the result of failure to maintain healthy epithelium.

Degeneration of germinal epithelium and seminiferous tubules and cessation of spermatogenesis in vitamin A-deficient rats are not prevented by retinoic acid. However, vitamin A-deficient rats had lowered testosterone levels that were restored by retinoic acid. Since retinoic acid cannot restore the germinal epithelium in vitamin A-deficient rats, these results indicate a role for retinoic acid in testosterone synthesis (Anonymous, 1977).

D. Bone Development

Vitamin A has a role in the normal development of bone through a control exercised over the activity of osteoclasts and osteoblasts of the epithelial cartilage (Mellanby, 1947). Disorganized bone growth and irritation of the joints are two manifestations of vitamin A deficiencies. In some cases, there is a constriction of the openings through which the optic and auditory nerves pass, thereby resulting in blindness and/or deafness.

Bone changes may also be responsible for the muscle incoordination and other nervous symptoms shown by vitamin A-deficient cattle, sheep, and swine. These changes may be involved in the increase in cerebrospinal fluid pressure shown to be characteristic of the deficiency. While the pathological basis is unknown, several studies have shown that a lack of vitamin A causes congenital malformation in certain soft tissues. Examples are the birth of pigs without eyeballs and hydrocephalus in rabbits.

E. Relationship to Immunological Response and Disease Conditions

Animals deficient in vitamin A will show increased frequency and severity of bacterial, protozoal, and viral infections as well as other disease conditions. Part of disease resistance, as a function of vitamin A, is related to maintenance of mucous membranes and normal functioning of the adrenal gland for production of corticosteroids needed to combat disease. An animal's ability to resist disease depends on a responsive immune system, with a vitamin A deficiency causing a reduced immune response.

In many experiments with laboratory and domestic animals, the effects of both clinical and subclinical deficiencies of vitamin A on the production of antibodies and on the resistance of the different tissues against microbial infection or parasitic infestation have frequently been demonstrated. An inadequate supply of vitamin A will increase the incidence of spontaneous infections in both humans and animals. Vitamin A-deficient chicks showed rapid loss of lymphocytes and deficient rats showed atrophy of the thymus and spleen and reduced response to diphtheria and tetanus toxoids (Krishnan *et al.*, 1974). Mortality from fowl typhoid *(Salmonella gallinarum)* was reduced in chicks fed vitamin A levels greater than the normal levels in a high-protein diet. Serum antibody levels in chicks were increased 2- to 5-fold by high dietary vitamin A concentrations.

A protective effect of dietary vitamin A supplementation against experimental *Staphylococcus aureus* mastitis in mice has been reported (Chew *et al.*, 1984). Harmon *et al.* (1963) studied the effect of a vitamin A deficiency on antibody production by baby pigs and found a high correlation coefficient ($R = 0.70$) between serum vitamin A and antibody titer.

Vitamin A is instrumental in curing ringworm *(Trichophyton verrucosum)* infestation in cattle. Worm infestation in dogs and chicks and occurrence of lungworms in sheep were aggravated by a deficiency of vitamin A. Supplementation with vitamin A improved the health of animals infected with roundworms, of hens infected with the genus *Capillaria,* and of rats infected with hookworms (Herrick, 1972).

In vitamin A-deficient animals, incidence of cancer has been shown to be higher than in animals receiving normal vitamin A intakes. In some special cases, administration of toxic doses of vitamin A caused a regression of tumors. Extreme toxicity of natural vitamin A makes it an ineffective drug for treatment of certain types of neoplasms. Synthetic analogs (retinoids) of vitamin A, however, have been successfully used to prevent cancer of the skin, lung, bladder, and breast in experimental animals. This is a pharmacological approach to prevention of cancer by enhancement of intrinsic epithelial defense mechanisms. Synthetic retinoids (retinyl methyl ether versus retinyl acetate or retinyl palmitate) are superior for this purpose (Goodman, 1980). In humans, studies suggest that diets rich in β-carotene provide a blocking or inhibition of certain types of cancer, including lung cancer (even in people who smoke). Also, skin diseases, including psoriasis and cystic acne, have responded exceptionally well to synthetic retinoids.

VII. REQUIREMENTS

Extensive research has been conducted to determine the vitamin A requirements of various species. Requirements have been published in the United States by the Committee on Animal Nutrition of the National Academy of Science–National Research Council. Vitamin A requirements can be expressed on an IU per kilogram body weight basis, on a daily basis, or as a unit of diet. In agreement with general practice, the requirements are normally expressed per unit of diet rather than per kilogram body weight.

Table 2.1 summarizes the vitamin A requirements for various species, with a more complete listing given in Appendix Table 1. These requirements are deemed sufficient to provide for optimal growth, satisfactory reproduction and milk, egg, or wool production, and prevention of deficiency signs. Presented requirements are designed to be adequate for these purposes under practical conditions of feeding and management as well as allow for a certain amount of storage.

In establishing a satisfactory vitamin A level for practical diets it is necessary to consider a number of factors that may alter the vitamin A requirement. Type and level of production are important as greater production rates increase requirements. Pregnancy, lactation, and egg production also result in higher requirements. Other practical factors are listed in Table 2.2.

Different species of animals convert β-carotene to vitamin A with varying degrees of efficiency. The conversion rate of the rat has been used as the standard value, with 1 mg of β-carotene equal to 1667 IU of vitamin A. Based on this standard, the comparative efficiencies of various species are shown in Table 2.3. Of the species studied, only poultry are equal to the rat in vitamin conversion, with cattle being only 24% as efficient.

Some factors that influence the rate at which carotenoids are converted to vitamin A are type of carotenoid, class and production level of animal, individual

TABLE 2.1

Vitamin A Requirements for Various Animals and Humans[a]

Animal	Purpose	Requirement	Reference
Beef cattle	Feedlot cattle	2,200 IU/kg	NRC (1984a)
	Pregnant heifers and cows	2,800 IU/kg	NRC (1984a)
	Lactating cows and bulls	3,900 IU/kg	NRC (1984a)
Dairy cattle	Growing	2,200 IU/kg	NRC (1978a)
	Lactating cows and bulls	3,200 IU/kg	NRC (1978a)
	Calf milk replacer	3,800 IU/kg	NRC (1978a)
Goat	All classes	5,000 IU/kg	Morand-Fehr (1981)
Chicken	Growing	1,500 IU/kg	NRC (1984b)
	Laying and breeding	4,000 IU/kg	NRC (1984b)
Turkey	Growing and breeding	4,000 IU/kg	NRC (1984b)
Geese	Growing	1,500 IU/kg	NRC (1984b)
	Breeding	4,000 IU/kg	NRC (1984b)
Sheep	Replacement ewes, 60 kg	1,567 IU/kg	NRC (1985b)
	Pregnancy, 70 kg	3,306 IU/kg	NRC (1985b)
	Lactation, 70 kg	2,380 IU/kg	NRC (1985b)
	Replacement rams, 80–100 kg	1,976 IU/kg	NRC (1985b)
Swine	Growing, 5–10 kg	2,200 IU/kg	NRC (1988)
	Growing, 20–100 kg	1,300 IU/kg	NRC (1988)
	Pregnant swine and boars	4,000 IU/kg	NRC (1988)
	Lactating	2,000 IU/kg	NRC (1988)
Horse	Growing	2,000 IU/kg	NRC (1978b)
	Maintenance and working	1,600 IU/kg	NRC (1978b)
	Pregnancy	3,400 IU/kg	NRC (1978b)
	Lactating	2,800 IU/kg	NRC (1978b)
Mink	Growing	5,930 IU/kg	NRC (1982a)
Fox	Growing	2,440 IU/kg	NRC (1982a)
Cat	Gestation	6,000 IU/kg	NRC (1986)
Dog	Growing	3,336 IU/kg	NRC (1985a)
Rabbit	Growing	580 IU/kg	NRC (1977)
	Gestation	1,160 IU/kg	NRC (1977)
Fish	Catfish	1,000–2,000 IU/kg	NRC (1983)
	Trout	2,500–5,000 IU/kg	NRC (1981)
Rat	Growing	4,000 IU/kg	NRC (1978c)
Nonhuman primate	All classes	10,000 IU/kg	NRC (1978d)
Human	Children	400–700 μg RE[b]	RDA (1980)
	Adults	800–1,000 μg/RE	RDA (1980)
	Lactating	1,200 μg/RE	RDA (1980)

[a]Expressed as per unit of animal feed either on an as fed (approximately 90% dry matter) or dry basis (see Appendix Table 1). Humans data are expressed as μg/day.

[b]Retinol equivalents (RE): 1 RE = 1 μg retinol or 6 μg β-carotene.

TABLE 2.2

Factors Influencing Vitamin A Requirements

Genetic differences (species, breed, strain)
Carryover effect of stored vitamin A (principally in the liver)
Conversion efficiency of carotene to vitamin A
Variations in level, type, and isomerization of carotenoid vitamin A precursors in feedstuffs
Presence of adequate bile *in vivo*
Destruction of vitamin A in feeds through oxidation, long length of storage, high temperatures of pelleting, catalytic effects of trace minerals, and peroxidizing effects of rancid polyunsaturated fats
Presence of disease and/or parasites
Environmental stress and temperature
Adequacy of dietary fat, protein, zinc, phosphorus, and antioxidants (including vitamin E and selenium)
Pelleting and subsequent storage of feed

TABLE 2.3

Conversion of β-Carotene to Vitamin A by Different Animals[a]

Animal	Conversion of mg of β-carotene to IU of vitamin A (mg) (IU)	IU of vitamin A activity (calculated from carotene) (%)
Standard	1 = 1667	100
Beef cattle	1 = 400	24
Dairy cattle	1 = 400	24
Sheep	1 = 400–450	24–30
Swine	1 = 500	30
Horse		
Growth	1 = 555	33.3
Pregnancy	1 = 333	20
Poultry	1 = 1667	100
Dog	1 = 833	50
Rat	1 = 1667	100
Fox	1 = 278	16.7
Cat	Carotene not utilized	—
Mink	Carotene not utilized	—
Human	1 = 556	33.3

[a]Adapted from Beeson (1965) and "United States–Canadian Tables of Feed Composition" (NRC, 1982b).

genetic differences in animals, and level of carotene intake (NRC, 1984a). Efficiency of vitamin A conversion from β-carotene is decreased with higher levels of intake: as β-carotene level is increased, conversion efficiency drops from a ratio of 2 : 1 to 5 : 1 for the chicken and from 8 : 1 to 16 : 1 for the calf (Bauernfeind, 1972).

Stress conditions, such as extremely hot weather, viral infections, and altered thyroid function, have also been suggested as causes for reduced carotene to vitamin A conversion. Vitamin A requirements are higher under stressful conditions such as abnormal temperatures or exposure to disease conditions. As an example with poultry, coccidiosis not only causes destruction of vitamin A in the gut but also injures the microvilli of the intestinal wall, thereby decreasing absorption of vitamin A and at the same time causing the chickens to stop eating for several days (Scott *et al.*, 1982).

Likewise, other factors may possibly affect the metabolism and increase requirements of vitamin A. These include free nitrates in feeds, inadequate protein, a zinc deficiency, and low dietary phosphorus (Harris, 1975). Considerable work and controversy have been reported on the relationship between nitrates and vitamin A nutrition. In a review of this subject by Rumsey (1975), it was concluded that although nitrates can be shown to have an adverse effect on vitamin A *in vitro*, this does not appear to translate into a significant effect under most feeding conditions.

The efficiency of β-carotene in meeting the vitamin A requirement of trout and salmon apparently is dependent on water temperature. Cold-water fishes utilize precursors of vitamin A at 12.4° to 14°C, but do not at 9°C (Poston *et al.*, 1977). Activity of β-carotene-15,15′-dioxygenase, which oxidizes β-carotene to retinal in the intestinal mucosa, may be restricted at cold temperatures.

The recommended human allowance for adult females is set at 80% of that for males or 800 retinol equivalents (4000 IU) (RDA, 1980). The allowance during pregnancy is increased to 1000 retinol equivalents to compensate for storage of the vitamin in the fetus, and an even greater allowance (1200 retinol equivalents) is recommended during lactation to provide for vitamin A secreted in milk. Daily vitamin A requirements for children (10 years or younger) vary between 400 and 700 μg retinol equivalents.

A more recent RDA publication than the 1980 edition was written but was rejected because of the controversy concerning requirements for vitamins A and C (Olson, 1986). The committee for the newer RDA edition had recommended reducing the requirements of both vitamins by approximately one-third.

VIII. NATURAL SOURCES

The richest sources of vitamin A are fish oils. Some swordfish liver oils contain as many as 250,000 IU units of vitamin A per gram. Halibut liver oil may run even higher. Thus, both are many times more potent than cod liver oils. Products from the same species, however, may be highly variable in potency, and so in their manufacture for use as a vitamin A supplement they are subjected to a biological assay so that the user may be assured of a certain minimum potency. Among the common foods of animal origin, milk fat, egg yolk, and liver are rated as rich sources, but this is not the case if the animal from which they came has been receiving a vitamin A-deficient diet for an extended period. Since the vitamin is present in the fat, skim milk contains very little. The effect of dietary vitamin A on egg yolk concentration was reported by Hill *et al.* (1961). Vitamin A concentrations of 1760, 4400, and 22,000 IU/kg feed resulted in vitamin A yolk levels of 0.9, 6.3, and 16.3 IU/g, respectively.

Sources of supplemental vitamin A are derived primarily from fish liver oils, in which the vitamin occurs largely in esterified form, and from industrial chemical synthesis. Before the era of the chemical production of vitamin A, the principal source of vitamin A concentrates was the liver and/or body oils of marine fish. Since industrial synthesis was developed in 1949, the synthetic form has become the major source of the vitamin to meet the requirements of domestic animals and humans. The synthetic vitamin usually is produced as the all-trans retinyl palmitate or acetate.

Provitamin A carotenoids, mainly β-carotene in green feeds, are the principal source of vitamin A for grazing livestock. All green parts of growing plants are rich in carotene and, therefore, have a high vitamin A value. In fact, the degree of green color in a roughage is a good index of its carotene content. Although the yellow color of carotenoids is masked by chlorophyll, all green parts of growing plants are rich in carotene and thus have a high vitamin A value. Good pasture always provides a liberal supply, and type of pasture plant, whether grass or legume, appears to be of minor importance. At maturity, however, leaves contain much more than stems, and thus legume hay is richer in vitamin content than grass hay (Maynard *et al.*, 1979). With all hays and other forage, vitamin A value decreases after the bloom stage. Plants at maturity can have 50% or less of the maximum carotenoid value of immature plants.

Both carotene and vitamin A are destroyed by oxidation, and this is the most common cause of any depreciation that may occur in the potency of sources. The process is accelerated at high temperatures, but heat without oxygen has a minor effect. Butter exposed in thin layers in air at 50°C loses all its vitamin A potency in 6 hr but in the absence of air there is little destruction at 120°C over the same period. Cod liver oil in a tightly corked bottle has shown activity after 31 years, but it may lose all its potency in a few weeks when incorporated in a feed mixture stored under usual conditions (Maynard *et al.*, 1979).

Much of carotene content is destroyed by oxidation in the process of field curing. Russell (1929) found that there may be a loss of more than 80% of the carotene of alfalfa during the first 24 hr of the curing process. It occurs chiefly during the hours of daylight, owing in part to photochemical activation of the destructive process. In alfalfa leaves, sunlight-sensitized destruction is 7–8% of the total pigment present, while enzymatic destruction amounts to 27–28% (Bauernfeind, 1972). Enzymatic destruction requires oxygen, is greatest at high temperatures, and ceases after complete dehydration.

Hays that are cut in the bloom stage or earlier and cured without exposure to rain or too much sun retain a considerable proportion of their carotene content, while those cut in the seed stage and exposed to rain and sun for extended periods lose most of it. Green hay curing in the swath may lose one-half its vitamin A activity in 1 day's exposure to sunlight and almost all of it if left exposed to rain as well as sunlight. Thus, hay usually has only a small proportion of the carotene content of fresh grass. Under similar conditions of curing, alfalfa and other legume hays are much richer than grass hays because of their leafy nature, but a poor grade of alfalfa may have less than a good grade of grass hay (Maynard *et al.*, 1979).

The carotene content of dried or sun-cured forages decreases on storage with the rate of destruction depending on factors such as temperature, exposure to air and sunlight, and length of storage. Under average conditions, carotene content of hay can be expected to decrease about 6–7% per month. In artificial curing of hay with a "hay drier," there is only a slight loss of carotene because of the rapidity of the process and protection against exposure to oxygen, with the final product having 2–10 times the value of field-cured hay. Severe heating of hay in the mow or stack reduces vitamin content, and there is a gradual loss in storage so that old hay is poorer than new. Aside from yellow corn and its by-products, practically all the concentrates used in feeding animals are devoid of vitamin A value, or nearly so. In addition, yellow corn contains a high proportion of non-β-carotenoids (i.e., cryptoxanthin, lutein, and zeacarotene) that contain much less vitamin A value than that of β-carotene.

The potency of yellow corn is only about one-eighth that of good roughage. Roots and tubers as a class supply practically no vitamin A, but carrots are a very rich source as are sweet potatoes, as might be expected from their yellow color. Pumpkins and squash also supply considerable amounts, and green leafy vegetables used in human nutrition are rich in carotene (Maynard *et al.*, 1979).

Tankage, meat scraps, and similar animal by-products have little if any vitamin A potency. Certain fish meals are fair sources, but variation in the raw material and in methods of processing may entirely destroy any potency originally present.

There is evidence that yellow corn may lose carotene rapidly during storage. For instance, a hybrid corn high in carotene lost about half of its carotene in 8 months' storage at 25°C and about three-quarters in 3 years. Less carotene was lost during storage at 7°C (Quackenbush, 1963).

A marked discrepancy exists between the carotene content of corn silage and the vitamin A status of ruminants fed corn silage. On the average, corn silage carotenes were found to be about two-thirds as effective as β-carotene for maintaining liver stores in rats (Miller *et al.*, 1969; Rumsey, 1975). Martin *et al.* (1971) reported 5-fold less carotenes in October and November corn silage than in September corn silage. More mature silages were not able to sustain liver vitamin A stores in beef steers, particularly if the ensiled corn plant was finely chopped. Diets high in corn silage harvested after a killing frost and fed to cattle would be marginal in their supply of both vitamins A and E. Miller *et al.* (1969) have reported that ethanol, sometimes found in corn silage as a product of fermentation, may reduce liver vitamin A stores as much as 26% by increasing mobilization of vitamin A from liver.

Wing (1969) reported carotene digestibility in plants to be greater during the warmer months. Variations were found in the digestibility of carotenes in plants according to year, species of plant, dry matter content, and form of forage; carotene digestibility was somewhat lower in silages than in pastures or hay. Table 2.4 presents typical vitamin A values of foods and carotene concentration of feeds. As noted earlier, degree of greenness in a roughage is a good index of its carotene content. The data in Table 2.4 are useful to indicate the order of the differences found among various roughages differing as to color, kind, and other factors. Average published values of carotene content can serve only as approximate guides in feeding practice because of many factors affecting actual potency of individual samples as fed (NCR, 1982b).

Cooking processes commonly used in human food preparation do not cause much destruction to the vitamin potency. The blanching and freezing process generally causes little loss of carotenoid content in vegetables and fruits. Heat, however, does isomerize the all-trans carotenoids to cis forms. In a report from Indonesia, isomerization during traditional cooking caused a loss of up to 9% of vitamin A potency (Van der Pol *et al.*, 1988). Hydrogenation of fats lessens their vitamin A value, while saponification does not destroy the vitamin if oxidation is avoided.

Several factors can influence the loss of vitamin A from feedstuffs during storage. The trace minerals in feeds and supplements, particularly copper, are detrimental to vitamin A stability. Dash and Mitchell (1976) reported the vitamin A content of 1293 commercial feeds over a 3-year period. The loss of vitamin A was over 50% in 1 year's time. Vitamin A loss in commercial feeds was evident even if the commercial feeds contained stabilized vitamin A supplements. The stability of vitamin A in feeds and premixes has been improved tremendously in recent years by chemical stabilization as an ester and by physical protection using antioxidants, emulsifying agents, gelatin, and sugar in spray-dried, beaded, or prilled products (Shields *et al.*, 1982). Nevertheless, vitamin A supplements should not be stored for prolonged periods prior to feeding.

Vitamin A and carotene destruction also occurs from processing of feeds with steam and pressure. Pelleting effects on vitamin A in feed are caused by die thickness and hole size, which produce frictional heat and a shearing effect that

TABLE 2.4

Vitamin A (Retinol) and β-Carotene Content of Feeds[a]

Vitamin A source	Vitamin A (IU/g)
Whale liver oil	400,000
Swordfish liver oil	250,000
Halibut liver oil	240,000
Herring liver oil	211,000
Tuna liver oil	150,000
Shark liver oil	150,000
Bonito liver oil	120,000
White sea bass liver oil	50,000
Barracuda liver oil	40,000
Dogfish liver oil	12,000
Seal liver oil	10,000
Cod liver oil	4,000
Sardine body oil	750
Pilchard body oil	500
Menhaden body oil	340
Butter	35
Cheese	14
Eggs	10
Milk	1.5
Carotene source	**Carotene (mg/kg)**
Fresh green legumes and grasses, immature (wet basis)	33–88
Dehydrated alfalfa meal, fresh, dehydrated without field curing, very bright green color	242–297
Dehydrated alfalfa meal after considerable time in storage, bright green color	110–154
Alfalfa leaf meal, bright green color	120–176
Legume hays, including alfalfa, very quickly cured with minimum sun exposure, bright green color, leafy	77–88
Legume hays, including alfalfa, good green color, leafy	40–59
Legume hays, including alfalfa, partly bleached, moderate amount of green color	20–31
Legume hays, including alfalfa, badly bleached, or discolored, traces of green color	9–18
Nonlegume hays, including timothy, cereal, and prairie hays, well cured, good green color	20–31
Nonlegume hays, average quality, bleached, some green color	9–18
Legume silage (wet basis)	11–44
Corn and sorghum silages, medium to good green color (wet basis)	4–22
Grains, mill feeds, protein concentrates, and by-product concentrates, except yellow corn and its by-products	0.02–0.44

[a]Adapted from Scott *et al.* (1982) and Maynard *et al.* (1979).

can break supplemental vitamin A beadlets and expose the vitamin. In addition, steam application exposes feed to heat and moisture. Running fines back through the pellet mill exposes vitamin A to the same factors a second time. Between 30 and 40% of vitamin A present at mixing may be destroyed during pelleting (Shields *et al.*, 1982).

IX. DEFICIENCY

A. Effects of Deficiency

Vitamin A is necessary for normal vision in animals, for maintenance of healthy epithelial or surface tissues, and for normal bone development. The vitamin A deficiency signs observed in various species vary somewhat but most relate to these three changes in tissues. A wide range of vitamin A deficiency signs are noted in Table 2.5.

1. Ruminants

For cattle, signs of vitamin A deficiency (Figs. 2.9–2.11) include reduced feed intake, rough hair coat, edema of the joints and brisket, lacrimation, xerophthalmia, night blindness, slow growth, diarrhea, convulsive seizures, improper bone growth, blindness, low conception rates, abortion, stillbirths, blind calves, abnormal semen, reduced libido, and susceptibility to respiratory and other infections (NRC, 1984b).

Clinical signs for vitamin A deficiency in sheep (Fig. 2.12) are similar to those of cattle, with the appearance of night blindness being the common means of determining the deficiency (NRC, 1985b). Vitamin A deficiency has detrimental effects on wool production and characteristics, with shorter wool fibers and decreases in fiber thickness, strength, and elongation (Farid and Ghanem, 1982).

The greatest need for vitamin A is during calving and breeding time. If it is inadequate, young ruminants may get pinkeye, pneumonia, or other illnesses related to the mucous membranes. If a vitamin A deficiency occurs in pregnant ruminants, they may either abort or produce blind, dead, or weak offspring that will not survive. These weak offspring have trouble getting to their feet and lack instinct to nurse. Newborn calves with a vitamin A deficiency may show a very severe diarrhea that may soon be followed by death. In young calves, signs of a vitamin A deficiency also include watery eyes, a nasal discharge, and sometimes muscular incoordination, staggering gait, and convulsive seizures.

Reduced libido and sterility in bulls with degeneration of seminiferous tubules has been reported (Larkin and Yates, 1964). Spermatozoa decrease in numbers and motility and there is a marked increase in abnormal forms. Vitamin A de-

TABLE 2.5

Deficiency Signs of Vitamin A[a]

- General
  - Cessation of growth
  - Cystic pituitary glands
  - Death
  - Decline in body weight
  - Diarrhea (scours)
  - Failure of appetite
  - General edema
  - Reduced resistance to parasite infections
  - Untidy hair or feathers
  - Xerosis of membranes
- Bone formation
  - Cancellous bone
  - Defective modeling
  - Narrowing of foramina
  - Restriction of brain cavity
- Congenital abnormalities
  - Anophthalmia
  - Aortic arch deformities
  - Cleft palate
  - Hydrocephalus
  - Kidney deformities
  - Microophthalmia
- Defective reproduction
  - Abnormal estrous cycle
  - Dead, weak, or blind offspring
  - Degeneration of testes
  - Reduced egg production and hatchability
  - Resorption of fetuses
- Eyes
  - Keratomalacia
  - Lacrimation
  - Night blindness (nyctalopia)
  - Xerophthalmia
  - Constriction of optic nerve
  - Loss of lens
  - Opacity of cornia
  - Papilledema
- Liver
  - Degeneration of Kupffer cells
  - Metaplasia of bile ducts
- Nervous system
  - Constriction at foramina
  - Convulsive seizure
  - Hydrocephalus
  - Incoordination and staggering gait
  - Paresis
  - Raised cerebrospinal fluid pressure
  - Twisting of nerve
- Respiratory system
  - Lung abscesses
  - Metaplasia of nasal passages
  - Nasal discharge
  - Pneumonia
- Urinary system
  - Cystitis
  - Nephrosis
  - Pus in ureters
  - Pyelitis
  - Thickening of bladder wall
- Urolithiasis

[a]Modified from Bauernfeind and DeRitter (1972).

Fig. 2.9 Vitamin A-deficient calf. Note the emaciated appearance and evidence of diarrhea. The calf also shows excessive lacrimation and nasal discharges characteristic of the deficiency. (Courtesy of G. Patterson, Chas. Pfizer Company.)

ficiency has resulted in low semen quality in rams (Lindley *et al.*, 1949). Feedlot cattle suffering from mild vitamin A deficiency reduce their feed intake and fail to make satisfactory weight gains. Lowered feed intake may result in other deficiencies when the diet is borderline in other nutrients.

The classic sign of vitamin A deficiency in ruminants is night blindness, with total and permanent blindness in younger animals resulting in stenosis of the optic nerve. Excessive eye lacrimation in cattle (rather than xerophthalmia) usually occurs; the corners of the eyes become keratinized and may with infection develop ulceration.

Low vitamin A and β-carotene plasma levels play an important role in regulating the incidence and severity of mastitis in dairy cows (Chew *et al.*, 1982). Potential pathogens exist regularly in the teat orifice and under suitable circumstances can invade and initiate clinical mastitis. Any deviation of the epithelium from a healthy state would increase susceptibility of a mammary gland to invasion by pathogens. Chew (1983) reports that cows receiving both supplemental vitamin A and β-carotene had a lower incidence of mastitis than did controls, with the combination of both vitamin A and carotene being most effective.

In the 1950s, it was discovered that cattle developed signs of vitamin A de-

Fig. 2.10 Advanced stage of anasarca in hindquarters of vitamin A-deficient steer. (Courtesy of L. L. Madsen, Washington State University.)

ficiency that was originally referred to as X-disease (hyperkeratosis). The disease was due to feeds that contained highly chlorinated naphthalene found in lubricating oil. Because of the depressed vitamin A levels in blood plasma, it was concluded that the toxic substance was interfering with the conversion of carotene to vitamin A (Maynard *et al.*, 1979). Removal of naphthalenes from oils eliminated X-disease.

Studies have shown that vitamin A-deficient cattle lack heat tolerance: deficient cattle stood and panted a great deal and their daily consumption of feed decreased (Perry 1980). The vitamin A-supplemented cattle tolerated the hot weather better and spent much of their time chewing their cud.

Fig. 2.11 Vitamin A-deficient calf, showing incoordination and weakness. (Courtesy of J. W. Thomas, Michigan State University.)

Fig. 2.12 Typical appearance of vitamin A-deficient lamb. Note the extreme weakness and swayed back. This was followed by the inability to stand. (Courtesy of T. J. Cunha and Washington State University.)

2. Swine

In pigs, the absence of vitamin A results principally in nervous signs such as unsteady gait, incoordination, trembling of the legs, spasms, and paralysis (Hentges *et al.*, 1952) (Fig. 2.13). Eye lesions are less common. An effect of vitamin A deficiency on appetite or rate of gain does not occur until eventual paralysis and weakness prohibit movement to the feeder (Cunha, 1977).

During reproduction and lactation, a vitamin A deficiency in the sow produces the following clinical signs: failure of estrus, resorption of young, wobbly gait, weaving and crossing of the hind legs while walking, dropping of the ears, curving with head down to one side, spasms, loss of control of hind and fore quarters and thus inability to stand up, and impaired vision (Cunha, 1977). Depending on degree of severity of vitamin A deficiency, fetuses were either resorbed, born dead, or carried to term. Fetuses carried to term showed a variety of defects, including various stages of arrested formation of the eyes as well as complete lack of eyeballs, harelips, cleft plate, misplaced kidneys, accessory earlike growths, some with one eye and some with one large and one small eye, and bilateral cryptorchidism (Guilbert *et al.*, 1937; Cunha, 1977). Vitamin A appears to improve reproductive performance of gilts by decreasing embryonic mortality, resulting in more pigs per litter (Brief and Chew, 1985).

3. Poultry

In poultry, lack of vitamin A in the diet causes slower growth (Fig. 2.14), lowered resistance to disease, eye lesions, muscular incoordination (Fig. 2.15), and other signs (Scott *et al.*, 1982). As deficiency progresses in adult poultry, the chickens become emaciated and weak and their features are ruffled. A marked decrease in egg production occurs and the length of time between clutches increases greatly. Hatchability is decreased and there is an increase in embryonic mortality in eggs from affected birds. A watery discharge from the nostrils and eyes is noted and eyelids are often stuck together.

When day-old chicks are given a vitamin A-free diet, clinical signs may appear at the end of the first week if the chicks are progeny of hens receiving a diet low in vitamin A. If chicks are progeny of hens receiving high levels of vitamin A, signs of deficiency may not appear until chicks are 6 or 7 weeks of age even though they are receiving a diet completely devoid of vitamin A (Scott *et al.*, 1982). Gross signs of vitamin A deficiency in chicks are characterized by anorexia, cessation of growth, drowsiness, weakness, incoordination, emaciation, and ruffled plumage. The mucous epithelium is replaced by a stratified squamous, keratinizing epithelium. As a result of mucous membrane breakdown, bacteria and other pathogenic microorganisms may invade these tissues and enter the body, thereby producing infections that are secondary to original vitamin A deficiency signs.

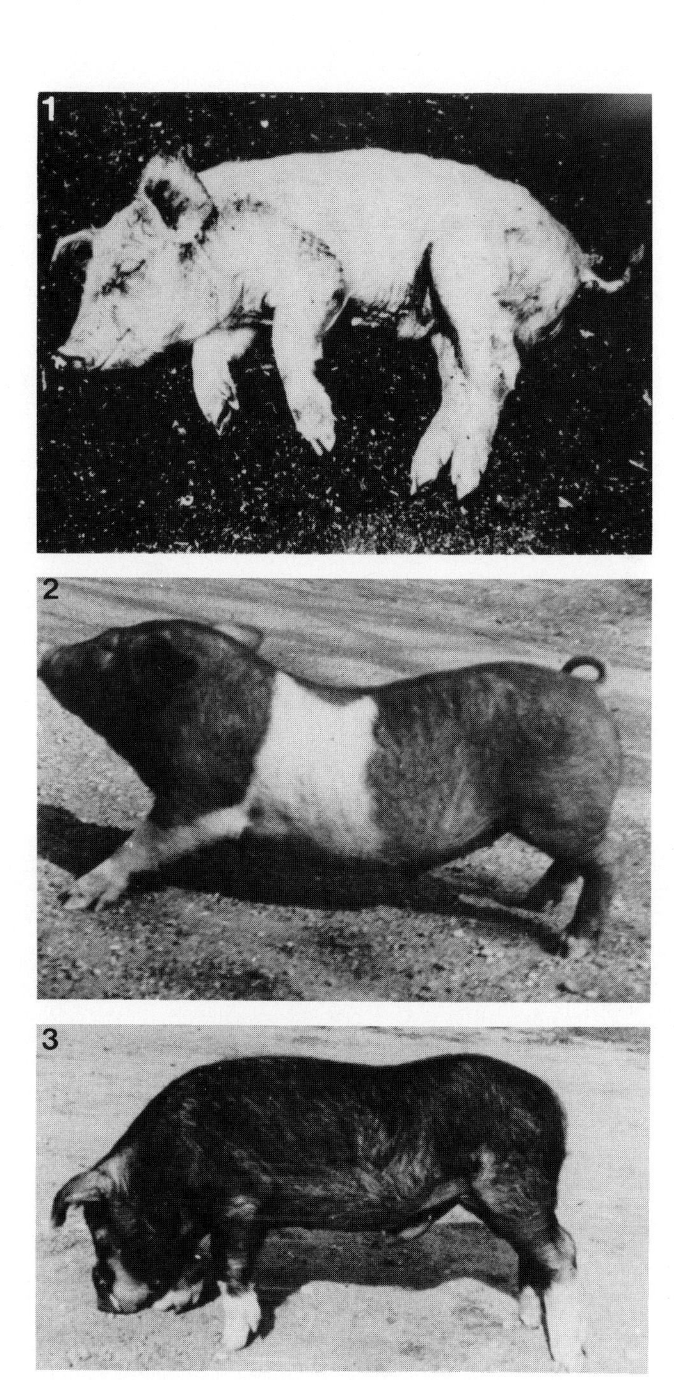

Fig. 2.13 Vitamin A deficiency in growing pigs. (1) shows a pig exhibiting partial paralysis and seborrhea; (2) shows a pig in the initial stage of spasm; (3) shows a pig exhibiting lordosis and weakness of hind legs. (Courtesy of J. F. Hentges, R. H. Grummer, and University of Wisconsin.)

Fig. 2.14 Vitamin A deficiency in a 6-week-old turkey. Right: fed a control diet; left: fed a vitamin A-deficient diet. (Courtesy of L. M. Potter, Virginia Polytechnic Institute.)

Fig. 2.15 Ataxia in chicks due to dietary deficiency of vitamin A. (Courtesy of M. L. Scott, Cornell University.)

4. Horses

The importance of vitamin A to the overall health and well-being of horses has been well documented. Night blindness, lacrimation, keratinization of the cornea and respiratory system, reproductive difficulties, capricious appetite, progressive weakness, and death occur in horses deficient in vitamin A (NRC, 1978b).

In a lengthy experiment with vitamin A-deficient horses, all horses developed night blindness that abated when dietary vitamin A was provided (Howell *et al.*, 1941). Upon withdrawal of vitamin A supplements, night blindness would again appear in the animals. Histological changes in the retinas of horses showing vitamin A deficiency have been shown. Copious lacrimation is characteristically associated with vitamin A deficiency along with varying degrees of corneal keratinization.

5. Other Animal Species

In avitaminosis A in the dog, xerophthalmia, corneal changes, nervous disturbances, and bone malformations are seen. Vitamin A deficiency in the fox is characterized mainly by nervous disorders. In some cases eye lesions, widespread epithelial changes, and degenerations in the nervous system are seen (Ensminger and Olentine, 1978). For cold-water fish, vitamin A deficiency results in impaired growth, exophthalmos, eye lens displacement, corneal thinning and expansion, degeneration of retina, edema, and depigmentation (NRC, 1981a).

6. Humans

Vitamin A deficiency in humans occurs in endemic proportions in many developing countries and is only seen occasionally in technologically developed societies in patients with severe malabsorption, transport disorders, or liver disease (McLaren, 1984). It has been reported, however, that alcohol consumption results in a significant hepatic vitamin A depletion, due perhaps to associated failure to consume an adequate diet. Also drugs such as phenobarbital or food additives such as butylated hydroxytoluene (BHT), when combined with ethanol, resulted in a striking depletion of hepatic vitamin A concentrations in rats (Leo *et al.*, 1987).

Vitamin A deficiency in humans under natural conditions is not wholly comparable to the human volunteers or experimental animals on an otherwise good diet lacking only vitamin A. Under field conditions, vitamin A deficiency is nearly always accompanied by protein-energy malnutrition, is exacerbated by parasitic infections, and is frequently complicated by intercurrent infections. Xerophthalmia is the most common cause of blindness in young children throughout the world (Fig. 2.16), with as many as 5 million Asian children developing xerophthalmia each year (Sommer *et al.*, 1981). One-tenth of children with xerophthalmia have severe corneal involvement and half of them become blind. Xerophthalmia has long been recognized to be endemic in many of the countries

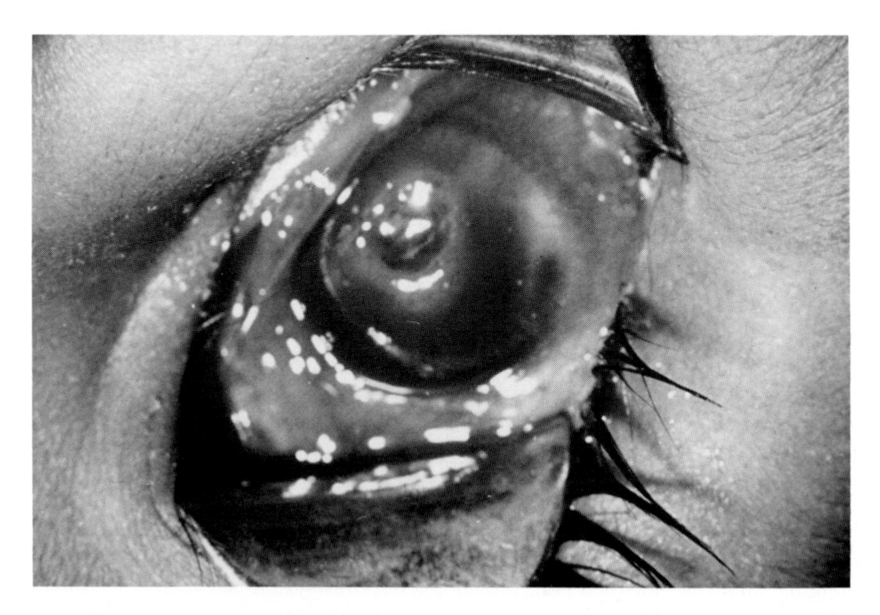

Fig. 2.16 Keratomalacia in a Jordanian infant with severe vitamin A deficiency. The central area of the cornea is undergoing typical colliquative necrosis. Sight is irrevocably lost at this advanced stage. (Courtesy of D. S. McLaren, Royal Infirmary, Edinburgh, Scotland.)

in southern and eastern Asia and parts of Latin America, but only recently has its frequent occurrence in many countries in Africa and the Middle East been recognized.

Lack of vitamin A has been associated with an increased susceptibility to infections, especially of the respiratory and digestive tracts. Immune mechanisms become impaired and antibody synthesis is diminished. Mortality rate among children with mild xerophthalmia (night blindness and/or Bitot's spots) was on average 4 times the rate, and in some age groups 8 to 12 times the rate, of children without xerophthalmia (Sommer *et al.*, 1983). In Indonesia, a 34% difference in mortality was observed between preschool children who received vitamin A supplements for 1 year and those who received none (Sommer *et al.*, 1986).

B. Deficiency Incidence and Relation to Vitamin A Body Stores

The circumstances most conducive to vitamin A deficiencies are (1) extended periods of drought, resulting in pastures becoming dry and bleached with no green color; (2) diets composed primarily of concentrates and no green pasture; (3) feeding mainly corn silage and a concentrate mixture low in vitamin A activity;

(4) in young animals fed milk from mothers on a low intake of vitamin A or carotene or poultry hatched where previous diets were low in vitamin A value; (5) when calves are fed relatively little whole milk or colostrum; and (6) when grains containing no carotene are substituted for yellow corn. Mild deficiencies of vitamin A, especially in winter (or during the dry season) and early spring, are probably fairly common.

When grazing livestock receive a modest amount of fresh green pasture forage, there is little likelihood of a deficiency. Likewise, with a substantial amount of good silage made from green forage, or with liberal feeding of fresh hay with a good green color, a deficiency will not occur. However, low-quality forages and weathered and leached hay may contain very little available carotene. Amounts of carotene in fresh green forages are very high relative to dietary requirements. Thus, a small quantity of fresh pasture forage will fully supply the need for vitamin A. For example, sun-cured full-bloom alfalfa hay contains 14,000 to 15,000 IU per kilogram dry matter, and sun-cured orchard grass hay contains 13,000 to 14,000 IU per kilogram.

One of the most frequent stress conditions for grazing livestock in many parts of the world is low intake of carotene under conditions of winter feeding or protracted drought. When normal seasonal cycles prevail and rainfall is sufficient to provide 4 months or more of green grazing, mature animals store sufficient vitamin A to carry them for many months on dry feed, including grain and roughage that is low in carotene. However, when severe and prolonged drought conditions prevail, especially during the normal green grazing period, vitamin A deficiency signs may occur in adult animals. Geographical locations where vitamin A deficiency might be expected would therefore be in regions with extensive dry seasons. Many tropical regions in Africa, Asia, and Latin America routinely have dry seasons 6 months or longer in duration. Under conditions in India, it has been reported (Ray, 1963) that fair grazing is possible for only 3 months of the year. With the cessation of the monsoon rains, grasses mature rapidly with a large drop in their carotene content. The value has been found to decrease from 100–200 mg per kilogram on a dry matter basis in the middle of the rainy season to as low as 0.5 to 1 mg per kilogram during the dry season. As a result, clinical signs like night blindesss, blindness in newborn calves, and birth of weak calves have been reported from all parts of India.

In the various vitamin A supplementation trials to improve ruminant livestock production, contradictions on the benefits of supplementation have been reported. Overall it has been shown beneficial to supplement grazing livestock with vitamin A, especially in middle to late dry season, and to feedlot cattle not receiving green forages. A general rule of thumb is that forages with a bright green color are always adequate in vitamin A value (carotene).

Use of concentrate feeds in place of forages is probably the largest single factor that has increased need for vitamin A supplements in feedlot cattle, poultry, and swine diets. Formerly poultry and swine diets contained alfalfa meal, a rich source of carotene. Inefficient utilization of corn carotene and the destruction of carotene and vitamin A that occurs in the rumen are reasons why the calculated total daily vitamin A requirement has increased (Rumsey, 1975).

Deficiencies of other nutrients interfere with vitamin A metabolism and can result in a vitamin A deficiency. Both protein and zinc have been shown to be required for mobilization of vitamin A from liver. Vitamin A is required for intestinal absorption of zinc (Berzin and Bauman, 1987) in poultry, while zinc influences vitamin A by affecting RBP synthesis and release from the liver (Brown *et al.*, 1976).

Cattle from tropical Northern Australia suffered a 12% annual mortality in part because of a slow release of liver vitamin A (Guerin, 1981). Apparently high calcium and low zinc forage concentrations contributed to this slow liver vitamin A release. Since tropical forages have been shown to be low in zinc (McDowell *et al.*, 1984), conditioned vitamin A deficiencies may be resulting even though liver vitamin A values indicate adequate concentrations of this vitamin.

Liver is the main vitamin A storage site for humans as well as for farm species. Although most of the vitamin A reserves are in the liver, when carotene intake is high, some is stored in fat. During periods of low dietary carotene, this stored vitamin A can be mobilized and utilized without signs of a vitamin A deficiency. At birth, the ruminant usually does not have sufficient vitamin A reserves to provide for its needs for any substantial time. Accordingly, it is important that young ruminants receive colostrum, which generally is rich in vitamin A, within a few days after birth. If the cow has received a diet low in vitamin A activity, the newborn calf is likely to be susceptible to a vitamin A deficiency because body reserves are low and colostrum will have a subnormal content (Miller *et al.*, 1969). Likewise, if the hen received a vitamin A-deficient diet, chicks will develop the deficiency.

Grazing livestock with access to green, high-quality pastures can store sufficient vitamin A in the liver to be adequate for periods of low intake during the winter or dry season, perhaps as long as 4 to 6 months. Cattle grazing good pasture will have 30 to 80 ppm of liver vitamin A (Rumsey, 1975), and cattle entering the feedlot with 20 to 40 ppm will have adequate liver stores for 3 to 4 months (Perry *et al.*, 1967). Intramuscular injection of 1 million IU of emulsified vitamin A apparently provides sufficient vitamin A to prevent deficiency signs for 2 to 4 months in growing or breeding beef cattle (NRC, 1984a).

Because of vitamin A storage, sheep that graze on green forage during the normal growing season may be able to do reasonably well on a low-carotene diet of dry feed for periods of 4 to 6 months (NRC, 1985b). Goats that have had access to good quality green feed can probably depend on vitamin A stores for a minimum of 3 months without detrimental effects (NRC, 1981b). The tendency of the goat to search out palatable green plant parts ensures it an advantage over other ruminant species. However, goats that are forced to consume more conventional cattle or sheep diets because of the unavailability of browse would not have such an advantage (NRC, 1981b).

Florida beef cattle finished on Roselawn St. Augustine grass during summer did not require supplemental vitamin A for production, however, 25,000 IU per animal daily increased weight gains by approximately 10% when cattle were pastured during the winter (Chapman *et al.*, 1964). In a study of nine cattle ranches from four regions in Florida, forage carotene and consequently liver vitamin A were lower during the winter than the summer (Kiatoko *et al.*, 1982). In this study, cattle in the northern most region, with fewer total grazing days, had significantly lower liver vitamin A than cattle in the other three regions and approached critical levels.

Storage of vitamin A is considerable for humans and farm species that consume diets with high vitamin A value. If sows have built up stores of vitamin A until they are 8 to 9 months of age and are then placed on a diet practically devoid of carotene or vitamin A, they still can produce at least two normal litters (Selke *et al.*, 1967). Well-nourished humans have at least several months' supply that the body can utilize (RDA, 1980).

C. Assessment of Status

A number of critera are available to evaluate the vitamin A status of livestock, including production response, liver vitamin A stores, plasma vitamin A, and cerebrospinal fluid pressure. Vitamin A level of blood may also reflect the nutritional status with respect to vitamin A (Green *et al.*, 1987), however blood vitamin A concentrations are governed by extent of liver stores as well as by current intake. In fact, in some species blood level tends to be maintained until liver stores are exhausted when the diet is devoid of the vitamin. For this reason a normal blood level cannot be interpreted to ensure that current intake is adequate, but a low level indicates a deficiency.

A plasma vitamin A level less than 20 μg/100 ml in Holstein calves suggests a deficiency (Eaton *et al.*, 1970), and a plasma level of 10 μg/100 ml indicates an advanced deficiency. For beef cattle, Perry *et al.* (1967) suggested that plasma vitamin A is probably the most accurate indicator of a borderline deficiency, with a level of less than 40 μg of vitamin A per 100 ml of blood serum indicating deficiency.

For dairy cattle, liver vitamin A values below 1 IU/kg are indicative of a critical deficiency (NRC, 1978a). Feedlot performance of beef cattle was good as long as liver vitamin A stores were above 2 IU/kg (Kohlmeier and Burroughs, 1970). In dairy calves, cerebrospinal fluid pressure rises at a rapid rate after liver concentrations of vitamin A fall below 1–2 ppm. In feeding trials of Perry *et al.* (1967), cattle showed a positive response to supplemental vitamin A when their liver concentrations were as low as 3–4 ppm.

One cannot determine the exact status of vitamin A storage in the pig by analyzing the blood for vitamin A. When vitamin A values drop below 10 μg/g in the liver and below 10 μg/100 ml in the plasma, the pig is quite deficient in vitamin A (Cunha, 1977). Liver biopsies and determination of liver vitamin A values would give the best indication of vitamin A storage or status in the pig as well as for most species.

The vitamin A content of liver of newly hatched chicks is an excellent measure of the vitamin A nutrition of the breeding hens. A deficiency will not occur in chickens having storage levels above 2–5 IU vitamin A per gram of liver. Thus, throughout the life of the chicken vitamin A content of liver serves as a good method of assessing vitamin A status (Scott *et al.*, 1982).

X. SUPPLEMENTATION

Available means of supplementing vitamin A are (1) as part of a concentrate or liquid supplement, (2) included with a free-choice mineral mixture, (3) as an injectable product, and (4) in drinking water preparations. The most convenient and often most effective means to provide vitamin A to livestock is inclusion with concentrate mixtures that will provide uniform consumption of the vitamin. For grazing livestock, however, providing vitamin A in concentrate mixtures generally is not economically feasible. Most vitamin A preparations used today in animal applications are administered orally as an ingredient to blend uniformly in dry feed. With the advent of low-cost synthetic vitamin A, fish oils are only used to a limited extent in various regions of the world.

In the technology of feed manufacturing, liquids such as oils are sprayed onto the feed in the batch or continuous blending operation, the amount being controlled volumetrically. In some cases the use of synthetic vitamin A with antioxidants in oil dilutions has continued; the oil addition plays a secondary role of reducing dustiness of feed and/or supplying an added source of calories (Bauernfeind and DeRitter, 1972).

Because of the lack of stability of vitamin A, particularly regarding exposure to oxygen, trace minerals, pelleting, feed storage, and other factors, the feed industry has readily accepted the dry stabilized forms of the vitamin. Stabilized and protectively coated (or beaded) forms of vitamin A slow destruction of the vitamin but for highest potency fresh supplies of the mixture should be available on a regular basis. Practical considerations that affect vitamin A stability are listed in Table 2.6. The gelatin beadlet in which the vitamin A ester (palmitate or acetate) is emulsified into a gelatin–plasticizer–antioxidant viscous liquid formulation and spray-dried into discrete dry particles results in products with good

TABLE 2.6

Practical Factors That Affect the Stability of Vitamin A[a]

Factors detrimental to stability:
- Long storage of the vitamin product before mixing or of the feed product after mixing
- Vitamin premixes containing minerals
- High environmental temperature and humidity
- Pelleting, blocking, and extrusion
- Hot feed bins that sweat inside upon cooling
- Rancid fat in the feed
- Rain leaks in feed bins or storage facilities

Factors promoting stability:
- Minimum time between manufacture of the vitamin product and consumption by the animal
- Store vitamins in a cool, dark, dry area in closed containers
- Do not mix vitamins and minerals in the same premix until ready to mix the feed
- Do not mix deliquescent substances such as choline chloride with the vitamins
- Use good quality feed ingredients and vitamins
- Good maintenance of storage bins and other equipment
- Coordination between production and purchasing to obtain a quality product that will not be stored an excessively long time before use

[a]Modified by personal communication (1985) with Dr. Charles H. McGinnis, Jr., Manager of Nutritional Services, Rhône-Poulenc, Inc., Atlanta, Ga.

chemical stability, good physical stability, and excellent biological availability (Bauernfeind and DeRitter, 1972).

Vitamin A is often included, along with vitamins D and E, in liquid feed supplements in addition to molasses, fat, urea, and selected minerals. Since the viscosity, pH, and solids content of liquid feed supplements vary considerably, the development of vitamin A forms that would blend uniformly and be stable in such a product of variable composition and characteristics represented a new challenge in the technology of vitamin A application. Products of choice are liquid emulsions of fat-soluble vitamins A, D, and E in a tested formulation of emulsifiers, antioxidants, synergists, preservatives, and carriers.

For grazing livestock, vitamin A can be provided as part of a free-choice mineral mixture as an alternative to mixing it with the feed. The greatest limitation of administering vitamin A with free-choice minerals is unknown consumption by individual animals and destruction of the vitamin with time. Vitamin A stability in a mineral mix is affected by abrasion, moisture, and prooxidant action of trace metals, particularly copper. Mitchell (1967) reported virtual destruction of gelatinized vitamin A when mixed with salt after 5 weeks. Apparently enough moisture was picked up from humid air and perhaps saliva from cattle to induce caking. This may have dissolved some of the gelatin coating, resulting in loss of vitamin A potency. Improved coatings of vitamin A are more insoluble in water and prevent exposure to air and the catalyzing effects of mineral contact.

In recent years, many livestock producers have followed the practice of intramuscular injection with a vitamin A concentrate. This has become a means of administering vitamin A either as a therapeutic measure or more frequently as a prophylactic approach when oral or dietary administration is either inconvenient or impossible. Feedlot cattle with an unknown history are often given an injection of vitamin A (i.e., 1,000,000 IU) as part of the adaptation or preconditioning process before the animal adjusts to the new environment and the high-energy fattening ration. Recommended high doses of vitamin A for ruminants are presented in Table 2.7.

Massive doses of vitamin A are also important under intensive housing conditions where there are increased requirements in early life or under conditions of stress. The practice has been found particularly valuable under the following circumstances (Hoffmann-LaRoche, 1967): for day-old chicks before transport; for laying hens when, for reasons unknown, laying performance suddenly drops; for piglets and calves during weaning or changeover to milk substitutes; as an adjuvant in treatment of attacks by intestinal parasites, lungworms, and other parasites; and for newly arrived feedlot cattle.

Administration of vitamin A in the form of a "megadose" in drinking water or by injection is recommended to support any specific measures used in treatment of diseased and convalescent animals. This is particularly true for animals in

TABLE 2.7

Recommendations for Periodic Administration of High Doses of Vitamin A to Ruminants[a]

Animal	IU per animal
Cattle	
Calves	200,000–500,000
Cattle (rearing)	1,000,000–1,500,000
Cattle (fattening)	1,000,000–2,000,000
Cattle (dairy)	2,000,000 (1 month before calving)
Breeding bulls	2,000,000–3,000,000 (2 months before and at the beginning of service)
Sheep	
Suckling and weaned lambs	100,000–500,000
Rams	1,000,000 (2 months before and at the beginning of service)
Ewes	1,000,000 (during service and a month before lambing)

[a]Adapted from Hoffmann-LaRoche (1967).

which vitamin A stores may have been depleted due to fever or in animals suffering from intestinal disorders when vitamin A absorption is seriously impaired.

The decision for vitamin A supplementation should be based mainly on whether or not a deficiency could be a practical problem. As with most nutrients, a borderline deficiency is much more likely than a severe deficiency. Likewise, a marginal deficiency adversely affecting performance by a few percentage points is not easily detected (Miller, 1979). Based on the positive results that may be derived and taking into account that vitamin A supplementation is inexpensive and no toxicity problems have been reported when given at recommended levels, it seems beneficial to supplement vitamin A at all times when livestock are not grazing or receiving green pastures (roughages) and for animals that are diseased or under stress. Vitamin A injection is one good way to determine if a deficiency exists by injecting half the animals and using the others as controls. This can be done in cases where it would be difficult to conduct complicated experiments.

XI. β-CAROTENE FUNCTION INDEPENDENT OF VITAMIN A

Since 1978, a number of studies have indicated that β-carotene has a function independent of vitamin A in dairy cattle (Bindas *et al.*, 1984). Dairy cattle receiving extra β-carotene have a higher intensity of estrus, increased conception rates, and reduced frequency of follicular cysts than controls. The corpus luteum of the cow has higher β-carotene concentrations than any other organ and it has been suggested that β-carotene has a specific effect on reproduction in addition to its role as a precursor of vitamin A. Graves-Hoagland *et al.* (1988) suggest a positive relationship between β-carotene and luteal cell progesterone during the winter when plasma β-carotene and vitamin A are decreased. Other researchers have found no effect (Folman *et al.*, 1979; Wang *et al.*, 1988) or adverse effects (Folman *et al.*, 1987) of β-carotene supplementation on fertility of dairy cattle.

Friesecke (1978) suggests that β-carotene intake should be sufficient to allow at least 300 μg/100 ml plasma for cattle to ensure maximum reproduction efficiency. He also summarized data indicating low β-carotene concentrations, as fed, in corn silage (0.5–3.4 mg/kg) as compared to grass silage (24.6–32.9 mg/kg) and hay (5.1–22.7 mg/kg). Adams and Zimmerman (1984) analyzed various feedstuffs from different regions of the United States and concluded that in the absence of fresh pasture or good quality forage, diets may not provide sufficient β-carotene. This investigation revealed that samples of moldy corn averaged 98% less carotene than sound corn.

Because of conflicting results relative to the beneficial effects of β-carotene on reproduction, further research using carefully designed studies are needed in a number of species. There has been suggestion in human nutrition that β-carotene may be effective as an inhibitor of some types of cancer.

XII. TOXICITY

In general, the possibility of vitamin toxicities for livestock and poultry is remote. However, of all vitamins, vitamins A is most likely to be provided in toxic concentrations to both humans and livestock, and excess vitamin A has been demonstrated to have toxic effects in most species studied. Presumed upper safe levels are 4 to 10 times the nutritional requirements for nonruminant animals, including birds and fishes, and about 30 times the requirements for ruminants (NRC, 1987). Most of the harmful effects have been obtained by feeding over 100 times the daily requirements for a period of time. Thus, small excesses of vitamin A for short periods of time should not exert any harmful effects. Recommended upper safe levels of vitamin A for livestock and poultry are presented in Table 2.8 (NRC, 1987).

TABLE 2.8

Presumed Upper Safe Levels of Vitamin A (IU/kg Diet) for Livestock and Poultry[a]

Animal	Presumed safe level[b]
Birds	
Chicken, growing	15,000
Chicken, laying	40,000
Duck	40,000
Goose	15,000
Quail	25,000
Turkey, growing	15,000
Turkey, breeding	24,000
Cat	100,000
Cattle, feedlot	66,000
Cattle, pregnant, lactating, or bulls	66,000
Dog	33,330
Fish	
Catfish	33,330
Salmon	25,000
Trout	25,000
Goat	45,000
Horse	16,000
Monkey	100,000
Rabbit	16,000
Sheep	45,000
Swine, growing	20,000
Swine, breeding	40,000

[a]Adapted from NRC (1987).
[b]For chronic dietary administration.

Large doses can be toxic and many cases of overdoses have been reported in various species. The most characteristic signs of hypervitaminosis A are skeletal malformations, spontaneous fractures, and internal hemorrhage (NRC, 1987). Other signs include loss of appetite, slow growth, loss of weight, skin thickening, suppressed keratinization, increased blood-clotting time, reduced erythrocyte count, enteritis, congenital abnormalities, and conjunctivitis. Degenerative atrophy, fatty infiltration, and reduced function of liver and kidney are also typical. A recent report on clinical signs of hypervitaminosis A in rainbow trout *(Salmo gairdneri)* listed growth depression, increased mortality, abnormal and necrotic anal, caudal, pectoral, and pelvic fins, and pale yellow, fragile livers (Hilton, 1983). Bone abnormalities may include extensive bone resorption and narrowing of the bone shaft, bone fragility, and short bones because of retarded growth. Abnormalities in bone modeling are the essential causes of fractures, and the cartilage matrix of bone may be destroyed. For example, high levels of vitamin A injected into rabbits may cause ears to curl because of the destruction of cartilage.

It is believed that release of lysosomal enzymes is responsible for degradative changes observed in tissues and intact animals suffering from hypervitaminosis A (Fell and Thomas, 1960). In hypervitaminosis A, retinol penetrates the lipid of the membrane and causes it to expand, and because the protein of the membrane is relatively inelastic, the membrane is therefore weakened. Thus, many phenomena in hypervitaminosis A can be explained in terms of damage to membranes either of cells or of organelles within cells.

Excess vitamin A affects metabolism of other fat-soluble vitamins with competition for absorption and transport demonstrated. Therefore in diets containing barely adequate levels of vitamins D, E, and K, a marked increase in dietary vitamin A may cause decreases in growth or egg production due to a deficiency of one or more of the other fat-soluble vitamins rather than a toxic effect of vitamin A. The mechanisms of absorption and transport apparently are similar for the carotenoid pigments and the fat-soluble vitamins. A marked increase in dietary vitamin A has been shown to interfere with absorption of carotenoids, thereby resulting in decreased pigmentation for poultry.

The efficiency of carotene conversion to vitamin A declines progressively with increasing intakes. This appears to be a natural homeostatic control mechanism that protects grazing livestock from any harmful effects due to great abundance of carotene present in high quality, fresh forages when they are the major feed for long periods (Miller *et al.*, 1979). Likewise, comparatively rapid disposal of very high levels of stored vitamin A is a protective mechanism. On a practical basis, toxicity is more easily caused by vitamin A than by carotene. Even so, vitamin A toxicity is not a practical problem for livestock, except when unreasonably large amounts are given accidentally (Miller, 1979).

The symptomatology of hypervitaminosis A in humans varies considerably depending on the age of the subject and the duration of the excessive intake.

The young child is especially susceptible, with toxicity occurring in children taking therapeutic doses prescribed for various skin problems. The onset of clinical vitamin A toxicity is insidious, and usually follows consumption of doses of the order of 100,000 IU per day for periods ranging from weeks to months. The varied clinical signs and symptoms are well documented and include headache and other symptoms attributable to increased intracranial pressure, dermatological changes, pruritic skin rash, irritability, pain in arms and legs, and hydrocephalus in children. Plasma vitamin A is always elevated, usually well above 200 μg per deciliter.

Arctic and Antarctic explorers have suffered from acute toxicity after eating polar bear or seal liver. It has been calculated that consumption of 500 g of polar bear liver (13,000–18,000 IU of vitamin A/g) would result in a toxic dose (Sebrell and Harris, 1967). Symptoms of this acute polar bear liver poisoning, which may appear within 2–4 hr, are drowsiness, sluggishness, irritability or an irresistible desire to sleep, severe headache, and vomiting. The medical literature records death following intake of a single dose of 1,000,000 IU in adults and 500,000 IU in children.

Hypervitaminosis A in humans is becoming a clinical problem of increasing frequency as a result of self-medication and overprescription. According to McClaren (1982), vitamin A accumulation in human liver with age signifies that hypervitaminosis A is also becoming a public health problem in Western countries. He points out that habitual intake of vitamin A in the United States and Europe is greater than the recommended daily allowance, the body content of vitamin A is in excess of that of other vitamins in relationship to requirements, and rate of catabolism of retinol is exponential, regardless of stores. Explicit warnings for both children and adults are made by the RDA (1980) regarding excessive vitamin A intakes of more than 25,000 IU on a regular basis.

3

Vitamin D

I. INTRODUCTION

Vitamin D is thought of as the "sunshine vitamin" because it is synthesized in various materials when they are exposed to sufficient sunlight. The two major natural sources of vitamin D are cholecalciferol (vitamin D_3, which occurs in animals) and ergocalciferol (vitamin D_2, which occurs predominantly in plants). In this chapter, the term vitamin D in the absence of a subscript will imply either vitamin D_2 or vitamin D_3. Under modern farming conditions many animals, particularly swine and poultry, are raised in total confinement with little or no exposure to natural sunlight. Even though with adequate sunlight exposure vitamin D is not needed in the diet, it still fits the definition of a vitamin in all respects for animals and humans who are confined indoors away from the sun.

Vitamin D's primary functions are enhancement of intestinal absorption and mobilization, retention, and bone deposition of calcium and phosphorus. Failure of bone mineralization results in rickets in young and osteomalacia in adults. From studies of vitamin D's metabolism and functions, it has been found that the vitamin functions as a hormone.

II. HISTORY

Historical aspects of vitamin D have been reviewed by DeLuca (1979), Miller and Norman (1984), Holick (1987), and Loosli (1988). Vitamin D deficiency rickets is a disease known since antiquity. Hippocrates described conditions that resembled rickets and Soranus Ephesius (born A.D. 130) provided a classic description referring to "the backbone bending" and "legs twisted at the thighs" in a disease noted to be more common in smoky cities than in the country (Arneil, 1975).

Since the middle ages, it was observed that sunlight seemed to have health-giving effects. During the early stages of the Industrial Revolution in the late eighteenth and early nineteenth centuries, there was a mass migration of population from the rural countryside into the industrial centers of Great Britain and Europe. Young children of the working class who lived in the densely populated cities were affected by a severe bone-deforming disease. The combination of industrial pollution and narrow, shaded alleyways prevented children who were reared in this environment from being exposed to sunlight. When necropsy studies were performed on children who died of various causes, it was found in Leyden, The Netherlands, that 85–90% of these children suffered from rickets (Holick, 1987).

In 1822 Sniadecki suggested that rickets was caused by lack of exposure to sunlight. He observed that children who lived in inner cities of Warsaw, Poland, had a very high incidence of rickets, whereas children living on the farms on the outskirts of the city essentially were free of this disease. Almost 70 years later Palm concluded from an epidemiological survey that the common denominator in rickets in children was lack of exposure to sunlight. He encouraged systematic sunbathing as a means of preventing and curing rickets. The majority of scientists and physicians at the time did not believe that simple exposure to sunlight could cure or prevent this bone-deforming disease.

In the early twentieth century, Sir Edward Mellanby began his work on the disease rickets, which was in epidemic proportions in the human population in his native country of England. Mellanby attempted to produce rickets by nutritional means, and in 1922 he was successful in producing the disease in dogs maintained on a diet of oatmeal and unplanned by him, in absence of sunlight. Although Mellanby incorrectly concluded that the healing of rickets was a property of the fat-soluble vitamin A, he placed the study of rickets on an experimental basis.

McCollum, who had discovered the fat-soluble vitamin A, realized that the antirachitic activity discovered by Mellanby was distinct from the antixerophthalmia activity in cod liver oil. McCollum bubbled oxygen through cod liver oil and heated it to destroy the vitamin A activity but the properties of cod liver oil in prevention and cure of rickets remained. Therefore, in 1922 he concluded that this unknown substance represented a new fat-soluble vitamin, which he called vitamin D.

Although it was known that ultraviolet light and vitamin D from cod liver oil were both equally effective in preventing and curing rickets, the close interdependence of these two factors was not immediately realized. Goldblatt and Soames (1923) conducted experiments in which they irradiated rachitic rats with ultraviolet light, removed their livers, and demonstrated that they contained the antirachitic substance, whereas if they were not irradiated, no antirachitic substance could be detected. Steenbock and Black (1924) then realized that ultraviolet irradiation was causing the alteration of some substance in animals and proceeded to demonstrate that ultraviolet irradiation of not only the animals but of their food could heal or prevent rickets.

These important discoveries led to the use of ultraviolet light irradiation of such foods as milk and butter to fortify them with vitamin D and thus eliminate rickets as a major medical problem. In addition, the demonstration that irradiation of food resulted in the production of an antirachitic factor provided the key for the isolation and chemical characterization of vitamin D_2 from the provitamin ergosterol and that irradiation of skin produced vitamin D_3 from the provitamin 7-dehydrocholesterol. In 1932, the structure of vitamin D_2 was simultaneously determined by Windaus in Germany, who named it vitamin D_2, and by Askew in England, who named it ergocalciferol. In 1936, Windaus succeeded in identifying the structure of vitamin D_3. Vitamin D_1 had been isolated earlier by Windaus and his colleagues but was later shown to be an adduct of vitamin D_2 and an irradiation side product, lumisterol.

A new phase of vitamin D research began in the late 1960s and has resulted in a complete change in attitude regarding the problems associated with this substance. It is now recognized that vitamin D is simply the precursor of at least one new steroid hormone, 1,25-dihydroxyvitamin D (1,25-$(OH)_2D$). In 1966 Lund and DeLuca demonstrated the disappearance of vitamin D_3 following administration and the appearance of several metabolites that possessed more potent antirachitic activity. The first of these metabolites, 25-hydroxyvitamin D_3 (25-OHD_3), was found to be produced in the liver. The 25-OHD_3 form of vitamin D was chemically synthesized by Blunt and co-workers in 1968. Using the radioactively labeled chemical, 25-OHD_3 was shown to be metabolically altered to at least three dihydroxy compounds, the most important thought to be 1,25-$(OH)_2D_3$. In the early 1970s it was determined that the kidney was the principal site of 1,25-$(OH)_2D_3$ production. From the discoveries that this biologically active vitamin D metabolite is produced exclusively in kidney and is found in the nuclei of intestinal cells came the concept that, in terms of its structure and mode of action, vitamin D is similar to steroid hormones. The discovery that the biological actions of vitamin D can be explained by a hormone-like mechanism of action marked the beginning of the modern era of vitamin D research.

III. CHEMICAL STRUCTURE, PROPERTIES, AND ANTAGONISTS

Vitamin D designates a group of closely related compounds that possess antirachitic activity. It may be supplied through the diet or by irradiation of the body. There are about 10 provitamins that, after irradiation, form compounds having variable antirachitic activity. The two most prominent members of this group are ergocalciferol (vitamin D_2) and cholecalciferol (vitamin D_3). Chemical structures of ergocalciferol and cholecalciferol and their precursors, ergosterol and 7-dehydrocholesterol, are shown in Fig. 3.1. All sterols possessing vitamin D activity have the same steroid nucleus; they differ only in the nature of the side chain attached to carbon 17.

Ergocalciferol is derived from a common plant steroid, ergosterol, and is the usual dietary source of vitamin D. Cholecalciferol is produced exclusively from

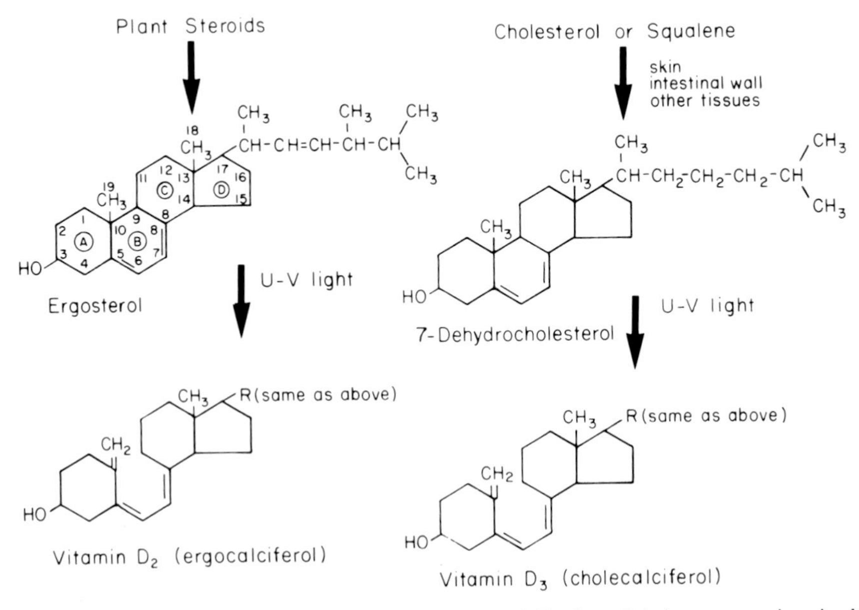

Fig. 3.1 Vitamin D_2 (ergocalciferol), vitamin D_3 (cholecalciferol), and their precursors in animal (7-dehydrocholesterol) and plant (ergosterol) tissues.

animal products. 7-Dehydrocholesterol is derived from cholesterol or squalene, which is synthesized in the body and present in large amounts in skin, intestinal wall, and other tissues. Vitamin D precursors have no antirachitic activity until the B-ring is opened between the 9 and 10 positions by irradiation and a double bond is formed between carbons 10 and 19 to form vitamin D.

Vitamin D occurs as colorless crystals that are insoluble in water but readily soluble in alcohol and other organic solvents. It is less soluble in vegetable oils. Cholecalciferol crystallizes as fine white needles from diluted acetone and has a melting point between 84° and 85°C. Vitamin D can be destroyed by overtreatment with ultraviolet light and by peroxidation in the presence of rancidifying polyunsaturated fatty acids. There is negligible loss of crystalline cholecalciferol over 1 year's storage in amber evacuated capsules at refrigerator temperatures, or of ergocalciferol for 9 months. Solutions in corn oil stored in amber bottles have shown no loss of activity during 30 months' storage. Overirradiation of ergocalciferol or cholecalciferol produces numerous irradiation products such as tachysterols, supra-sterol$_1$, supra-sterol$_2$, and others. Some of these compounds have partial vitamin D activity, some are toxic, and some may be potent antagonists of vitamin D_3 (Scott *et al.*, 1982).

IV. ANALYTICAL PROCEDURES

Vitamin D activity can be expressed in units based on a bioassay in vitamin D-deficient rats or chicks. When assayed in the rachitic rat, ergocalciferol and cholecalciferol are equally active with a potency of 40,000 units/mg of pure steroid. However, in chicks and other birds, ergocalciferol has only about one-tenth the activity of cholecalciferol (Chen and Bosmann, 1964). One international or USP unit of vitamin D activity is defined as the activity of 0.025 μg of vitamin D_3 contained in the USP vitamin D reference standard. For poultry, the term ICU (International Chick Unit) is employed with reference to use of D_3 versus D_2.

Methods of analysis for vitamin D are complex and somewhat difficult because there are so many isomers, and not all are biologically active. The standard method for assay of vitamin D supplements for feeds is a biological assay. Vitamin D is the only vitamin in which a biological method has not been largely replaced by chemical, physical, or microbiological assay. The rat is often used to assay products for human and animal use, but the chick is the assay animal of choice in assessing supplements intended for poultry feeding because of unequal activity of vitamin D_2 and D_3 for this species as compared to most mammals. Vitamin D_3 biological assay using young chicks has been standardized by the Association of Analytical Official Chemists and involves feeding a standard rachitogenic ration to young chicks for a 21-day period and determining the ash content of dry fat-free tibia of either chicks that have recieved the supplements containing unknown quantities of vitamin D or those receiving standard quantities of a vitamin D reference standard (Scott *et al.*, 1982).

A curative method involves developing rickets in young rats and chicks, then graded increments of unknown samples and standard vitamin D are added to diets for 7 days, followed by a "line test." The "line test" entails staining a section of the metaphysis of the proximal end of the tibia with silver nitrate ($AgNO_3$) to show deposition of calcium salts. The silver precipitates the PO_4 and, on exposure to light, Ag_3PO_4 is reduced to Ag, which deposits in a black line. Bones are then graded numerically according to degree of calcification. Also x-rays may be taken instead of using the line test. Chemical and physical methods to analyze vitamin D generally lack sensitivity of biological assays. Thus, they are not adequate for measuring samples that contain low concentrations of vitamin D. However, these physical and chemical means of vitamin D determination have the advantage of being less time-consuming than biological assays and are frequently used on samples known to contain high levels of vitamin D (Miller and Norman, 1984).

Physical and chemical methods of vitamin D analysis include ultraviolet absorption, colorimetric procedures, fluorescence spectroscopy, gas chromatography–mass spectroscopy, competitive binding assays, and high-pressure liquid chromatography (HPLC). The HPLC procedure is very promising, with the separation process resulting in an exceedingly high resolving capability and increased sensitivity (Miller and Norman, 1984). A major advantage of HPLC is that compounds are not altered by the heat of gas–liquid chromatography, so may be detected as the actual known compound. Compounds may be identified by the retention time of either an internal or external standard and a new technique of stop flow in which the UV spectrum of the molecule separated may be examined.

V. METABOLISM

A. Absorption and Conversion from Precursors

Vitamin D obtained from the diet is absorbed from the intestinal tract, with conflicting reports as to which portion of the small intestine serves as the primary absorption site. It has also been suggested that the largest amount of dietary vitamin D is more likely to be absorbed in the ileum because of longer retention time of food in the distal portion of the intestine (Norman and DeLuca, 1963).

Vitamin D is absorbed from the intestinal tract in association with fats, as are all the fat-soluble vitamins. Like the others, it requires the presence of bile salts for absorption. Because it is fat soluble, vitamin D is absorbed with other neutral lipids via chylomicra into the lymphatic system of mammals or the portal circulation of birds and fishes. It has been reported that only 50% of a dose of vitamin D is absorbed. However, considering that sufficient amounts of vitamin D are usually produced by daily exposure to sunlight, it is not surprising that the body has not evolved a more efficient mechanism for dietary vitamin D absorption (Miller and Norman, 1984).

Cholecalciferol is produced by irradiation of 7-dehydrocholesterol with ultraviolet light either from the sun or from an artificial source. Cholecalciferol is synthesized in the outer skin layers. Presence of the provitamin 7-dehydrocholesterol in the epidermis of the skin and sebaceous secretions is well recognized. For poultry, Koch and Koch (1941) reported that skin of legs and feet of the chicken contains about eight times as much 7-dehydrocholesterol (provitamin D_3) as the body skin.

During exposure to sunlight, the high-energy UV photons (290–315 nm) penetrate the epidermis and photolyze 7-dehydrocholesterol (provitamin D_3) to previtamin D_3. Once formed, previtamin D_3 undergoes a thermally induced isomerization to vitamin D_3 that takes 2–3 days to reach completion. Approximately 15% of provitamin D_3 in human skin exposed to 10 min of simulated sunlight is converted in the stratum basale to vitamin D_3 (Holick *et al.*, 1981). Longer exposure times do not significantly increase D_3 concentrations in the epidermis.

Heuser and Norris (1929) showed that 11–45 min of sunshine daily were sufficient to prevent rickets in growing chicks, and that no further improvements in growth were obtained under these conditions by adding cod liver oil. During initial exposure to sunlight, provitamin D_3 in the human epidermis is efficiently converted to previtamin D_3. However, because previtamin D_3 is also labile to sunlight, once it is formed it begins to photolyze to additional photoproducts, principally lumisterol and tachysterol. The net result is that prolonged exposure to sunlight does not significantly increase the previtamin D_3 concentration above about 15% of the initial provitamin D_3 concentrations (Holick, 1987).

More than 90% of previtamin D_3 synthesis in skin occurs in the epidermis. The cholecalciferol formed by irradiation of the 7-dehydrocholesterol in the skin is absorbed through the skin and transported by the blood to the lipids throughout the body. It is clear that absorption can take place from the fact that rickets can be successfully treated by rubbing cod liver oil on the skin. Once vitamin D_3 is formed, it is transported in the blood, primarily bound to an α-globulin, and becomes immediately available for further metabolism (Imawari *et al.*, 1976). Some of the vitamin D_3 formed in and on the skin ends up in the digestive tract as many animals consume the vitamin as they lick their skin and hair.

B. Conversion to Metabolically Active Forms

Prior to 1968, it was almost universally thought that cholecalciferol and ergocalciferol (D_3 and D_2) were the circulating antirachitic agents in the living animal system. Starting in 1968 DeLuca and a number of other researchers demonstrated that vitamin D_3 undergoes a multiple series of transformations and multisite interactions in the living system (DeLuca, 1979). Whether the vitamin is ingested orally or produced photochemically in the skin, these chemical changes occur before any significant biological response is registered in the intestine or bone. Production and metabolism of the vitamin D_3 necessary to activate the target organs are illustrated in Fig. 3.2.

Once vitamin D (D_2, D_3, or both) enters the blood, it circulates at relatively low concentrations. This phenomenon is probably a result of its rapid accumulation in the liver. Once in the liver, the first transformation occurs in which a microsomal system hydroxylates the 25 carbon in the side chain to produce 25-OH vitamin D. This metabolite is the major circulating form of vitamin D under normal conditions and during vitamin D excess (Littledike and Horst, 1982).

Conversion to 25-OHD_3 takes place predominantly in the microsomes but also in the mitochondria. The microsomal enzyme is an enzyme of low capacity and high affinity and is probably the one of greatest physiological importance (Madhok and DeLuca, 1979). The mitochondrial enzyme is a high-capacity, low-affinity enzyme and is thought to hydroxylate vitamin D under conditions of high substrate concentrations, such as vitamin D_3 toxicity. Liver is the major site of 25-hy-

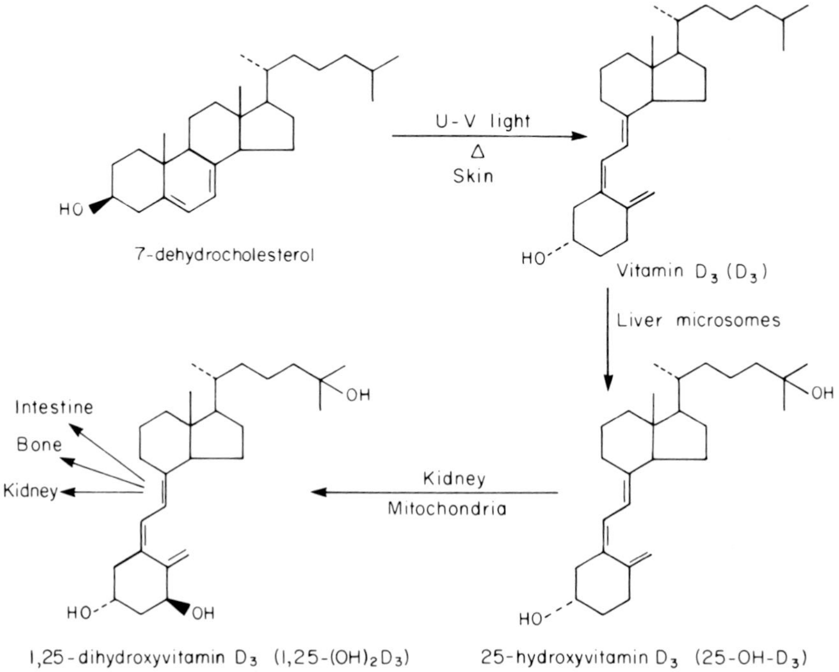

Fig. 3.2 The functional metabolism of vitamin D_3 necessary to activate the target organs of intestine, bone, and kidney.

droxylation of vitamin D_3, however, the intestine and kidney can also produce 25-OHD_3, although the amount of 25-hydroxylation taking place in these organs is small (Tucker *et al.*, 1973).

The 25-OHD_3 is then transported to the kidney on the vitamin D transport globulin, where it can be converted to a variety of compounds, of which the most important appears to be 1,25-$(OH)_2D_3$. This reaction occurs in the mitochondrial fraction and is catalyzed by a three-component, mixed-function monoxygenase involving NADPH, molecular oxygen, a flavoprotein, an iron–sulfur protein, and cytochrome P-450 (Ghazarian *et al.*, 1974). The 1,25-$(OH)_2D_3$ that is formed in the kidney is then transported to the intestine, bones, or elsewhere in the kidney, where it is involved in the metabolism of calcium and phosphorus. The hormonal form is the metabolically active form of the vitamin that functions in intestine and bone, whereas 25-OHD_3 and vitamin D_3 do not function at these specific sites.

Production of 1,25-$(OH)_2D_3$ is very carefully regulated by parathyroid hormone in response to serum calcium and phosphate concentrations. Ribovitch and DeLuca

(1975) reported that removal of parathyroid glands resulted in an animal's inability to adapt to varying calcium demands by increasing intestinal calcium absorption. It is now known that the most important point of regulation of the vitamin D endocrine system occurs through the stringent control of the activity of the renal 1-hydroxylase. In this way, the production of the hormone 1,25-$(OH)_2D$ can be modulated according to the calcium needs of the organism (Miller and Norman, 1984). Factors known to affect the activity of the 1-hydroxylase include calcium, parathyroid hormone, and vitamin D status.

Studies have confirmed that both intestinal and serum 1,25-$(OH)_2D$ concentrations and *in vitro* activity of 1-hydroxylase are inversely related to dietary and serum calcium concentrations. Further evidence that calcium exerts an effect on activity of 1-hydroxylase came from experiments showing that an animal's ability to adapt to dietary calcium could be eliminated by administration of an exogenous source of 1,25-$(OH)_2D$ (Miller and Norman, 1984).

Besides 1,25-$(OH)_2D$, the kidney also converts 25-OHD_3 to other known compounds, including 24,25-$(OH)_2D_3$, 25,26-$(OH)_2D_3$, and 1,24,25-$(OH)_3D_3$. The role of these compounds in function or inactivation of vitamin D has not been fully evaluated, and significant physiological roles are yet to be discovered. Although controversy exists, it has been suggested that 24,25-$(OH)_2D_3$ is responsible for the mineralization of bone, for the suppression of parathyroid hormone secretion, for the production of cartilaginous components, and for the development of chick embryos. Henry and Norman (1978) have shown that hatchability of eggs from hens is severely depressed in vitamin D-deficient hens even though they are fed 1,25-$(OH)_2D_3$. Evidently, 1,25-$(OH)_2D_3$ is effective in maintaining blood calcium levels so that egg production and egg shell thickness remain normal. However, without vitamin D, the upper mandible of the chicks fails to develop and consequently the chicks cannot crack the shell, resulting in mortality. The reason for this has been suggested to be the requirement of another vitamin D metabolite, 24,25-$(OH)_2D_3$. However, Hart *et al.* (1984) were unable to find a role for 24,25-$(OH)_2D_3$ in hatchability. Bordier *et al.* (1978) showed that 24,25-$(OH)_2D_3$ is required with 1,25-$(OH)_2D_3$ for normal healing of osteomalacia in humans. Although it has been shown that 24-hydroxylation appears to have a role in bone mineralization, it is also believed to be involved in the elimination pathway for 25-hydroxy and 1,25-dihydroxy vitamin D_3.

C. Transport

In mammals, vitamin D, 25-OHD, and possibly 24,25-$(OH)_2D$ and 1,25-$(OH)_2D$ are all transported on the same protein. This protein has a molecular weight of 50,000–60,000 and in humans appears to be a single-chain polypeptide. It is called transcalciferin, or vitamin D-binding protein (DBP), and binds 25-

OHD (the major circulating metabolite) with a higher affinity than it binds vitamin D or 1,25-$(OH)_2D_3$ (Haddad and Walgate, 1976). With radioactively labeled metabolites, Hay and Watson (1977) observed in 65 of the 72 species of mammals studied that the vitamin D metabolites associate with a protein of α-globulin mobility as determined by disk gel electrophoresis. This was also observed in bony fish, reptiles, and some species of birds. However, for several mammalian species and in four species of birds, 25-OHD was carried on albumin or on a protein with albumin-like electrophoretic mobility.

D. Storage and Placental Transfer

In contrast to aquatic species, which store significant amounts of vitamin D in liver, land animals and humans do not store appreciable amounts of the vitamin. The body has some ability to store the vitamin, although to a much lesser extent than is the case for vitamin A. Principal stores of vitamin D occur in blood and liver, but it is also found in lungs, kidneys, and elsewhere (Maynard *et al.*, 1979). Since it was known that liver serves as a storage site for vitamin A, another fat-soluble vitamin, it was thought that the liver also functioned as a storage site for vitamin D. However, it has since been shown that blood has the highest concentration of vitamin D when compared with other tissues; in pigs, the amount of vitamin D in blood is several-fold higher than that in liver (Quaterman *et al.*, 1964).

The persistence of vitamin D in animals during periods of vitamin D deficiency may be explained by slow turnover rate of vitamin D in certain tissues, such as skin and adipose tissue. During times of deprivation, vitamin D in these tissues is released slowly, thus meeting vitamin D needs of the animal over a long period of time (Miller and Norman, 1984).

Transplacental movement of calcium increases dramatically during the last trimester of gestation in all species observed. It is well established that in most mammalian species, fetal plasma calcium levels are higher than maternal levels at term. In sheep, passive diffusion accounts for a minor component of placental calcium movement with active transport mechanisms responsible for greater than 90% (Braithwaite *et al.*, 1972). In the pregnant rat (and perhaps human), 1,25-$(OH)_2D$ is a critical factor in the maintenance of sufficient maternal calcium for transport to the fetus and may play a role in normal skeletal development of the neonate (Lester, 1986). While there is no large transfer to the fetus, a liberal intake during gestation does provide a sufficient store in newborn to help prevent early rickets. For example, newborn lambs can be provided enough in this way to meet their needs for 6 weeks. Parenteral cholecalciferol treatment of sows before parturition proved an effective means of supplementing young piglets with cholecalciferol (via the sow's milk) and its more polar metabolites via placental transport (Goff *et al.*, 1984).

E. Excretion

Excretion of absorbed vitamin D and its metabolites occurs primarily in feces with the aid of bile salts. Very little vitamin D appears in urine. Ohnuma *et al.* (1980) suggested that the newly discovered metabolite 1,25-$(OH)_2D_3$-26,23-lactone may represent one of the first steps in the catabolism of 1,25-$(OH)_2D_3$. Because the half-life of 25,26-$(OH)_2D$ in serum is only 10 days, this metabolite might have an excretory role.

VI. FUNCTIONS

A. Relationship to Calcium and Phosphorus Homeostasis

The general function of vitamin D is to elevate plasma calcium and phosphorus to a level that will support normal mineralization of bone as well as other body functions. Recently, evidence also suggests a regulatory role of vitamin D (1,25-$(OH)_2D$) in immune cell functions (Reinhardt and Hustmyer, 1987). The possible use of vitamin D analogs to bring about differentiation of myelocytic-type leukemias and in treatment of psoriasis has been an important new development (DeLuca, 1988).

Tetany in humans and animals results if plasma calcium levels are appreciably below normal. Two hormones, thyrocalcitonin (calcitonin) and parathyroid hormone (PTH), function in a delicate relationship with 1,25-$(OH)_2D_3$ to control blood calcium and phosphorus levels. Production rate of 1,25-$(OH)_2D_3$ is under physiological control as well as dietary control (see Section V, B). Calcitonin, contrary to the other two, regulates high serum Ca levels by (1) depressing gut absorption, (2) halting bone demineralization, and (3) reabsorption in the kidney. Vitamin D brings about an elevation of plasma calcium and phosphorus by stimulating specific pump mechanisms in the intestine, bone, and kidney. These three sources of calcium and phosphorus thus provide reservoirs that enable vitamin D to elevate calcium and phosphorus levels in blood to levels that are necessary for normal bone mineralization and for other functions ascribed to calcium.

1. Intestinal Effects

It is well known that vitamin D stimulates active transport of calcium and phosphorus across intestinal epithelium. This stimulation does not involve the parathyroid hormone directly but involves the active form of vitamin D. Parathyroid hormone indirectly stimulates intestinal calcium absorption by stimulating production of 1,25-$(OH)_2D_3$ under conditions of hypocalcemia.

The mechanism whereby vitamin D stimulates calcium and phosphorus absorption is still not completely understood. Current evidence (Wasserman, 1981) indicates that 1,25-$(OH)_2D_3$ is transferred to the nucleus of the intestinal cell, where it interacts with the chromatin material. In response to the 1,25-$(OH)_2D_3$, specific RNAs are elaborated by the nucleus and when these are translated into specific proteins by ribosomes, the events leading to enhancement of calcium and phosphorus absorption occur (Scott *et al.*, 1982).

In the intestine, 1,25-$(OH)_2D_3$ promotes synthesis of calcium-binding protein (CaBP) and other proteins and stimulates calcium and phosphorus absorption. Administration of 1,25-$(OH)_2D_3$ to rachitic animals has been shown to stimulate the incorporation of [^{3}H]leucine into several proteins of the intestinal mucosa. This apparent increase in protein synthesis was at least in part accounted for by the discovery that 1,25-$(OH)_2D$ induces synthesis of a specific intestinal protein that has been identified as CaBP. This CaBP is not present in the intestine of rachitic chicks but appears following vitamin D treatment.

Intestinal calcium transport relies on the integrated effects of both genomic and nongenomic mechanisms of hormone action. Two kinds of mucosal proteins are dependent on vitamin D: (1) CaBP and (2) intestinal membrane calcium-binding protein (IMCal). IMCal was only recently discovered and it is a membrane component of the translocation mechanism rather than a cytosol constituent (Schachter and Kowarski, 1982). It is proposed that the primary nongenomic mechanism by which 1,25-$(OH)_2D_3$ regulates calcium transport across the luminal membrane of the enterocyte involves inducing a specific alteration in membrane phosphatidylcholine content and structure, which leads to an increase in membrane fluidity and thereby to an increase in calcium transport rate. In addition to inducing CaBP and IMCal, 1,25-$(OH)_2D_3$ has been shown to increase levels of several other proteins in the intestinal mucosa. These include alkaline phosphatase, calcium-stimulated ATPase, and phytase enzyme activities (Miller and Norman, 1984).

Originally, it was felt that vitamin D did not regulate phosphorus absorption and transport, but in 1963 it was demonstrated, through the use of an *in vitro* inverted sac technique, that vitamin D does in fact play such a role (Harrison and Harrison, 1963). Little is known about the actual mechanism of phosphate transport, but phosphate is transported against an electrochemical potential gradient involving sodium in response to 1,25-$(OH)_2D_3$.

2. Bone Effects

In young animals during bone formation, minerals are deposited on the matrix. This is accompanied by an invasion of blood vessels that gives rise to trabecular bone. This process causes bones to elongate. During a vitamin D deficiency, this organic matrix fails to mineralize, causing rickets in the young and osteomalacia in adults. 1,25-$(OH)_2D_3$ brings about mineralization of the bone matrix, and Weber *et al.* (1971) provided evidence that 1,25-$(OH)_2D_3$ is localized in the nuclei of bone cells. Also, there is some indication that 24,25-$(OH)_2D_3$ and possibly 25-OHD_3 may have unique actions on bone. The 24,25-$(OH)_2D_3$ appears to be accumulated in bone, where it promotes normal development (Bar *et al.*, 1982).

Vitamin D plays another role in bone, that is, in the mobilization of calcium from bone to the extracellular fluid compartment. This function is shared by PTH (Garabedian *et al.*, 1974). However, little is known about the mechanism of bone reabsorption in response to these factors, although it may be similar or identical to the intestinal transport system. It is an active process requiring metabolic energy, and presumably it transports calcium and phosphate across the bone membrane by acting on osteocytes and osteoclasts.

Another role of vitamin D has been proposed in addition to its involvement in bone, namely, in the biosynthesis of collagen in preparation for mineralization (Gonnerman *et al.*, 1976). A vitamin D deficiency causes inadequate cross-linking of collagen as a result of low lysyl oxidase activity, which is involved in a condensation reaction for the collagen cross-linking. This may be a direct effect of vitamin D or a result of mineral changes in blood, it is not considered a major function of vitamin D.

3. Kidney Effects

There is evidence that vitamin D functions in the distal renal tubules to improve calcium reabsorption. It is known that 99% of the filtered load of calcium is reabsorbed in the absence of vitamin D and the parathyroid hormone. The remaining 1% is under control of these two hormonal agents, although it is not known whether they work in concert. It has been shown that 1,25-$(OH)_2D_3$ functions in improving renal reabsorption of calcium (Sutton and Dirks, 1978). With intact parathyroids and without vitamin D, renal tubular resorption of inorganic phosphate decreases, thereby increasing phosphate clearance and resulting in hypophosphatemia, although the parathyroids maintain a normal plasma calcium level. With adequate vitamin D, greater resorption of phosphorus by the renal tubules occurs. Without intact parathyroids, vitamin D actually increases renal loss of phosphorus.

B. Calcium and Phosphorus Absorption by Ruminants

1. Calcium

There is now clear evidence that sheep and cattle absorb calcium from their gut according to need and that they can alter the efficiency of absorption to meet a change in requirement. For example, Braithwaite and Riazuddin (1971) have shown that young sheep with a high calcium requirement absorb calcium at a higher rate and with greater efficiency than mature animals with a low requirement. An increase in both absorption and efficiency of absorption also occurs in mature sheep when their requirement for calcium is increased through pregnancy or lactation or after a period of calcium deficiency (Braithwaite, 1974).

Studies in cattle have given similar results. Thus the efficiency of absorption of calcium in the small intestine of the dairy cow has been shown to rise in response to a reduction in dietary calcium intake and to the onset of lactation (Scott and McLean, 1981). Calcium absorption has also been shown to be directly related to milk production, though in early lactation when the demand for calcium is greatest, the increase in absorption falls short of the requirement, with the deficit being met by increased bone resorption (Braithwaite *et al.*, 1969).

The mechanism by which calcium is adjusted in response to requirement has received much attention. In this sequence, a fall in plasma calcium concentration resulting from an increase in demand leads in turn to an increase in parathyroid hormone release. This then stimulates the increased production by the kidney of 1,25-$(OH)_2D_3$, which acts on the gut to increase the production of CaBP and so accelerates calcium absorption. In a reverse manner, an increase in plasma calcium concentration causes suppression of parathyroid hormone release, a reduction in 1,25-$(OH)_2D_3$ production, and reduced calcium absorption. Although all aspects of this system have not yet been fully examined in ruminants, it does appear that the same mechanism operates in that an increase in circulatory 1,25-$(OH)_2D_3$ level has been found to precede the increase in calcium absorption that occurs in cattle soon after parturition (Horst *et al.*, 1978).

2. Phosphorus Absorption

Sheep fed on roughage diets usually excrete little phosphorus in their urine (Scott and McLean, 1981), so control of phosphorus balance must therefore be achieved within the gut through control of either absorption or secretion or both. Saliva is the main contributor of phosphorus in the gut. Little or no net absorption of phosphorus appears to occur from either the forestomach or large intestine, and it is generally accepted that the upper small intestine is the major absorptive site. Using sheep fitted with reentrant cannulas, several workers have shown that the secretion of phosphorus before the pylorus (salivary) is closely matched by net absorption in the small intestine. Ability to balance absorption against secretion has been shown to be unaffected by wide variations in dietary calcium : phosphorus values (Scott and McLean, 1981).

Until recently, secretion of phosphorus in saliva has usually been viewed in the context of its role as a buffer against the volatile fatty acids produced in the rumen. However, studies by Australian workers (Tomas, 1974) have suggested that salivary glands, apart from their role as a source of buffer, may also play an important role in phosphorus homeostasis by controlling the amount of phosphorus secreted into the gut. Evidence for this function was provided by sheep studies in which both parotid salivary ducts were ligated, a procedure that led to a small increase in urinary phosphorus excretion and a proportional reduction in fecal phosphorus excretion (Tomas and Somers, 1974). Similar changes in the pathway of phosphorus excretion were also seen in sheep in which part of the parotid salivary flow was collected and returned directly to the circulation.

Ruminants have a higher renal threshold for phosphorus excretion than do monogastric species, and it is important to consider what advantage this confers. Poor quality roughage diets, apart from their low digestibility, also tend to contain little phosphorus (McDowell, 1985a). Therefore, the ratio between the amount of phosphorus required for saliva production and dietary intake is wide. If the renal threshold for phosphorus excretion in ruminants were as low as in monogastric species, then at times when the concentration of inorganic phosphorus in the plasma rises in response to reabsorption, phosphorus would be excreted in the urine. This phosphorus would not, however, in any real sense be surplus to requirement and its loss would have to be met from a diet low in available phosphorus. Under such conditions, there is clearly an advantage to the ruminant in maintaining the high renal threshold for phosphorus excretion.

Concentrate diets, especially those that include fish meal, contain much larger quantities of phosphorus than do roughage diets, to the point where intake may equal or exceed the amount secreted in the saliva. Under these conditions, need to control phosphorus absorption is clearly less critical and, as a result, a different level of control may operate. Increasing dietary phosphorus intake leads to increased absorption and increased urinary phosphorus excretion.

Administration of large amounts of parathyroid hormone over several days has been shown to reduce fecal phosphorus excretion in cattle (Mayer *et al.*, 1968), though whether this was due to reduced secretion or increased absorption is not clear. In sheep, parathyroidectomy has been shown to result in a negative balance for both calcium and phosphorus, and it has been suggested that the effect on phosphorus balance was the result of a reduction in amount of salivary phosphorus reabsorbed (McIntosh and Tomas, 1978). 1,25-$(OH)_2D_3$ has also been suggested as a possible regulator of intestinal phosphorus absorption in ruminants (Scott and McLean, 1981), though whether this was a primary effect or secondary to its effects on calcium absorption and deposition in bone is not clear.

VII. REQUIREMENTS

Animals and humans do not have a nutritional requirement for vitamin D when sufficient sunlight is available, since vitamin D_3 is produced in skin through action of ultraviolet light on 7-dehydrocholesterol. In addition to sunlight, other factors influencing dietary vitamin D requirements include (1) amount and ratio of dietary calcium and phosphorus, (2) availability of phosphorus and calcium, (3) species, and (4) physiological factors.

Vitamin D becomes a nutritionally important factor in the absence of sufficient sunlight. Sunlight that comes through ordinary window glass is ineffective in producing vitamin D in skin since glass does not allow penetration of ultraviolet rays, and its effectiveness is dependent on length and intensity of UV rays that reach the body. The radiation that reaches the earth contains only a small part of the UV range that has an antirachitic effect. Sunlight is more potent in the tropics than in the temperate or Arctic zones, more potent in summer than in winter, more potent at noon than in the morning or evening, and more potent at high altitudes (Maynard *et al.*, 1979). Sunlight provides most of its antirachitic powers during the 4 hr around noon.

Besides geographical and seasonal considerations, sunlight may be blocked by many means. Potency of UV light varies greatly with differences in atmospheric conditions and because of clouds, mist, or smoke. Air pollution screens out many UV rays. Air pollution became prevalent during the Industrial Revolution and at the same time the incidence of rickets became widespread in industrial cities. It is now known that this epidemic of rickets was partly due to the lack of sunlight because of the presence of air pollution. Thus, rickets has been called the first air pollution disease (Miller and Norman, 1984). Animals housed for much of the year must depend on their feed for the vitamin D they need; in a modern agricultural economy this applies particularly to intensive swine and poultry production.

The colors of the coat and skin are important in determining response to irradiation. Ultraviolet irradiation is more effective on exposed skin than through a heavy coat of hair. Irradiation is less effective on dark-pigmented skin. This has been shown to be true for white and black breeds of hogs as well as for people. White pigs have been shown to resist vitamin D deficiency signs about twice as long as colored pigs; an average of 45 min of daily exposure to January sunshine for 2 weeks was sufficient to cure rickets in pigs in Minnesota (Cunha, 1977).

In white humans, 20–30% of the UV radiation is transmitted through the epidermis, while in black people less than 5% of the UV radiation penetrates the epidermis (Holick, 1987). When specimens of surgically obtained white and black skin were exposed to simulated sunlight under the same conditions, longer exposure times were needed to maximize vitamin D_3 formation in the samples of black skin. As skin pigmentation increased, the time required to maximize vitamin D_3 formation increased from 30 min to 3 hr.

Aging has an effect on the cutaneous production of vitamin D_3. In humans older than about 20 years, skin thickness decreases linearly with time. A comparison of amounts of previtamin D_3 generated in skin of young subjects and elderly subjects showed that aging can decrease more than 2-fold the capacity of the skin to produce previtamin D_3 (Holick, 1986).

Amounts of dietary calcium and phosphorus, and the physical and chemical forms in which they are presented, must be considered when determining requirements for vitamin D. High dietary calcium concentrations can precipitate phosphates as insoluble calcium phosphate. If calcium is given in the form of a relatively insoluble compound, or even a calcium compound that is normally easily soluble but is too coarsely ground, it may be comparatively unavailable, for example, coarse limestone (calcium carbonate) (Franklin and Johnstone, 1948). Soluble calcium salts are more readily absorbed and oxalates tend to interfere with absorption, but some of this interference can be overcome by dietary vitamin D or irradiation.

Correspondingly, while the phosphorus of inorganic orthophosphate tends to be well absorbed, other factors being favorable, that of phytic acid (see Chapter 16), which is the predominant phosphorus compound of unprocessed cereal grains and oilseeds, seems to be poorly available except to species, such as ruminants, with massive populations of microorganisms in the gut that synthesize phytase enzymes (Abrams, 1978). Phosphorus absorption is mostly independent of vitamin D intake, with the inefficient absorption in rickets being secondary to failure of calcium absorption and the improvement upon vitamin administration being a result of improving calcium absorption.

The need for vitamin D depends to a large extent on the ratio of calcium to phosphorus. As this ratio becomes either wider or narrower than the optimum, the requirement for vitamin D increases, but no amount will compensate for severe deficiencies of either calcium or phosphorus. The dietary dry matter of rapidly growing young stock should approximately contain between 0.6 and 1.2% calcium, with a calcium/phosphorus ratio in the range of about 1.2:1 to 1.5:1; for adult maintenance, lower calcium levels and wider calcium/phosphorus ratios are possible. In these situations vitamin D requirements seem to be at a minimum and risks of vitamin D deficiency are less probable. The young rat needs very little vitamin D, indeed, to make the species suitably responsive in the biological assay of the vitamin, the dietary phosphorus content has to be kept low and the calcium/phosphorus ratio high (Abrams, 1978).

Intestinal pH as well as other dietary nutrients influence calcium and phosphorus requirements, and thus vitamin D requirements. Fermentation of excess carbohydrates makes intestinal contents more acid, which favors both calcium and phosphorus absorption, probably by converting less soluble alkaline salts to the more soluble acid salts. High intakes of fats containing higher fatty acids increase highly insoluble calcium soaps. Potassium may increase phosphorus absorption but cations that form insoluble phosphates such as iron and aluminum interfere.

Vitamin D requirements of various species (Table 3.1) are sufficiently high to produce normal growth, calcification, production, and reproduction in the

TABLE 3.1

Vitamin D Requirements for Various Animals and Humans[a]

Animal	Purpose	Requirement	Reference
Beef cattle	All classes	275 IU/kg	NRC (1984a)
Dairy cattle	Calf	300 IU/kg	NRC (1978a)
	Adult	300 IU/kg	NRC (1978a)
Goat	All classes	1400 IU/kg	Morand-Fehr (1981)
Chicken	Leghorn, 0–20 weeks	200 ICU/kg	NRC (1984b)
	Leghorn, laying and breeding	500 ICU/kg	NRC (1984b)
	Broilers, 0–8 weeks	200 ICU/kg	NRC (1984b)
Turkey	All classes	900 ICU/kg	NRC (1984b)
Duck (Pekin)	0–7 weeks	220 ICU/kg	NRC (1984b)
Japanese quail	All classes	1200 ICU/kg	NRC (1984b)
Sheep	Adult	555 IU/100 kg liveweight	NRC (1985b)
Swine	Growing–finishing, 1–10 kg	220 IU/kg	NRC (1988)
	Growing–finishing, 20–110 kg	150 IU/kg	NRC (1988)
	Adult	200 IU/kg	NRC (1988)
Horse	Adult	275 IU/kg	NRC (1978b)
Cat	Growing	500 IU/kg	NRC (1986)
Dog	Growing	22 IU/kg body wt	NRC (1985a)
Fish	Catfish	500–1000 mg/kg	NRC (1983)
	Trout	1600–2400 IU/kg	NRC (1981a)
Rat	Growing	1000 IU/kg	NRC (1978c)
Mouse	Growing	1000 IU/kg	NRC (1978c)
Guinea pig	Growing	1000–2000 IU/kg	NRC (1978c)
Hamster	Growing	2484 IU/kg	NRC (1978c)
Human	Children	400 IU/day	RDA (1980)
	Adults	200–600 IU/day	RDA (1980)

[a]Expressed as per unit of animal feed (except for sheep and dog) either as fed (approximately 90% dry matter) or dry basis (see Appendix Table 1). Human requirements are expressed as IU/day (1 IU = 0.025 μg vitamin D).

absence of sunlight provided that diets contain recommended levels of calcium and available phosphorus. Species differences can be illustrated by the fact that adequate intakes of calcium and phosphorus in a diet that contains only enough vitamin D to produce normal bone in the rat or pig will quickly cause the development of rickets in chicks. Turkeys and pheasants have higher requirements than chicks. Surprisingly, the human baby is more like birds in this respect than like the other mammals mentioned. In full-term infants, greater absorption of calcium and greater growth rates occurred in those children given 400 IU per day compared to those given 100 IU/day, although 100 IU is enough to prevent rickets (Miller and Norman, 1984).

VIII. NATURAL SOURCES

Cholecalciferol and ergocalciferol contents of natural materials are shown in Table 3.2. Sources of vitamin D are natural foods, irradiated sebaceous material licked from skin or hair, or directly absorbed products of irradiation formed on

TABLE 3.2

Vitamin D Concentrations of Various Foods and Feedstuffs[a]

Food or feedstuff[a]	IU/100 g
	Ergocalciferol (D_2)
Alfalfa pasture	4.6
Alfalfa hay, sun cured	142
Alfalfa silage	12
Alfalfa wilted silage	60
Birdsfoot trefoil hay, sun cured	142
Barley straw	60
Cocoa shell meal, sun dried	3,500
Corn grain	0
Corn silage	13
Molasses (sugar beet)	58
Red clover, fresh	4.7
Red clover, sun cured	192
Sorghum grain	2.6
Sorghum silage	66
	Cholecalciferol (D_3)
Blue fin tuna liver oil	4,000,000
Cod liver oil meal	4,000
Cod liver oil	10,000
Eggs	100
Halibut liver oil	120,000
Herring, entire body oil	10,000
Menhaden, entire body oil	5,000
Milk, cow's whole (summer)	4
Milk, cow's whole (winter)	1
Sardine, entire body oil	8,000
Sturgeon liver oil	0
Swordfish liver oil	1,000,000

[a]Adapted from NRC (1982b) and Scott *et al.* (1982). Concentrations are on a fed basis.

or in the skin. The distribution of vitamin D is very limited in nature although provitamins D occur widely. Of feeds for livestock, grains, roots, and oilseeds, as well as their numerous by-products, contain insignificant amounts of vitamin D; green fodders are equally poor. However, when plants, especially pasture species, begin to die and the fading leaves are favorably exposed to UV light, some vitamin D_2 is formed. So arises the vitamin D of hay, the potency of which depends on local climatic conditions, for if it is made very quickly in the absence of direct sunlight and is baled when still quite green, its potency will be low (Abrams, 1952). The principal source of the antirachitic factor in the diets of farm animals is thus provided by the action of radiant energy upon ergosterol in forages. Legume hay that is cured to preserve most of its leaves and green color contains considerable amounts. Alfalfa, for example, will range from 650 to 2200 IU/kg (Maynard *et al.*, 1979); timothy and other grass hays contain less. Stemmy hay lacking in leaves and color that has been exposed to a minimum of sunlight may contain none, whether legume or nonlegume. The antirachitic value of the alfalfa crop increases with state of maturity because of the increase in dead leaves.

Artificially dried and barn-cured hay contains less vitamin D than hay that is properly sun cured. Even hay dried in the dark immediately after cutting has some of the vitamin present. This is because the dead or injured leaves on the growing plant are responsive to irradiation even though the living tissues are not. This fact is also largely responsible for the vitamin D found in corn silage (Maynard *et al.*, 1979). Under normal conditions, even wilted legume silage furnishes ample vitamin D for dairy calves.

For non-forage-consuming species, the vitamin D that occurs naturally in unfortified food is generally derived from animal products. Saltwater fish, such as herring, salmon, and sardines, contain substantial amounts of vitamin D, and fish liver oils are extremely rich sources. However, eggs, veal, beef, unfortified milk, and butter supply only small quantities of this vitamin. Among animal products, eggs, especially the yolks, are a good source if the diet of the hen is rich in the vitamin. Milk contains a variable amount in its fat fraction (5 to 40 IU in cow's milk per quart), but neither cow's milk nor human milk contains enough to protect the baby against rickets (Maynard *et al.*, 1979). Cow's milk is reportedly higher in vitamin D when produced during the summer compared to the winter.

It has generally been assumed that for all but a few species, vitamin D_2 and vitamin D_3 are equally potent. For poultry and other birds and a few of the rarer mammals that have been studied, including some New World monkeys, vitamin D_3 is many times more potent than D_2 on a weight basis. Vitamin D_3 may be 30 times more effective than the D_2 form for poultry, therefore plant sources (vitamin D_2) of the vitamin should not be provided to these species. Recently, the dogma that mammals (other than the New World monkey) do not discriminate between the two sources has been proven incorrect. Data for the pig (Horst and Napoli, 1981) and for ruminants (Sommerfeldt *et al.*, 1981) suggest that these species discriminate in the metabolism of vitamin D_2 and vitamin D_3, with vitamin D_3 being the preferred substrate. Sommerfeldt *et al.* (1983) report that the amount of 1,25-$(OH)_2$D in the plasma of ergocalciferol-treated dairy calves was one-half to one-fourth the amount for cholecalciferol-treated calves. Discrimination against vitamin D_2 by ruminants may be in part a result of its preferred degradation by rumen microbes or its less efficient absorption by the intestine. Similarly, Harrington and Page (1983) found vitamin D_3 to be more hypercalcemic and overtly toxic to horses than vitamin D_2, likewise suggesting preference for D_3. Vitamin D_3 is reported to be at least three times as effective as vitamin D_2 in satisfying the requirement for vitamin D in trout (NRC, 1981a). Although the recent data suggest a preference for D_3 by a number of animals, in practice D_2 is still relatively comparable to D_3 in antirachitic function except for poultry and certain monkeys.

IX. DEFICIENCY

A. Effects of Deficiency

The outstanding disease of vitamin D deficiency is rickets, generally characterized by a decreased concentration of calcium and phosphorus in the organic matrices of cartilage and bone. A deficiency of vitamin D results in signs and symptoms similar to those of a lack of calcium or phosphorus or both, as all three are concerned with proper bone formation. In the adult, osteomalacia is the counterpart of rickets and, since cartilage growth has ceased, is characterized by a decreased concentration of calcium and phosphorus in the bone matrix.

Clinical signs of vitamin D deficiency are seen mainly in the young. General consequences of a deficiency can appear in the form of an inhibition of growth, a loss of weight, and reduced or lost appetite before characteristic signs that relate primarily to the bone system become apparent. The role of vitamin D in the adult appears to be much less important except during reproduction and lactation. Congenital malformations in newborn result from extreme deficiencies in the diet of the mother during gestation, and the mother's skeleton is injured as well.

The same disruption of the orderly processes of bone formation with vitamin D deficiency occurs in animals as it does in humans and includes (Kramer and Gribetz, 1971):

1. Failure of calcium salt deposition in the cartilage matrix.
2. Failure of cartilage cells to mature, leading to their accumulation rather than destruction.

3. Compression of the poliferating cartilage cells.
4. Elongation, swelling, and degeneration of proliferative cartilage.
5. Abnormal pattern of invasion of cartilage by capillaries.

Outward signs of rickets include the following skeletal changes, varying somewhat with species depending on anatomy and severity:

1. Weak bones cause curving and bending of bones.
2. Enlarged hock and knee joints.
3. Tendency to drag hind legs.
4. Beaded ribs and deformed thorax.

Although there appear to be differences between species in the susceptibility of different bones to such degenerative changes, differences that probably reflect bodily conformation (e.g., pig compared with sheep) and stance (e.g., humans compared with the common quadrupeds), there is nevertheless an apparent common pattern (Abrams, 1978). Spongy parts of individual bones, and bones relatively rich in such tissue, are first and worst affected. As in simple calcium deficiency, the vertebrae and the bones of the head suffer the greatest degree of resorption. Next come the scapula, sternum and ribs. The most resistant bones are metatarsals and shafts of long bones.

1. RUMINANTS

Clinical signs of vitamin D deficiency in ruminants are decreased appetite and growth rate, digestive disturbances, stiffness in gait, labored breathing, irritability, weakness, and occasionally tetany and convulsions. There is enlargement of joints, slight arching of back, and bowing of legs, with erosion of joint surfaces causing difficulty in locomotion (NRC, 1984a). Young ruminants may be born dead, weak, or deformed.

Clinical signs involving bones begin with thickening and swelling of the metacarpal or metatarsal bones. As the disease progresses, the forelegs bend forward or sideways: In severe or prolonged vitamin D deficiency, tension of the muscles will cause a bending and twisting of long bones to give the characteristic deformity of bone. There is enlargement at ends of bones from deposition of excess cartilage, giving the characteristic "beading" effect along the sternum where ribs attach. The lower jaw bone, the mandible, becomes thick and soft; in the worst cases eating is then difficult. In calves so affected there can be slobbering, inability to close the mouth, and protrusion of the tongue (Craig and Davis, 1943). Joints (particularly the knee and hock) become swollen and stiff, the pastern straight, and the back humped. In more severe cases synovial fluid accumulates in the joints (NRC, 1978a).

In older animals with a vitamin D deficiency, bones become weak and fracture easily, and posterior paralysis may accompany vertebral fractures. For daily cattle, milk production may be decreased and estrus inhibited by inadequate vitamin D (NRC, 1978a).

Vitamin D should be supplied to growing animals that are denied sunlight over extended periods because of cloud cover or confinement housing. In more northern latitudes during winter months, photochemical conversion of provitamin D to its active compound in the skin of ruminants could be limited because of insufficient ultraviolet radiation. Hidiroglou *et al.* (1979) reported that 25-OHD was higher in plasma of cattle in summer than in winter. Vitamin D deficiency may be observed in young ruminants that are closely confined and do not consume sun-cured roughage. Hidiroglou *et al.* (1978) reported the clinical history of a flock of sheep kept under total confinement that showed a high incidence (8%) of an osteodystrophic condition, a vitamin D-responsive disease. A form of osteodystrophia has also been produced experimentally in goats (NRC, 1981b).

When grazing ruminants have normal exposure to direct sunlight or are fed normal amounts of sun-cured forage, little chance for vitamin D deficiency exists. However, seasons of minimum sunlight, artificially cured forages, sheep with full fleece, feedlot animals without access to sunlight or sun-cured forages, and high-producing dairy cows with limited access to sunlight or sun-cured forage are situations that may require dietary supplementation.

2. Swine

During growth, a deficiency of vitamin D causes poor growth, stiffness, lameness (Fig. 3.3), stilted gait, a general tendency to "go down" or lose the use of the limbs (posterior paralysis), frequent cases of fractures, softness of bones, bone deformities, beading of the ribs, enlargement and erosion of joints, and unthriftiness (Cunha, 1977). Bones may also be deformed by the weight of the animal and the pull of body muscles.

The trend toward confinement of swine in completely enclosed houses through the life cycle increases the importance of adequate dietary vitamin D fortification. Goff *et al.* (1984) conclude that subclincal rickets may become more of a problem as swine producers convert to swine confinement operations, which deprive sows and piglets of the ultraviolet irradiation needed for endogenous production of cholecalciferol. Research has shown that sunshine cannot always be depended on to meet vitamin D requirements of growing or finishing pigs during winter months in northern latitudes.

3. Poultry

In addition to retarded growth, the first sign of vitamin D deficiency in chicks is rickets, which is characterized by a severe weakness of the legs. In young, growing chickens and turkeys there is a tendency to rest frequently in a squatting position, a disinclination to walk, and a lame, stiff-legged gait. These are dis-

Fig. 3.3 Vitamin D deficiency. Photo shows pig with advanced rickets caused by lack of vitamin D. The pig was fed indoors and was not exposed to sunlight. It shows leg abnormalities and was unable to walk; it later responded to supplementary vitamin D. (Courtesy of J. M. Bell, University of Saskatchewan, Canada.)

tinguished from the clinical signs of vitamin A deficiency in that birds with a vitamin D deficiency are alert rather than droopy and walk with a lame rather than a staggering gait (ataxia). The beaks and claws become soft and pliable (Fig. 3.4), usually between 2 and 3 weeks of age. The most characteristic internal sign of vitamin D deficiency in chicks is a beading of the ribs at their juncture with the spinal column (Scott *et al.*, 1982).

In chronic vitamin D deficiency, marked skeletal distortions become apparent (Scott *et al.*, 1982) in which the spinal column may bend downward in the sacral and coccygeal region. The sternum usually shows both a lateral bend and an acute dent near the middle of the breast. These changes reduced the size of the thorax with consequent crowding of the vital organs.

As in many other nutritional diseases of poultry, the feathers soon become ruffled. In red or buff color breeds of chickens, a deficiency of vitamin D causes an abnormal black pigmentation of some of the feathers, especially those of the wings. If the deficiency is very marked, the blackening becomes pronounced and nearly all the feathers may be affected (NRC, 1984b). When vitamin D is

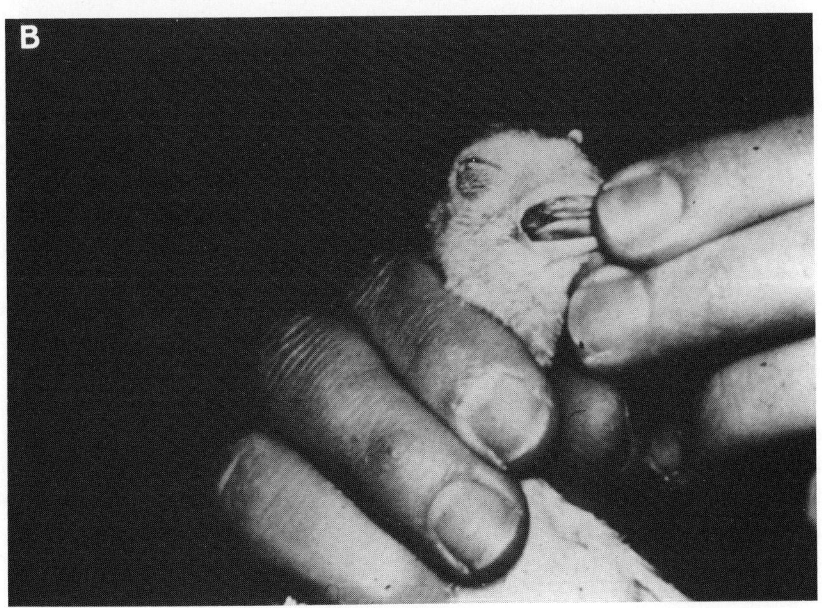

Fig. 3.4 Vitamin D deficiency in turkeys. Comparison is made between vitamin D-deficient and normal poult (A). Note rubbery beak from vitamin D deficiency (B). (Courtesy of L. S. Jensen and Washington State University.)

supplied in adequate quantity, the new feathers and newer part of older feathers are normal in color, although the discolored portion remains black.

Signs of vitamin D deficiency begin to occur in laying hens in confinement about 1–2 months after they are deprived of vitamin D. When laying chickens are fed a diet deficient in vitamin D, the first sign of the deficiency is a thinning of the shells of their eggs. Commercial layers will continue to lay eggs with reduced shell quantity for weeks. If the diet is also completely devoid of vitamin D_3, egg production may decrease rapidly and eggs with very thin shells or no shell will be produced. Hatchability is markedly reduced, with embryos frequently dying at 18–19 days of age. These embryos show a short upper mandible or incomplete formation at the base of the beak. Eventually breast bones become noticeably less rigid and there may be beading at the ends of the ribs. Individual hens may show temporary loss of use of the legs, with recovery after laying an egg (usually shell-less) (Scott *et al.*, 1982). During the periods of extreme leg weakness, hens show a characteristic posture that has been described as a "penguin-like squat."

Of all vitamins provided in poultry feeds, vitamin D is one of two (the other being vitamin B_{12}) that is most likely to be deficient. Typical grain–soybean-based diets contain virtually none of the vitamin. Also the trend to complete confinement will eliminate UV light as a source of the vitamin, and therefore supplemental vitamin D must be provided for all poultry operations in which birds remain in confinement.

4. Horses

A deficiency of calcium, phosphorus, or vitamin D can cause bone deformities in the horse caused by the weight of the animal and the pull of the body muscles on weak, porous bones. Vitamin D deficiency is characterized by reduced bone calcification, stiff and swollen joints, stiffness of gait, bone deformities, frequent cases of fractures, and reduction in serum calcium and phosphorus (NRC, 1978b). El Shorafa *et al.* (1979) observed early stages of rickets in ponies to include decreased bone ash, decreased cortical area and bone density, and delayed epiphyseal closure.

Grazing horses or horses that exercise regularly in sunlight and consume sun-cured hay normally will have their requirements for vitamin D met. However, if exposure to sunlight is restricted by confinement, hay may not always supply the requirement. Sunlight provides most of its antirachitic powers during the 4 hr around noon. This is something to bear in mind when racehorses are briefly exercised in the early morning before being housed for the rest of the day (Abrams, 1978).

5. Other Animal Species

a. Dogs and Cats. Rickets with typical bone lesions is readily produced in dogs but clinical signs are frequently confounded by a simultaneous deficiency or imbalance of calcium and phosphorus (NRC, 1985a). Severe rickets in kittens resulted in enlarged costochondral junctions (''rachitic rosary'') with disorganization in new bone formation and excessive osteoid (NRC, 1986). Classic signs of rickets are rare in kittens and confined to those born in winter, kept permanently in dark quarters, or from queens fed vitamin D-deficient diets.

b. Laboratory Animals. Rickets is classically brought about in rats by a diet lacking vitamin D, adequate in calcium, and low in phosphorus. Bones of rachitic rats show decreased or absent calcification with wide areas of uncalcified cartilage at the junction of diaphysis and epiphysis. Bone ash may be less than half normal (NRC, 1978c).

Guinea pigs housed without access to UV light and fed a low-vitamin D purified diet with 0.028% calcium and 0.2% phosphorus did not grow normally and developed rickets. Typical lesions occurred in the zone of cartilage proliferation at the epiphyseal plate of long bones and ribs. Also, incisors exhibited a high degree of enamel hypoplasia, while enamel and dentin were disorganized and irregular with poor calcification (NRC, 1978c). Hamsters do not require dietary vitamin D for prevention of rickets when the dietary calcium to phosphorus ratio is about 2 : 1 and calcium is at 0.6% (NRC, 1978c).

c. Nonhuman Primates. Softening, demineralization, and fibrous dysplasia of bone, which are compatible with a diagnosis of rickets and which are responsive to vitamin D treatment, have been observed in many species of nonhuman primates (NRC, 1978d). Rickets also has been induced unintentionally in New World primates kept under laboratory conditions when it was incorrectly assumed that they could utilize the vitamin D_2 form as efficiently as D_3 (see Section VIII).

d. Foxes and Mink. Rickets can be produced in both foxes and mink by feeding diets having low vitamin D content and abnormal calcium-to-phosphorus ratios (1982a). Signs of rickets in growing kits are generally seen between 2 and 4 months of age.

e. Fish. Channel catfish raised in aquaria and fed a vitamin D-deficient diet for 16 weeks showed reduced weight gain, lower body ash, lower body phosphorus, and lower body calcium compared with controls (NRC, 1983). Signs of vitamin D deficiency in trout include slow growth, an impairment of calcium homeostasis manifested by clinical signs of tetany of the white skeletal muscles, and ultrastructural changes in the white muscle fibers of the musculature (NRC, 1981a; Barnett *et al.,* 1982). No changes in bone ash have been detected in this species of fish.

6. Humans

Vitamin D deficiency in children leads to the pathological bone condition of rickets, which is characterized by disordered cartilage cell growth and enlargement of the epiphyseal growth plates in the long bones (Fraser, 1984). There is also a prominent accumulation of unmineralized bone matrix (osteoid) on trabecular bone surfaces. Classic bone symptoms associated with rickets, such as bowlegs (Fig 3.5), knock-knees, curvature of the spine, and pelvic and thoracic deformities, result from the application of normal mechanical stress to demineralized bone (Miller and Norman, 1984). Enlargement of bones, especially in the knees, wrists, and ankles, and changes in the costochondral junctions also occur. With these defects the bones become structurally weak, they bend under weight of the child, and are liable to fracture. Rickets also results in inadequate mineralization of tooth enamel and dentin. If the disease occurs during the first 6 months of life, convulsions and tetany can develop.

When humans changed from a hunting and gathering culture to an agricultural one, they also moved from tropical and subtropical climates to temperate zones in which protection from the cold by houses and clothing was necessary for a considerable part of the year (Harrison, 1978). This limited the production of vitamin D in skin, particularly of infants during those months when exposure of large areas of skin to sunshine was not possible. When a further change to urban living occurred, children in crowded city slums had little opportunity for exposure

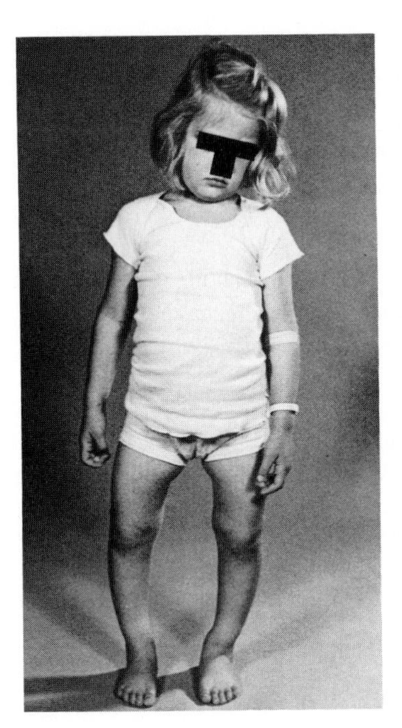

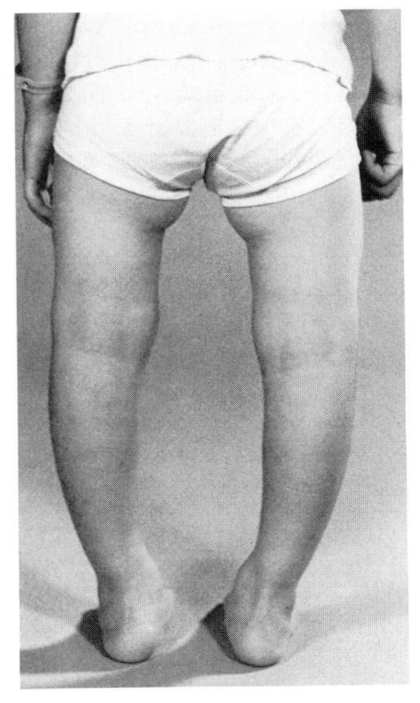

Fig. 3.5 Child with rickets with marked bowlegs. Note the angle of the feet. (Courtesy of Alan T. Forrester, "Scope Manual on Nutrition," The Upjohn Company, Kalamazoo, Mich.)

to sunshine. As noted earlier, this was accentuated by the Industrial Revolution, during which the use of fossil fuels contaminated the atmosphere and thus blocked out UV radiation.

Rickets resulting from vitamin D deficiency became widespread in cities of northern latitudes, particularly in the British Isles, Europe, the United States, and Canada. The disease usually appeared in the first year of life and was characterized by muscle weakness, deformities of long bones including bowed legs, knuckle-like projections along the rib cage known as the rachitic rosary, and deformities of the pelvis, which were often permanent. Consequences of this disease were quite profound, especially for young women, in whom a deformed pelvis would cause difficulty with childbirth and result in a high incidence of infant and maternal morbidity and mortality (Holick, 1987).

Osteomalacia occurs after skeletal development is complete. As in rickets, even though bone mineralization has ceased, collagen formation continues; this results in formation of uncalcified bone matrix. In adults, the bones no longer grow in length but are constantly remodeled. The main symptoms of osteomalacia are muscular weakness and bone pain. As the disease progresses, bone fractures occur.

Although it is accepted that vitamin D is absolutely essential for growing children, it is not well appreciated that it is also essential for maintenance of a healthy mineralized skeleton in adults. One reason for vitamin D deficiency in older populations is that individuals do not get enough exposure to sunlight. Older persons require more sunlight to get the same vitamin effect as do young individuals (see Section VII). Holick (1987) observed that one of the primary causes of poor vitamin D nutrition in the elderly in the United States is a decrease in or complete abstinence from consumption of milk, which is the principal food source that is fortified with vitamin D. Also, milk and milk products are the principal sources of calcium and phosphorus in many diets. In Great Britain, where dairy products are not routinely fortified with vitamin D, it was found that 30% of women and 40% of men who developed hip fracture were deficient in vitamin D (Doppelt *et al.*, 1983).

B. Assessment of Status

Several methods have been used to assess nutritional status of individuals for vitamin D. Poor production rates in livestock as well as bone abnormalities in both animals and humans are the chief indications when vitamin D deficiency is substantially advanced. The incomplete calcification of the skeleton is easily detectable with x-rays, but, like other production-related signs, would not be specific for vitamin D deficiency versus other nutrient inadequacies.

Low serum calcium levels in the range of 5–7 mg/100 ml and high serum alkaline phosphatase activity can be used to diagnose rickets and osteomalacia.

Also, a marked reduction in circulating 1,25-$(OH)_2D$ levels in individuals with osteomalacia has been reported. The plasma concentration of 25-OHD is related to and therefore is an index of the vitamin D supply for both animals and humans. The concentration of 25-OHD in plasma, which is the principal form of vitamin D in plasma, is about 10 times and 500 to 1000 times higher than those of 24,25-$(OH)_2D$ and 1,25-$(OH)_2D$, respectively (Fraser, 1984).

The circulating concentration of 25-OHD is a good index for determining the cumulative effect of sunlight on synthesis of vitamin D_3 in skin and also dietary sources of the vitamin. In plasma of the vitamin D-deficient child, a low concentration of 25-OHD (less than 5 ng/ml is found (Fraser, 1984). Toxicity caused by excess vitamin D administration is also associated with plasma 25-OHD; concentrations of more than 400 ng/ml are reported (Hughes *et al.,* 1976). Patients suffering from hypervitaminosis D have been shown to exhibit a 15-fold increase in plasma 25-OHD concentration as compared with normal individuals.

X. SUPPLEMENTATION

For animals that are kept indoors or that live in climates where the sunlight is not adequate for vitamin D production, the vitamin D content of the diet becomes important. In modern poultry and swine operations, animals are often entirely without sunlight. Under these intensive conditions supplemental vitamin D must be provided.

Animals that are on pasture during the summer never suffer from the lack of the antirachitic factor even though their diet may be practically devoid of it. In the wintertime, however, animals are outside only a part of the time, there are generally fewer sunny days, and the sunlight that reaches the animal in much less effective than in summer. Under most conditions in northern latitudes, it is unsafe to rely on exposure to sunlight to provide the antirachitic factor during the winter months. Milk is not especially rich in vitamin D but the calf or lamb can obtain adequate amounts by skin irradiation if exposed to the sunlight for 1–2 hr per day. Sun-cured forage is the best natural source of vitamin D. Even silages generally have dead leaves (thus sun cured) that will provide some vitamin D to growing animals that are housed indoors. However, it is probably prudent to provide a vitamin D supplement to young calves in their milk replacer and starter diets until they are turned out or are consuming adequate forage (e.g., 6–8 weeks).

Supplemental vitamin D has been used to prevent parturient hypocalcemia (milk fever) in dairy cows for a number of years. Treatment with high levels of vitamin D has been successful, but toxicity problems have sometimes resulted and for some animals the disease has been induced by the treatment. Because of the extreme toxicity of vitamin D_3 in pregnant cows and the low margin of safety between doses of vitamin D_3 that prevent milk fever and doses that induce milk fever, Littledike and Horst (1982) concluded that vitamin D_3 cannot be used practically to prevent milk fever when injected several weeks prepartum. However, a more recent report from the same laboratory has provided data that suggest that injection of 24-F-1,25 $(OH)_2D_3$ (fluoridation at the 24R position) delivered at 7-day intervals prior to parturition can effectively reduce incidence of parturient paresis (Goff *et al.,* 1988). Parturient paresis can be prevented effectively by feeding a prepartum low-calcium and adequate-phosphorus diet. Prepartal low-calcium diets are associated with increased plasma PTH and 1,25-$(OH)_2D_2$ plus 1,25-$(OH)_2D_3$ concentrations during the prepartal period. Green *et al.* (1981) suggested that these increased PTH and 1,25-$(OH)_2D$ concentrations resulted in "prepared" and effective gut and bone calcium homeostatic mechanisms at parturition that prevented parturient paresis.

Vitamin D is available in two forms, vitamin D_3 of animal origin and vitamin D_2 of plant origin. In recent years, vitamin D_3 has been the primary source of supplemental vitamin D for domestic animals, whereas vitamin D_2 has been the chief source of supplemental vitamin D in foods and pharmaceuticals (Adams, 1978). Vitamin D_3 is commercially available as a resin, usually containing 24 to 26 million IU per gram, for use as the vitamin D source in various vitamin products. Vitamin D_3 products for feed include gelatine beadlets (with vitamin A), oil dilutions, oil absorbates, emulsions, and spray- and drum-dried powders. Test results have shown that the gelatin beadlet offers optimum vitamin D_3 stability.

Pure vitamin D crystals or vitamin D resin is very susceptible to degradation upon exposure to heat or contact with mineral elements. In fact the resin is stored at refrigerated temperatures under nitrogen gas. Dry, stabilized supplements retain potency much longer and can be used in high mineral supplements. It has been shown that vitamin D_3 is much more stable than D_2 in feeds containing heavy minerals and trace metals. Continued irradiation eventually destroys the vitamin D that it produced, but the chief cause of loss from foods is doubtless oxidation, as was recognized long ago (Fritz *et al.,* 1942). When the unprotected vitamin is thinly spread over the surface of free-flowing powders, as in some of the mineral components of compound animal feeds, its life in storage may be no more than 1 month.

Stabilization of the vitamin can be achieved by (1) rapid compression of the mixed feed, for example, into cubes, so that air is excluded, (2) storing feed under cool, dry, dark conditions, (3) preventing close contact between the vitamin and potent metallic oxidation catalysts, for example, manganese, and (4) including natural or synthetic antioxidants in the mix. The vitamin can also be protected by enclosing it in tough, gelatin micropellets. But all stability is relative; warmth and moisture will soften gelatin, while temperatures below about 10°C cause it to become hard, brittle, and friable (Abrams, 1978).

Stability of dry vitamin D supplements is affected most by high temperature, high moisture content, and contact with trace minerals such as ferrous sulfate, manganese oxide, and others. Hirsh (1982) reports the results of a ''conventional'' or nonstabilized vitamin D_3 product being mixed into a trace mineral premix or into animal feed and stored at ambient room temperature (20° to 25°C) for up to 12 weeks. The mash feed had lost 31% of its vitamin D activity after 12 weeks and the trace mineral premix had lost 66% of its activity after only 6 weeks in storage.

Natural sources of vitamin D in feeds must likewise to be protected from loss. Poor handling of hay, which can otherwise be an important source of vitamin D (and other nutrients) for cattle, sheep, goats, and horses, can lead to extensive fragmentation and loss of the leaf, which is much richer in vitamin D than the stem. Animals fed on grain, silage, and hay that is badly made or kept are extremely liable to suffer dietary deficiency of the vitamin (Abrams, 1978).

Cost of vitamin D supplementation to animal and poultry feeds is very low (Rowland, 1982). In contrast, the cost of not adding enough vitamin D to prevent deficiencies is very high. Factors that increase the amount of vitamin D needed to maximize productive and reproductive responses often are not reflected in the National Research Council (NRC) requirements. Consequently, responsible formulators often use more than the minimum levels of vitamin D in feeds. Successful nutrition programs may contain two to five times the NRC minimum of vitamin D. However, no amount of vitamin D can make up for lack of enough calcium or phosphorus in the diet.

Rowland (1982) notes that diets for young, rapidly growing chickens must contain liberal amounts of vitamin D to prevent field problems. He further observed that the NRC level (200 IU/kg of feed) for young chickens is extremely unrealistic for broilers. Even under low-stress research conditions, 3 to 5 times the NRC level is required to support maximum weight gain of broilers, and under commercial conditions, 10 times the NRC level is prudent for broiler feed under field conditions. The NRC vitamin D levels for laying hens and turkeys are somewhat more realistic, with a factor of five times the NRC level generally supporting optimum performance and providing some margin of safety (Rowland, 1982). The most logical approach is to adjust supplemental vitamin D levels to expected production conditions.

Besides inadequate quantities of dietary vitamin D, deficiencies may result from (1) errors in vitamin addition to diets, (2) inadequate mixing and distribution in feed, (3) separation of vitamin D particles after mixing, (4) instability of the vitamin content of the supplement, or (5) excessive length of storage of diets under environmental conditions causing vitamin D loss (Hirsch, 1982).

Supplementation considerations are dependent on other dietary ingredients. The requirements for vitamin D_3 are increased several-fold by inadequate levels of calcium and/or phosphorus or by improper ratios of these two elements in the diet (see Section VII). A number of reports have indicated that molds in feeds interfere with vitamin D_3 (Cunha, 1977), for example, when corn contains the mold *Fusarium roseum,* a metabolite of this mold prevents vitamin D_3 in the intestinal tract from being absorbed by the chick. Other molds may also be involved, and they result in a large percentage of birds with bone disorders. A number of flocks have been successfully treated by adding water-dispersible forms of vitamin D to drinking water at three to five times the normally recommended vitamin D levels.

Other factors that influence vitamin D status are diseases of the endocrine system, intestinal disorders, liver malfunction, kidney disorders, and drugs. Liver malfunction limits production of the active forms of the vitamin, while intestinal disorders reduce absorption. Persons with kidney failure are unable to synthesize 1,25-$(OH)_2D$, and for patients in renal dialysis who are waiting for a compatible kidney transplant donor, 1,25-$(OH)_2D$ has been found to alleviate the painful and serious bone disease that goes with chronic renal failure. The prolonged use of anticonvulsant drugs can result in an impaired response to vitamin D (Miller and Norman, 1984), and T. S. Cunha (personal communication, 1987) suggested the possibility that livestock with certain diseases or heavy infestation of internal parasites might be unable to synthesize the metabolically active forms of vitamin D as a result of liver or kidney damage.

Widespread fortification of human diets with vitamin D and provision of oral supplements over the past 50 years have greatly reduced the incidence of rickets in children. However, most of the world's population still depends on exposure to sunlight for their vitamin D nutritional needs because very few natural foods contain sufficient quantities to meet the daily requirement. It has been assumed that in countries where food is fortified with vitamin D, such as the United States, exposure to sunlight is no longer necessary for vitamin D nutrition. Although this may be true for young children who drink milk that is fortified with vitamin D, many elderly persons who do not drink milk and do not take a vitamin D supplement are prone to vitamin D deficiency when they are not exposed to sufficient sunlight. Among the fortified foods are milk, both fresh and evaporated, margarine and butter, cereals, and chocolate mixes. Milk is fortified to supply 400 IU vitamin D per quart, and margarine usually contains 4400 or more IU per kilogram (Miller and Norman, 1984).

XI. TOXICITY

Besides the toxicity resulting from excess vitamin A, vitamin D is the vitamin most likely to be consumed in concentrations toxic to both humans and livestock. Excessive intake of vitamin D produces a variety of effects, all associated with abnormal elevation of blood calcium. Elevated blood calcium is caused by greatly stimulated bone resorption, as well as increased intestinal calcium absorption.

The main pathological effect of ingestion of massive doses of vitamin D is widespread calcification of soft tissues. Pathological changes in these tissues are observed to be inflammation, cellular degeneration, and calcification. Diffuse calcification affects joints, synovial membranes, kidneys, myocardium, pulmonary alveoli, parathyroids, pancreas, lymph glands, arteries, conjunctivae, and cornea. More advanced cases interefere with cartilage growth. The abnormal calcification is grossly observed as a whitish chalky material, and kidney insufficiency is the most critical development of these processes. Initial kidney damage is due to calcium deposition in distal tubules, causing inflammation and later obstruction, which in turn causes hypertension and pathology related to it. As would be expected, the skeletal system undergoes a simultaneous demineralization that results in the thinning of bones.

Other common observations of vitamin D toxicity are anorexia (loss of appetite), extensive weight loss, elevated blood calcium, and lowered blood phosphate. Many investigators have described the clinical signs of hypervitaminosis D in mammals (NRC, 1987). Cows receiving 30 million IU of vitamin D_2 orally for 11 days developed anorexia, reduced rumination, depression, premature ventricular systoles, and bradycardia. Toxicosis in monkeys resulted in weight loss, anorexia, elevated blood urea nitrogen, diarrhea, anemia, and upper respiratory infections. In pigs, signs of toxicity were anorexia, stiffness, lameness, arching of the back, polyuria, and aphonia. Symptoms of vitamin D intoxication for humans include hypercalcemia, hypercalciuria, anorexia, nausea, vomiting, thirst, polyuria, muscular weakness, joint pains, diffuse demineralization of bones, and disorientation (Miller and Norman, 1984).

Severity of the effects and pathogenic lesions in vitamin D intoxication depend on such factors as the type of vitamin D (D_2 versus D_3), the dose, the functional state of the kidneys, and the composition of the diet (NRC, 1987). Studies indicate that vitamin D_3 is 10 to 20 times more toxic than vitamin D_2. When equal amounts of vitamin D_3 and vitamin D_2 are provided together in diets of mammals, the predominant circulating form of the vitamin is usually 25-OHD_3 rather than 25-OHD_2. Similarly, in toxicity experiments where vitamin D_2 was less toxic than vitamin D_3, the metabolite 25-OHD_2 was found to be present at lower plasma concentrations than was 25-OHD_3 (Harrington and Page, 1983).

Vitamin D toxicity is enhanced by elevated supplies of dietary calcium and phosphorus and is reduced when the diet is low in calcium. Toxicity is also reduced when the vitamin is accompanied by intakes of vitamin A or by thyroxin injections (Payne and Manston, 1967). Route of administration also influences toxicity. Parenteral administration of 15 million IU of vitamin D_3 in a single dose caused toxicity and death in many pregnant dairy cows (Littledike and Horst, 1982). On the other hand, oral administration of 20 to 30 million IU of vitamin D_2 daily for 7 days resulted in little or no toxicity in pregnant dairy cows (Hibbs and Pounden, 1955). Rumen microbes are capable of metabolizing vitamin D to the inactive 10-keto-19-nor vitamin D, which may partially explain the difference in toxicity between oral and parenteral vitamin D. The toxic dose of vitamin D is quite variable, with an important factor being duration of intake, as this is a cumulative toxicity. However, there is marked variation among individuals in tolerance to excessive doses of vitamin D, the mechanism of which is unknown. Some species are less affected by toxicity due to vitamin D metabolism differences. For example, in the hooded seal there is increased conversion of 25-OHD to 24,24 $(OH_2)D$ and a high capacity for vitamin D storage in their large blubber mass, which provides resistance for this species to toxicity (Keiver *et al.*, 1988).

Although it is usually assumed that living plants do not contain vitamin D_2, some plants contain compounds that have vitamin D activity. Grazing animals in several parts of the world develop calcinosis, a disease characterized by the deposition of calcium salts in soft tissues (Carrillo, 1973; Morris, 1982). The ingestion of the leaves of the shrub *Solanum malacoxylon* by grazing animals causes enzootic calcinosis in Argentina and Brazil, where the disease is referred to as "enteque seco" and "espichamento," respectively. As few as 50 fresh leaves per day (200 g of fresh leaves per week) over a period of 8–20 weeks are enough to develop the disease in cows (Fig. 3.6) (Okada *et al.*, 1977).

Another solanaceous plant, *Cestrum diurnum* (a large ornamental plant), causes calcinosis in grazing animals in Florida, while the grass *Trisetum flavescens* is the causative agent in the Alpine regions in Europe. *Solanum torvum* is suspected of causing calcinosis in cattle in Papua, New Guinea. A condition known as "humpy-back," in which clinical symptoms reminiscent of calcinosis occur, may be caused by sheep grazing the fruits of *S. esuriale* in Australia. In Jamaica "Manchester wasting disease" and in Hawaii "Naalehu disease" are conditions seen in cattle that are virtually identical to "enteque seco" in relation to clinical and pathological signs (Wasserman, 1975).

The calcinogenic factor in *S. malacoxylon* and *C. diurnum* is a water-soluble glycoside of 1,25-$(OH)_2D_3$ (Wasserman, 1975). The digestive system of the animal releases the sterol, which promotes a massive increase in the absorption of dietary calcium and phosphate such that accommodation of these by the normal physiological processes is ineffective and soft tissue calcification results. Vitamin D_3 has been identified in *T. flavescens* but at concentrations that would not be calcinogenic. Evidence is emerging that the grass may also contain 1,25-$(OH)_2D_3$ or a substance able to mimic its actions as well as an aqueous soluble factor that promotes phosphate absorption.

During development of plant-induced calcinosis diseases, destruction of connective tissues occurs and this precedes mineralization in which magnesium is involved as well as calcium and phosphate. Clinical signs of the disease are stiffened and painful gait with progressive loss of weight. If the animals are removed in the early stages from the affected areas they recover quickly, but they may die if they are not removed. In advanced cases, joints cannot be extended

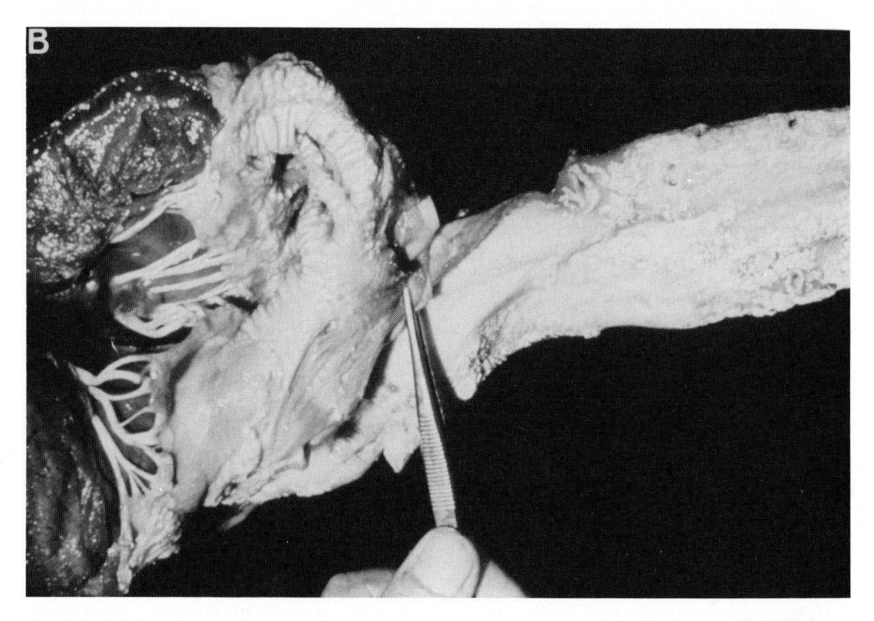

Fig. 3.6 Vitamin D toxicity (''enteque seco'') in Argentina. (A) Animal had consumed the shrub *Solanum malacoxylon*. (B) Photograph illustrates calcium deposits in soft tissue. (Courtesy of Bernardo Jorge Carrillo, CICV, INTA, Castelar, Argentina.)

completely and animals tend to walk with an arched back, carrying the weight on the forepart of the hooves.

Animals with calcinosis may also show signs of acute cardiac and pulmonary insufficiency. Postmortem examination shows a widespread metastatic calcification of the vascular system and the soft tissues. The heart and aorta exhibit the most marked effects, and calcification in the lung develops later than in the vascular system. Cartilages of the appendicular skeleton are eroded in advanced cases, with the joints being almost denuded of cartilage. The kidney may also show some calcification. Animals grazing in areas where the disease occurs show high serum levels of calcium and inorganic phosphorus. All clinical signs are similar to those found in vitamin D intoxication (Wolker and Carrillo, 1967).

It is clear that the calcinogenic plants are economically important, particularly because of their toxic effects that result in enormous losses in meat and milk production (Morris, 1982). In some fields in Argentina, between 10 and 30% of cattle show signs of "enteque seco," and *S. malacoxylon* is now regarded as one of the most important poisonous plants of that country. A recent survey conducted in Bavarian slaughterhouses revealed mineral deposits in soft tissues in 22–52% of cattle from the south of Bavaria that had been grazing at an altitude above 500 m. *Trisetum flavescens* is also an important toxic plant in these alpine pastures. In 1977, it was estimated that the annual loss in revenue resulting from *S. malacoxylon*-induced calcinosis in Argentina was of the order of U.S. $10 million. No assessment of the economic losses arising from plant-induced calcinosis is available for any other country.

TABLE 3.3

Estimation of Safe Upper Dietary Levels of Vitamin D_3 (IU/kg in Diet) for Animals[a]

Animal	Dietary requirement	Exposure time	
		< 60 days	> 60 days
Birds			
Chicken	200	40,000	2,800
Japanese quail	1,200	120,000	4,700
Turkey	900	90,000	3,500
Cow	300	25,000	2,200
Fish			
Catfish	1,000		20,000
Rainbow trout	1,800		1,000,000
Horse	400		2,200
Sheep	275	25,000	2,200
Swine	220	33,000	2,200

[a]Modified from NRC (1987).

With the exception of grazing animals consuming certain plants containing 1,25-$(OH)_2D_3$ in several parts of the world, excessive amounts of vitamin D are not available from natural sources. Although certain fish liver and body oils are good sources of vitamin D, concentrations are not so high that toxic amounts of vitamin D would likely be ingested. Therefore, danger of vitamin D toxicosis is from dietary supplementation. Extensive whole-body irradiation with ultraviolet light will not result in vitamin D toxicity. Vitamin D toxicosis from dietary sources can result in 400 ng/ml of plasma 25-OHD while extensive UV light exposure can only result in 40–80 ng of this transport form of vitamin D (Davie and Lawson, 1980). Therefore, supplying vitamin D by mouth bypasses protective mechanisms that prevent excessive 25-OHD formation.

For most species the presumed maximal safe level of vitamin D_3 for long-term feeding conditions (more than 60 days) is 4 to 10 times the recognized dietary requirement (NRC, 1987). Catfish and rainbow trout, on the other hand, can tolerate as much as 50 and 500 times their requirements, respectively. Under short-term feeding conditions (less than 60 days), most species can tolerate as much as 100 times their apparent dietary requirements. Table 3.3 provides recommendations for safe upper dietary levels of vitamin D_3 for animals.

Rodents are sensitive to vitamin D toxicity and have an LD_{50} of 43.6 mg/kg body weight for cholecalciferol compared to 88 and 200 mg/kg for dogs and humans, respectively (Tindall, 1985). Cholecalciferol is being marketed as an effective rodenticide, with low oral toxicity to nonrodent species.

In the early stages of vitamin D intoxication, the effects are usually reversible. Treatment consists merely of withdrawing vitamin D and perhaps reducing dietary calcium intake until serum calcium levels fall. However, it is usually not immediately successful because of the long plasma half-life of vitamin D (5–7 days) and 25-OHD (20–30) days). This is in contrast to the short plasma half-life of 1α-OHD_3 (1–2 days) and 1,25-$(OH)_2D_3$ (4–8 hr). Sodium phytate, an agent that reduces intestinal calcium absorption, has also been used successfully in vitamin D toxicity management in monogastrics. This treatment would be of little benefit to ruminants because of the presence of rumen microbial phytases. There have also been reports that calcitonin, glucagon, and glucocorticoid therapy reduces serum calcium levels resulting from vitamin D intoxication (NRC, 1987).

4

Vitamin E

I. INTRODUCTION

Vitamin E is recognized as an essential nutrient for all species of animals, including humans. However, opinions differ among research workers as well as practical livestock producers regarding conditions under which vitamin E supplementation is required and at what levels it should be fed. For years, vitamin E in human nutrition was described as "a vitamin looking for a disease." Vitamin E deficiency conditions that occurred in animals were not seen in humans, however, there are a number of medical claims for physiological benefits that still are largely subjective.

II. HISTORY

Excellent reviews of vitamin E history have been provided by Scott (1980) and Ullrey (1982). By the early 1920s, existence of vitamins A, B (thiamin), and C was established, and that of vitamin D was virtually assured.

As a result of the stimulus to experiment with purified diets that followed the discovery of the first vitamins, it was frequently observed that on certain diets, which were satisfactory for growth and health, rats failed to reproduce. Vitamin E was discovered in 1922 by Evans and Bishop of the University of California, Berkeley, as an unidentified factor in vegetable oils required for reproduction in female rats. In their experiments, estrus, mating, and all detectable phases of beginning of pregnancy were normal, but fetuses soon died and were resorbed unless the diet was supplemented with small amounts of wheat germ, dried alfalfa leaves, or fresh lettuce.

At first this substance was known as factor X, but Sure (1924) and Evans (1925) soon proposed the name vitamin E, since this was the next serial alphabetical designation. Vitamin E was isolated as α-tocopherol. The name tocopherol is derived from the Greek "tokos" meaning childbirth or offspring, the Greek "pherein" meaning to bring forth, and "ol" to designate an alcohol.

It also became apparent that male rats were affected by a deficiency of this nutrient that resulted in testicular degeneration. Throughout the 1920s, vitamin

E was recognized only as a factor required for reproduction in rats. Many clinical studies were undertaken to determine the effects of vitamin E on various reproductive problems in humans, though in most cases it was found to have little or no effect.

In 1931, Pappenheimer and Goettsch conducted a series of what became classic experiments showing that vitamin E is also required for prevention of encephalomalacia of chicks and of nutritional muscular dystrophy in rabbits and guinea pigs. By 1944, it was found that a multiplicity of clinical signs occur in animals suffering from vitamin E deficiency. In a single species, the chick, three distinct vitamin E deficiency diseases were documented: exudative diathesis, encephalomalacia, and muscular dystrophy.

The first controlled study of vitamin E deficiency in swine was that of Adamstone *et al.* (1949). These workers reported a decline in reproductive efficiency and signs of locomotor incoordination and muscular necrosis. Obel (1953) described a naturally occurring dietary disease in vitamin E-deficient swine that was characterized by hepatic necrosis, fibrinoid degeneration of blood vessel walls, and muscular dystrophy. In 1957, Kalus Schwarz and associates, studying dietary liver necrosis of rats receiving a diet low in vitamin E, showed that dried brewer's yeast, which contains no vitamin E, was as effective as vitamin E in preventing liver necrosis. Shortly after this discovery, selenium was found to be the active ingredient in brewer's yeast and able to replace vitamin E for prevention of exudative diathesis in poultry, tissue degeneration in swine, and muscular degeneration in young ruminants. Much confusion existed because earlier discoveries had showed that synthetic antioxidants as well as sulfur amino acids were as effective as vitamin E for prevention of some vitamin E deficiency diseases. In recent years, considerable effort has been made to clarify the mechanisms of vitamin E, selenium, sulfur amino acids, and antioxidants in relation to vitamin E-responsive diseases.

III. CHEMICAL STRUCTURE AND PROPERTIES

Vitamin E activity in food derives from a series of compounds of plant origin, the tocopherols and tocotrienols. Eight forms of vitamin E are found in nature: four tocopherols (α, β, γ, and δ) and four tocotrienols (α, β, γ, and δ). The structures of α-tocopherol and the commercially available α-tocopheryl acetate are presented in Fig. 4.1, while different active forms of vitamin E are shown in Fig. 4.2. Differences between α, β, γ, and δ are due to the placement of methyl groups on the ring. The difference between tocopherols and tocotrienols is due to unsaturation of the side chain in the latter.

The *dl* α-tocopheryl acetate (also called all-rac-α-tocopheryl acetate) is accepted as the International Standard (1 mg = 1 international unit). Synthetic free to-

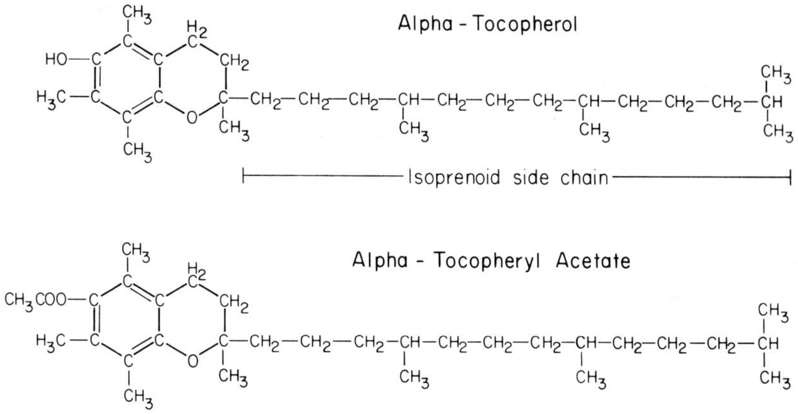

Alpha - Tocopherol

Isoprenoid side chain

Alpha - Tocopheryl Acetate

Fig. 4.1 Structure of α-tocopherol and α-tocopheryl acetate.

copherol, *dl*-α-tocopherol, has a potency of 1.1 IU/mg. Activity of naturally occurring α-tocopherol, *d*-α-tocopherol (also called RRR-tocopherol), is 1.49 IU/mg, and of its acetate, 1.36 IU/mg.

α-Tocopherol is a yellow oil that is insoluble in water but soluble in organic solvents. Tocopherols are extremely resistant to heat but readily oxidized. Natural vitamin E is subject to destruction by oxidation, which is accelerated by heat,

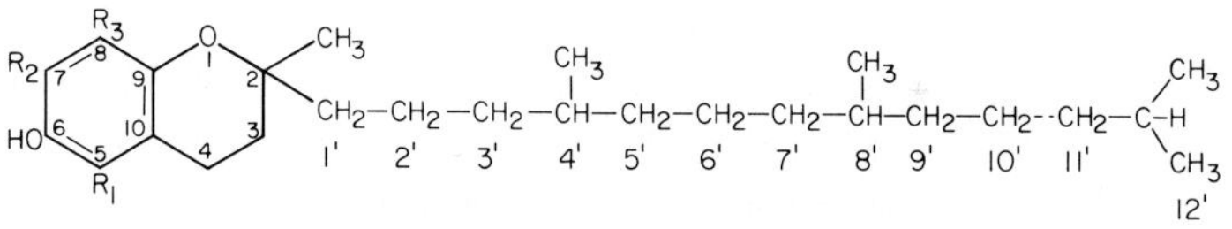

Tocopherol	$R_1(5)$	$R_2(7)$	$R_3(8)$	Side chain double bonds (3', 7', 11', positions)
alpha	CH_3	CH_3	CH_3	
beta	CH_3	H	CH_3	
gamma	H	CH_3	CH_3	
delta	H	H	CH_3	
alpha tocotrienol	CH_3	CH_3	CH_3	+
beta tocotrienol	CH_3	H	CH_3	+
gamma tocotrienol	H	CH_3	CH_3	+
delta tocotrienol	H	H	CH_3	+

Fig. 4.2 Structural differences among vitamin E forms.

moisture, rancid fat, and certain trace minerals. α-Tocopherol is an excellent natural antioxidant that protects carotene and other oxidizable materials in feed and in the body. However, in the process of acting as an antioxidant it is destroyed. Since esterification of the vitamin improves its stability, commercial supplements usually contain *d*-α-tocopheryl acetate or *dl*-α-tocopheryl acetate.

IV. ANALYTICAL PROCEDURES

Many methods for the determination of tocopherols in feedstuffs and animal tissues have been introduced, most of them based on separation of the tocopherols by column, paper, or thin-layer chromatography, followed by a colorimetric reaction. Separation steps, however, are usually laborious and time consuming, and colorimetric reactions such as that of Emmerie and Engel are often subject to interference from other compounds (McMurray and Blanchflower, 1979). A relatively new procedure of high-pressure liquid chromatography (HPLC) offers the possibility of combining rapid analysis with separation of tocopherols from interfering substances. The method consists of three main steps via extraction, saponification, and chromatography (McMurray and Blanchflower, 1979).

Biological assay methods for vitamin E are common. International units have been established on the basis of the most common biological assay, the rat fetal resorption test. Biological variations are inherent in bioassays, considerably more so than in chemical assays as a rule. They involve variations in response between

TABLE 4.1

Relative Biopotency of Vitamin E Forms

	Rat antisterility	Rat weight gain	Rabbit, cure of muscular dystrophy	Hemolysis of erythrocytes	
				in vivo	*in vitro*
Tocopherols					
d-α	135	—	100	130	100
dl-α	100	100	90	100	100
d-β	54	—	30	30	40
dl-β	27	25	—	25	54
d-γ	1	—	20	4–22	30
dl-γ	1–11	19	6	18	67
d-δ	1	—	—	3	20
Tocotrienols					
d-β	5	—	—	1–5	133
d-α	29	—	—	23	106
d-γ	3	—	—	—	88

individuals, families, strains, and species of animals, as well as management skill, stability of the test material, diet standardization and purity, and control of other environmental factors. Dietary history of parent stock (particularly previous vitamin E intakes) can have an influence. Table 4.1 illustrates a comparison among vitamin E forms related to biological assays.

V. METABOLISM

A. Absorption and Transport

Absorption is related to fat digestion and is facilitated by bile and pancreatic lipase (Ullrey, 1981; Sitrin *et al.*, 1987). Whether presented as free alcohol or as esters, most vitamin E is absorbed as the alcohol. Esters are largely hydrolyzed in the gut wall, and the free alcohol enters the intestinal lacteals and is transported via lymph to the general circulation.

Balance studies indicated that much less vitamin E is absorbed, or at least retained, in the body than is vitamin A. Vitamin E recovered in feces from a test dose was found to range from 65 to 80% in human, rabbit, and hen, although in chicks it was reported at about 25%. It is not known how much fecal vitamin E represents unabsorbed tocopherol and how much may come via secretion in the bile. The latter usually has a tocopherol content similar to that of blood plasma (Anonymous, 1972).

The tocopherol form, which is the naturally occurring one, is subject to destruction in the digestive tract to some extent, whereas the acetate ester is not. Much of the acetate is readily split off in the intestinal wall and the alcohol is reformed and absorbed, thereby permitting the vitamin to function as a biological antioxidant. Any acetate form absorbed or injected into the body evidently is converted there to the alcohol form.

Vitamin E in plasma is attached mainly to lipoproteins in the globulin fraction. Rates and amounts of absorption of the various tocopherols and tocotrienols are in the same general order of magnitude as their biological potencies. α-Tocopherol is absorbed best, with γ-tocopherol absorption 85% that of α forms, but with a more rapid excretion. Thus, non-α-tocopherol forms tend to be discriminated against, and one can generally assume that most of the vitamin E activity within plasma and other animal tissues is α-tocopherol (Ullrey, 1981). In humans, whose natural diet contains a high percentage of non-α forms, blood serum tocopherols identified consited of about 87% α-, 11% γ-, and 2% β-tocopherol (Anonymous, 1972).

Tocopherols pass through placental membranes and also the mammary gland; thus the diet of the female influences store of the young at birth and the amount obtained from mother's milk. However, less than 2% of dietary vitamin E is transferred from feed to milk.

B. Storage and Excretion

Vitamin E is stored throughout all body tissues, with highest storage in the liver. However, liver contains only a small fraction of total body stores, in contrast to vitamin A, for which about 95% of the body reserves are in the liver. The extent of storage is shown by the fact that females born of mothers whose diets contained a liberal supply frequently have enough in their bodies at birth to carry them through a first pregnancy. Rats reared on natural foods rich in the vitamin and then placed on a deficient diet may produce three or four litters before exhausting their reserves (Maynard *et al.*, 1979). However, Gallo-Torres (1980) reports that unlike vitamin A, lower body stores of vitamin E are available for periods of low dietary intake.

Small amounts of vitamin E will persist tenaciously in the body for a long time. However, stores are exhausted rapidly by polyunsaturated fatty acids (PUFA) in the tissues, the rate of disappearance being proportional to the intake of PUFA. A major excretion route of absorbed vitamin E is bile, in which tocopherol appears mostly in the free form.

VI. FUNCTIONS

It is well established that some functions of vitamin E can be fulfilled in part or entirely by traces of selenium or by certain synthetic antioxidants. Even sulfur-bearing amino acids, cystine and methionine, affect certain vitamin E functions. Much evidence points to undiscovered metabolic roles for vitamin E that may be paralleled biologically by roles of selenium and possibly other substances. The most widely accepted functions of vitamin E are discussed in this section.

A. Vitamin E as a Biological Antioxidant

Vitamin E has a number of different but related functions. One of the most important functions is its role as an intercellular and intracellular antioxidant. In this capacity, it prevents oxidation of unsaturated lipid materials within cells, thus protecting fats within the cell membrane from breaking down. If lipid hydroperoxides are allowed to form in the absence of adequate tocopherols, direct cellular tissue damage can result, in which peroxidation of lipids destroys structural integrity of the cell and causes metabolic derangements. Morphological damage to muscle is common in cases of vitamin E–selenium deficiency, and dystrophic tissue membranes may leak creatine and transaminases (e.g., glutamic-oxaloacetic transaminase) into plasma.

Vitamin E reacts or functions as a chain-breaking antioxidant, thereby neutralizing free radicals and preventing oxidation of lipids within membranes. At least one important function of vitamin E is to interrupt production of free radicals at the initial stage. Consequences of lipid peroxidation include perturbation of membrane microarchitecture, inhibition of membrane enzyme activity, and accumulation of reaction products (e.g., lipofuscin associated with brown spots of aging) that are not readily degraded to harmless metabolic debris (Ullrey, 1981). The more active the cell (e.g., the cells of skeletal and involuntary muscles), the greater is the inflow of lipids for energy supply and the greater is the risk of tissue damage if vitamin E is limiting. This antioxidant property also ensures erythrocyte stability and maintenance of capillary blood vessel integrity.

Interruption of fat peroxidation by tocopherol explains the well-established observation that dietary tocopherols protect or spare body supplies of such oxidizable materials as vitamin A and the carotenes. Certain deficiency signs of vitamin E (i.e., encephalomalacia, experimental muscular dystrophy) can be prevented by diet supplementation with other antioxidants, thus lending support to the antioxidant role of tocopherols. Antioxidant properties of the vitamin explain earlier observations (dating back to 1926) that large intakes of cod liver oil, which is high in unsaturated acids but low in vitamin E, caused muscular dystrophy in various herbivora that did not occur when hydrogenated oil was fed. It is clear that highly unsaturated acids in the diet increase vitamin E requirements (Maynard *et al.*, 1979). When acting as an antioxidant, vitamin E supplies become depleted, thus furnishing an explanation for the often observed fact that the presence of dietary unsaturated fats (susceptible to peroxidation) augments or precipitates a vitamin E deficiency.

B. Membrane Structure and Prostaglandin Synthesis

α-Tocopherol may be involved in the formation of structural components of biological membranes, thus exerting a unique influence on architecture of membrane phospholipids (Ullrey, 1982). It is reported that α-tocopherol stimulated the incorporation of ^{14}C from linoleic acid into arachidonic acid in fibroblast phospholipids. Also, it was found that α-tocopherol exerted a pronounced stimulatory influence on formation of prostaglandin E from arachidonic acid, while a synthetic antioxidant had no effect.

C. Blood Clotting

Vitamin E is an inhibitor of platelet aggregation, and may play a role by inhibiting peroxidation of arachidonic acid, which is required for formation of prostaglandins involved in platelet aggregation (Panganamala and Cornwell, 1982). The inhibition of platelet aggregation by prostaglandin E and the observation that vitamin E, but not an antioxidant, stimulates prostaglandin E synthesis suggest that vitamin E may have a role other than as an antioxidant in the blood-clotting mechanism.

D. Disease Resistance

Considerable attention is presently being directed to the role vitamin E and selenium play in protecting leukocytes and macrophages during phagocytosis, the mechanism whereby mammals immunologically kill invading bacteria. Both vitamin E and selenium may help these cells to survive the toxic products that are produced in order to effectively kill ingested bacteria (Badwey and Karnovsky, 1980). Mice fed vitamin E-deficient diets are unable to produce a vigorous humoral response (Gebremichael *et al.*, 1984), and this decreased immune reactivity undoubtedly contributes to increased susceptibility to bacterial infections associated with vitamin E deficiencies. Large doses of vitamin E protected chicks against *Escherichia coli* with increased phagocytosis and antibody production (Tengerdy and Brown, 1977). When pigs were fed minced colon from swine with dysentery or with a pure culture of *Treponema hyodysenteriae,* vitamin E and selenium supplementation to a deficient diet increased resistance to the disease (Tiege *et al.*, 1978). Calves receiving 125 IU of vitamin E daily were able to maximize their immune responses compared to calves receiving low dietary vitamin E (Reddy *et al.*, 1987b).

E. Electron Transport and Deoxyribonucleic Acid

There is limited evidence that vitamin E is involved in biological oxidation–reduction reactions (Anonymous, 1972). It may act as a cofactor in the cytochrome reductase portion of the nicotinamide–adenine dinucleotide (NAD) oxidase and the succinate oxidase systems. Restoration of the specific activity of cytochrome C reductase by vitamin E has been shown. This vitamin alone can reactivate this enzyme system following its inactivation by isolation and aging or freezing and thawing. Vitamin E also appears to regulate the biosynthesis of deoxyribonucleic acid (DNA) within cells.

F. Relationship to Toxic Elements

Both vitamin E and selenium provide protection against toxicity with three classes of heavy metals (Whanger, 1981). One class consists of metals like cadmium and mercury, in which selenium is highly effective in altering toxicities but where vitamin E has little influence. In the second group, which includes silver and arsenic, vitamin E is highly effective, with selenium also effective but only at relatively high levels. A third group of metals, of which lead is an example, is counteracted by vitamin E, but selenium has little effect.

G. Relationship with Selenium in Tissue Protection

There is a close working relationship between vitamin E and selenium within tissues (see Section VII). Selenium has a sparing effect on vitamin E and delays onset of deficiency syndromes. Likewise, vitamin E and sulfur amino acids partially protect against or delay onset of several forms of selenium deficiency syndromes. Tissue breakdown occurs in most species receiving diets deficient in both vitamin E and selenium, mainly through peroxidation. Peroxides and hydroperoxides are highly destructive to tissue integrity and lead to disease development. It now appears that vitamin E in cellular and subcellular membranes is the first line of defense against peroxidation of vital phospholipids, but even with adequate vitamin E, some peroxides are formed. Selenium, as part of the enzyme glutathione peroxidase, is a second line of defense that destroys these peroxides before they have an opportunity to cause damage to membranes. Therefore selenium, vitamin E, and sulfur-containing amino acids, through different biochemical mechanisms, are capable of preventing some of the same nutritional diseases. Vitamin E prevents fatty acid hydroperoxide formation, sulfur amino acids are precursors of glutathione peroxidase, and selenium is a component of glutathione peroxidase (Smith *et al.*, 1974).

To some extent vitamin E and selenium are mutually replaceable, but there are lower limits below which substitution is ineffective. In diets severely deficient in selenium, vitamin E does not prevent or cure exudative diathesis, whereas addition of as little as 0.05 ppm selenium completely prevents this disease (Scott, 1980).

H. Other Functions

Additional functions of vitamin E that have been reported (Scott *et al.*, 1982) include (1) normal phosphorylation reactions, especially of high-energy phosphate compounds such as creatine phosphate and adenosine triphosphate; (2) in synthesis of ascorbic acid; (3) in synthesis of ubiquinone; and (4) in sulfur amino acid metabolism. Vitamin E is reported to have a role in vitamin B_{12} metabolism (Pappu *et al.*, 1978). A deficiency of vitamin E interfered with conversion of vitamin B_{12} to its coenzyme 5′-deoxyadenosylcobalamin and concomitantly metabolism of methylmalonyl-CoA to succinyl-CoA.

VII. REQUIREMENTS

Estimated vitamin E requirements for selected animals and humans are presented in Table 4.2. Scott (1980), after reviewing the literature, concluded that the minimum vitamin E requirement of normal animals and humans is approximately 30 ppm of diet. Vitamin E requirements are exceedingly difficult to determine because of the interrelationships with other dietary factors, therefore, its requirement is dependent on dietary levels of polyunsaturated fatty acids (PUFA), antioxidants, sulfur amino acids, and selenium (see Sections VI and IX). The requirement may be increased with increasing levels of PUFA, oxidizing

TABLE 4.2

Vitamin E Requirements for Various Animals and Humans[a]

Animal	Purpose	Requirement	Reference
Beef cattle	Growing	15–60 IU/kg	NRC (1984a)
Dairy cattle	Milk replacer	300 IU/kg	NRC (1978a)
Goat	All classes	100 IU/kg	Morand-Fehr (1981)
Chicken	Growing, 0–6 weeks	10 IU/kg	NRC (1984b)
	Growing, 6–20 weeks	5 IU/kg	NRC (1984b)
	Laying	5 IU/kg	NRC (1984b)
	Broilers, 0–8 weeks	10 IU/kg	NRC (1984b)
Turkey	Growing, 0–8 weeks	12 IU/kg	NRC (1984b)
	Growing, 8–24 weeks	10 IU/kg	NRC (1984b)
	Breeding	25 IU/kg	NRC (1984b)
Sheep	All classes	15–20 IU/kg	NRC (1985b)
Horse	Growing	233 μg/kg body weight	NRC (1978b)
Swine	All classes	11–22 IU/kg	NRC (1988)
Mink	Growing	25 IU/kg	NRC (1982a)
Cat	All classes	30 IU/kg	NRC (1986)
Dog	Growing	22 IU/kg	NRC (1985a)
Rabbit	All classes	40 IU/kg	NRC (1977)
Fish	Salmon	30 IU/kg	NRC (1981a)
Rat	All classes	30 IU/kg	NRC (1978c)
Human	Infants	3–4 mg/day	RDA (1980)
	Children	5–7 mg/day	RDA (1980)
	Adults	8–13 mg/day	RDA (1980)

[a]Expressed as per unit of animal feed (except for horses) either on an as fed (approximately 90% dry matter) or dry basis (see Appendix Table 1). Data for humans are expressed as mg α-tocopherol equivalents/day.

agents, vitamin A, carotenoids, and trace minerals and decreased with increasing levels of fat-soluble antioxidants, sulfur amino acids, and selenium. On otherwise adequate diets containing sufficient cystine and methionine and containing a minimum of PUFA, vitamin E requirements appear to be low. This is evidenced by difficulties in producing deficiency signs on such diets under optimum environmental conditions.

The levels of PUFA found in unsaturated oils such as cod liver oil, corn oil, soybean oil, sunflower seed oil, and linseed oil all increase the vitamin E requirements. This is especially true if these oils are allowed to undergo oxidative rancidity in the diet or are in the process of peroxidation when consumed by the animal. If they become completely rancid before ingestion, the only damage is the destruction of the vitamin E present in the oil and in the feed containing the rancidifying oil. But if they are undergoing active oxidative rancidity at the time of consumption, they apparently cause destruction of body stores of vitamin E as well (Scott *et al.*, 1982). The vitamin E requirement for dogs is 5-fold higher under conditions of high PUFA intake (NRC, 1985a). The amount of vitamin E required per gram of PUFA is dependent on experimental conditions, species differences, levels and kinds of PUFA, and test used. Nevertheless, for a number of species, 0.6 IU of vitamin E per gram of PUFA is inadequate and 1 IU is a realistic minimum (Anonymous, 1972).

A combination of stress of infection and presence of oxidized fats in swine diets was reported to exaggerate vitamin E needs still further (Tiege *et al.*, 1978). These researchers reported that supplements of 100 IU vitamin E and 0.1 ppm selenium did not entirely prevent deficiency lesions in weanling pigs afflicted with dysentery and fed 3% cod liver oil.

Requirements of both vitamin E and selenium are greatly dependent on the dietary concentrations of each other (see Sections VI and IX). As noted earlier, they are mutually replaceable above certain limits. Chicks consuming a diet containing 100 ppm vitamin E required 0.01 ppm selenium, while those receiving no added vitamin E required 0.05 ppm selenium (Thompson and Scott, 1969).

Vitamin E is known to reduce the selenium requirement in at least two ways (McDowell, 1985b): (1) by maintaining body selenium in an active form, or preventing loss from body, and (2) by preventing destruction of membrane lipids within the membrane, thereby inhibiting the production of hydroperoxides, and reducing the amount of glutathione peroxidase needed to destroy peroxides formed in the cell. Selenium is known to spare vitamin E in at least three ways: (1) it is required to preserve the integrity of the pancreas, which allows normal fat digestion and thus normal vitamin E absorption; (2) it reduces the amount of vitamin E required to maintain the integrity of lipid membranes via glutathione peroxidase; and (3) it aids in some unknown way in retention of vitamin E in the blood plasma.

Determination of vitamin E requirements is further complicated because the body has a fairly large ability to store both vitamin E and selenium. Sows maintained on a diet deficient in vitamin E and selenium produced normal piglets during the first reproductive cycle of the deficiency and clinical deficiency signs occurred only after five such cycles (Glienke and Ewan, 1974). A number of studies to establish requirements for both nutrients have underestimated the requirements by failing to account for their augmentation from both body stores as well as experimental dietary concentrations.

In humans it has been concluded that a daily intake of between 3 and 15 mg of tocopherol is required from natural diets. However, the allowances, will not be adequate in individuals who, for a variety of reasons, do not absorb fat efficiently or who have medical conditions that result in an abnormal vitamin E status in the blood and tissues. As in other species, the human requirement for vitamin E is related to dietary intake of PUFA. However, in normal diets in the United States, this relationship is probably of little significance, inasmuch as primary dietary sources of PUFA—vegetable oils, margarine, and shortening—are also the richest sources of vitamin E. This situation is not true when foods consumed contain the longer-chained fatty acids (i.e., fish oils).

VIII. NATURAL SOURCES

Many vitamin E analyses of foods and feedstuffs have been reported using a variety of analytical techniques, however, there is a lack of characterization of individual tocopherols in the majority of analyses. Total tocopherol analysis of a food or feedstuff is of limited value in providing a reliable estimate of the biological vitamin E value. The occurrence of tocopherols (and tocotrienols) other than alpha (Table 4.3) as well as the prevalence of non-tocopherol-reducing substances in natural products has led to analytical examination of these materials by techniques capable of precise quantitation of individual species.

Because α-tocopherol is the most active form of vitamin E, many nutritionists prefer listing this form in feeds versus the unreliable total tocopherol values. Some of these less active tocopherols, particularly γ-tocopherol, are present in mixed diets in amounts two to four times greater than that of α-tocopherol. For purposes of calculating total vitamin E activity of mixed diets, milligrams of β-tocopherol should be multiplied by 0.5, those of γ-tocopherol by 0.1, and those

TABLE 4.3

Tocopherols in Selected Feedstuffs (ppm)[a]

Feedstuff	α	β	γ	δ
Barley	4	3	0.5	0.1
Corn	6	—[b]	38	Tr[c]
Oats	7	2	3	—
Rye	8	4	6	—
Wheat	10	9	—	0.8
Corn oil	112	50	602	18
Cottonseed oil	389	—	387	—
Palm oil	256	—	316	70
Safflower oil	387	—	174	240
Soybean oil	101	—	593	264
Wheat germ oil	1330	710	260	271

[a]Modified from Ullrey (1981).
[b]No value reported.
[c]Trace.

of δ-tocopherol by 0.3 (RDA, 1980). These three forms in addition to α-tocopherol provide the only significant vitamin E activity for typical diets. If only α-tocopherol in a mixed diet is reported, the value in milligrams should be increased by 20% (multiply by 1.2) to account for other tocopherols that are present, thus giving an approximation of total vitamin E activity as milligrams of α-tocopherol equivalents. The α-tocopherol concentrations of various foods and feedstuffs are listed in Table 4.4.

TABLE 4.4

α-Tocopherol Content of Feeds (ppm)[a,b]

Source	Mean	Range
Alfalfa meal, dehydrated 17% protein	73	28–121
Alfalfa meal, sun cured 13% protein	41	18–61
Alfalfa hay	53	23–102
Barley, whole	36	22–43
Beef, meat	6	5–8
Brewer's grains, dried	27	17–48
Butter	24	10–33
Chicken, meat	3	2–4
Corn, whole	20	11–35
Cottonseed meal	9	2–16
Distiller's grains, dehydrated	30	17–40
Eggs	11	8–12
Fat, animal	8	2–16
Fish, halibut	9	4–13
Fish, shrimp	9	6–19
Fish meal, herring	17	8–31
Fish meal, Peruvian	2	1–3
Lard	12	2–30
Linseed meal	8	3–10
Meat and bonemeal	1	1–2
Milo	12	10–16
Molasses, cane	5	3–9
Oats, whole	20	18–24
Pork, meat	5	4–6
Poultry by-products meal	2	1–4
Rice, brown	13.5	13–14
Rice, bran	61	34–87
Sorghum, grain	12	10–16
Soybean meal, solvent process	3	1–5
Wheat, whole	11	3–15
Wheat, bran	17	15–19

[a]Adapted from Bauernfeind (1980) and Ullrey (1981).

[b]When only α-tocopherol in a mixed diet is available, multiply value by 1.2 to account for the other tocopherols present.

Vitamin E is widespread in nature with the richest sources being vegetable oils, cereal products containing the oils, eggs, liver, legumes, and, in general, green plants. In nature the synthesis of vitamin E is a function of plants and thus their products are by far the principal sources. It is abundant in whole cereal grains, particularly in germ, and thus in by-products containing the germ. There is wide variation in vitamin content of particular feeds, with many feeds having a 3- to 10-fold range in reported α-tocopherol values. As an example, there can be a 5-fold seasonal difference in the α-tocopherol content of cow's milk.

Animal by-products supply only small amounts, and milk and dairy products are poor sources. Eggs, particularly the yolk, make a significant contribution depending on the diet of the hen. Wheat germ oil is the most concentrated natural source, and various other oils such as soybean, peanut, and particularly cottonseed are also rich. Unfortunately, most of the oil meals now marketed are almost devoid of these oils because of their removal by solvent extraction (Maynard *et al.*, 1979). Green forage and other leafy materials, including good quality hay, are very good sources, with alfalfa being especially rich. Concentration of tocopherols per unit dry matter in fresh herbage is between 5 and 10 times as great as that in some cereals or their by-products (Hardy and Frape, 1983).

Stability of all naturally occurring tocopherols is poor and substantial losses of vitamin E activity occur in feedstuffs when processed and stored, as well as in manufacturing and storage of finished feeds. Vitamin E sources in these ingredients are unstable under conditions that promote oxidation of feedstuffs—heat, oxygen, moisture, oxidizing fats, and trace minerals. For concentrates, oxidation increases following grinding, mixing with minerals, the addition of fat, and pelleting. When feeds are pelleted, destruction of both vitamins E and A may occur if the diet does not contain sufficient antioxidants to prevent their accelerated oxidation under conditions of moisture and high temperature. Iron salts (i.e., ferric chloride) can completely destroy vitamin E. Both nitrogen trichloride and chlorine dioxide, at concentrations usually used to bleach flour, will destroy much of the vitamin E activity in flour. According to Moore *et al.* (1957), baking destroyed 47% of remaining tocopherols in treated flour.

Artificial dehydration or processing of forages and grains will reduce availability of tocopherol as well as that of selenium. For example, in one study 80% of the vitamin E was lost in hay making (King *et al.*, 1967), whereas ensiling or rapid dehydration of forages retains most of the vitamin. Vitamin E content in forage is affected by stage of maturity at time of forage cutting and the period of time from cutting to dehydration. Storage losses can reach 50% in 1 month and losses during drying in the swath can amount to as much as 60% within 4 days. Vitamin E losses of 54–73% have been observed in alfalfa stored at 33°C for 12 weeks and 5–33% losses have been obtained with commercial dehydration of alfalfa.

Artificial drying of corn results in a much lower vitamin E content. Young *et al.* (1975) reported 9.3 ppm α-tocopherol in artificially dried corn versus an average of 20 ppm in undried corn (Table 4.4). Preservation of moist grains by ensiling caused almost complete loss of vitamin E activity. Corn stored as acid-treated (propionic or acetic-propionic mixture), high-moisture corn contained approximately 1 ppm dry matter of α-tocopherol, whereas similar corn artificially dried following harvesting contained approximately 5.7 ppm of α-tocopherol (Young *et al.*, 1978). Apparently damage is not due to moisture alone but to the combined propionic acid/moisture effect (McMurray *et al.*, 1980). Further decomposition of α-tocopherol occurs over a more extended period of time until the grain eventually has α-tocopherol levels of less than 1 ppm, which is commonly found in propionic acid-treated barley.

IX. DEFICIENCY

Vitamin E displays the greatest versatility of all vitamins in the range of deficiency signs. Deficiency signs differ among species and even within the same species. The amount of vitamin E required in diets can vary depending on such factors as levels of polyunsaturated fatty acids, selenium, antioxidants, and sulfur amino acids in feed. These deficiency diseases and compounds preventing them are shown in Table 4.5.

Blaxter (1962) reported that muscular dystrophy seemed to be the one syndrome commonly encountered in vitamin E deficiency in all species. He cited references indicating that vitamin E-preventable muscular degeneration occurs in all laboratory and farm animals, including camels, buffalos, kangaroos, and quokkas. In some 20 different animal species, a deficiency of tocopherol leads to muscular dystrophy. Fundamentally this is Zenker's degeneration of both skeletal and cardiac muscle fibers. Connective tissue replacement that follows is observed grossly as white striations in the muscle bundles (Smith, 1970).

Occurrence of muscular dystrophy is worldwide, but its incidence or at least diagnosis, particularly in a mild or subclinical form, varies widely in different countries and regions within countries (McDowell *et al.*, 1983). Considerable research has revealed the positive relationship between selenium content in soil and geographical occurrence of vitamin E–selenium-responsive muscular dystrophy.

A. Effects of Deficiency

1. Ruminants

White muscle disease (WMD, also known as nutritional muscular dystrophy), a serious muscle degeneration disease in young ruminants, is caused by a selenium deficiency, but is influenced by vitamin E status. It may develop intrauterine or

TABLE 4.5

Vitamin E Deficiency Diseases as Influenced by Other Factors[a]

Disease	Experimental animal	Tissue affected	PUFA influence	Prevented by			Sulfur amino acids
				Vitamin E	Se	Antioxidant	
Reproductive failure							
Embryonic degeneration							
Type A	Rat, hamster, mouse, hen, turkey	Vascular system of embryo	X	X		X	
Type B	Cow, ewe			—[b]	X[c]		
Sterility (male)	Rat, guinea pig, hamster, dog, cock, rabbit, monkey	Male gonads		X			
Neuropathy	Chick, human	Brain	X	X		X	
Liver, blood, brain, capillaries, pancreas							
Necrosis	Rat, pig	Liver		X	X		X
Fibrosis	Chick, mouse	Pancreas			X		
Erythrocyte hemolysis	Rat, chick, human (premature infant), dog, monkey	Erythrocytes	X	X		X	
Plasma protein loss	Chick, turkey	Serum albumen		X	X		
Anemia	Monkey	Bone marrow		X		X	
Encephalomalacia	Chick	Cerebellum	X	X		X	
Exudative diathesis	Chick, turkey	Vascular system		X	X		
Kidney degeneration	Rat, mouse, monkey, mink	Kidney tubular epithelium	X	X	X		

Steatitis (ceroid)	Mink, pig, chick	Adipose tissue	X	X		X	
Depigmentation	Rat	Incisors	X	X		X	
Nutritional myopathies							
Type A (nutritional muscular dystrophy)	Rabbit, guinea pig, monkey, duck, mouse, mink, dog	Skeletal muscle		X		?	
Type B (white muscle disease)	Lamb, calf, kid, foal	Skeletal and heart muscles		—[b]	X[c]		
Type C	Turkey	Gizzard, heart		—[b]	X		
Type D	Chicken	Skeletal muscle	—[d]	X			X
Retinopathy	Dog, monkey, rat	Retinal pigment epithelium (photoreceptor cells)	X	X			
Dermatosis	Dog	Skin					
Immunodeficiency	Dog, chick, mouse, sheep, pig	Reticulo endothelial	X	X	X		

[a]Modified from Scott (1980) and Sheffy and Williams (1981).
[b]Not effective in diets severely deficient in selenium.
[c]When added to diets containing low levels of vitamin E.
[d]A low level (0.5%) of linoleic acid is necessary to produce dystrophy; higher levels did not increase vitamin E required for prevention.

extrauterine and is characterized by generalized weakness, stiffness, and deterioration of muscles, with affected animals having difficulty standing (Figs. 4.3 and 4.4). In calves the tongue musculature may be affected and prevent suckling (NRC, 1984a). Often death occurs suddenly from heart failure as a result of severe damage to heart muscle. In milder cases with calves where the chief clinical signs are stiffness and difficulty standing, dramatic, rapid improvement can result with vitamin E–selenium injections.

An acute and chronic as well as a peracute form of the disease can be distinguished in older calves, usually already in the finishing period. In particular,

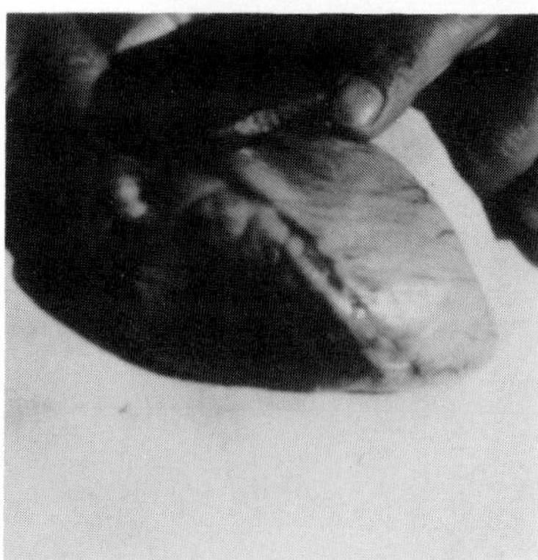

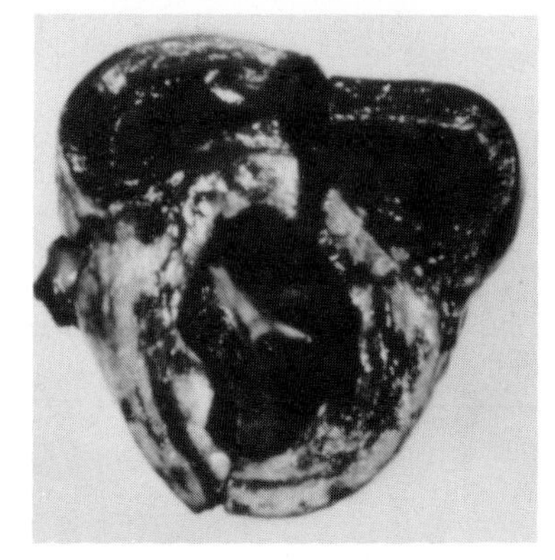

Fig. 4.3 White muscle disease. The calf in the top photograph is about 3 months old; lameness and generalized weakness of muscles can be seen. Bottom photographs show abnormal white areas in the heart muscle. (Courtesy of O. H. Muth, College of Veterinary Medicine, Oregon State University.)

Fig. 4.4 Selenium–vitamin E deficiency in sheep known as stiff-lamb disease or white muscle disease (WMD). The lamb is unable to stand as a result of tissue degeneration. (Courtesy of O. H. Muth, College of Veterinary Medicine, Oregon State University.)

stress situations such as transport, regrouping, or abrupt changes in feed composition are generally considered as precipitating factors. Sudden death without previous unmistakable signs of disease is the main feature of the peracute condition (Bostedt, 1980). The cause is usually found in advanced degeneration of the myocardium, and motor disturbances such as an unsteady gait or stiff-calf disease, hard lumbar, neck, and forelimb muscles, muscle tremor, and perspiration are encountered in the acute form.

In Florida the condition is most common in ''buckling'' calves that come off the truck or out of the processing chute with weakness of rear legs, buckling of fetlocks, and frequently a generalized shaking or quivering of muscles (Figs. 4.5–4.7). Many calves become progressively worse until they are unable to rise and may appear to be paralyzed. Many animals will be down or continue to buckle for extended periods, and death loss is high in severe cases. Calves with excitable temperaments appear to be most affected. Postmortem examination of affected calves reveals pale, chalky streaks in muscles of the hamstring and back, and the heart, rib muscles, and diaphragm may also be affected (McDowell *et al.*, 1985).

In lambs, WMD (also known as stiff-lamb disease) takes a course similar to

Fig. 4.5 Vitamin E–selenium deficiency illustrated as flexion of the hock and fetlock joints as a result of decreased support by the gastrocnemius muscle that is severely affected by myodegeneration. (Courtesy of Bob Mason and University of Florida.)

that found in calves. A gradual swelling of the muscles, particularly in the lumbar and rear thigh regions, gives the erroneous impression of an especially muscular young animal. In addition to the peracute form encountered in calves (changes primarily in the myocardium), chronic cardiac muscle degeneration is also found in the lamb. Despite good initial development, affected lambs quickly lose weight after the third week of life and are unthrifty. They try to avoid any strain and usually stand apart from the herd. Cardiac arrhythmia and increased heart rate can result even after slight exercise. In the advanced stage, animals eat little feed and rapidly waste away.

Absence of vitamin E and/or selenium in the goat, as in other ruminants, results mainly in WMD. Kids especially suffer from this disturbance of muscle metabolism, as they are born with little or no reserves of the fat-soluble vitamins A, D, and E. Sudden death of young kids under 2 weeks of age may reveal postmortem evidence of muscle disease and degeneration in the heart muscle or the diaphragm. In older kids and mature animals, the disease may occur after sudden exercise and the animals show bilateral stiffness, usually in their hind legs.

Fig. 4.6 Vitamin E–selenium deficiency illustrated as flexion of the hock and fetlock joints as a result of decreased support by the gastrocnemius muscle that is severely affected by myodegeneration. (Courtesy of Bob Mason and University of Florida.)

In preruminant calves, white muscle disease has been easily induced by feeding polyunsaturated oils, however, it was thought unlikely that unsaturated fats would be responsible for the disease in ruminating calves because of the apparent near 100% hydrogenation of all unsaturated fatty acids by the rumen microflora (Noble *et al.*, 1974). Recent research indicates that unsaturated fatty acids from lipids in grasses can act as substitutes for the peroxidative challenge in nutritional muscular dystrophy in calves (McMurray and Rice, 1984). Nutritional degenerative myopathy in older calves occurs most frequently at turnout to spring pasture (Anderson *et al.*, 1976). McMurray *et al.* (1980) showed that polyunsaturated fatty acids were capable of escaping ruminal hydrogenation at turnout, resulting in a 3-fold increase of plasma linolenic acid within 3 days of turnout. Rice *et al.* (1981) showed that linolenic acid, if protected from ruminal hydrogenation, rapidly reaches high levels in blood and is associated with a rise in plasma creatine phosphokinase, indicating muscular degenerative myopathy.

Most nutritional myopathy cases have involved young ruminants, with effects less fully described for adult animals. However, degenerative myopathy in adult cattle has been reported (Van Vleet *et al.*, 1977; Gitter and Bradley, 1978;

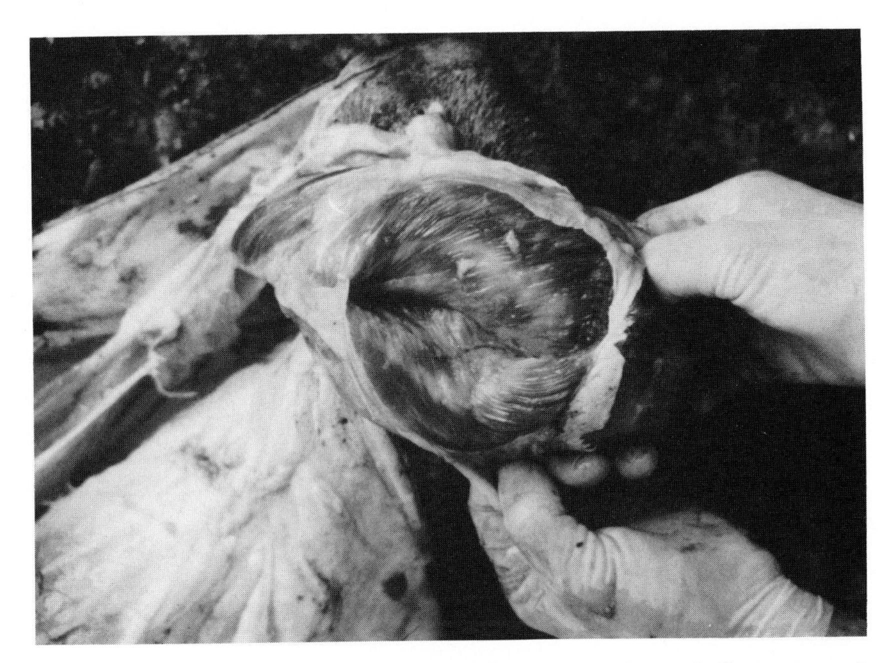

Fig. 4.7 Selenium–vitamin E deficiency in cattle as illustrated by white muscle disease or necrosis of the gastrocnemius muscle, as evidenced by chalky white streaks in the belly of the muscle. (Courtesy of Bob Mason and University of Florida.)

Hutchinson *et al.*, 1982), and a group of yearling Chianina heifers experienced abortion, stillbirth, and periparturient recumbency (Hutchinson *et al.*, 1982). Necropsy and tissue analyses revealed myodegeneration and a combined deficiency of vitamin E and selenium. Rapid growth in these heifers coupled with stresses of late pregnancy and parturition may have contributed to this vitamin E deficiency. A myopathic condition affecting yearling cattle was reported by Barton and Allen (1973) and was associated with animals fed grains treated with propionic acid, which is known to destroy vitamin E.

Attempts to establish a practical role for vitamin E in ruminant reproductive deficiencies of both males and females have been largely unsuccessful (NRC, 1984a). The relationship to reproduction is of special interest since early rat research demonstrated that reproductive failure was a key feature of vitamin E deficiency. In one experiment, four generations of female and male dairy cattle were fed low-vitamin E diets (Gullickson *et al.*, 1949). Although growth, reproduction, and milk production were normal, several cattle died suddenly of apparent heart failure between 21 months and 5 years of age.

From a different aspect of reproduction in dairy cattle, Harrison *et al.* (1984)

reported that supplemental vitamin E was required in addition to selenium for prevention of retained placenta. Groups administered vitamin E alone, selenium alone, and controls had a retained placenta incidence of 17.5% compared to 0% for animals receiving both vitamin E and selenium. Other research indicates that the incidence of retained placenta (22.1%) was not reduced by a combination of vitamin E and selenium or selenium alone (Hidiroglou *et al.*, 1987). In high-producing dairy goats, deficiency manifests itself in poor involution of the uterus with accompanying retained placenta and metritis following kidding (Guss, 1977).

Adequate amounts of vitamin E in the diet are needed to prevent oxidative flavors in milk. However, the cost is high, with efficiency of transfer into milk less than 2% (NRC, 1978a).

2. Swine

Most vitamin E deficiency signs for the pig have been associated with selenium deficiency, and scientists usually refer to a vitamin E and/or selenium deficiency since it is not clear which is involved and generally dietary levels of both must be low to bring about deficiency signs and lesions. Since the early 1950s, reports in the European literature have revealed tissue degeneration signs in swine under field conditions associated with vitamin E deficiency, with the significance of selenium deficiency not realized until 1957.

Muscular dystrophy and hepatosis dietetica (toxic liver dystrophy) were particularly widespread in the swine industry in Sweden. Obel (1953) reported that records from the State Veterinary Medical Institute of Stockholm from 1947 to 1952 revealed that of a total of 4382 pigs autopsied, over 10% suffered from hepatosis dietetica.

Selenium–vitamin E deficiencies have been readily produced in swine diets through use of both highly unsaturated fats (i.e., cod liver oil) and rancid fats. However, naturally occurring vitamin E–selenium deficiencies were not reported in the United States until the late 1960s (Michel *et al.*, 1969), and in the 1970s they became widespread. High incidence of vitamin E–selenium deficiencies in swine was believed to be due to a number of factors (Trapp *et al.*, 1970), including (1) swine raised in complete confinement, without access to pasture, (2) low selenium content in Midwestern U.S. feeds, (3) solvent-extracted protein supplements low in vitamin E, (4) limited feeding programs for sows, (5) loss of vitamin E and selenium from corn due to oxidation as a result of air and heat drying or storing high-moisture grains, and (6) selection of meatier-type pigs that require more selenium. Evidence also suggests that moldy feed in bulk-holding bins may produce mycotoxins that either inhibit the uptake of vitamin E in the small intestine or affect the antioxidant balance of cells. Orstadius *et al.* (1963) reported that vitamin E content of corn was reduced from 30 to 50 ppm to about 5 ppm of dry weight as a result of artificial drying at 100°C for 24 hr under a continuous flow of air.

An increase in confinement rearing of swine on concrete floors or slats has been accompanied by a decrease in the utilization of pasture and forages. Such crops are not only excellent sources of vitamin E but also provide the more highly available form of the vitamin, α- versus γ-tocopherol.

Vitamin E–selenium deficiency in swine is often associated with sudden death. In most cases clinical signs of the condition were not observed prior to death (Michel *et al.*, 1969; Trapp *et al.*, 1970), although occasionally pigs were observed with clinical signs of icterus, difficult locomotion, reluctance to move, and weakness. Clinical signs also include peripheral cyanosis (particularly the ears), dyspnea (abdominal respiration), and a weak pulse, all occurring shortly before death. In many cases the faster-growing, more thrifty-appearing pigs died suddenly.

The most common pathological lesions include massive hepatic necrosis (hepatosis dietetica) (Figs. 4.8 to 4.10), degenerative myopathy of cardiac (Fig. 4.11) and skeletal muscles (Fig. 4.12), edema, esophagogastric ulceration, icterus, nephrosis, hemoglobinuria, acute congestion, hemorrhaging (Fig. 4.13) in various

Fig. 4.8 Lesions in growing pigs fed diets low in vitamin E. (7.0 IU α-toxopherol/kg diet) and selenium (0.061 ppm). Liver with severe acute lesions is characteristic of hepatosis dietetica, consisting of a mosaic pattern of deep red and yellow lobules of massive coagulation necrosis. (Courtesy of L. R. McDowell, R. C. Piper, and Washington State University.)

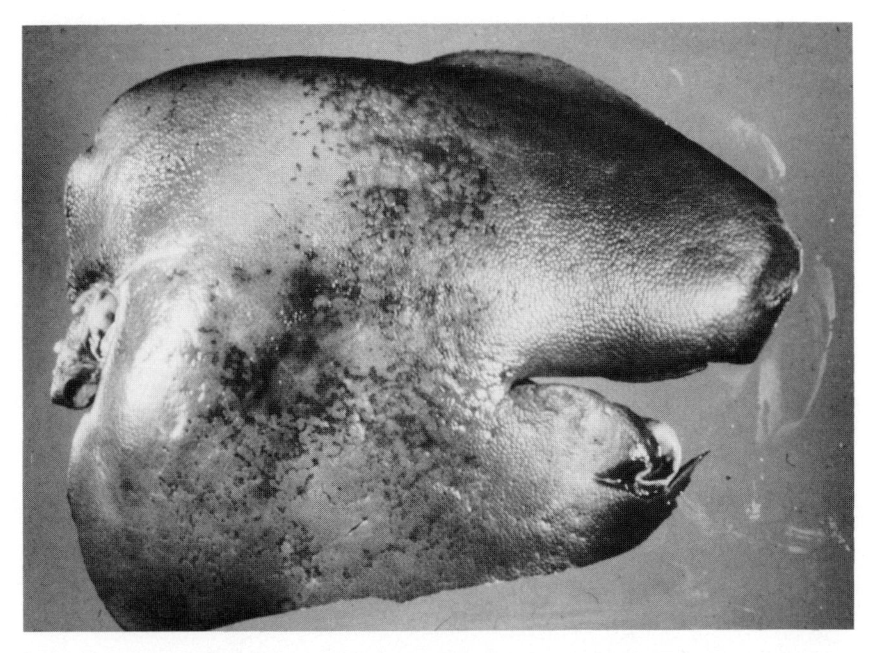

Fig. 4.9 Lesions in growing pig fed diet low in vitamin E (7.0 IU α-tocopherol/kg diet) and selenium (0.061 ppm). In subacute dietary massive hepatic necrosis, the lesions are more chronic than those seen in liver of pig in Fig. 4.8, in that many hepatic lobules are atrophic and collapsed, causing a pitted appearance on the surface of the liver. (Courtesy of L. R. McDowell, R. C. Piper, and Washington State University.)

tissues (Trapp *et al.*, 1970; Piper *et al.*, 1975), and yellowish discoloration of adipose tissue (''yellow fat'').

Many pathological reports of vitamin E–selenium deficiency note that the most striking lesion was liver necrosis (Trapp *et al.*, 1970), however, bilateral paleness of skeletal muscles was the gross lesion most commonly found. In some pigs, microscopic lesions in liver were either absent or minimal, whereas changes in skeletal muscles were extensive. In other cases, the reverse was true. Other conditions reported in swine herds with vitamin E–selenium deficiency include mastitis–metritis–agalactia syndrome in sows, spraddled rear legs in newborn pigs, gastric ulcers, infertility, and poor skin condition. These conditions were believed initially to be unrelated to pig deaths from a vitamin E–selenium deficiency. However, after supplementation with dietary vitamin E or injections of selenium and vitamin E, a noticeable reduction in these conditions occurred (Trapp *et al.*, 1970).

In swine not only are newborn pigs affected but also young growing animals.

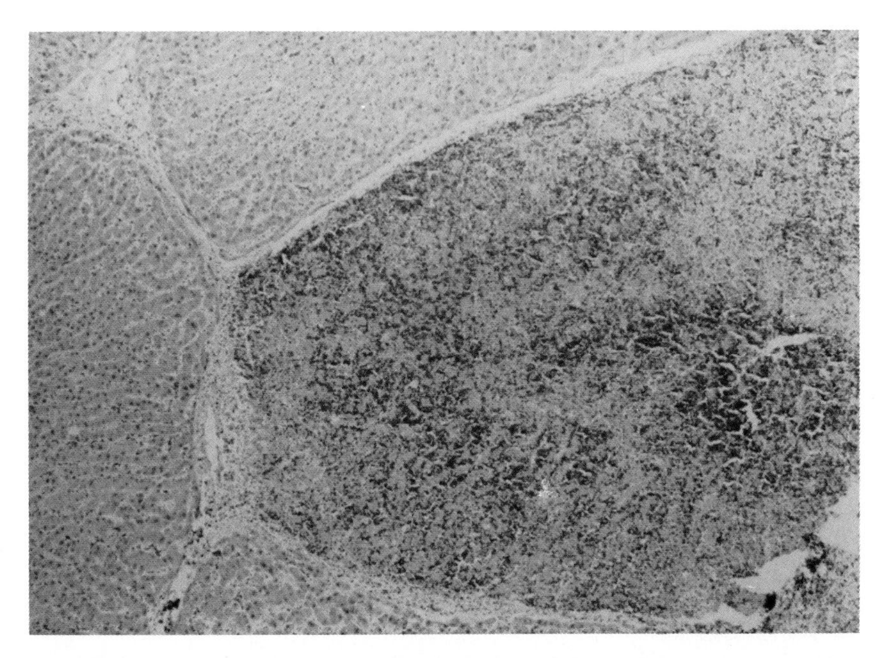

Fig. 4.10 Lesions in liver of growing pig fed diet low in vitamin E (7.0 IU α-tocopherol/kg diet) and selenium (0.061 ppm). Shown is an individual hepatic lobule that has undergone acute massive coagulation necrosis and the necrosis cells are being replaced by blood. Note the normal adjacent lobule. (Courtesy of L. R. McDowell, R. C. Piper, and Washington State University.)

The vitamin E–selenium deficiency syndrome (nutrition-related microangiopathy, nutritional hepatic dystrophy, muscle degeneration in the back, pelvis, and upper thigh) can, however, also be found in the fattening and reproductive stages of pig production (Bostedt, 1980). At an early age it is particularly myocardial damage, also known as nutritional microangiopathy or mulberry heart disease, that may cause substantial losses within a litter. This is the most serious of disorders, since when heart muscle tissue is damaged the result is usually sudden death. There may be hemorrhagic lesions within the heart that give the characteristic "mulberry" appearance of mulberry heart disease (MHD).

Some researchers have demonstrated a low tolerance of vitamin E- and selenium-deficient baby pigs to intramuscular injections of iron dextrose for prevention of anemia. At 2 or 3 days, piglets die from iron shock if given routine treatment with iron, with death resulting from an iron-induced lipid peroxidation in tissues. Pretreatment with vitamin E, selenium, or ethoxyquin was protective against toxic effects of injectable iron (Tollerz and Lannek, 1964).

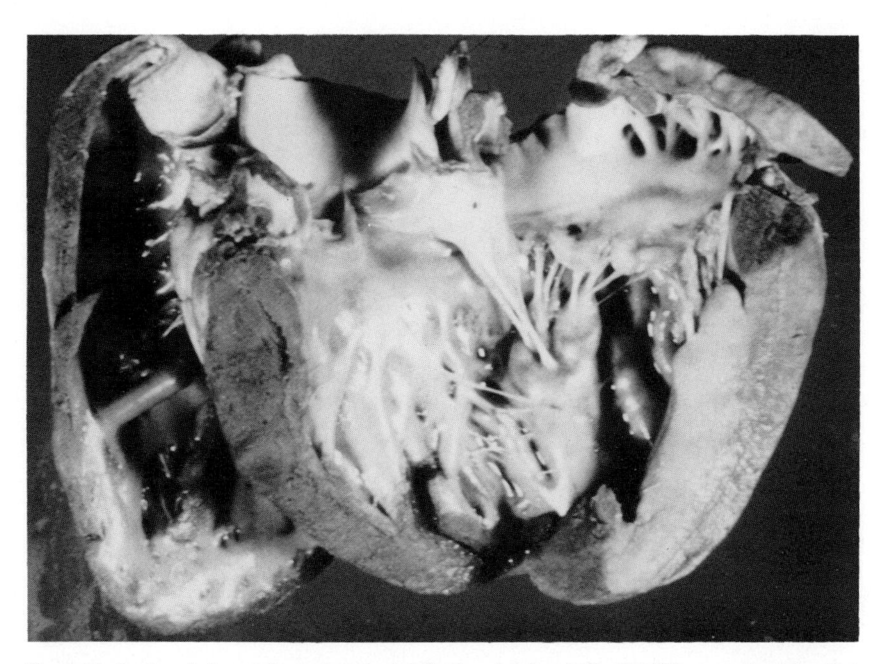

Fig. 4.11 Lesions in heart of growing pig fed diet low in vitamin E. (7.0 IU α-tocopherol/kg diet) and selenium (0.061 ppm). Note the degenerative nutritional cardiac myopathy and the large pale areas due to degeneration and necrosis of myocardial fibers that are most severe along the inner border of the left ventricle. (Courtesy of L. R. McDowell, R. C. Piper, and Washington State University.)

Maximum incidence of death due to selenium–vitamin E deficiency generally occurs at 6–8 weeks of age with the incidence declining up to the sixteenth week of life. A Michigan survey diagnosed selenium–vitamin E deficiencies in swine herds with mortality ranging from 3 to 10% (Michel *et al.*, 1969; Trapp *et al.*, 1970). One producer, however, lost approximately 300 of 800 pigs weaned.

It has been realized for many years that vitamin E-deficient animals are more subject to the effects of "stress" than normal animals. The concept of stress is, in itself, difficult to define, but experience has shown that dietary and environmental abnormalities of various kinds can lead to clinical signs of disease and death in animals deprived of vitamin E. Death is often associated with unaccustomed muscular activity. Incidence of death in baby pigs is greatly increased because of fighting when animals are weaned and mixed with different litters. Castration is an additional stress that has been implicated to cause early death in selenium- and vitamin E-deficient pigs (Piper *et al.*, 1975).

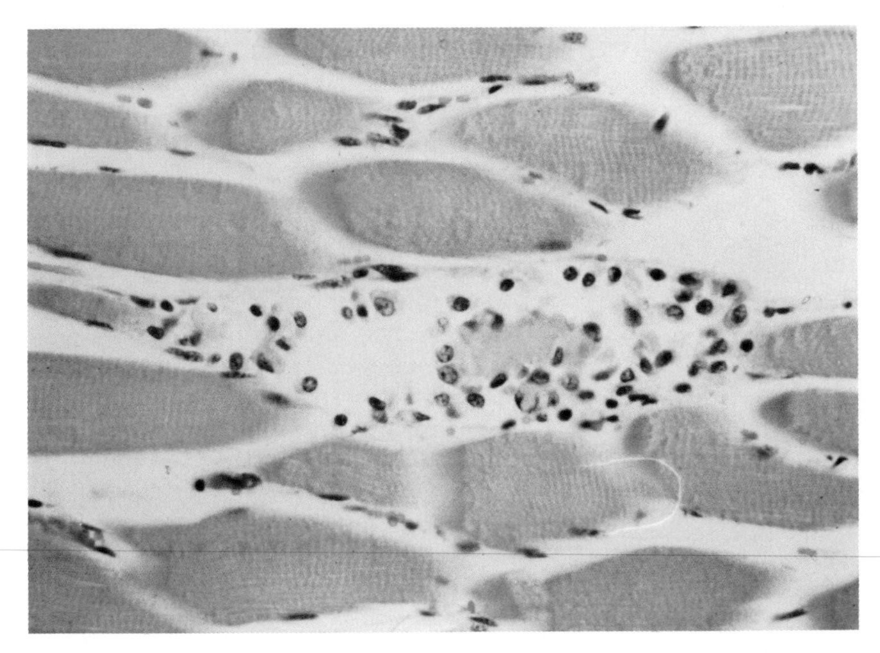

Fig. 4.12 Lesions in skeletal muscle of pig fed diet low in vitamin E. (7.0 IU α-tocopherol/kg diet) and selenium (0.061 ppm). Shown is a muscle fiber (low power) of biceps fermoris undergoing degeneration, surrounded by normal fibers. Cytoplasm is breaking up into granules from enzymatic digestion as the fiber is invaded by macrophages. (Courtesy of L. R. McDowell, R. C. Piper, and Washington State University.)

3. Poultry

Vitamin E deficiency in poultry can result in at least three conditions: exudative diathesis (Fig. 4.14), with signs of subcutaneous edema and, in severe cases, blackening of the affected parts, apathy, and inappetance; encephalomalacia (''crazy chick disease'') (Fig. 4.15) characterized by ataxia, head retraction, and ''cycling'' with legs; and muscular dystrophy (Fig. 4.16) (Scott *et al.*, 1982). Vitamin E deficiency is also known to reduce hatchability in turkey eggs (Jensen and McGinnis, 1957).

Ataxia of encephalomalacia generally affects chicks from 2 to 6 weeks of age and results from hemorrhages and edema within the cerebellum. At least one important function of vitamin E is to interrupt the production of free radicals at the initial stage of encephalomalacia. The quantitative need for vitamin E for this function depends on the amount of linoleic acid in the diet. Selenium is ineffective in preventing encephalomalacia, while synthetic antioxidants are highly effective. The fact that low concentrations of antioxidants are capable of pre-

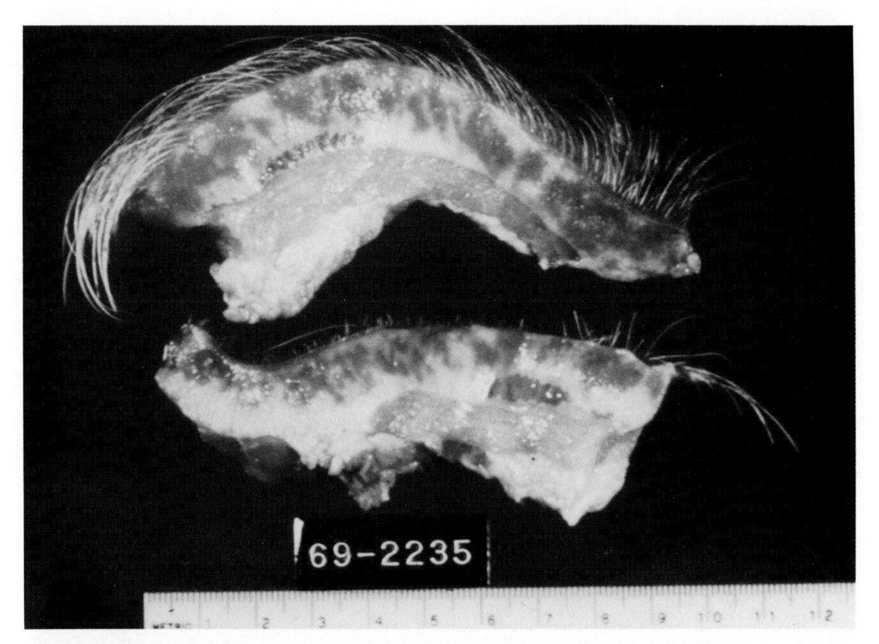

Fig. 4.13 Skin lesion in growing pig fed diet low in vitamin E (7.0 IU α-tocopherol/kg diet) and selenium (0.061 ppm). Shown is the skin from a pig with severe congestion and hemorrhage. The skin hemorrhage extended througout the epidermis, dermis, subcutaneous fat, and down to the cutaneous musculature. (Courtesy of L. R. McDowell, R. C. Piper, and Washington State University.)

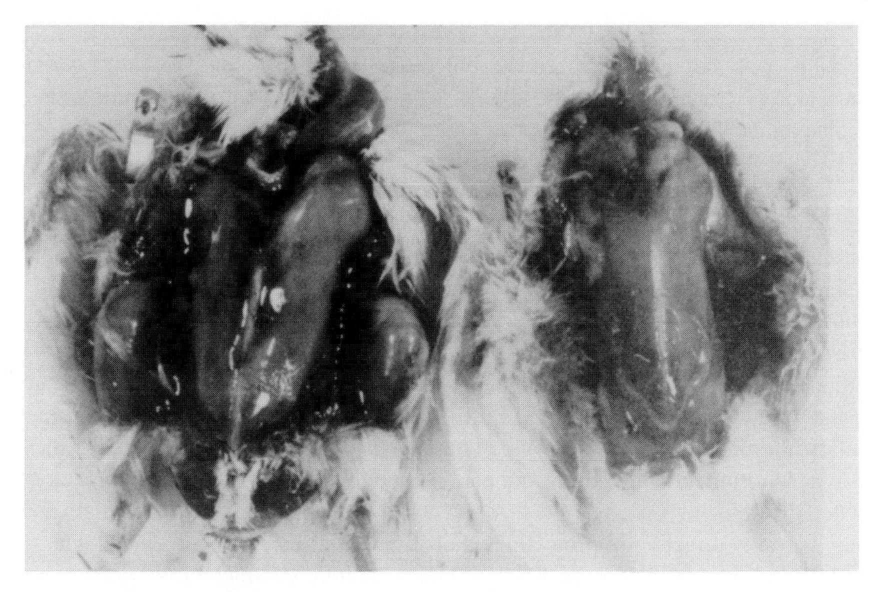

Fig. 4.14 Vitamin E–selenium deficiency in chicks resulting in exudative diathesis. Note profuse subcutaneous edema. (Courtesy of L. E. Krook, Cornell University.)

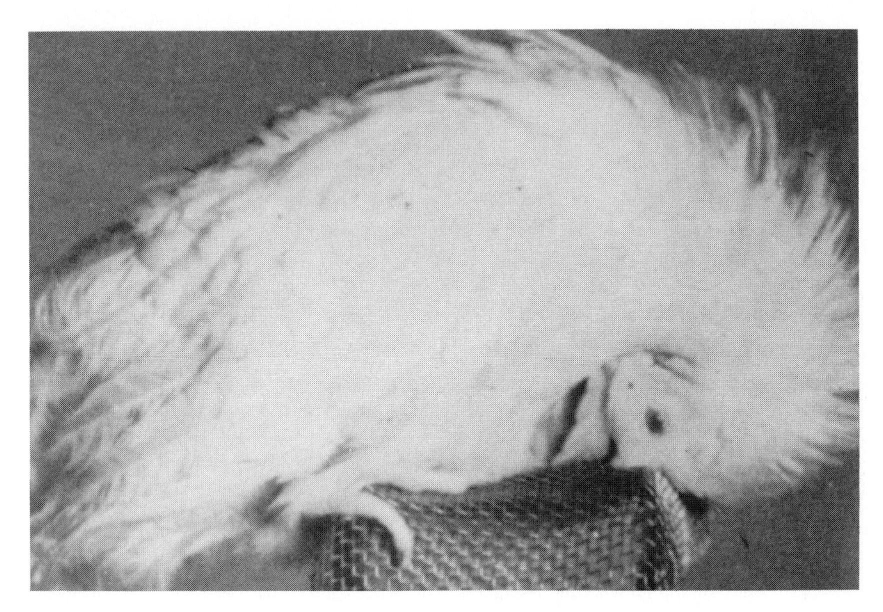

Fig. 4.15 Encephalomalacia in a chick fed a vitamin E-deficient diet. This disease is caused by a deficiency of vitamin E or antioxidants. Selenium supplementation of the diet will not prevent it. (Courtesy of M. L. Scott, Cornell University.)

venting encephalomalacia in chicks but fail to prevent exudative diathesis or muscular dystrophy in the same chicks strongly suggests that in preventing encephalomalacia vitamin E acts as an antioxidant.

Exudative diathesis in chicks, depicted in Fig. 4.14, is a severe edema produced by a marked increase in capillary permeability. When vitamin E deficiency is accompanied by a sulfur amino acid deficiency, chicks show a severe nutritional muscular dystrophy, especially of breast muscle, at about 4 weeks of age. Both selenium and vitamin E are involved in prevention of exudative diathesis and nutritional muscular dystrophy. In diets severely deficient in selenium, however, vitamin E does not prevent or cure exudative diathesis, whereas addition of as little as 0.05 ppm of dietary selenium completely prevents this disease. Cystine is likewise effective in preventing nutritional muscular dystrophy in vitamin E-deficient chicks. Cystine, however, is apparently ineffective in preventing the dystrophic condition in other animals. Although vitamin E and selenium are generally both highly effective in preventing exudative diathesis, selenium is only partially effective in protecting against muscular dystrophy in chicks when added in the presence of a low level of dietary vitamin E. Much larger quantities of selenium are required to reduce the incidence of muscular dystrophy in chicks

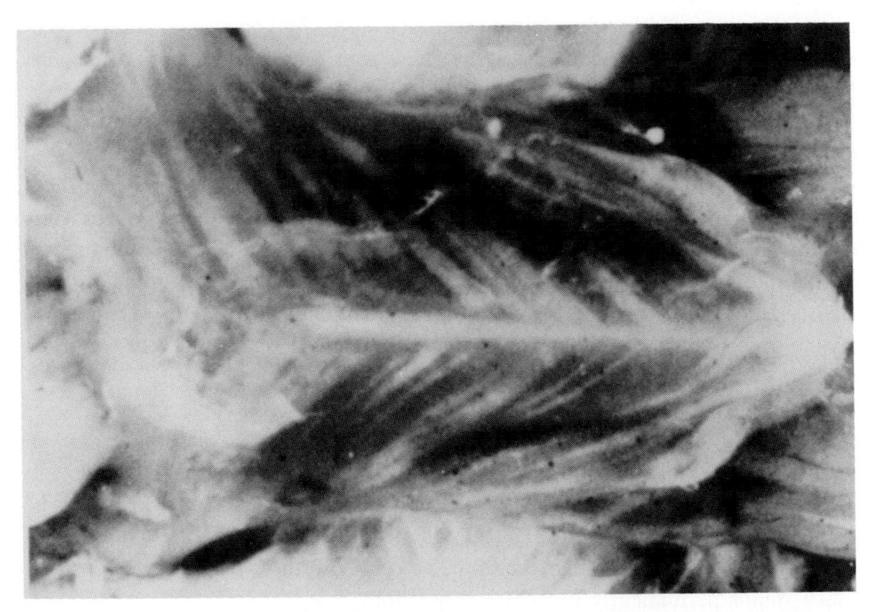

Fig. 4.16 Nuritional muscular dystrophy in chick fed a vitamin E-deficient diet low in the sulfur amino acids. The diet contained an antioxidant and 0.1 ppm of selenium to prevent encephalomalacia and exudative diathesis. Note the white degenerated muscle fibers in both the breast and thigh. (Courtesy of M. L. Scott, Cornell University.)

receiving a vitamin E-deficient diet low in methionine and cystine (Scott *et al.*, 1982).

Prolonged vitamin E deficiency can result in permanent sterility. Hatchability of eggs from vitamin E-deficient hens is reduced (NRC, 1984b), and embryonic mortality may be high during the first 4 days of incubation as a result of circulatory failure. Turkey embryos deficient in vitamin E may have eyes that protrude with a bulging of the cornea.

4. Horses

The newborn foal that is born with nutritional muscular dystrophy usually exhibits various clinical signs in this acute phase. Animals are hardly able to stand up and give the impression of general weakness. After laborious attempts to struggle to its feet, the foal stands rather awkwardly and stiffly. If neck muscles are affected, suckling is substantially impaired, with the foal unable to raise its head to the mother's udder, although it gives the impression of wishing to suckle (Bostedt, 1980). Changes affect predominantly pectoral, intercostal, and diaphragm muscles and result in an accelerated, intermittent, primarily abdominal respiration that can be incorrectly diagnosed as bronchial pneumonia or dry pleurisy. Movement of the thorax appears to cause pain. Cardiac arrhythmia is another clinical sign that occurs as a result of muscle changes in the heart.

As with calves, the musculature of the tongue may be affected. In spite of their obvious appetite, the animals are unable to swallow milk and slightly opened mouth allows milk to leak out or trickle from their nostrils. Their stomach is empty and their belly drawn up, so that the person in charge of the mare frequently consults a veterinarian for what is believed to be a lack of milk. Foals affected in this manner usually die during the first few days of life; the animal's death is accelerated by hypostatic or deglutition pneumonia or by neonate infections arising from immunological disorders caused by an inadequate intake of colostrum or from cardiovascular insufficiency (Bostedt, 1980).

In older foals (6–12 weeks), the progressive degeneration of motor muscles rather than changes in the head and thoracic muscles are the first indications of the disease. Typical signs are an increasingly clumsy gait and unsteady movements of the hindquarters. These foals also lie down a great deal and can hardly be made to stand (Bostedt, 1980). The urine may be coffee colored because of myoglobin released from damaged muscle cells. As the condition worsens, the foal remains permanently in the lying position.

Combinations of vitamin E and selenium have been used in the treatment of the "tying-up" syndrome in horses (NRC, 1978b), which is characterized by lameness and rigidity of the loin muscles. However, no experimental evidence to confirm the value of vitamin E on the condition has been provided. Likewise, studies published concerning the influence of vitamin E on reproduction have been contradictory.

5. Other Animal Species

a. Rabbits. Muscular dystrophy in rabbits is recognized as caused primarily by vitamin E deficiency. Signs of this syndrome include degeneration of the skeletal and cardiac muscles, paralysis, and fatty liver (Bragdon and Levine, 1949). Ringler and Abrams (1970) encountered widespread signs of vitamin E deficiency in a commercial herd of rabbits fed a natural diet that provided 16.7 ppm vitamin E. The rabbit appears to be unusual in that selenium has neither a protective nor sparing effect on muscular dystrophy that is preventable by vitamin E (Hove *et al.*, 1958).

b. Foxes and Mink. Diets containing rancid fats or high in unsaturated fat cause a "yellow fat" disease (steatitis) in foxes and mink. The most frequent clinical sign with an uncomplicated vitamin E deficiency was sudden death due to minor stress (NRC, 1982a). Selenium had some but not a complete vitamin E-sparing effect.

Signs of "yellow fat" disease in both foxes and mink are most evident among fast-growing male pups. Acute and subacute cases frequently occur. In fur-bearing animals, anemia is often found in chronic cases of "yellow fat." A pronounced fragility of the red blood cells and an increased number of leukocytes and thrombocytes are frequent. In the postmortem examination, the musculature often appears light and musty, and fatty degeneration is usually discovered in the parenchymatous organs and in the musculature. The skin is of poor quality and loses hair easily, both during and after tanning (Helgebostad and Ender, 1973).

c. Dogs. Clinical signs of vitamin E deficiency in dogs include dystrophy of skeletal muscle and associated muscle weakness. The interrelationship of selenium and vitamin E requirements has not been sufficiently studied (NRC, 1985a).

d. Cats. Steatitis was noted when sources of highly unsaturated fatty acids were fed to cats in the absence of supplemental vitamin E. Cats fed a diet containing 5% tuna oil without supplemental vitamin E exhibited severe steatitis, focal interstitial myocarditis, focal myositis of the skeletal muscle, and periportal mononuclear infiltration in the liver (NRC, 1986).

e. Monkeys. When deprived of vitamin E for prolonged periods, usually a year or more, rhesus and cebus monkeys develop a characteristic anemia and muscular dystrophy (NRC, 1978d), Anemia results in the production of defective erythrocytes, with muscular dystrophy closely resembling vitamin E deficiency–muscular dystrophy in other species.

f. Fish. Channel catfish fed a vitamin E-deficient diet containing oxidized menhaden oil exhibited reduced growth, muscular dystrophy, fatty livers, anemia, exudative diathesis, and depigmentation in 16 weeks (NRC, 1983). "Sekoke" disease is a condition in common carp characterized by a marked loss of flesh. The disease in carp resulted from feeding oxidized silk worm pupae and was completely prevented with supplemental vitamin E. Signs of vitamin E deficiency in trout and salmon either with or without selenium are numerous and include dystrophic skeletal muscle lesions, impaired erythropoiesis, extreme anemia, susceptibility to stress of handling, high mortality, yellow-to-brown serous fluid in the body cavity, and increased content of body water (exudative diathesis) (NRC, 1981a).

6. Humans

In the United States and other developed countries, vitamin E intake of most human populations is considered adequate for maintenance, growth, and reproduction in normal individuals. Exceptions are newborn infants, particularly those born premature or otherwise of low body weight, and some members of low-income groups or individuals practicing bizarre food consumption habits that entail consumption of low-vitamin E foods.

Plasma vitamin E concentration in the newborn infant is about one-third that of adults, with that in the low-birth-weight (LBW) infant even lower. This is primarily a reflection of lower concentration of blood lipids in newborn infants but may also be due to inefficient placental transfer of the vitamin. In the United States, edema and anemia attributed to vitamin E deficiency have been reported in LBW infants fed low-vitamin E commercial formulas made with polyunsaturated fat (Oski and Barness, 1967). Premature infants fed these formulas with inadequate vitamin E develop hemolytic anemia, resulting in shortened life span of red blood cells.

Human diseases that are alleviated by vitamin E include thrombophlebitis and intermittent claudication, both of which involve blood flow, particularly in the extremities of elderly persons (Haeger, 1974). A number of studies have indicated that vitamin E has a beneficial effect in preventing cardiac diseases, eye disorders, skin diseases, ulcers, and even cancer.

Low levels of vitamin E also have been found in plasma of many persons with absorptive defects (Sitrin *et al.*, 1987). Patients suffering from sprue, cystic fibrosis of the pancreas with accompanying steatorrhea, or any other disease that causes malabsorption of fat also show a marked decrease in plasma tocopherol levels.

Effects of vitamin E on human diseases have been reviewed by numerous research workers (Scott, 1980). These reviewers presented two extreme points of view: one group claims that vitamin E is a cure for almost every disease known to man; the other group holds that vitamin E has not been proved scientifically to have any of the effects being claimed for it. The most universally recognized disease in animals due to vitamin E deficiency is muscular dystrophy. Attempts to demonstrate improvement from vitamin E therapy in humans with various forms of muscular dystrophy have failed for the most part (Horwitt, 1980). Many medical claims for physiological benefits in healthy persons are still largely subjective, and controlled experiments are needed to elucidate the need for supplemental vitamin E.

B. Assessment of Status

Confirmation of a low vitamin E and/or selenium status in animals is obtained when specific deficiency diseases associated with lack of these nutrients are present. Likewise, gross lesions and histopathological examinations provide definite evidence of vitamin E and/or selenium deficiency.

Muscular damage as a result of vitamin E and/or selenium deficiencies causes leakage of intercellular contents into the blood. Thus elevated levels of selected enzymes, above normal concentrations for particular species, serve as diagnostic aids in detecting tissue degeneration. Serum enzyme concentrations used to follow incidence of nutritional muscular dystrophy include serum glutamic-oxaloacetic-transaminase (SGOT), aspartate amino transferase (ASPAT), lactic dehydrogenase (LDH), creatine phosphokinase (CPK), and malic dehydrogenase (MDH). A distinction regarding type of tissue degeneration can sometimes be made. For example, in swine SGOT and (ornithine carbamyl transferase) OCT are useful indicators of muscular dystrophy and liver necrosis, respectively (Wretlind *et al.*, 1959). Enzyme tests are very sensitive, and an elevation of enzyme activity in serum is usually discovered before any pathological changes or clinical signs appear (Tollersrud, 1973).

In addition to serum enzyme determinations, other laboratory tests devised to assist in diagnosis of vitamin E and selenium deficiency include (1) vitamin E and selenium analyses of feeds, blood, and tissues and (2) electrocardiogram changes that reflect heart muscle injury. Low tissue concentrations of glutathione peroxidase, a selenium-containing enzyme, is a relatively good status indicator of this element. However, an *in vitro* hemolysis test is an indicator of vitamin E but not selenium status.

Nutritional status with respect to vitamin E is commonly estimated from plasma (or serum) concentration. There is a relatively high correlation between plasma and liver levels of α-tocopherol (and also between amount of dietary α-tocopherol administered and plasma levels). This has been observed in rats, chicks, pigs, lambs, and calves within rather wide ranges of intake. There is a much higher correlation between blood and liver concentrations for vitamin E than for vitamin A. Plasma tocopherol concentrations of 0.5–1 μg/ml are considered low in most animal species, with less than 0.5 μg/ml generally considered a vitamin E deficiency. Thus, plasma α-tocopherol concentrations can be used for assay purposes without the necessity of liver biopsy or animal slaughter (Ullrey, 1981).

In normal adult human populations of the United States, the range of total plasma tocopherols is 0.5–1.2 μg/ml, with values for α-tocopherol 10–15% lower (Bieri and Prival, 1965). It is generally accepted that a plasma level of total tocopherols below 0.5 μg/ml is undesirable, although it has not been shown that lower concentrations in adults, unless of a duration of a year or longer, are associated with inadequate tissue concentrations. Plasma tocopherol concentrations are highly correlated with total lipid, with less than 0.6–0.8 mg/g lipid considered deficient in humans (Machlin, 1984).

For many species, selenium concentrations in liver, renal cortex, and blood (as well as other tissues) each adequately portray selenium status. For example, swine hepatic, renal cortex, and blood selenium concentrations of 0.25, 2.5, and 0.1 ppm (dry basis), respectively, were determined to be critical levels below which clinical illness, death, or lesions of selenium–vitamin E deficiency could be expected (McDowell *et al.*, 1977). Serum or plasma selenium is considered a good status indicator, with less than 0.03 ppm being critical for cattle (McDowell, 1985a).

X. SUPPLEMENTATION

Methods of providing supplemental vitamin E are (1) as part of a concentrate or liquid supplement, (2) included with a free-choice mineral mixture, (3) as an injectable product, and (4) in drinking water preparations.

The most active form of natural vitamin E found in feed ingredients is *d*-α-tocopherol. For many years the primary source of vitamin E in animal feed was the natural tocopherols found in green plant materials and seeds. The *dl*-α-tocopherol (all-rac) form of vitamin E does not exist in nature and is less than the natural *d*-α-tocopherol (RRR). For sheep (Hidiroglou and McDowell, 1988) and cattle (Hidiroglou *et al.*, 1988) a higher bioavailability and higher plasma α-tocopherol for intramuscularly administered *d*-α-tocopherol versus *dl*-α-tocopherol is reported. The principal commercially available forms of vitamin E used in the food, feed, and pharmaceutical industries are acetate and hydrogen succinate esters of RRR α-tocopherol and the acetate ester of all-rac-α-tocopherol (Table 4.6). During commercial synthesis of *dl*-α-tocopherol, it is esterified to acetate to stabilize it, with the ester extremely resistant to oxidation. Thus, *dl*-α-tocopherol acetate does not act as an antioxidant in the feed and only has antioxidant activity after it is hydrolyzed in the intestine and free *dl*-α-tocopherol is released and absorbed.

The acetate forms of α-tocopherol are commercially available from two basic sources (Anonymous, 1972): (1) *d*-α-tocopheryl acetate is made by extraction of natural tocopherols from vegetable oil refining by-products, molecular distillation to obtain the alpha form, and then acetylation to form the acetate ester, and (2) *dl*-α-tocopheryl acetate is made by complete chemical synthesis, producing

TABLE 4.6

Forms of α-Tocopherol Commercially Available

Form	IU/mg
dl-α-Tocopheryl acetate (all-rac)	1.00
dl-α-Tocopherol (all-rac)	1.10
d-α-Tocopheryl acetate (RRR)	1.36
d-αTocopherol (RRR)	1.49
dl-α-Tocopheryl acid succinate (all-rac)	0.89
d-α-Tocopheryl acid succinate (RRR)	1.21

a racemic mixture of equal parts of *d* and *l* isomers. The *d* and *l* forms differ only in spatial placement of the isoprenoid side chain.

Commercially, the *dl*- and *d*-α-tocopheryl acetates are available in purified form or in various dilutions and include (Anonymous, 1972) (1) a highly concentrated oily form, for further processing; (2) emulsions incorporated in powders or beadlets for use in dry, water-dispersible preparations; (3) beadlets or powders consisting of the tocopheryl acetate incorporated in oil, or in emulsifiable form mixed with gelatin and sugar, gum acacia, soy grits, or dextrin as carriers (such beadlets or powders may be further diluted to lower potencies in a feed ingredient or water-soluble material and are primarily for use in feed); and (4) adsorbates of the oily tocopheryl acetate on selected absorbent carriers, in free-flowing "dry" powder, meal, or granules. This type is for use in feeds only.

Even though the ester is more stable than the natural free or alcohol form, it is desirable to further stabilize it by a gelatin coating or adsorption technique that reduces it to a beadlet, granule, or powder form to be added more easily and uniformly to animal feeds (Bauernfeind, 1969). Both types of products are quite stable. Stabilized *dl*-α-tocopheryl acetate gelatin beadlets have been blended in mash feed that also has been pelleted and stored with satisfactory stability results.

A water-miscible injectable vitamin E preparation is also available that contains free *dl*-α-tocopherol. Liquid emulsions of appropriate types are used in drinking water and liquid feeds, in regular feed, and for injection. In this form, α-tocopherol is more efficiently absorbed from intramuscular injection sites than a water-miscible preparation containing α-tocopheryl acetate or either form dissolved in an oily base (Machlin, 1984).

The need for supplementation of vitamin E is dependent on the requirement of individual species, conditions of production, and in relation to available vitamin E in food or feed sources (see Section VIII). The primary factors that influence the need for supplementation include (1) vitamin E- and/or selenium-deficient concentrates and roughages; (2) excessively dry ranges or pastures for grazing livestock; (3) confinement feeding where vitamin E-rich forages are not included or only forages of poor quality are provided; (4) diets that contain predominantly non-α-tocopherols and thereby are less biologically active; (5) diets that include ingredients that increase vitamin E requirements (i.e., unsaturated fats, waters high in nitrates); (6) harvesting, drying, or storage conditions of feeds that result in destruction of vitamin E and/or selenium (see Section VIII); (7) accelerated rates of gain, production, and feed efficiency that increase metabolic demands for vitamin E; and (8) intensified production that also indirectly increases, vitamin E needs of animals by elevating stress, which often increases susceptibility to various diseases.

Vitamin E–selenium deficiencies are found in specific world regions and are characterized by low concentrations of selenium in feedstuffs. Regions that rely on concentrate importations from these areas deficient in selenium and/or vitamin E (i.e., midwestern United States) likewise must provide these nutrients to livestock. Adverse conditions such as poor weather (drought and early frost), molds, and insect infestation will reduce the vitamin E value of feedstuffs. The vitamin E activity in blighted corn was 59% lower than that in sound corn, and the vitamin E activity in light-weight corn averaged 21% below that in sound corn. Feed spoilage will also promote selenium–vitamin E deficiencies, therefore, to prevent loss of vitamin E in diets, the producer should use fresh feed at all times because the vitamin is rapidly destroyed under hot, humid conditions. Also, the producer should use an antioxidant in the diet to prevent the destruction of the vitamin E (as well as other vitamins, i.e., vitamin A). Losses during storage increase as the duration and temperature of storage increase.

The most convenient and often most effective means to provide vitamin E to livestock is inclusion with the concentrate mixtures, which provide uniform consumption of the vitamin on a daily intake basis. For grazing livestock, however, providing vitamin E in concentrate mixtures generally is not economically feasible and of doubtful value if pasture is of good quality. Most of the vitamin E used today in animal applications is administered orally as an ingredient blended uniformly in the dry feed or ration. There are occasions, however, where a larger quantity single dose is needed, and then one has the choice of (1) giving a single oral dose or a drench or (2) injecting intramuscularly. The parenteral application of a proven formulation appears to be the favored choice over a limited period of time. Particularly in the ruminant there may be some ruminal destruction of a single large dose of α-tocopherol given as a drench (Bauernfeind, 1969).

Products containing vitamin E and selenium are often given intramuscularly to animals exhibiting clinical signs of muscular degeneration. Response to treatment of this condition is extremely variable depending on degree of muscular degeneration. For cattle, animals that are down are less likely to survive, while animals showing moderate weakness of the rear limbs or slight buckling of fetlocks usually respond rapidly. Total recovery may require several days to 1 month (Mason *et al.*, 1985). However, affected animals are often responsive to treatment and generally recover the ability to walk unassisted 3–5 days following selenium–vitamin E therapy.

Some cattle ranchers make it a practice to inject newborn calves intramuscularly with a combination of vitamin E and selenium. For dystrophic lambs, an oral therapeutic dose of 500 mg of *dl*-α-tocopherol followed by 100 mg on alternate days until recovery is successful (Rumsey, 1975). Reddy *et al.* (1987a) conclude that supplementation of conventional dairy calf diets with 125–250 IU vitamin E per animal daily increases performance compared to controls. Stuart (1987) recommends that weaned calves with transit shrink should receive 400 IU of vitamin E daily during arrival to a feedlot finishing program, while those with minimal shrink receive 100–200 IU. Yearling cattle were recommended to receive 200–400 IU vitamin E, depending on previous dietary history in relation to vitamin E and selenium. Most preventive preparations for WMD in ruminants contain a combination of both vitamin E and selenium. As a preventive measure, they should be administered to pregnant cows during the second trimester and again 30 days prior to calving.

For humans, good experimental evidence indicates a need for supplemental vitamin E in the diets of pregnant and lactating women, newborn infants, particularly premature infants, and older persons suffering from circulatory disturbances and intermittent claudication. Higher levels may be indicated for persons exposed to oxygen and environmental pollutants, such as ozone, nitrites, and heavy metals. Feeding low-birthweight infants entails problems somewhat different from those of normal-weight infants. Because of their reduced absorption of fat, utilization of α-tocopherol is impaired, with the result that special effort is required to assure an adequate intake. The Committee on Nutrition of the American Academy of Pediatrics (AAP, 1977) recommended that formulas for these infants should provide 5 IU of water-soluble α-tocopherol daily.

XI. TOXICITY

Compared with vitamin A and vitamin D, both acute and chronic studies with animals have shown that vitamin E is relatively nontoxic, but not entirely devoid of undesirable effects. Hypervitaminosis E studies in rats, chicks, and humans indicate maximum tolerable levels in the range of 1000–2000 IU/kg of diet (NRC, 1987).

Massive doses of vitamin E (5000 mg of *dl*-α-tocopherol/kg of diet) caused reduced packed-cell volumes in trout (NRC, 1981a). Administration of vitamin E to vitamin K-deficient rats, dogs, chicks, and humans exacerbates the coagulation defect associated with a vitamin K deficiency (Corrigan, 1982). In humans, isolated but inconsistent reports have appeared of adverse effects from high intakes (400–1000 IU) of *dl*-α-tocopheryl acetate, but most adults appear to tolerate these doses. Large doses of α-tocopherol in anemic children suppress normal hematological response to parenteral iron administration (RDA, 1980).

5

Vitamin K

I. INTRODUCTION

Vitamin K was the last fat-soluble vitamin to be discovered. In contrast to the other fat-soluble vitamins A, D, and E, which have multiple functions and wide biological importance, vitamin K appeared to be limited in its function to the liver for the normal blood-clotting mechanism. However, vitamin K-dependent proteins have recently been identified that suggest roles for the vitamin in addition to blood coagulation. Because of the blood-clotting function, vitamin K was previously referred to as the "coagulation vitamin," "antihemorrhagic vitamin," and "prothrombin factor."

Vitamin K is indispensible for maintaining the function of the blood coagulation system in humans and all investigated animals. Even though vitamin K is synthesized by intestinal microorganisms, deficiency signs have been observed under field conditions. Poultry, and to a lesser degree pigs, are susceptible to vitamin K deficiency. In ruminants a deficiency can be caused by ingestion of spoiled sweet clover hay, which is a natural source of dicumarol (a vitamin K antagonist). Vitamin K is most required in human nutrition in infants because of insufficient intestinal synthesis and in adults under conditions where fat absorption is impaired.

II. HISTORY

The history of vitamin K can be found in Olson and Suttie (1978), Suttie (1980), and Loosli (1988). The presence of a dietary antihemorrhagic factor was first suspected in 1929, when Henrik Dam of Denmark fed chickens a purified low-fat diet in an attempt to determine whether they were able to synthesize cholesterol. He noted that chickens became anemic and developed subcutaneous and intermuscular hemorrhages and that blood taken from these animals clotted slowly. Hemorrhagic signs were reported by other workers using ether-extracted fish meal for chicks (McFarlene *et al.*, 1931). Since the condition was prevented by unextracted fish meal, the curative factor was fat soluble. However, studies in a number of laboratories soon demonstrated that this disease could not be cured by administration of any of the known fat-soluble vitamins (A, D, and E) or other known physiologically active lipids.

Dam continued to study the distribution and lipid solubility of the active component in vegetable and animal sources, and in 1935 proposed that the antihemorrhagic vitamin of the chick was a new fat-soluble vitamin that he called vitamin K, from the Danish word for coagulation (koagulation). At the same time Almquist and Stokstad (1935) described their success in curing the hemorrhagic disease with ether extracts of alfalfa and clearly pointed out that microbial action in fish meal and bran preparations could lead to development of antihemorrhagic activity. Of historical note, it is interesting that Almquist and Stokstad of California actually discovered what were later referred to as vitamins K_1 and K_2 in 1928. The paper reporting their results was delayed by university administrators and finally when it was submitted to the journal *Science*, it was rejected. About that time Dam's research was published. When Dam (with Doisey) later received the Nobel prize for the discovery of vitamin K, it is reported that he had expected Almquist to share the prize. Doisey contributed to the knowledge of the role of vitamin K in blood clotting (Loosli, 1988). Dam *et al.* (1936) demonstrated that prothrombin activity decreased in vitamin K-deficient chicks. At about the same time, the hemorrhagic condition resulting from obstructive jaundice (a deficiency of bile) was shown to be due to poor absorption of vitamin K, and bleeding episodes were attributed to lack of plasma prothrombin (Suttie, 1984). This hemorrhagic condition was originally thought to be due solely to a lowered concentration of plasma prothrombin (factor II), but it was later shown during the 1950s that the synthesis of clotting factors VII, IX, and X was also depressed in the deficient state. Therefore, four blood-clotting proteins are dependent on vitamin K for their synthesis.

A number of groups were involved in attempts to isolate and characterize this new vitamin, and Dam's collaboration with Karrer of the University of Zurich resulted in isolation of the vitamin from alfalfa as a yellow oil (Suttie, 1984). Research proceeded to show that a large number of chemical compounds possess some degree of vitamin K activity. The main source present in plants was referred to as K_1 and the form synthesized by microflora as K_2. The simplest source of vitamin K_3 is menadione, produced by laboratory synthesis. Vitamin K_1 was synthesized by three laboratories in 1939, while the structure of K_2 was elucidated in 1940 but was not synthesized until 1958 by Isler and co-workers in Switzerland.

III. CHEMICAL STRUCTURE, PROPERTIES, AND ANTAGONISTS

The general term vitamin K is now used to describe not a single chemical entity but a group of quinone compounds that have characteristic antihemorrhagic effects. Vitamin K is a generic term for a homologous group of fat-soluble vitamins consisting of a 2-methyl-1,4-naphthoquinone derivatives, commonly called menadione. The basic molecule is a naphthoquinone and the various isomers differ in the nature and length of the side chain (Fig. 5.1). Vitamin K extracted from plant material was named phylloquinone or vitamin K_1. Phylloquinone has a phytyl side chain composed of four isoprene units, the first of which contains a double bond.

Vitamin K-active compounds from material that had undergone bacterial fermentation were named menaquinones or vitamin K_2. The simplest form of vitamin K is the synthetic menadione (K_3), which is composed of the active nucleus (2-methyl-1,4 naphthoquinone) and has no side chain. The menaquinone family of K_2 homologs is a large series of vitamins containing unsaturated side chains that differ in number of isoprenyl units. Numerous natural analogs have been isolated, almost all of which are variations of the side chain at position 3. Some of these chains are quite long, with as many as 65 carbon atoms in some bacterial vitamin K forms, but none is specifically required for activity. Menaquinone-4 is synthesized in liver from ingested menadione or changed to a biologically active menaquinone by intestinal microorganisms.

Vitamin K is a golden yellow viscous oil. Natural sources of vitamin K are fat soluble, stable to heat, and labile to oxidation, alkali, strong acids, light, and irradiation. Vitamin K_1 is slowly degraded by atmospheric oxygen but fairly rapidly destroyed by light. In contrast to natural sources of vitamin K, some of the synthetic products, such as salts of menadione, are water soluble.

A number of vitamin K antagonists exist that increase the need for this vitamin. A deficiency of vitamin K is brought about by ingestion of dicumarol (Fig. 5.2), an antagonist of vitamin K, or by the feeding of sulfonamides (in monogastric

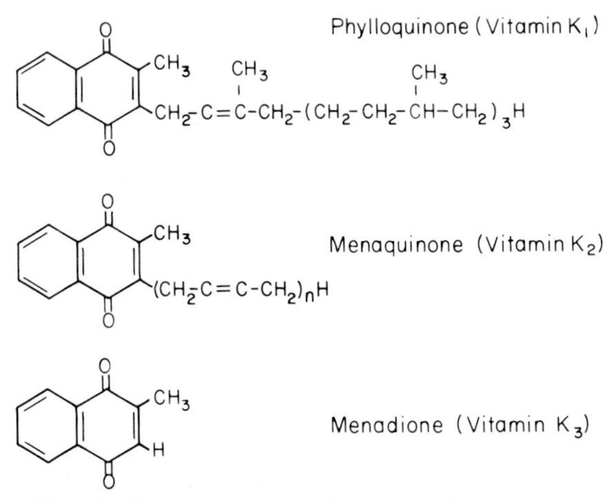

Fig. 5.1 Chemical structures of the vitamin K compounds.

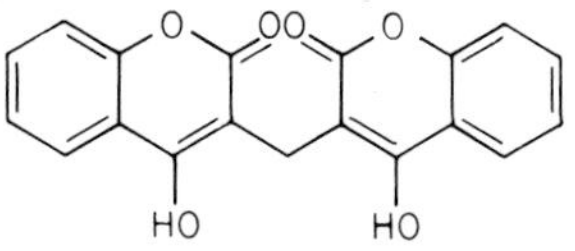

Dicumarol

3,3'-methyl-bis-(4-hydroxycoumarin)

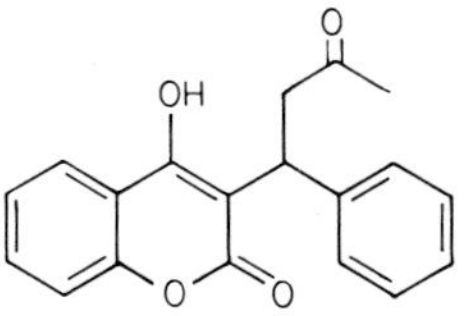

Warfarin

3-(α-acetonylbenzyl)-4-hydroxycoumarin

Fig. 5.2 Antagonists of vitamin K include coumarin derivatives.

species) at levels sufficient to inhibit intestinal synthesis of vitamin K. Mycotoxins, toxic substances produced by molds, are also antagonists causing vitamin K deficiency. A hemorrhagic disease of cattle that was traced to consumption of moldy sweet clover hay was described in the 1920s. The destructive agent was found to be dicumarol, a substance produced from natural coumarins. Dicumarols are produced by molds, particularly those that attack sweet clover, thus giving rise to the term sweet clover disease (see Section IX). During the process of spoiling, harmless natural coumarins in sweet clover are converted to dicumarol (bis-hydroxycoumarin), and when toxic hay or silage is consumed by animals, hypoprothrombinemia results, presumably because dicumarol combines with the proenzyme to prevent formation of the active enzyme required for the synthesis of prothrombin. It probably also interferes with synthesis of factor VII and other coagulation factors.

Dicumarol serves as an anticoagulant in medicine to slow blood coagulation in people afflicted with cardiovascular disease to avoid intravascular blood clots, just as vitamin K under other conditions increases the coagulation time. Thus additional vitamin K will overcome this action by dicumarol. Goplen and Bell (1967) have shown in cattle that vitamin K_1 is much more potent as an antidote to dicumarol than is vitamin K_3. The most successful dicumarol for the long-term lowering of the vitamin K-dependent clotting factors is warfarin (Fig. 5.2); which is widely used as a rodenticide. Concern has been expressed in recent years because of the identification of anticoagulant-resistant rat populations. Spread of this resistance has led to renewed interest in synthesis of new coumarin derivatives that are more effective (Hadler and Shadbolt, 1975).

Because of their use as clinical anticoagulants, investigation of the mechanism of action of dicumarol has been of interest to researchers in the vitamin K field (Suttie, 1984). Recent investigations have centered on interconversion of vitamin K and its 2,3-epoxide as the site of coumarin action. The current hypothesis is that metabolic effects of these compounds are the consequence of their inhibition of the microsomal epoxide reductase (Bell, 1978). The epoxide apparently acts as a competitive inhibitor of vitamin K at its site of action and a coumarin such as warfarin was an inhibitor of vitamin K action only to the extent that it increased the cellular ratio of oxide to the vitamin.

IV. ANALYTICAL PROCEDURES

Vitamin K can be analyzed by a variety of color reactions or by direct spectroscopy. The chemical reactivity is a function of the naphthoquinone nucleus, and other quinones also react with many of the available colorimetric assays. For spectroscopy to be successful, a significant amount of separation is often required to eliminate interfering substances present in crude extracts.

Improved procedures for vitamin K analyses have used high-pressure liquid chromatography (HPLC) as an analytical tool to investigate the vitamin as well as the interconversion of vitamin K to its 2,3-epoxide. The HPLC method is highly suitable for vitamin K analysis because of its high sensitivity, specificity, and accuracy (Manz and Maurer, 1982).

The classic biological assay for the amount of vitamin K in an unknown source is a determination of the whole-blood clotting time of the chick. This assay measures prothrombin time since prothrombin is the most limiting vitamin K-dependent blood-clotting factor in chicks receiving vitamin K-deficient diets. Young chicks are raised on a vitamin K-deficient diet until their whole-blood clotting time is increased to four to seven times normal. The usual one-stage method for measuring prothrombin time consists of adding excess thromboplastin (obtained from chick brain or other source) and calcium to the blood and then measuring the time for the blood to clot. The response of chicks receiving a standard vitamin K preparation is compared to the prothrombin time obtained when a supplement containing an unknown quantity of vitamin K is added to the vitamin K-deficient basal diet (Scott *et al.*, 1982; Weiser and Tagwerker, 1982).

V. METABOLISM

A. Absorption

Like all fat-soluble vitamins, vitamin K is absorbed in association with dietary fats and requires the presence of bile salts and pancreatic juice for adequate uptake from the alimentary canal. Absorption of vitamin K depends on its incorporation into mixed micelles, and optimal formation of these micellar structures requires the presence of both bile and pancreatic juice. Thus, any malfunction of the fat absorption mechanism, for example, biliary obstruction, will reduce availability of vitamin K. Unlike phylloquinone and the menaquinones, menadione bisulfites and phosphates are relatively water soluble and therefore are absorbed satisfactorily from low-fat diets.

The lymphatic system is the major route of transport of absorbed phylloquinone from the intestine. Shearer *et al.* (1970) demonstrated the association of phylloquinone with serum lipoproteins, but little is known of the existence of specific carrier proteins. The absorption of various forms of vitamin K has been studied and found to differ significantly. Ingested phylloquinone is absorbed by an energy-dependent process from the proximal portion of the small intestine (Hollander, 1973). In contrast to the active transport of phylloquinone, menaquinone is absorbed from the small intestine by a passive noncarrier-mediated process.

Efficiency of vitamin K absorption has been measured from 10 to 70%, depending on the form in which the vitamin is administered. Some reports have indicated menadione to be completely absorbed, but phylloquinone only at a rate of 50%. Rats were found to excrete about 60% of ingested phylloquinone in the feces within 24 hr of ingestion, but only 11% of ingested menadione (Griminger and Donis, 1960; Griminger, 1984a). However, 38% of ingested menadione but only a small amount of phylloquinone were excreted via the kidneys in the same period of time. The conclusion was that although menadione is well absorbed, it is poorly retained, while just the opposite was true for phylloquinone. Poor retention of menadione can be explained by the need to add, in the liver, a difarnesyl chain and thus transform it into menaquinone (vitamin K) with a 20-carbon chain (MK-4). Apparently there are quantitative limitations in this biosynthetic step. The menadione not converted is rapidly detoxified and excreted.

Konishi *et al.* (1973) administered radioactive menadione, phylloquinone, or menaquinone-4 to rats and found that radioactive menadione was spread over the whole body much faster than the other two compounds, but the amount retained in the tissues was low. Martius and Alvino (1964) utilized radioactive menadione to establish that it could be converted to a more lipophilic compound that, on the basis of their limited characterization, appeared to be menaquinone-4. It was therefore concluded that the vitamin K form of animal tissues was menaquinone-4. They found that when either a menaquinone or a phylloquinone was given to animals, the side chain was removed, probably by the microorganisms in the gut.

On the contrary, Griminger and Brubacher (1966) showed that a major portion of the phylloquinone that they fed to chicks was absorbed and deposited in the liver intact, and that as such it had equally as good biological activity upon prothrombin synthesis as the menaquinone-4, which they found in the chick's liver following feeding of menadione. Therefore, menaquinone-4 is most likely produced only if menadione is fed, or if the intestinal microorganisms degrade the dietary K_1 or K_2 to menadione, and the formation of menaquinone-4 is not obligatory for metabolic activity, since phylloquinone is equally active in bringing about synthesis of the K-dependent blood-clotting proteins (Scott *et al.*, 1982).

B. Storage and Excretion

A number of studies have shown that phylloquinone is specifically concentrated and retained in the liver but that menadione is poorly retained in this organ. Menadione is found to be widely distributed in all tissues and to be very rapidly excreted. Although phylloquinone is rapidly concentrated in liver, it does not have a long retention time in this organ (Thierry *et al.*, 1970). The inability to rapidly develop a vitamin K deficiency in most species results, therefore, from the difficulty in preventing absorption of the vitamin from the diet or from intestinal synthesis rather than from a significant storage of the vitamin.

Some breakdown products of vitamin K are excreted in the urine. One of the principal excretory products is a chain-shortened and oxidized derivative of vitamin K, which forms a γ-lactone and is probably excreted as a glucuronide. Vitamin K oxide has also been identified as a metabolite of vitamin K in rats (Matschiner *et al.*, 1970). The principal metabolites of menadione are the sulfate and glucuronide of dihydromenadione (Losito *et al.*, 1967). Some vitamin K is reexcreted into the intestine with bile, part of which is excreted in the feces. In humans, 20% of injected phylloquinone was excreted in the urine and 40–50% was excreted in the feces via the bile (Shearer and Barkhan, 1973).

VI. FUNCTIONS

Coagulation time of blood is increased when vitamin K is deficient because the vitamin is required for the synthesis of prothrombin (factor II). Plasma clotting factors VII, IX, and X also depend on vitamin K for their synthesis. Vitamin K, while required for the synthesis of prothrombin and the other proteins, does not become part of the prothrombin molecule, nor has an enzyme system been found that includes vitamin K as a coenzyme. These four blood-clotting proteins are rather synthesized in the liver in inactive precursor forms and then converted to biologically active proteins by the action of vitamin K (Suttie and Jackson, 1977). In deficiency, administration of vitamin K brings about a prompt response and return toward normal of depressed coagulation factors in 4–6 hr. In the absence of the liver, this response does not occur.

The blood-clotting mechanism can apparently be stimulated by either an intrinsic system in which all the factors are in the plasma or an extrinsic system. In the extrinsic system of coagulation, injury to the skin or other tissue frees tissue thromboplastin that, in the presence of various factors and calcium, changes prothrombin in the blood to thrombin. The enzyme thrombin facilitates the conversion of the soluble fibrinogen into insoluble fibrin (Fig. 5.3). Fibrin polymerizes into strands and enmeshes the formed elements of the blood, especially the red blood cells, to form the blood clot (Griminger, 1984a). The final active component in both the intrinsic and extrinsic systems appears to activate the Stuart factor, which in turn leads to activation of prothrombin. The various steps involved in blood clotting are presented in Fig. 5.3. The action of vitamin K is required at four different sites in these reactions.

It is now recognized that vitamin K-deficient animals synthesize vitamin K-dependent proteins but in an inactive form, and vitamin K is needed to convert these inactive protein precursors to biologically active proteins (Suttie, 1980). Investigations related to vitamin K mechanisms revealed that prothrombin contained 10 residues of a previously unrecognized amino acid, γ-carboxyglutamic acid (Gla). Likewise, the Gla residues were found in the three other vitamin K-dependent proteins. The amino-terminal regions of these proteins are homologous, and the Gla residues are in essentially the same position in all of these clotting factors (Suttie and Olson, 1984).

The action of converting inactive precursor proteins to biological activity in-

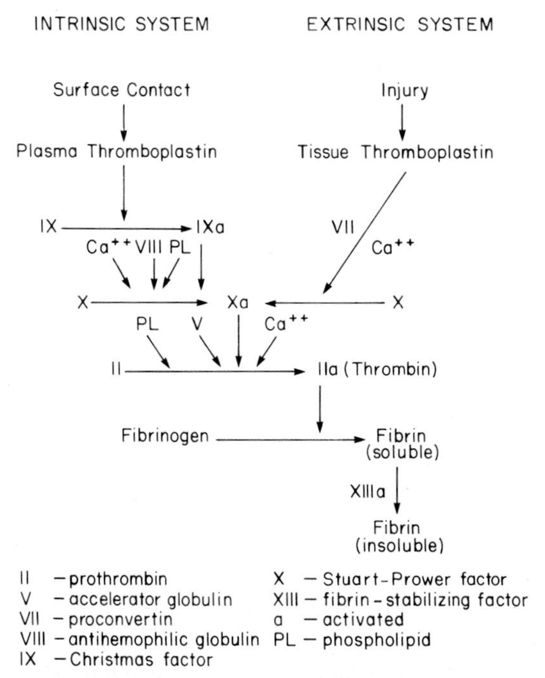

Fig. 5.3 Scheme involving blood clotting. The vitamin K-dependent clotting factors (synthesis of each is inhibited by dicumarol) include: factor IX, plasma thromboplastin components (PTC); factor X, Stuart factor; factor VII, proconvertin; and factor II, prothrombin.

volves the carboxylation of glutamic acid residues in the inactive molecules. Carboxylation allows prothrombin and the other procoagulant proteins to participate in a specific protein–calcium–phospholipid interaction that is a necessity for their biological role (Suttie and Jackson, 1977). To participate in this reaction, vitamin K is converted to hydroquinone and is then reconstituted to the quinone form via vitamin K-epoxide.

Liver microsomes contain enzymes that oxidize the vitamin to its 2,3-epoxide and reduce the epoxide back to the reduced vitamin. The carboxylase activity and epoxidase activity appear to share a common oxygenated intermediate, and available data suggest that this may be a hydroperoxide of the vitamin. Current evidence would indicate that the role of vitamin K is to labilize the γ-hydrogen of the substrate for CO_2 attack rather than to activate or transfer the CO_2 (Suttie, 1980). Warfarin and other anticoagulants of the coumarin type interfere with 2,3-epoxide reductase and therefore with the reconstitution of the active form of vitamin K (see Section III).

The γ-carboxyglutamyl residues have recently been shown to be present in a number of proteins (i.e., bone, kidney, lung, placenta, skin, and spleen) other than the long-recognized vitamin K-dependent clotting factors. Therefore, vitamin K-dependent carboxylase enzyme systems may have other roles in addition to blood clotting, such as in bone formation. Osteocalcin is a vitamin K-dependent protein found in bone, and it appears in embryonic chick bone and rat bone matrix at the beginning of mineralization of the bone (Gallop *et al.*, 1980). A vitamin K-dependent carboxylase system has been identified in skin, which may be related to calcium metabolism in skin (de Boer-van den Berg *et al.*, 1986).

In animals, discoveries of active proteins that depend on vitamin K for their synthesis are becoming common. Discovery of a vitamin K-dependent protein in skeletal tissue suggests that distribution of γ-carboxyglutamic acid is much more widespread than was once realized. A number of pathological states result in an ectopic calcification of various tissues. Residues of Gla have been found in these lesions as well as in calcium-containing kidney stones. Microsomes from kidney cortex possess a vitamin K-dependent carboxylase activity. Proteins containing Gla residues all interact with Ca^{2+} or other divalent cations but their physiological roles are not well understood. Although much is unknown concerning vitamin K function apart from blood coagulation, it is apparent that a reasonably large number of proteins may be involved.

VII. REQUIREMENTS

Vitamin K requirement of mammals is met by a combination of dietary intake and microbial biosynthesis in the gut, which may involve intestinal microorganisms (such as *Escherichia coli*) as well as ruminal microbes. Ruminal microorganisms in particular synthesize large amounts of vitamin K, explaining why ruminants do not appear to need a dietary source of the vitamin. Animals that practice coprophagy, such as the rabbit, can utilize much of the vitamin K that is eliminated in the feces. In rats, the majority of menaquinone absorbed resulted from fecal ingestion compared to dietary sources or from direct synthesis and absorption from the intestine (Kindberg *et al.*, 1987). Animal feces contain substantial amounts of the vitamin even when none is present in feed. Despite the intestinal synthesis, animals can be rendered deficient when fed vitamin K-free diets and coprophagy is prevented if animals are maintained germ-free or if a vitamin K antagonist is given. Difficulties in demonstrating dietary requirement in many species include the varying degrees to which they utilize vitamin K synthesized by intestinal bacteria and the degree to which different species practice coprophagy. Contrary to most animals, poultry have a limited ability for intestinal synthesis, and so adequate dietary supplies are of greater importance (see Section IX).

Rats caged on litter obtain vitamin K, as well as B vitamins, by feces eating. Thus the practice was established of keeping experimental animals on wire screens to prevent this complication in vitamin feeding experiments. The practice was assumed to be successful until 1957, when Barnes and co-workers showed, by the use of a new technique, that rats may eat 50–65% of their feces even when maintained on wire screens. In a later experiment when rats were prevented by this new technique from feces eating, a vitamin K deficiency uniformly developed from a vitamin K-free diet.

Because of microbial synthesis, a precise expression of vitamin K requirements is not feasible. However, attempts to determine the contribution of microbial synthesis have been made. In conventional rats, the vitamin requirement is 0.05–0.10 ppm, whereas in germ-free rats the requirement is more than doubled to about 0.25 ppm (Suttie and Olson, 1984). In humans, vitamin K homologs stored in liver indicate that about 40–50% of the daily requirement is derived from plant sources (vitamin K_1) and the remainder from microbiological synthesis.

Estimated vitamin K requirements for various animals and humans are presented in Table 5.1. The daily requirement for most species falls in a range of 2–200 μg vitamin K per kilogram body weight. It should be remembered that this requirement can be altered by age, sex, strain, anti-vitamin K factors, disease conditions, and any condition influencing lipid absorption or altering intestinal flora (see Sections IX and X).

The adult human requirement for vitamin K is extremely low, and there seems little possibility of a simple dietary deficiency developing in the absence of complicating factors (Suttie, 1984). Since intestinal flora are insufficiently developed in humans during the first days after birth to provide the vitamin K requirement of the infant, and because of the low vitamin k content of the mother's milk, prothrombin level is very low in the newborn. A daily dose of 1–2 mg vitamin

TABLE 5.1

Vitamin K Requirements for Various Animals and Humans[a]

Animal	Purpose	Requirement	Reference
Beef cattle	Adult	Microbial synthesis	NRC (1984a)
Dairy cattle	Adult	Microbial synthesis	NRC (1978a)
Chicken	All classes	0.5 mg/kg	NRC (1984b)
Duck	All classes	0.4 mg/kg	NRC (1984b)
Turkey	0–8 weeks	1.0 mg/kg	NRC (1984b)
	12–24 weeks	0.8 mg/kg	NRC (1984b)
	Breeding	1.0 mg/kg	NRC (1984b)
Sheep	Adult	Microbial synthesis	NRC (1985b)
Swine	All classes	0.5 mg/kg	NRC (1988)
Horse	Adult	Microbial synthesis	NRC (1978b)
Goat	Adult	Microbial synthesis	NRC (1981b)
Rabbit	Adult	2 mg/kg	NRC (1977)
Cat	Adult	0.1 mg/kg	NRC (1986)
Dog	Growing	1 mg/kg	NRC (1985a)
Fish	Trout	0.5–1 mg/kg	NRC (1981a)
Rat	All classes	0.05 mg/kg	NRC (1978c)
Guinea pig	Growing	5 mg/kg	NRC (1978c)
Human	Infants	12 μg/day	RDA (1980)
	Adults	70–140 μg/day	RDA (1980)

[a]Exact requirements are impossible to determine because intestinal synthesis occurs in all species. Animal requirements are expressed as per unit of feed either on an as fed (approximately 90% dry matter) or dry basis (see Appendix Table 1), and for humans as μg/day.

K for newborn infants, or 2–5 mg daily to the prepartum mother,is therefore recommended. The daily requirement of adults is estimated at about 1 mg (Marks, 1975; RDA, 1980). On the basis of studies in patients sustained on intravenous fluids who were given antibiotics, a minimal daily requirement as low as 0.03 μg/kg body weight has been suggested (Frick *et al.*, 1967).

VIII. NATURAL SOURCES

There are two major natural sources of vitamin K, phylloquinones (vitamin K_1) in plant sources and menaquinones (vitamin K_2) produced by bacterial flora in animals. Vitamin K concentrations of various foods and feedstuffs are presented in Table 5.2. Green leaves are the richest natural source of vitamin K_1. Light is important for its formation and parts of plants that do not normally form chlorophyll contain little vitamin K. However, natural loss of chlorophyll as in the fall yellowing of leaves does not bring about a corresponding change in vitamin K.

Vitamin K is present in fresh dark-green vegetables and also in an extract of pine needles. Cauliflower, broccoli, spinach, lettuce, and brussels sprouts are excellent sources. Most feeds containing very high levels of vitamin K are not usually fed to domestic animals, with the exception of alfalfa leaf meal, which is frequently included in small amounts in poultry and other animal diets. Cereals and oil cakes contain only small amounts of vitamin K, while liver and fish meal are good animal sources of the vitamin. All feeds or foods of animal origin, including fish meal and fish liver oils, are much higher in vitamin K after they have undergone extensive bacterial putrefaction. Milk from cattle has been reported to be 10 times as rich as that of the human, and vitamin K content of cow's milk is known to vary significantly with the month of sampling. The effect of food processing and cooking on vitamin K content has not been carefully considered, however, limited data suggest that food vitamin K is relatively stable,

TABLE 5.2

Vitamin K Concentration of Various Foods and Feedstuffs (on as Fed Basis)[a]

Food or feedstuff	Vitamin K level (ppm)
Alfalfa hay, sun cured	19.4
Alfalfa meal, dehydrated (20% protein)	14.2
Barley, grain	0.2
Cabbage (green)	4.0
Carrots	0.1
Corn, grain	0.2
Eggs	0.2
Fish meal, herring (mechanically extracted)	2.2
Liver (cattle)	1–2
Liver (swine)	4–8
Meat (lean)	1–2
Milk (cattle)	0.02
Milk (human)	0.2
Peas	0.1–0.3
Potatoes	0.8
Sorghum, grain	0.2
Soybean, protein concentrate (70.0% protein)	0.0
Spinach	6.0
Tomatoes	4.0

[a]From NRC (1982b) and Marks (1975).

with the greatest loss resulting from oxidation. Irradiation of beef has resulted in destruction of the vitamin.

The menaquinones (vitamin K_2) are produced by the bacterial flora in animals and are especially important in providing the vitamin K requirements of humans and most other mammals. However, the chick does not receive sufficient vitamin K from intestinal microbial synthesis (Scott *et al.*, 1982). Vitamin K production in the rumen of ruminants and subsequent passage along the small intestine, a region of active absorption, make such synthesized vitamins highly available to the host. In nonruminants, site of synthesis is in the lower gut, an area of poor absorption, and thus availability to the host is limited unless the animal practices coprophagy, in which case the synthesized vitamin K is highly available.

IX. DEFICIENCY

The major clinical sign of vitamin K deficiency in all species is impairment of blood coagulation. Other clinical signs include low prothrombin levels, increased clotting time, and hemorrhaging. In its most severe form, a lack of vitamin K will cause subcutaneous and internal hemorrhages, which can be fatal. Vitamin K deficiency can result from dietary deficiency, lack of microbial synthesis within the gut, inadequate intestinal absorption, or inability of the liver to use the available vitamin K.

A. Effects of Deficiency

1. Ruminants

Microorganisms in the rumen synthesize large amounts of vitamin K and a deficiency is seen only in the presence of a metabolic antagonist, such as dicumarol from moldy sweet clover. This condition is referred to as "hemorrhagic sweet clover disease" and has been responsible for large animal losses. Ruminants may die from hemorrhage following a minor injury, or even from apparently spontaneous bleeding. Dicumarol passes through the placenta in pregnant animals and newborn animals may become affected immediately after birth. All species of animals studied have been shown to be susceptible, but cases of poisoning have involved mainly cattle and, to a very limited extent, sheep, swine, and horses.

All clinical signs relate to the hemorrhages caused by blood coagulation failure. First appearance of clinical disease after consumption of spoiled sweet clover varies greatly and depends to a large extent on dicumarol content of the particular sweet clover fed and animal age. If dietary dicumarol is low or variable, animals may consume it for months before signs of disease appear. Initial signs may be stiffness and lameness from bleeding into the muscles and articulations. Hematomas, epistaxis, or gastrointestinal bleeding may be observed. Death may occur suddenly with little preliminary evidence of disease and is caused by spontaneous massive hemorrhage or bleeding after injury, surgery, or parturition.

2. Swine

Schendel and Johnson (1962) were able to produce a vitamin K deficiency in the baby pig by using a sulfa drug and an antibiotic and by carefully minimizing coprophagy by cleaning the feces from wire bottom cages where pigs were housed. Clinical and subclinical signs of vitamin K deficiency include both increased prothrombin and blood-clotting time, internal hemorrhage, and anemia due to blood loss. Newborn pigs may be pale with loss of blood from the umbilical cord. Until recently, vitamin K deficiency under natural conditions was not expected as it was thought that the pig synthesized most if not all of the vitamin that was required. However, in the late 1960s and early 1970s there were prevalent reports of a bleeding disease of young pigs on commercial diets that was successfully overcome by vitamin K medication. Observations from Australia and New Zealand were of hemorrhaging in the navel of newborn pigs (Cunha, 1977).

A number of field trials in the United States have reported a hemorrhagic syndrome for growing pigs. In one study, hemorrhagic syndrome occurred 9 days after pigs were fed a standard diet, while those receiving either 2.5% dehydrated alfalfa meal or supplemental vitamin K remained in good health (Fritschen *et al.,* 1970). Gross visible signs for hemorrhagic syndrome were large subcutaneous hemorrages, blood in urine, and abnormal breathing. Additional clinical signs from field observations are that some pigs will develop enlarged blood-filled joints and become lame, whereas others may have swelling along the body wall that are filled with unclotted blood. Hematomas (or blood swellings) in the ears also occur (Cunha, 1977). Hemorrhagic conditions in the growing pig have in some cases been associated with ingestion of molds, such as aspergillus or moldy materials, and have usually responded to vitamin K therapy.

The exact causes of more recent needs for vitamin K supplementation are not definitely known. Cunha (1977) has summarized the following likely reasons for vitamin K deficiency under field conditions:

1. As confinement feeding has increased, less pasture and alfalfa, both good sources of vitamin K, are used, so diets are now lower in the vitamin. Likewise there has been a trend toward the use of solvent-extracted soybean meal and other seed meals and better quality (less putrefied) fish meals, which are lower in vitamin K than the originial expeller meals and somewhat putrid fish meals.
2. Hemorrhaging gastric ulcers, which occur frequently, may increase vitamin K needs.

3. A mycotoxin produced by certain molds that may be present in the feed might cause the disease.

4. An antimetabolite (antivitamin K) may be in the feed and thus increase vitamin K needs.

5. Use of slatted floors lessens the opportunity for coprophagy since feces was an excellent source of this vitamin.

6. Use of sulfa drugs, antibiotics, and other medications that have reduced intestinal synthesis of the vitamin.

7. Breeding of strains of pigs that possibly require more vitamin K. Increased litter size and rate of gain may also be increasing the need for vitamin K.

3. Poultry

A deficiency of vitamin K causes a reduction in the prothrombin content of the blood and, in the chick, may reduce the quantity in the plasma to less than 2% of normal. Since the prothrombin content of the blood of normal, newly hatched chicks is only about 40% that of adult birds, very young chicks are readily affected by a deficiency of vitamin K. A carryover from the parent hen to the chick has been demonstrated (Almquist, 1971). Laying hens fed a diet deficient in vitamin K produce eggs low in the vitamin, and when the eggs are incubated, the chicks produced have low reserves and a prolonged blood-clotting

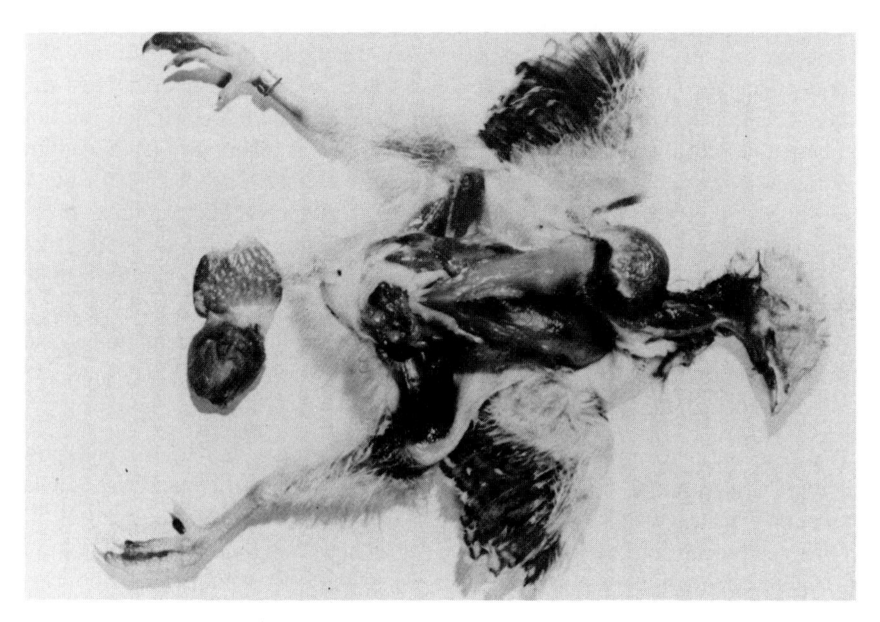

Fig. 5.4 Generalized hemorrhage due to severe vitamin K deficiency in a young chick. (Courtesy of M. S. Scott, Cornell Universtiy.)

time. As a consequence, the chicks may bleed to death from an injury as slight as that caused by wing banding (Fig. 5.4).

In very young chicks deficient in vitamin K, blood coagulation time begins to increase after 5–10 days of age, with clinical signs occurring most frequently in chicks 2–3 weeks after they begin consuming a vitamin K-deficient diet. Hemorrhages often occur in any part of the body, either spontaneously or as the result of an injury or bruise. Postmortem examination usually reveals accumulations of blood in various parts of the body; sometimes there are petechial hemorrhages in the liver and almost invariably there is erosion of the gizzard lining.

Borderline deficiencies of vitamin K often cause small hemorrhagic blemishes on the breast, legs, wings, in the abdominal cavity, and on the surface of the intestine (Fig. 5.5). Chicks show an anemia that in part may be due to loss of blood, but also to the development of a hypoplastic bone marrow. Even a borderline deficiency of vitamin K is of economic importance in broiler production because the hemorrhagic areas that occur in the legs or throughout the body may result in a high percentage of condemnations during inspection at the processing plant (Scott *et al.*, 1982). A condition manifested by numerous small hemorrhages scattered throughout all tissues has been reported frequently in the commercial broiler industry (Almquist, 1978).

A number of considerations influence the likelihood of a vitamin K deficiency

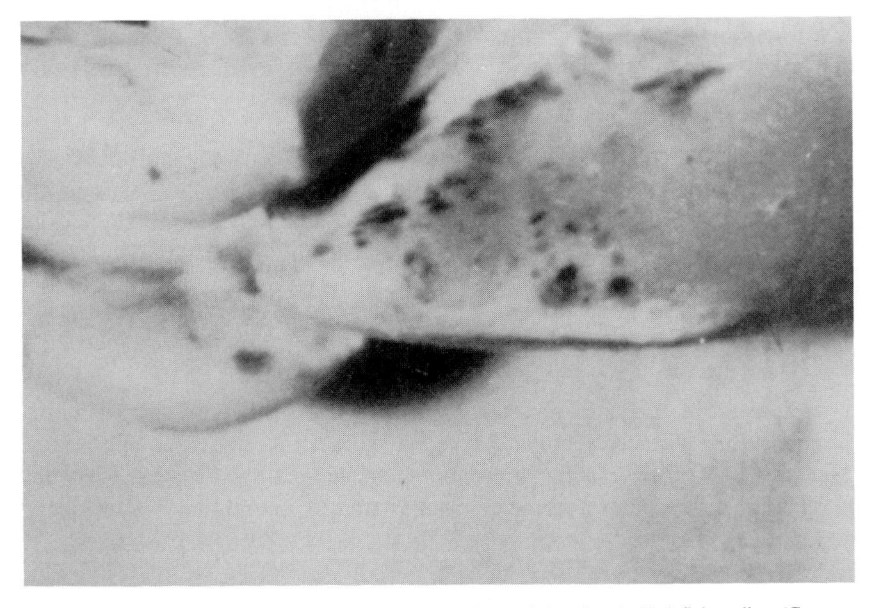

Fig. 5.5 Hemorrhagic blemishes in the muscle of a chicken fed a vitamin K-deficient diet. (Courtesy of M. L. Scott, Cornell University.)

in poultry, including dietary sources of the vitamin, level of vitamin K in the maternal diet, intestinal synthesis, coprophagy, presence of sulfa drugs and other nonnutrients in the diet, and disease conditions. Chicks attacked by coccidiosis, a disease that causes severe damage along the intestinal tract, may bleed excessively or fatally. When sulfaquinoxaline or certain other drugs are present in the feed or in the drinking water or when coccidiosis is being treated, supplementary vitamin K is needed at levels up to 10 times that needed in the absence of these drugs (Scott *et al.*, 1982). Antimicrobial agents suppress intestinal bacteria that synthesize vitamin K and in their presence the bird may be entirely dependent on dietary vitamin K (NRC, 1984b). Arsanilic acid increases the need for dietary vitamin K in both breeder and chick diets.

In poultry, little intestinal synthesis occurs because of the short digestive tract. The young chicken's large intestine or colon, a major area of bacterial activity, comprises less than 6% of the total length of the intestinal tract, while the figure for the adult of the species is 7% (Griminger, 1984b). In other domestic animals, the relative length varies from 13% for the dog to 28% for the rabbit. Also poultry cannot utilize the vitamin K synthesized by intestinal flora because the synthesis is taking place too close to the distal end of the intestinal tract to permit significant absorption.

Rapid rate of food passage through the digestive tract may also influence vitamin K synthesis in poultry. First defecation in pigs, for a specific portion of diet, may occur about 15 hr after feeding, but most of a given meal will be retained in the tract appreciably longer. A comparable time period for chicken would be approximately 3 hr (Griminger, 1984b).

4. Horses

There are no conclusive reports in the literature relating to vitamin K deficiency in the equine species (NRC, 1978b). Clinical responses of some "bleeders" to vitamin K therapy suggest that some horses may require additional supplies of vitamin K (Wakeman *et al.*, 1975), but it is generally assumed that vitamin K is synthesized by microorganisms of the cecum and colon in sufficient quantities to meet requirements.

5. Other Animal Species

As a result of adequate dietary supplies and intestinal synthesis, vitamin K deficiency has not been reported in the cat, guinea pig, and mink and is little studied. Other species, including trout and channel catfish, have been reported to have limited response to supplemental vitamin K but no dietary recommendations are available.

a. Dogs. Vitamin K deficiency has been demonstrated in adult dogs following diversion of bile from the intestine by means of a cholecystonephrostomy (NRC,

1985a). Vitamin K absorption from both diet and intestinal bacterial synthesis was apparently reduced. Some reports indicate that newborn pups suspected of vitamin K deficiency occasionally respond to vitamin K therapy.

b. Foxes. On farms where silver and blue foxes were born with subcutaneous hemorrhages, enrichment of diets of pregnant females with vitamin K was beneficial (NRC, 1982a).

c. Fish. Hemorrhages in channel catfish fed a vitamin K-deficient diet have been reported while other studies with this species have found no benefit with supplemental dietary vitamin K (NRC, 1983). For trout, no growth or survival data for vitamin K deficiency have been reported (NRC, 1981a). However, dietary sulfaguanidine, as well as cold water temperatures, have caused prolonged blood coagulation times.

d. Rabbits. A vitamin K-deficient diet has been fed to pregnant rabbits with the result of placental hemorrhage and abortion of young (NRC, 1977).

e. Monkeys. An increase in prothrombin time of rhesus monkeys has been demonstrated when vitamin K-deficient diets (soy protein based) were fed (Hill *et al.*, 1964).

f. Laboratory Animals. Vitamin K deficiency is produced (prolonged blood-clotting time and hemorrhage) within 2 weeks in rats fed low dietary quantities of the vitamin, with the process accelerated by feeding sulfonamides or preventing coprophagy (NRC, 1978c). In mice, supplemental vitamin K has corrected vaginal hemorrhages and fetal resorptions caused by feeding a vitamin K antagonist. Hypoprothrombinemia can be produced in adult male hamsters either by feeding a vitamin K-deficient diet or by treatment with vitamin K antagonists.

6. Humans

Primary vitamin K deficiency is uncommon in humans because of the widespread distribution of vitamin K in plant and animal tissues and the microbial flora of the normal gut, which synthesize the menaquinones in amounts that may supply the bulk of the requirement for vitamin K. Cases of an acquired vitamin K deficiency do, however, occur in the adult population, and although relatively rare they do present a significant problem for some individuals. Hazell and Baloch (1970) have observed that a relatively high percentage of an older adult hospital-admitted population has a hypoprothrombinemia that responds to administration of oral vitamin K.

Deficiency of vitamin K may result from inadequate intake, absorption, or utilization, or as a result of drugs that interfere with its activity. Deficiency can occur rapidly in patients with a poor food intake who are on antibiotics, particularly in the postoperative period (Pineo *et al.*, 1973). Vitamin K as a fat-soluble vitamin requires the presence of bile salts for its absorption in the upper small intestine, thus a deficiency state can be caused by biliary obstruction. In sprue, idiopathic steatorrhea, ulcerative colitis, and other conditions in which fats are not effectively absorbed, bleeding due to deficiency of vitamin K may occur. Severe bleeding during or, more frequently, a day or two after an operation for the relief of jaundice due to obstruction of the common bile duct was once a complication much feared by surgeons. Today it is recognized that this danger can be circumvented by giving vitamin K by injection prior to the operation.

Hemorrhagic disease of the human newborn can result from dietary lack of vitamin K. Newborn infants represent a special case of vitamin K nutrition because (1) the placenta is a relatively poor organ for maternal–fetal transmission of lipids and (2) the newborn baby has a sterile intestinal tract and furthermore is apt to be fed on foods relatively free from bacterial contamination. Breast milk and clean cow's milk are very poor sources of the vitamin.

Infants in the first week of life have less prothrombin in their blood than normal adults and have a prolonged prothrombin time. In normal infants, plasma prothrombin concentration and that of other vitamin K-dependent factors may decrease to levels as low as 30% of adult levels in the second and third days of life. Then, as food is taken and the gastrointestinal tract is colonized, these levels gradually climb to normal adult values over a period of weeks (Anonymous, 1985c). Mild vitamin K deficiency is seen in all newborn infants and is particularly severe in premature infants. It is due to inadequate liver synthesis and is aggravated by decreased dietary intake and probably inadequate production by intestinal bacterial flora. The tendency to correlate the general low levels of prothrombin seen in all infants, and particularly in premature or low-birth-weight infants, with an insufficient hepatic level of vitamin K is only partially correct as part of the prothrombin deficiency is due to an inability of the immature liver to synthesize sufficient clotting factors (Suzuki, 1979).

Breast-fed infants are at higher risk of hemorrhage than are formula-fed babies because human milk contains only 1–2 μg vitamin K/liter, whereas cow's milk contains 5–17 μg/liter (Haroon *et al.*, 1982). Furthermore, breast milk is sterile and delays (through the flora it encourages) colonization of the gut with vitamin K-synthesizing bacteria.

B. Assessment of Status

Measurement of blood-clotting time has been used to evaluate the body status of vitamin K (see Section IV), and although blood-clotting time is a fairly good measure of vitamin K deficiency, a more accurate measure is obtained by determining the "prothrombin time." A determination of prothrombin time is indicative of efficiency of conversion of prothrombin to thrombin and fibrinogen to fibrin. Prothrombin times in severely deficient chicks, for example, may be extended from a normal of 17–20 sec to 5–6 min or longer.

Prolongation of the prothrombin time in the absence of liver disease indicates vitamin K deficiency, with further clarification by specific vitamin K-dependent factor assays or by the rapid response to administration of vitamin K. The currently available method for measuring an inadequate intake of vitamin K is to measure the plasma concentration of one of the vitamin K-dependent clotting factors, prothrombin (factor II), factor VII, factor IX, or factor X (Suttie, 1984). There are a number of one- and two-stage modified assays for prothrombin determination. Also, snake venom preparations (i.e., *Oxyuranus echis* have been used to develop one-stage clotting assays for prothrombin (Carlisle *et al.*, 1975).

X. SUPPLEMENTATION

As long as natural dietary vitamin K sources (i.e., green leafy plants) are sufficiently high and/or bacterial synthesis in the intestinal tract remains functional, supplementary dietary vitamin K is not necessary. However, high sources of vitamin K such as green leafy plants are not usually fed to nongrazing domestic animals. An exception is alfalfa meal, which is sometimes included in small amounts in chicken and other animal feed. Therefore, a source of vitamin K needs to be added to the diets of animals that are not getting fresh greens or their dried equivalent and are not synthesizing sufficient amounts of this vitamin in their gastrointestinal tract.

Scott *et al.* (1982) reported that natural ingredients used in poultry diets some years ago probably contained sufficient vitamin K while present diets do not. The common feedstuffs used in past years, such as high levels of alfalfa meal, high-fat soybean and other oilseed meals, and putrefied fish meals, supplied ample vitamin K. Recent trends toward (1) reducing levels of alfalfa for production of higher energy, higher efficiency diets, (2) solvent extraction of soybean meals and other oilseed meals, (3) improved processing of fish meals, resulting in lower menaquinone levels because of less putrefaction, and (4) use of vitamin K-inhibiting drugs in feed and drinking water have had a combined effect that makes supplementation of most present-day poultry feeds a necessity. Dietary changes affecting poultry vitamin K needs are similar for swine (see Section IX) in confinement, however, it is generally accepted that swine better utilize vitamin K from microbial synthesis.

Vitamin K antagonists will increase the vitamin K needs of livestock. In adjusting dietary vitamin K fortification levels, an appropriate margin of safety is needed to prevent deficiency and a low optimum performance in livestock. Vitamin K antagonists include:

1. Use of certain antibiotics and sulfa drugs. By altering the ruminal and/or intestinal microflora, an excellent source of vitamin K is lost.
2. Ingestion of an antimetabolite. Sweet clover contains high levels of dicumarol, which can decrease prothrombin levels. Warfarin, a synthetic antimetabolite, is a highly effective rat poison. Rats ingest large amounts of this extremely palatable antimetabolite and eventually die from internal hemorrhaging. Marks (1975) observed that the most common cause of vitamin K deficiency in veterinary practice is the accidental poisoning of domestic animals with warfarin. Mycotoxins (aflatoxins) are toxic substances produced by molds. Vitamin K may be helpful in correcting vitamin K deficiency in aflatoxicosis.

The level of supplemental vitamin K should be adequate to meet the requirements under the wide variety of stress conditions encountered in practical poultry and swine production. Squibb (1964) obtained increased prothrombin times, indicating a higher vitamin K requirement in chicks during the early stages of Newcastle disease. Studies have shown an interrelationship between the severity of coccidiosis and vitamin K requirement and indicated that as much as 8 mg of vitamin K per kilogram of diet was needed at times for maximum growth and feed efficiency (Scott *et al.*, 1982). Field reports with swine indicate that hemorrhaging in stressed animals occurs at birth in the navel and following castration. Various reports indicate that levels of 2 to as high as 16 g vitamin K per ton of feed were needed because the lower levels were not effective under certain farm conditions (Cunha, 1977). Scott *et al.* (1982) concluded that coccidiosis possibly produces a triple stress on the vitamin K requirement by (1) reducing feed intake and thereby the supply of vitamin K, (2) injuring the intestinal tract and reducing absorption of the vitamin, and (3) treatment with sulfaquinoxaline or other coccidiostats that cause an increased requirement for vitamin K.

Vitamin K_1 is not presently available to the feed industry as it is too expensive for this purpose, instead water-soluble menadione (vitamin K_3) salts are used to provide vitamin K activity in feeds. Because of stability, menadione is not used in feed as the pure vitamin but is formulated as an addition product with sodium bisulfite and derivatives thereof. Water-soluble derivatives of menadione, including menadione sodium bisulfite (MSB), menadione sodium bisulfite complex (MSBC), and menadione dimethyl-pyrimidinol bisulfite (MPB), are the principal forms of vitamin K included in commercial diets. The greatest menadione activity is 50% for MSB followed by 45.4% for MPB and 33% for MSBC (Schneider and Hoppe, 1986). Sometimes MSB is coated with gelatin to increase stability, resulting in a 25% menadione activity.

Stability of the naturally occurring sources of vitamin K is poor. However, stability of the water-soluble menadione salts is excellent in multivitamin premixes unless trace minerals are present (Frye, 1978). Basic pH conditions also accelerate the destruction of menadione salts, thus soluble or slightly soluble basic mineral substances should not be included in multivitamin premixes containing menadione. Stability of vitamin K_3 derivatives is likewise impaired by moisture and choline chloride in feeds and premixes. Less water soluble forms or coated K_3 forms exhibit superior stability as compared to uncoated MSB. Heat, moisture, and trace minerals increase the rate of destruction of menadione salts in both pressure-pelleted and extruded feeds (Anonymous, 1981a). For these reasons, greater quantities of vitamin K_3 are recommended in premixes that contain large quantities of choline chloride and certain trace elements, and especially in all cases where plain MSB is used or when premixes are exported or stored for an extended period of time (Schneider and Hoppe, 1986).

Clinical use of vitamin K in human nutrition is largely limited to two forms, a water-soluble form of menadione (menadiol sodium diphosphate) and the more expensive phylloquinone. Some danger of hyperbilirubinemia has been associated with menadione usage (Suttie, 1984), whereas vitamin K in the form of phylloquinone is biologically more available. Griminger and Brubacher (1966) found four times as much vitamin K in eggs of hens fed phylloquinone as in eggs of hens fed an equivalent level of menadione. Vitamin K supplementation has been found to have a definite value in human therapy (1) as a preoperative and post-operative measure to prevent risk of bleeding, (2) in cases where absorption is impaired, as in obstructive jaundice, because bile is necessary for the absorption of vitamin K, and (3) in hemorrhagic diseases of the newborn.

XI. TOXICITY

Toxic effects of the vitamin K family are manifested mainly as hematological and circulatory derangements. Not only is species variation encountered, but profound differences are observed in the ability of the various vitamin K compounds to evoke a toxic response (Barash, 1978). The natural forms of vitamin K, phylloquinone and menaquinone, are nontoxic at very high dosage levels. The synthetic menadione compounds, however, have shown toxic effects when fed to humans, rabbits, dogs, and mice in excessive amounts. The toxic dietary level of menadione is at least 1000 times the dietary requirement (NRC, 1987). Menadione compounds can be safely used at low levels to prevent the development of a deficiency but should not be used as a pharmacological treatment for a hemorrhagic condition.

In animal studies, oral ingestion of large amounts of vitamin K_1 (25 g/kg body weight) produced no fatalities, whereas menadione had an LD_{50} (in mice) equal to 500 mg/kg of diet (Molitor and Robinson, 1940). Anemia, hemoglobinuria, urobilinuria, and urobilinoguria were observed with oral doses of 25–50 mg/kg. These doses are approximately 125 times the recommended clinical dose (0.05

mg/kg) in humans. The anemia seen following excessive vitamin K_3 administration appears to be reversible following withdrawal (Richards and Shapiro, 1945).

Full-term infants appear to have a greater tolerance to vitamin K than premature infants. It is concluded that all vitamin K analogs are safe when administered in the proper dosage; however, the formulation with the greatest margin of safety is vitamin K_1.

6

Thiamin

I. INTRODUCTION

Thiamin (also called thiamine, aneurin(e), and vitamin B_1) was the first of the water-soluble vitamins to be discovered. There is little chance of thiamin deficiency for monogastric animals, including humans, when diets contain ample quantities of whole cereal grains or starchy roots. Thiamin deficiency in humans has been a problem mostly in Asian countries where highly milled (polished) rice is consumed, thus eliminating the thiamin-rich bran fraction of the grain. For years, it was accepted that ruminants did not require B-vitamin supplementation because of adequate rumen microflora synthesis. Intensification of ruminant feeding, involving high-concentrate diets, and management systems with increased levels of production have resulted in nervous disorders that are responsive to thiamin supplementation.

II. HISTORY

Thiamin is considered to be the ''oldest'' vitamin, with the deficiency disease beriberi probably the earliest documented deficiency disorder. The early history of thiamin can be found in Sebrell and Harris (1973) and Loosli (1988). Beriberi was recognized in China as early as 2600 B.C. No cure was found for beriberi until Takaki studied sailors in the Japanese Navy in the early 1880s, when the incidence of beriberi was 32%. By substituting some of the polished rice with other foods, Takaki was able to dramatically reduce the incidence of beriberi. Takaki incorrectly thought that added dietary protein was responsible for preventing beriberi.

In the 1890s Eijkman, a Dutch investigator, produced a paralysis in chickens fed boiled polished rice. He called the condition ''polyneuritis'' and observed that clinical signs were similar to beriberi symptoms in humans. In the course of experiments Eijkman noticed by chance that both polyneuritis and beriberi could be prevented and cured if rice bran was consumed with the polished rice.

He mistakenly believed that the polished rice diet produced a toxin that was counteracted by feeding rice polishings. The conclusion that beriberi was caused by a vitally important food constituent was formulated in 1901 by Grijns, who was Eijkman's successor.

It was 10 years later (1911) that Funk of the Lister Institute in London obtained a potent anti-beriberi substance from rice bran (thiamin) that had the character of an amine. However, it was later found that many vitamins are not amines, and it was Funk who coined the term "vitamin(e)" ("vital amine"). In 1926, Jansen and Donath, successors to Eijkman and Grijns in the Indonesian laboratory, succeeded in crystallizing vitamin B in pure form. In 1936, R. R. Williams and colleagues after many years of work, determined the chemical structure of thiamin and were able to synthesize the vitamin.

III. CHEMICAL STRUCTURE, PROPERTIES, AND ANTAGONISTS

Thiamin consists of a molecule of pyrimidine and a molecule of thiazole linked by a methylene bridge (Fig. 6.1); it contains both nitrogen and sulfur atoms. Thiamin is isolated in pure form as the white thiamin hydrochloride. The vitamin has a characteristic sulfurous odor and a slightly bitter taste. Thiamin is very soluble in water, sparingly so in alcohol, and insoluble in fat solvents. It is very sensitive to alkali, in which the thiazole ring opens at room temperature when pH is above 7. In a dry state thiamin is stable at 100°C for several hours, but moisture greatly accelerates destruction and thus it is much less stable to heat in fresh than in dry foods. Under ordinary conditions, thiamin hydrochloride takes up moisture and, therefore, should be kept in a sealed container. Autoclaving destroys thiamin, an observation that played an important role in the discovery that what was originally considered to be a single vitamin was actually a member of the vitamin B complex (Maynard *et al.*, 1979).

Substances with an anti-thiamin activity are fairly common in nature and include

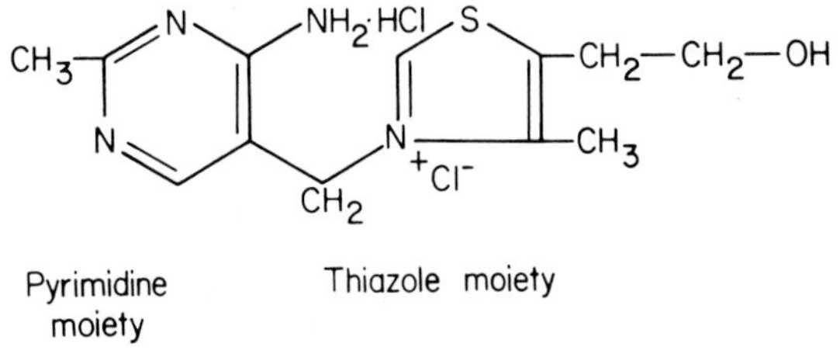

Pyrimidine moiety

Thiazole moiety

Fig. 6.1 Structure of thiamin hydrochloride.

structurally similar antagonists as well as structure-altering antagonists. The synthetic compounds pyrithiamine, oxythiamine, and amprolium (coccidiostate) are structurally similar antagonists and their mode of action is competitive inhibition with biologically inactive compounds, thus interfering with thiamin at different points in metabolism. Synthetic thiamin antagonists are often used in studies on the pharmacodynamics of thiamin (Barclay and Gibson, 1982). Pyrithiamine blocks chiefly the esterification of thiamin with phosphoric acid, resulting in inhibition of the thiamin coenzyme cocarboxylase. Oxythiamin likewise displaces cocarboxylase from important metabolic reactions. Amprolium inhibits the absorption of thiamin from the intestine and also blocks the phosphorylation of the vitamin.

Thiaminase activity destroys thiamin activity by altering the structure of the vitamin. The disease of ''Chastek paralysis'' in foxes and other animals fed certain types of raw fish results from a thiaminase that splits the thiamin molecule into two components and thus renders it inactive.

Since thiaminase is heat labile, the problem can be avoided by cooking the fish at 83°C for at least 5 min. Many different kinds of fish contain thiaminase, with thiamin deficiency being reported in penguin, seals, and dolphins fed primarily fish diets in zoos (Maynard *et al.*, 1979). Thiaminase is found mainly in herrings, sprats, stints, and various carp species, a total of some 50 species most of which live in fresh water. Wild aquatic animals apparently do not suffer thiamin deficiency even though they eat a diet primarily of fish, because fish must undergo some putrefaction to release the enzyme (Evans, 1975). *In vitro* and *in vivo* experiments have shown that 1 kg of fish can destroy up to 25 mg thiamin. This degradation takes place within the first 30 min after ingestion, when still in the stomach. Certain microorganisms (bacteria and molds) also have been shown to produce thiaminases. A disease in horses known as ''bracken fern poisoning'' results from antagonism to thiamin.

IV. ANALYTICAL PROCEDURES

Thiamin activity can be analyzed by biological, microbiological, and chemical methods. Biological methods are based on curative ability for polyneuritis in pigeons, bradycardia in the rat, or growth in the chick, pigeon, and rat. Microbiological tests are fairly rapid, inexpensive, and very sensitive, but some organisms lack specificity for thiamin compared to the pyrimidine and thiazole moieties or the enzyme form of the vitamin. Chemical analytical analysis for thiamin is conducted by oxidation to thiochrome, which shows a characteristic blue fluorescence in ultraviolet light. The fluorometric thiochrome method is the most widely used to estimate thiamin in foodstuffs and feeds.

V. METABOLISM

A. Digestion, Absorption, and Transport

Thiamin appears to be readily digested and released from natural sources. A precondition for normal absorption of thiamin is sufficient production of stomach hydrochloric acid. Phosphoric acid esters of thiamin are split in the intestine. Free thiamin is soluble in water and is easily absorbed, especially in the duodenum. Ruminants can also absorb free thiamin from the rumen, but the rumen wall is not permeable for bound thiamin or for thiamin contained in rumen microorganisms. The horse can also absorb thiamin from the cecum.

The mechanism of thiamin absorption is not yet fully understood, but apparently both active transport and simple diffusion are involved (Bräunlich and Zintzen, 1976). At low concentrations there is an active sodium-dependent transport against the electrochemical potential, whereas at high concentrations it diffuses passively through the intestinal wall. Absorption from the rumen is also believed to be an active mechanism. Absorbed thiamin is transported via the portal vein to the liver with a carrier plasma protein.

B. Phosphorylation

Thiamin phosphorylation can take place in most tissues but particularly in the liver. Four-fifths of thiamin in animals is phosphorylated in liver under the action of adenosine triphosphate (ATP) to form the metabolically active enzyme form thiamin pyrophosphate (TPP or cocarboxylase). Of total body thiamin, about 80% is TPP, about 10% is thiamin triphosphate (TTP), and the remainder is thiamin monophosphate (TMP) and free thiamin.

C. Storage and Excretion

Although thiamin is readily absorbed and transported to cells throughout the body, it is not stored to any great extent. Thiamin content in individual organs varies considerably, with the vitamin preferentially retained in organs with a high metabolic activity. During deficiencies thiamin is retained in greatest quantities in important organs such as heart, brain, liver, and kidneys. Intakes in excess of current needs are rapidly excreted. This means that the body needs a regular supply and also that unneeded intakes are wasted. The pig is somewhat of an exception, however, and for some unknown reason its tissues contain several times as much thiamin as is the case with other species studied, and thus has a store that can meet body needs on a thiamin-deficient diet for as long as 2 months (Heinemann *et al.*, 1946).

Absorbed thiamin is excreted in both urine and feces, with small quantities excreted in sweat. Fecal thiamin may originate from feed, synthesis by microorganisms, or endogenous synthesis (e.g., via bile or excretion through the mucosa of the large intestine). When thiamin is administered in large doses, urinary excretion first increases then reaches a saturation level, and with additional thiamin the fecal concentration increases considerably (Bräunlich and Zintzen, 1976).

VI. FUNCTIONS

A. Coenzyme

A principal function of thiamin in all cells is as the coenzyme cocarboxylase or TPP. The citric acid cycle (Krebs cycle or tricarboxylic acid cycle) is responsible for production of energy in the body. In this cycle, breakdown products of carbohydrates, fats, and proteins are brought together for further breakdown and for synthesis. The vitamins riboflavin and niacin, as well as thiamin, play roles in the cycle. Thiamin is the coenzyme for all enzymatic decarboxylations of α-keto acids. Thus it functions in the oxidative decarboxylation of pyruvate to acetate, which in turn is combined with coenzyme A for entrance into the tricarboxylic cycle.

In mammals, thiamin is essential in two oxidative decarboxylation reactions in the citric acid cycle that take place in cell mitochondria and one reaction in the cytoplasm of the cells (Fig. 6.2). These are essential reactions for utilization of carbohydrates to provide energy. Decarboxylation in the citric acid cycle removes carbon dioxide and the substrate is converted into the compound having the next lower number of carbon atoms:

$$\text{Pyruvate} \rightarrow \text{acetyl-CoA} + CO_2 \quad (1)$$

$$\alpha\text{-Ketoglutaric acid} \rightarrow \text{succinyl-CoA} + CO_2 \quad (2)$$

TPP is a coenzyme in the transketolase reaction that is part of the direction oxidative pathway (pentose phosphate cycle) of glucose metabolism that occurs not in mitochondria but in the cell cytoplasm in liver, brain, adrenal cortex, and kidney, but not skeletal muscle. Transketolase catalyzes transfer of C_2 fragments, hence with ribulose 5-phosphate as donor and ribose 5-phosphate as acceptor, sedoheptulose 7-phosphate and triose phosphate are formed. This is the only mechanism known for synthesis of ribose, which is needed for nucleotide formation, and also results in formation of NADPH, which is essential for reducing intermediates from carbohydrate metabolism to form fatty acids.

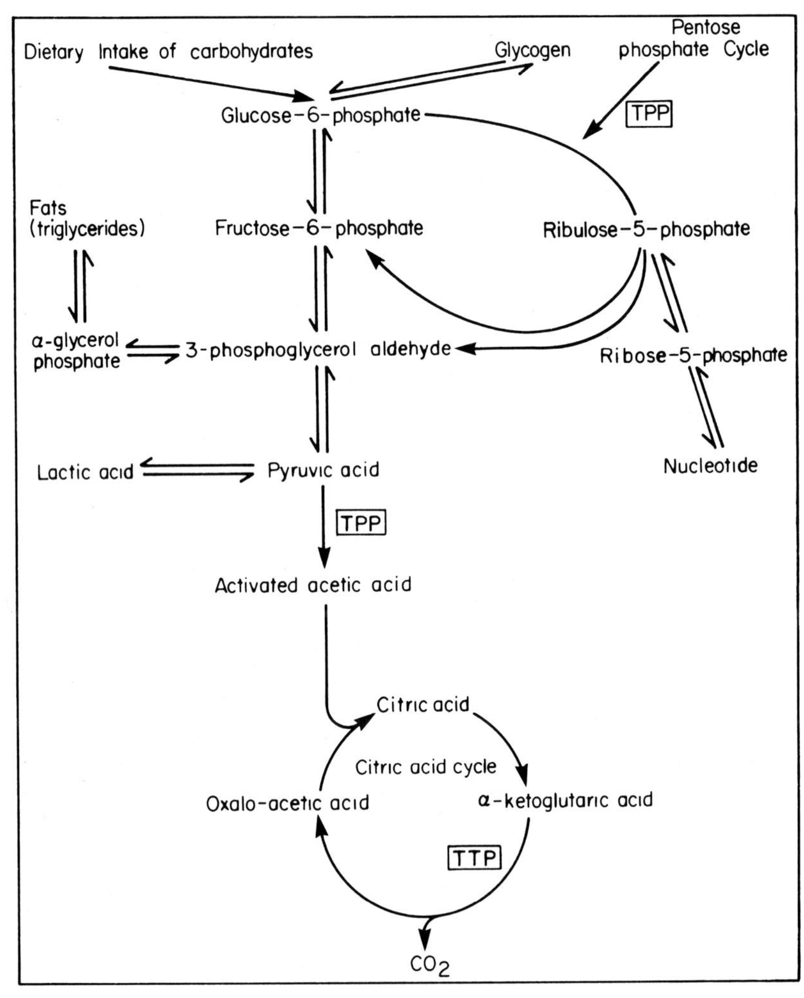

Fig. 6.2 Thiamin as thiamin pyrophosphate (TPP) in the metabolism of carbohydrates (Modified from Bräunlich and Zintzen, 1976.)

B. Neurophysiology

Little is known of thiamin functions in nervous tissue. However, evidence has accumulated for a specific role of thiamin in neurophysiology that is independent of its coenzyme function. Fatty acids and cholesterol are the major constituents of cell membranes, and effects on their synthesis would affect membrane integrity and function. Thiamin deficiency in cultured glial cells impairs their ability to synthesize fatty acids and cholesterol. The defect is related to reduced formation of key lipogenic enzymes. These changes could be the basis of degenerative changes seen in glial cells in early thiamin deficiency.

Possible mechanism of action of thiamin in nervous tissue include the following (Muralt, 1962; Cooper *et al.*, 1963): (1) thiamin is involved in the synthesis of acetylcholine, which transmits neural impulses; (2) thiamin participates in the passive transport of sodium of excitable membranes, which is important for the transmission of impulses at the membrane of ganglionic cells; and (3) the reduction in the activity of transketolase in the pentose phosphate pathway that follows a deficiency of thiamin reduces the synthesis of fatty acids and the metabolism of energy in the nervous system.

VII. REQUIREMENTS

Many factors influence requirements for thiamin (Sebrell and Harris, 1973). Thiamin requirements in some species are difficult to establish because of vitamin synthesis by microflora in ruminants and most likely for all species in the lower intestine. More important than dietary effects on synthesis is the general conclusion that most of the synthesis apparently occurs too far down the intestinal tract for absorption. Intestinal synthesis, therefore, would be of greatest benefit to animals that practice coprophagy. However, with the horse, there is considerable synthesis of thiamin in the digestive tract with an estimation of 25% of the free thiamin in the cecum absorbed (Linerode, 1966). For humans and most livestock species, other than ruminants and horses, it is doubtful whether the amount of thiamin thus produced by intestinal synthesis and absorbed is large enough to make a significant contribution to body needs.

For ruminants, total feed thiamin and that available from ruminal synthesis have to be considered together, and the total need is yet to be defined. Animals with functional rumens are generally considered to have no dietary thiamin requirements because of microbial synthesis. However, thiamin deficiency can be produced in lambs and calves and other young ruminants that do not have a functional rumen. Deficiency can be produced as late as the sixth week of life, the preruminant period. Mixed rations, consisting of milk, sources of high energy, and hay, stimulate growth of the villi in the rumen, but milk alone retards development of the rumen mucosa and thus thiamin synthesis. Self-synthesis of thiamin becomes significant in calves only from the sixth week of life on, and this synthesis is still relatively weak at weaning after 5 months (Zintzen, 1974).

The self-synthesis of thiamin is subject to dietary composition in the rumen; it is significantly enhanced by carbohydrates such as molasses, which are easily soluble, and by supplies of nitrogen such as urea. When monogastric animals receive only marginal amounts of thiamin, they will develop deficiency signs earlier if their feed is also low in protein (Abe, 1969). For the chicken, the amount of intestinal synthesis would depend on carbohydrate type present in the diet, synthesis being favored by cooked (dextrinized) starches as compared with glucose or sucrose (Scott *et al.,* 1982). Concentration of thiamin in the gastrointestinal tract (especially in the rumen) is more uniform than it is in the feed. The self-synthesis of thiamin is therefore high when the vitamin level in feed is low, and less when feed is rich in thiamin. Therefore, there must be some regulating mechanism that governs the concentrations of the vitamin in the rumen and maintains this within certain limits (Zintzen, 1974).

Diet composition can dramatically influence thiamin requirements. Since thiamin is specifically involved in carbohydrate metabolism (see Section VI), level of dietary carbohydrate relative to other energy-supplying components influences thiamin requirement. The need for thiamin increases as consumption of carbohydrate increases. When thiamin is deficient, body reserves become depleted more rapidly when animals are being maintained on a feed rich in carbohydrates than when they are receiving a diet rich in fat and protein. The "thiamin-sparing" effect of fats and protein has long been known.

Size, genetic factors, and metabolic status affect thiamin requirements. Thiamin requirement is also proportional to size. Light poultry breeds (Leghorn) seem to have higher thiamin requirements than heavy breeds (Thornton and Schutze, 1960), and Leghorn hens deposit more thiamin in eggs than do heavy hens. Need for thiamin increases during gestation, lactation, and in hyperthyroidism; during lactation, the rat needs five times the normal level. As an animal ages, its need for thiamin increases because efficiency of vitamin utilization likely diminishes.

Thiamin requirements are obviously higher if feeds contain raw materials (e.g., fish) or additives with anti-thiamin action (see Section III). Spoiled and moldy feeds may contain such antagonists or thiaminases. Chicks kept on a feed infected with *Fusarium moniliforme* developed polyneuritis that could be cured with thiamin injections (Fritz *et al.,* 1973). Moldy feed analyses showed a thiamin content of less than 0.1 mg/kg, whereas the same feed not contaminated with *Fusarium* had a thiamin content of 5.33 mg/kg. The antagonistic factor could be destroyed by treatment with steam.

Disease conditions also result in increased thiamin requirements. When dietary thiamin is marginal, typical deficiency signs of thiamin are more likely to develop in infected animals than in normal animals. Endoparasites such as strongylids and coccidia compete with the host for thiamin contained in food. It has been shown experimentally that infection with these coccidia results in considerable reduction in thiamin blood levels. Thiamin levels found were directly correlated to infection severity (McManus and Judith, 1972). Likewise, conditions such as diarrhea and malabsorption increase the requirement.

Table 6.1 summarizes thiamin requirements for various livestock species and humans, with a more complete listing given in Appendix Table 1. The minimum human requirement for thiamin appears to be between 0.20 and 0.23 mg per

TABLE 6.1

Thiamin Requirements for Various Animals and Humans[a]

Animal	Purpose	Requirement	Reference
Beef cattle	Adult	Microbial synthesis	NRC (1984a)
Dairy cattle	Calf	65 μg/kg	NRC (1978a)
	Adult	Microbial synthesis	
Chicken	Leghorn, 0–6 weeks	1.8 mg/kg	NRC (1984b)
	Leghorn, 6–20 weeks	1.3 mg/kg	NRC (1984b)
	Laying–breeding	0.8 mg/kg	NRC (1984b)
	Broilers, 0–8 weeks	1.8 mg/kg	NRC (1984b)
Turkey	All classes	2.0 mg/kg	NRC (1984b)
Japanese quail	All classes	1.6–3.2 mg/kg	Shim and Boey (1988)
Sheep	Adult	Microbial synthesis	NRC (1985b)
Swine	Growing–finishing, 1–5 kg	1.5 mg/kg	NRC (1988)
	Growing–finishing, 5–110 kg	1.0 mg/kg	NRC (1988)
	Adult	1.0 mg/kg	NRC (1988)
Horse	Growing	3.0 mg/kg	NRC (1978b)
	Adult	Microbial synthesis(?)	
Goat	Adult	Microbial synthesis	NRC (1981b)
Mink	Growing	1.2 mg/kg	NRC (1982a)
Fox	Growing	2.0 mg/kg	NRC (1982a)
Cat	Adult	5.0 mg/kg	NRC (1986)
Dog	Growing	0.75 mg/kg	NRC (1985a)
Fish	Catfish	1.0 mg/kg	NRC (1983)
	Shrimp	120 mg/kg	NRC (1983)
Rat	All classes	4.0 mg/kg	NRC (1978c)
Human	Infants	0.3–0.5 mg/day	RDA (1980)
	Children	0.7–1.2 mg/day	RDA (1980)
	Adults	1.0–2.0 mg/day	RDA (1980)

[a]Expressed as per unit of animal feed (except for calf) either on an as fed (approximately 90% dry matter) or dry basis (see Appendix Table 1). Human data are expressed as mg/day.

1000 kcal (RDA, 1980). On a calorie basis, infant requirements are similar to those of adults. Patients with thiamin-responsive inborn errors of metabolism respond to pharmacological doses of thiamin (Scriver, 1973).

VIII. NATURAL SOURCES

Cereal grains and their by-products, soybean meal, cottonseed meal, and peanut meal are relatively rich sources of thiamin. Brewer's yeast is the richest known natural source of thiamin. Since the vitamin is present primarily in the germ and seed coats, by-products containing the latter are richer than the whole kernel, while highly milled flour is very deficient. Whole rice may contain 5 ppm thiamin, with much lower and higher concentrations for polished rice (0.3 ppm) and rice bran (23 ppm), respectively (Marks, 1975). Wheat germ ranks next to yeast in thiamin concentration. Reddy and Pushpamma (1986) studied the effects of 1 year's storage and insect infestation on thiamin content of feeds. Thiamin losses were high in different varieties of sorghum, pigeonpea, and green gram (40–70%) and lower in rice and chickpea (10–40%), with insect infestation causing further loss. Lean pork, liver, kidney, and egg yolk are rich animal products. The content in lean pork can be doubled by increasing the thiamin intake of the pig (Heinemann *et al.*, 1946). Thiamin content of typical feedstuffs is shown in Table 6.2.

The level of thiamin in grain rises as the level of protein rises; it depends on species, strain, and use of nitrogenous fertilizers (Aitken and Hankin, 1970; Zintzen, 1974). The content in hays decreases as plants mature and is lower in cured than in fresh products. Thiamin concentration is correlated with leafiness

TABLE 6.2

Thiamin in Foods and Feedstuffs[a]

Energy feed sources	mg/kg	Protein sources	mg/kg
		Plant	
Barley grain, dried	5.7	Alfalfa meal	3.9
Beans	6.0	Brewer's grains, dried	0.8
Corn (maize), yellow grain	3.5	Brewer's yeast, dried	95.2
Corn (maize), germ meal	10.9	Coconut meal, dried	0.8
Corn (maize), dried gluten meal	2.1	Cottonseed meal, solvent extracted	6.4
Distiller's dried solubles	6.8	Linseed meal, expeller extracted	5.1
Millet, dried	4.5	Peanut meal	12.0
Oat grain	5.2	Peas, dried	9.0
Potatoes, sweet	0.9	Sesame meal	10.0
Potatoes, white	1.0	Animal	
Rice, bran	23.0	Blood meal, dried	0.2
Rice, grain	5.0	Chicken	0.4
Rice, polished	0.3	Eggs, whole	3.4
Rye, dried	4.4	Fish meal, with solubles	2.0
Sorghum grain	3.9	Liver and glandular meal	2.6
Sugarbeet pulp, dried	0.4	Meat and bone meal	0.1
Sugarcane molasses	1.2	Milk, cow's	0.4
Wheat bran	8.0	Skim milk, dried	3.5
Wheat grain	5.5		
Wheat standard middlings, dried	12.0		
Whey, dried	8.0		

[a]Modified from Bräunlich and Zintzen (1976), Marks (1975), and Scott *et al.* (1982).

and greenness as well as protein content. In general, good quality hay is a substantial source, and in a dry climate there is practically no loss in storage. Milk is not a rich source, and pasteurization for 30 min at 63°C destroys 25% of its content (Maynard *et al.*, 1979). Since thiamin is water soluble as well as unstable to heat, large losses in certain cooking operations result (McDowell, 1985b). There was up to 40% loss after 8 months' storage of frozen meat. Roasting beef and pork and boiling vegetables resulted in 40–50% loss. With electronic cooking of meat there is little loss, but with ionizing radiation, up to 88% of thiamin is destroyed (NRC, 1983).

IX. DEFICIENCY

The classic diseases, beriberi in humans and polyneuritis in birds, represent a late stage of the deficiency resulting from a peripheral neuritis, perhaps caused by accumulation of intermediates of carbohydrate metabolism. Comparing thiamin deficiency signs in various species, it is seen that disorders affecting the central nervous system are the same in all species. This is explained by the fact that, in all mammals, the brain covers its energy requirement chiefly by the degradation of glucose and is therefore dependent on biochemical reactions in which thiamin plays a key role.

In addition to neurological disorders, the other main group of disorders involves cardiovascular damage. Clinical signs involved with heart function are slowing of heart beat (bradycardia), enlargement of heart, and edema. Less specific clinical signs include gastrointestinal troubles, muscle weakness, easy fatigue, hyperirritability, and lack of appetite (anorexia). Of all nutrients, a deficiency of thiamin has the most marked effect on appetite. Animals consuming a low-thiamin diet soon show severe anorexia, lose all interest in food, and will not resume eating unless given thiamin. If the deficiency is severe, thiamin must be force-fed or injected to induce animals to resume eating. Table 6.3 illustrates clinical signs (and symptoms) associated with thiamin deficiency in various species. Animals developing thiamin deficiency most readily are chickens and pigeons; pigs, rats, mice, and rabbits are less readily depleted, probably because of endogenous thiamin production.

A. Effects of Deficiency

1. Ruminants

Young animals that do not have a fully developed rumen can suffer from a thiamin deficiency as shown in experiments with calves and lambs. Clinical signs include an apparent weakness that is usually first exhibited by poor leg coor-

TABLE 6.3

Clinical Signs Associated with Thiamin Deficiency in Various Species[a]

General signs	Cardiovascular signs	Nervous disorders	Other signs
Human			
Loss of appetite, tiredness/apathy, lack of concentration, irritable, anxiety states, depression	ECG changes, falling blood pressure, arrhythmia, edema, hypotension, tachycardia	Muscular weakness, calf cramp, weakened reflexes, loss of feeling, neuritic symptoms, coma	Dyspnea
Hen			
Loss of appetite, loss of weight, weakness	Bradycardia (from 300 down to 90–100/min), cyanosis, edema	Opisthotonos, leg weakness, muscular weakness, extension spasms, apathy, paresis	Sudden death, gonad damage, gut atony, diarrhea, hypothermia
Pigeon			
Loss of weight	Heart failure, tachycardia	Leg weakness, opisthotonos, ataxia, paresis	Crop voiding (vomiting)
Mouse			
Loss of appetite, loss of weight, weakness	Pathological changes in the myocardium	Hyperexcitability, ataxia, circumduction, hemorrhages in the mesencephalon, opisthotonos	Pathological changes in liver, testicles, skeletal muscle, diarrhea, hypothermia
Rat			
Loss of appetite, disturbances of growth, general weakness	Bradycardia (from 500 down to 250–300/min), slackened pulse and reduced respiratory rate	Ataxia and cramps, paralysis, circumduction, nodding of the head	Diarrhea, skin atony, hypothermia
Rabbit			
		Paralysis, cramps, opisthotonos	Coma
Cat			
Loss of appetite, loss of weight	Myocardial damage	Ataxia, ventral flexion of the head	
Dog			
Loss of appetite, loss of weight, weakness	Slow pulse	Paresis, cramps, opisthotonos	Gastrointestinal disorders, vomiting, hypothermia

(continued)

TABLE 6.3 (*Continued*)

General signs	Cardiovascular signs	Nervous disorders	Other signs
Fox			
Loss of appetite, loss of weight	Pathological changes in the heart	Polyneuritis, ataxia, irritability, cramps, spastic paresis and paralysis	Groaning, pathological changes in the liver
Mink			
Loss of appetite, loss of weight		Polyneuritis, ataxia, cramps, convulsions, spastic paresis, and paralysis, opisthotonos	Diarrhea, poor quality of pelt
Pig			
Loss of appetite, loss of weight, weakness, premature births, high mortality among the young	Slow pulse, cyanosis, heart failure, tachycardia, dilatation of the heart, edema, hemorrhages in the gastrointestinal wall and myocardium		Vomiting, dyspnea, diarrhea, skin atony, hypothermia, sudden death
Young ruminant			
Loss of appetite, loss of weight, weakness	Cardia arrhythmia, tachycardia	Hyperesthesia, polyneuritis, ataxia, convulsions, cramps, opisthotonos, spastic paresis and paralysis	Diarrhea, dyspnea, lachrymation, gnashing of the teeth, sometimes disorders of vision
Horse			
Loss of appetite, loss of weight, lusterless coat	Heart failure, tachycardia	Nervousness, ataxia, apathy, muscle tremors, cramps, coma	Diarrhea, constipation, dyspnea

[a]Modified from Bräunlich and Zintzen (1976).

dination, especially of the forelimbs, and by inability to rise and stand. The head is frequently retracted (Fig. 6.3) along the shoulder and the heart may develop arrhythmia. Specific signs are usually accompanied by anorexia and severe diarrhea, followed by dehydration and death (NRC, 1978a). Signs in the calves can be either acute or chronic. Acutely affected calves were anorectic, had severe diarrhea, and died within 24 hr. These signs appeared after 2–4 weeks on the low-thiamin diet (Johnson *et al.*, 1948a).

Fig. 6.3 Sheep with thiamin deficiency. Characteristics of the condition are head bent backward (opisthotonos), cramp-like muscular contractions, disturbance of balance, and aggressiveness. (Courtesy of Michel Hidiroglou, Animal Research Center, Ottawa, Ontario, Canada.)

Because of extensive ruminal thiamin synthesis, the general conclusion is that ruminants possessing a normally functioning rumen have no dietary thiamin requirement. Despite the fact that rumen microbes synthesize thiamin and also feeds, particularly whole grains, contain thiamin, deficiencies do develop in ruminants. From a number of world areas a thiamin-responsive disease, which occurs sporadically in cattle, sheep, and goats, is known as polioencephalomalacia (PEM). Other names for PEM in different regions include cerebrocortical necrosis, cerebral necrosis, and forage poisoning.

The condition affects mainly calves and young cattle aged between 4 months and 2 years and in lambs, young sheep, and goats between 2 and 7 months old. The incidence of PEM is reported to be between 1 and 20% and mortality may reach 100%. Clinical signs in mild cases include dullness, blindness, muscle tremors, especially of the head, and opisthotonos. The condition is characterized by circling, head pressing, and convulsions and in severe cases the animal collapses within 12–72 hr after the onset of the disease (Fig. 6.4). The ears drop and in the final stages the limbs and head are extended. General twitching of the musculature of the ears and eyelids, waving of the head and neck, and grinding of the teeth with groaning sometimes occur. Without treatment, death usually occurs within a few days. The main lesions in these animals are necrotic areas in both cerebral hemispheres.

The PEM condition may run a short, acute course, or it may occur in a milder form and run a more protracted course. It is probable that, particularly in its mild form, the condition is often not diagnosed and may in fact occur more frequently than is recognized. Without treatment mortality is about 50% of animals with the mild form and may be up to 100% in animals with acute PEM. The incidence and death rate are higher in young animals from 2 to 5 months of age than for animals older than 1 year.

Biochemical changes indicating that PEM is associated with thiamin deficiency include reduced tissue thiamin contents, dramatic elevation of blood pyruvate and lactate, and markedly reduced transketolase activity (Bräunlich and Zintzen, 1976). Moreover, sick animals react so promptly to treatment with thiamin (sometimes within hours) that early treatment with thiamin is even used for confirming the diagnosis of PEM.

Zintzen (1974) concludes that PEM can be definitely established if the following four situations exist:

1. Case history—animals have been maintained on high-energy feeds rich in carbohydrates, and other animals on the same farm have, from time to time, died after exhibiting central nervous system disorders.
2. Biochemical evidence—Blood pyruvate has steeply increased and the activity of erythrocyte-transketolase has been reduced.
3. Diagnostic therapy—animals thought to be suffering from PEM will react promptly to treatment with thiamin, provided they are treated in the early stages of the disease.
4. Pathological changes—necropsy shows typical pathological anatomical changes—bilateral cortical necrosis—in the brain.

Polioencephalomalacia generally occurs in feedlot cattle frequently about 3 weeks after a ration change. Research suggests that PEM is associated with lactic acid acidosis with both conditions related to adaptation to grain diets. Oltjen *et al.* (1962) reported that thiamin in the rumen is decreased by a reduction in rumen pH; a low ruminant pH is characteristic of cattle fed high-concentrate diets. Although little information is available on the direct addition of thiamin to finishing cattle diets, Brethour (1972) reported that in two trials, a combination of thiamin and sodium carbonate supplement increased feed intake by 5% and daily gain by 8%. In a third trial, thiamin administered alone gave an intermediate response to calves immediately after weaning.

PEM has caused significant economic losses in tropical countries, not only in feedlots where high-grain diets are provided but also where high levels of molasses are fed. When molasses is provided *(ad libitum)* together with rations containing little crude fiber, a disease referred to as "molasses toxicity" or "molasses drunk-

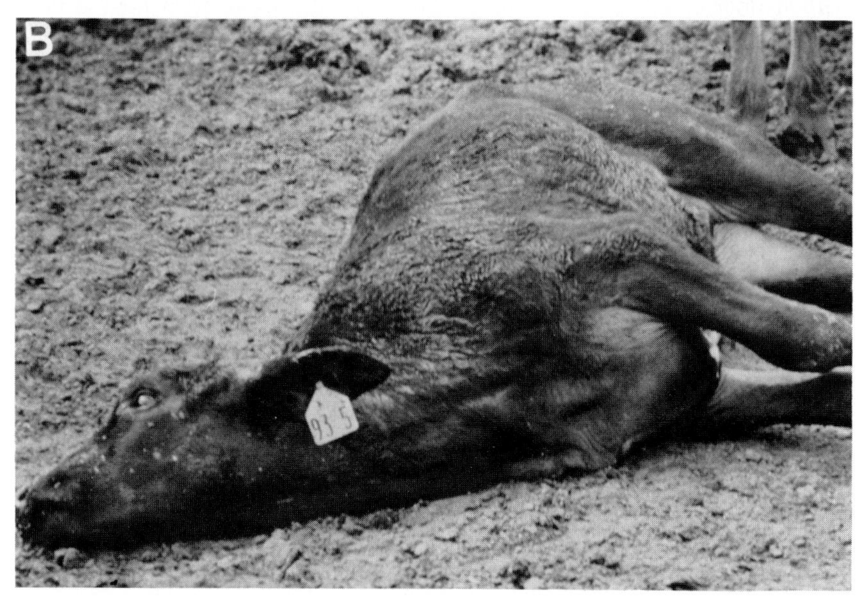

Fig. 6.4 An animal (A and B) with polioencephalomalacia, a thiamin deficiency. Feedlot cattle suffering from this condition show dullness and sometimes blindness with a series of nervous disorders such as circling, head pressing, and convulsions. After injections of thiamin the same animal (C) was able to stand after 6–8 hr. With continued thiamin treatment, in 3–5 days the animal returned to almost normal with slight brain damage. (Courtesy of B. Bock, University of Florida.)

Fig. 6.4 (*Continued*)

enness'' appears (Losada *et al.*, 1971). Clinical signs of this condition closely resemble PEM and some studies completed in Cuba have suggested that thiamin treatment, together with additional roughage, may be an effective cure.

A number of experiments have shown that PEM is caused by naturally occurring thiamin antagonists, in connection with a reduced thiamin synthesis or destruction of the vitamin (see Section III). Clinical reports of PEM have shown that under high-concentrate feeding systems of beef cattle and lambs, thiaminase, an enzyme that destroys thiamin, may become active in the rumen and cause a thiamin deficiency in animals with functional rumens (Edwin and Lewis, 1971). These thiaminases may be produced by microorganisms (bacteria and fungi) in contaminated feeds (Davies *et al.*, 1968). Thiaminases are found in certain plant species and are produced by some microorganisms believed to be responsible for PEM. This has been a special problem in Australia, where PEM occurs under pasture conditions, apparently being derived from some of the fern species.

From Colombia, a wasting disease known as ''secadera'' (''drying up'') has been reported as a thiamin deficiency because the condition has been alleviated within thiamin injections (Mullenax, 1983). Mullenax (1983) suggests that a fungus associated with native forages contains a thiaminase. On the contrary, Miles and McDowell (1983) report that the wasting disease ''secadera'' (Fig. 6.5) can be successfully controlled with a highly fortified complete mineral supplement. It is possible that supplementation of this wasting disease is controlled by either thiamin or minerals through different mechanisms (McDowell, 1985a).

Fig. 6.5 A wasting disease (''secadera'') of cattle in the llanos of Colombia. Animals are characterized by an emaciated condition in spite of good quality available forage. This has been reported as a thiamin deficiency as the condition has been alleviated with thiamin injections. However, ''secadera'' has also been controlled with a highly fortified complete mineral supplement. (L. R. McDowell, University of Florida.)

2. Swine

Thiamin deficiency in swine reveals itself particularly in a decrease of appetite and body weight, vomiting, a slow pulse, subnormal body temperature, nervous signs, postmortem heart changes, and sudden death. Heinemann *et al.* (1946) reported that the pig can utilize stored thiamin over a long period of time, as 56 days were required for the pigs to lose their appetites after being placed on a thiamin-deficient diet. Death has been reported 74 days after pigs were placed on a thiamin-free but otherwise adequate diet (Loew, 1978). For young pigs, severe thiamin deficiency has resulted in death at the age of 3–4 weeks.

First signs of thiamin deficiency in pigs are reduced feed consumption and vomiting with a sharp reduction in weight gains (Miller *et al.*, 1955). Functional and structural cardiac changes are the main findings in experimentally deficient swine; in contrast to clinical reports, nervous system lesions were not detected. Electrocardiographically demonstrable changes in heart tissue are also seen, with enlarged hearts obtained from pigs on thiamin-deficient diets (Fig. 6.6). Microscopically, it is possible to recognize inflammations and necrotic changes in the myocardial fibers.

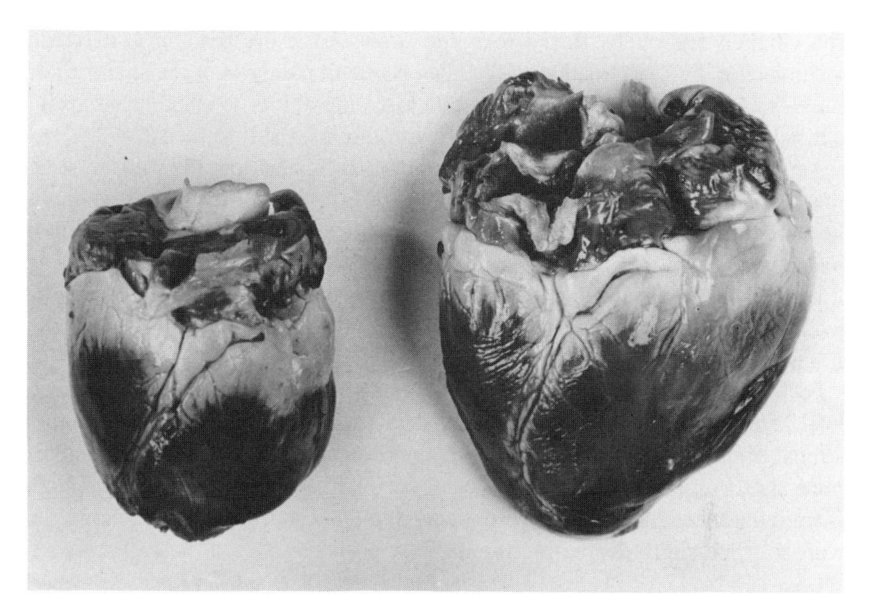

Fig. 6.6 Enlarged heart on right is due to a deficiency of thiamin. Heart on left is from a similar pig fed the same diet plus thiamin. (Courtesy of T. J. Cunha and Washington State Universtiy.)

3. Poultry

Poultry are more susceptible to neuromuscular effects of thiamin deficiency than most mammals. In chickens and turkeys, there is a loss of appetite, emaciation, impairment of digestion, a general weakness, opisthotonos or stargazing, and frequent convulsions, with polyneuritis as an extreme clinical sign. Early signs are lethargy and head tremors. Chicks fed very low thiamin (0.4 ppm) survived for only 7–10 days, apparently only a few days after the supply of thiamin in the yolk sac was exhausted (Gries and Scott, 1972). Some chicks developed nervous disorders, apathy, and tremor as early as the third or fourth day of life. These signs increased in severity up to ataxia, inability to stand, and high-grade opisthotonos or twisting of the neck. Severity of the spasms increased when the chicks were frightened. Chicks that showed these high-grade nervous disorders died within a few hours. Cardiac abnormalities have also been reported in acutely thiamin-deficient chicks (Sturkie *et al.*, 1954). A paralysis of the crop, manifested as delayed emptying, accompanies the general neuropathy of experimental thiamin deficiency in chicks (Naidoo, 1956).

In mature chickens, polyneuritis is observed approximately 3 weeks after they are fed a thiamin-deficient diet (Scott *et al.*, 1982). As the deficiency progresses, paralysis of the muscles occurs, beginning with the flexors of the toes and progressing upward, affecting the extensor muscles of the legs, wings, and neck.

The chicken sits on its flexed legs and draws back the head in a stargazing position (Fig. 6.7). Retraction of the head is due to paralysis of the anterior neck muscles. At this stage, the chicken soon loses the ability to stand or sit upright and falls to the floor, where it may lie with the head still retracted.

Acutely deficient pigeons developed vomiting, emaciation, leg weakness, and opisthotonos, the last of which appears between 7 and 12 days after beginning the thiamin-free diet (Swank, 1940). Chronic deficiency due to a diet partially inadequate in thiamin resulted in leg weakness but no opisthotonos. Evidence of cardiac failure was also noted. The lesions produced in thiamin-deficient pigeons are reported to be identical to those found in Wernicke's polioencephalitis in humans (Lofland *et al.*, 1963).

For chickens with thiamin deficiency, body temperature drops to as low as 36°C and respiratory rate progressively decreases (Scott *et al.*, 1982). There is adrenal gland hypertrophy that apparently results in tissue edema, particularly in the skin. Atrophy of genital organs also occurs in chickens affected with chronic thiamin deficiency, being more pronounced in the testes than in the ovaries. The heart shows a slight degree of atrophy.

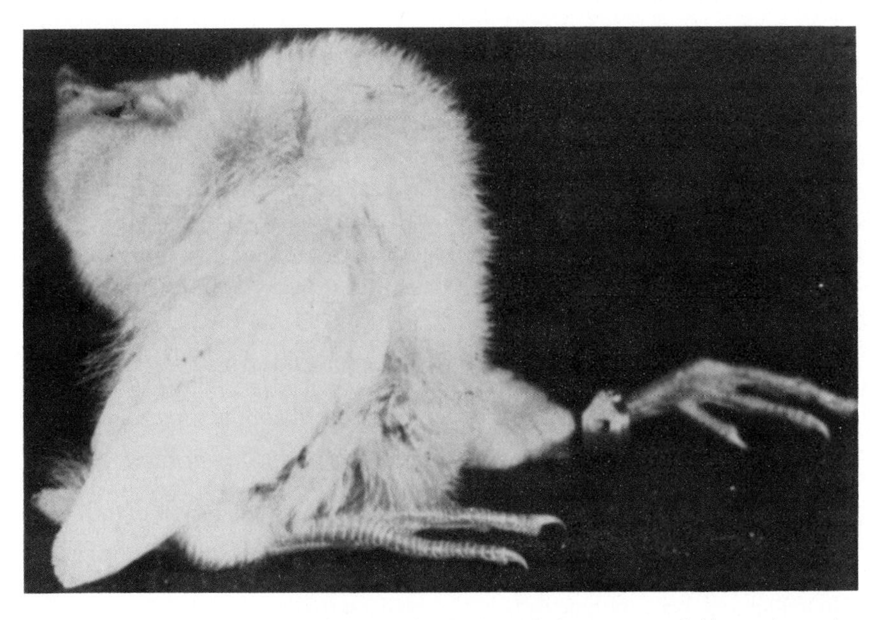

Fig. 6.7 Polyneuritis in a thiamin-deficient chick. Muscle paralysis causes extended legs and retraction of the head. (Courtesy of M. L. Scott, Cornell University.)

4. Horses

Horses fed experimental diets low in thiamin exhibited anorexia, loss of weight, incoordination (especially in the hind legs), lower blood thiamin, elevated blood pyruvic acid, and dilated and hypertrophied heart (Carroll *et al.*, 1949). Incoordination and other nervous signs were alleviated by feeding thiamin, indicating that this species requires a dietary source of this vitamin (Maynard *et al.*, 1979). A lack of thiamin causes reproductive failure in both sexes.

Because of generally adequate thiamin dietary supplies and intestinal synthesis, deficiency would result under practical conditions only when thiaminase-containing plants are provided (see Section III). Ingestion of sufficient quantities of the thiaminase-containing bracken fern *Pteridium aquilinum* was shown to readily cause thiamin deficiency in horses (Kingsbury, 1964). The signs in affected horses included weight loss, incoordination, lethargy, cardiac irregularities, muscular twitching and tremors, and prostration; death usually followed convulsions. Anorexia was not apparent until other signs became severe. Treatment with thiamin was successful if given before terminal stages. The horsetail *Equisetum* appears to act in the same way as bracken fern.

5. Other Animal Species

a. Foxes and Mink. Thiamin deficiency is also a problem in pelt animal farming and is referred to as "Chastek paralysis." This disease occurs in mink and foxes and is induced by feeding these animals on certain types of raw fish (see Section III). Clinical signs of thiamin deficiency in mink are reported as anorexia, weight loss, and emaciation, followed later by nervous disorders such as involuntary movements, unsteady gait, muscular spasms, and attacks of cramp or general paralysis. When the animals are touched they develop extension spasms in the hind limbs. The terminal phase is characterized by opisthotonos and unconsciousness, and animals die in a characteristic contorted position. Histological examination shows brain hemorrhages and lesions of the neural pathways (Bräunlich and Zintzen, 1976). For the fox, clinical signs include anorexia and abnormal gait, followed by severe ataxia, inability to stand, hyperesthesia, constant moaning, and convulsions (Long and Shaw, 1943).

b. Dogs and Cats. The main clinical signs for thiamin deficiency in dogs and cats are anorexia and nervousness, leading to spasms and finally to death (NRC, 1985a). Of all the domestic animals, the cat is most often reported to be clinically thiamin deficient. This might well be expected as domesticated cats often consume fish, with the possibility of thiaminase being present in many cat foods (NRC, 1986). However, these are usually heat treated, which would be expected to inactivate thiaminase.

c. Rabbits. Rabbits fed a thiamin-free diet, along with a thiamin antagonist (neopyrithamin), developed ataxia, flaccid paralysis, convulsions, coma, and death (Reid *et al.*, 1963). The rabbit differed from nearly all other species in not developing anorexia. Apparently, intestinal synthesis and coprophagy prevented the usual signs from developing.

d. Fish. Thiamin deficiency signs in fish are similar to those observed in mammals and birds. The principal signs are anorexia, arrest of growth, and neurological disturbances. Fish swim restlessly, twist around with spasmodic jerks of the body, and often collide with the walls of the tank (Bräunlich and Zintzen, 1976). The slightest disturbances trigger fright reactions that may result in sudden death. The surface of the body and the fins are discolored and the liver is pale.

Chinook salmon reared on a thiamin-inadequate diet exhibited poor appetite, muscle atrophy, instability, and loss of equilibrium, followed by convulsions and death. Freshwater eels developed deficiency after consuming thiaminase-containing clams (Hashimoto *et al.*, 1970). The main cause of thiamin deficiency in commercial fish production is dietary provision of raw fish.

6. Humans

Classic beriberi has been known since the earliest recorded times in the Far East, where it is endemic because of prevalent consumption of polished rice. Beriberi was the primary health problem in Indonesia, Malaysia, Japan, and the Philippines even as recently as the 1940s. In 1947 the Philippine mortality rate from beriberi was 132 per 100,000 population, being second only to tuberculosis. Through rice enrichment with thiamin, this mortality was reduced to 14 out of 100,000 by 1960.

Beriberi is a state in which usually both cardiac and nervous functions are disturbed. Beriberi occurs in a wet form characterized by edema (Fig. 6.8) and cardiovascular symptoms and a dry form characterized by peripheral neuritis, paralysis, and muscular atrophy, but the forms merge into one another so it is hard to differentiate. The chronic form may last for years; cardiac symptoms may appear suddenly (shoshin beriberi) and result in death in a short time. It is generally believed that the more serious the nervous lesions, the greater are the muscular pain and weakness and the less likely is the development of acute beriberi. The heart is saved from extreme insufficiency by the patient being forced into complete rest at an early stage in the attack.

General symptoms of beriberi include anorexia, heart enlargement, tachycardia (as contrasted to bradycardia in animals), lassitude and muscle weakness, paresthesia, loss of knee and ankle reflex with subsequent foot and wrist drop, ataxia due to muscle weakness, and dyspnea on exertion (Gubler, 1984). The "squat test" illustrates the neurological damage, as a beriberi patient is unable to rise from a squatting position.

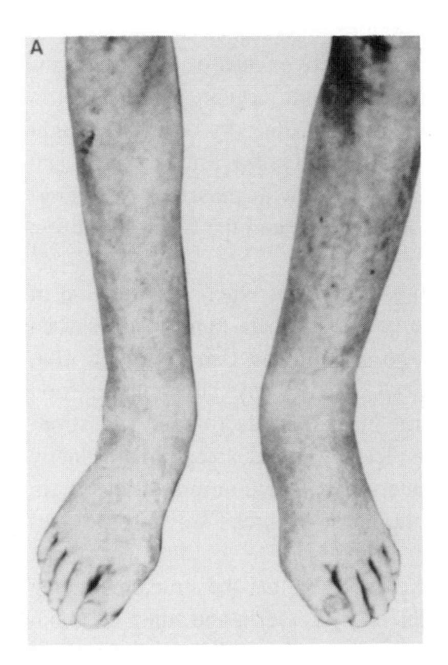

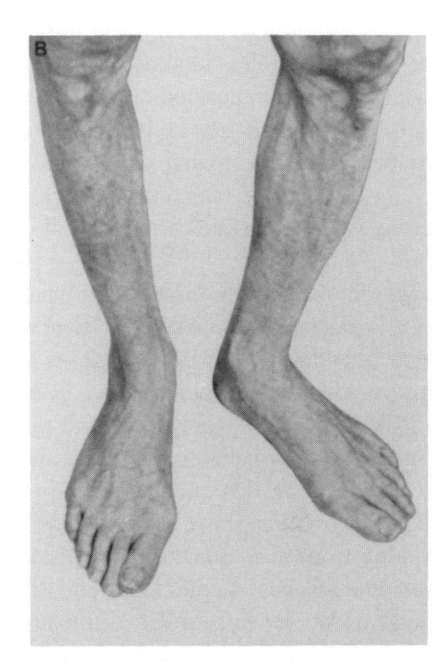

Fig. 6.8 Thiamin deficiency in a man. (A) Swelling of the legs with pitting in ankle region marks beginning of so-called wet beriberi. (B) Same patient is shown 4 days after a single intravenous injection of 50 mg thiamin. (During this period, the patients's excretion of fluid exceeded intake by 10½ lbs. The general nutritional state improved as a result of increased appetite. (Courtesy of Alan T. Forrester, "Vitamin Manual," The Upjohn Company, Kalamazoo, Mich.)

The signs and symptoms vary depending on age, individual, diet, duration and severity of the deficiency, and abruptness of onset. Peripheral neuropathy (dry beriberi) is a symmetrical impairment or loss of sensory, motor, and reflex function affecting the distal segments of limbs more severely than the proximal ones, with less cardiac involvement. Disturbances of the higher nervous centers, nystagmus, and ophthalmoplegia are frequent. In advanced stages there are general muscular atrophy, ataxia, mental confusion, and defective short-term memory.

The sequence of symptoms is variable but beriberi often begins with numbness in feet, heaviness of legs, prickly sensations, and itching. Muscles become tender and there is pain on squeezing the calf of the leg. Cardiac symptoms develop at some stage and are characterized by palpitation, epigastric pain, coldness of extremities, and enlargement of heart. Later symptoms and clinical signs include edema of ankles, puffiness of face, digestive disturbances, anorexia, nausea, and vomiting.

Infantile beriberi, common in the Orient, is an acute form. It is probably an important cause of the high infant mortality in Southeast Asia. The nursing mother, who is providing milk deficient in thiamin, may or may not show mild signs of deficiency. The condition usually occurs in the first few months of life and begins with anorexia, regurgitation, abdominal distention, and colicky pain. Oliguria is followed by edema, and there is dyspnea, with a peculiar cry or grunt thought to be caused by edema of the vocal cords. Later, cardiovascular signs and congestive failure increase and nervous signs appear, with muscular twitching, coma, and death. Each phase may last only a few hours and the whole condition only a day or two.

Beriberi has been largely eliminated in most countries where fortification of rice is practiced, but is still prevalent in those countries where unfortified polished rice is widely used. In countries where large amounts of fish are eaten raw, human thiamin deficiency may occur. Elsewhere, poverty, alcoholism, food faddism, or poorly prepared food may result in inadequate thiamin ingestion. Studies suggest that some degree of thiamin deficiency is widespread in the elderly (Marks, 1975). The most frequently encountered type of thiamin deficiency in developed countries is associated with alcoholism (Wernicke's syndrome), resulting from inadequate intakes of thiamin.

Some Japanese studies (Zintzen, 1974) have indicated that the structure-altering antagonists (see Section III) of thiamin produced by bacteria and fungi are main causes of beriberi, involving as many as 70% of all cases. Bacillary and clostridial species that produce these antagonists have been isolated from intestinal flora of Japanese people under test, or they may be consumed with contaminated food, such as moldy rice.

B. Assessment of Status

In thiamin deficiency determination, there is no question about the specificity of such signs as polyneuritis and Chastek paralysis, but these are terminal stages. The problem of identifying the early stages by any specific sign is an unsolved one for thiamin and for several other of the B vitamins. When a deficiency of thiamin is acute, moreover, the specific effects or the deficiency cannot be distinguished from some unspecific effects such as those of anorexia (Zintzen, 1974). For humans, thiamin deficiency can be tentatively diagnosed on a clinical basis, particularly on careful inquiry into the patient's dietary habits or other conditions that might predispose to a deficiency, and confirmed by positive results of thiamin supplementation.

Some specific criteria that are suitable for the assessment of the thiamin status and that are also susceptible to reasonably simple techniques of measurement are the metabolic dysfunctions that result from a deficiency of the vitamin. Thiamin status criteria include (1) concentration of thiamin in blood and urine, (2) levels of products of intermediate metabolism that are dependent on the enzyme function of thiamin, and (3) activities of enzymes in which thiamin is a cofactor.

Thiamin concentrations are decreased in blood and urine with a deficiency of this vitamin. Urinary excretion of thiamin reflects thiamin saturation or depletion because the vitamin is excreted promptly when ingested in excess of needs. Both blood levels and urinary excretion of thiamin are really only a reflection of the immediately preceding intake and other factors and may not be a reliable index of tissue stores, distribution, or actual biochemical functioning and hence are of limited value for interpretation of actual thiamin status.

Thiamin is needed for pyruvate metabolism and with a deficiency, abnormally high concentrations of pyruvic and lactic acids in the blood indicate thiamin inadequacy. For a calf on a deficient diet, urinary excretion of thiamin drops to very low levels in 20–25 days, and increased pyruvate excretion follows. Blood pyruvate and lactate levels increase suddenly to 400 and 500% of normal as the deficiency develops (NRC, 1978a). However, the measurements of these levels often cannot be used to detect mild deficiencies of thiamin. Also these tests are not entirely specific, because toxicity by minerals such as arsenic and antimony will also inhibit the utilization of pyruvate. Increased pyruvate can also result from a number of pathological conditions such as those arising from increased adrenal gland activity. Acute polyneuritis has been shown to develop before rise of blood pyruvate in pigeons.

The best criterion, so far, for determining thiamin status is activity of enzymes that depend on the vitamin as a coenzyme. These enzymes include pyruvate decarboxylase, α-ketoglutarate decarboxylase, and transketolase. Transketolase activity measurement is the most convenient, feasible, specific, and sensitive of tests of thiamin deficiency. Brin (1962) was able to show that blood (particularly the red cell) transketolase activity is a reliable index of the availability of coenzyme TPP, and thus correlated well with the degree of deficiency in both animals and humans. Transketolase is an excellent indicator in that it is useful in detecting a marginal thiamin deficiency.

X. SUPPLEMENTATION

Thiamin is found in most feedstuffs, but in widely differing concentrations. Good thiamin sources are cereals, milling by-products, oil extraction residues, and yeast (see Section VIII). The thiamin content of most common feeds should be three to four times greater than requirements for most species (Brent, 1985). For swine and poultry consuming typical diets (e.g., corn–soybean meal), thiamin is one of the vitamins least likely to be deficient. Likewise, under normal feeding and management conditions, and in the absence of antimetabolites, thiamin deficiency should theoretically not occur in either young or adult ruminants.

Nevertheless, utlization of available thiamin in feedstuffs may be limited and may also be impaired by thiamin antagonists, therefore it is common practice to add supplemental thiamin to poultry and pig feeds principally as a low-cost insurance. Thiamin supplementation should normally not be considered for grazing ruminants but rather for animals that have a potential to develop PEM as a result of consuming high-concentrate diets.

Thiamin supplementation should be greatly modified if diets contain anti-thiamin substances, such as thiaminases from fish or moldy feeds. As an example, in free-range farming, pigs may occasionally suffer from bracken poisoning, as the roots contain anti-thiamin substances. The animals can be saved by timely thiamin injections.

For mink thiamin requirement is estimated to be 1.2 mg per kilogram of dry feed. In Scandinavia, however, owing to the risk of destruction of thiamin when mink are fed on fish, much higher dosages (up to 6 mg/kg) are recommended (Bräunlich and Zintzen, 1976). Other fish eaters, for example, seals, may also develop thiamin deficiency (Geraci, 1974). It has been calculated and confirmed by determinations of blood transketolase activity that a seal weighing 80 kg and consuming 4–6 kg fish per day has a daily thiamin requirement of 100–150 mg.

Intentionally added nonnutrient substances to diets are sometimes of concern such as the coccidiostat amprolium, a thiamin antimetabolite. A mild thiamin deficiency from amprolium added to a standard commercial hen feed caused a reduction in the feed intake and in egg-laying performance and an increase in the mortality of embryos and chicks. These phenomena could be prevented or effectively counteracted by high thiamin doses in the feed. At recommended levels, apparently, amprolium does not interfere with the thiamin metabolism of the chicken (Scott *et al.*, 1982). Sometimes other thiamin antagonists, such as the free bisulfite present in certain menadione sulfite forms, are added to the feed.

Animals with clinical signs of thiamin deficiency and/or other indicators of thiamin insufficiency (i.e., transketolase activity) should be provided thiamin at therapeutic doses. Since thiamin deficiency causes anorexia, injection of the vitamin is preferred to oral doses with a severe deficiency. Clinical signs in calves weighing less than 50 kg were prevented with 0.65 mg thiamin–HCl per kilogram of liquid diet fed at 10% of liveweight (65 μg/kg liveweight) (Johnson *et al.*, 1948a). Animals with PEM need to be rapidly provided with supplemental thiamin. Levels of thiamin to be administered intravenously or intermuscularly for 3 days have been recommended for lambs and calves (100–400 mg/day) and for sheep and cattle (500–2000 mg/day) (Zintzen, 1974).

For general maintenance following the treatment of mild cases (or as a prophylactic measure when a herd is at risk), 5–10 mg of thiamin should be added to 1 kg of dry feed. Feeds should be enriched with thiamin in a concentration such that each animal will daily receive 100–500 mg. Likewise, roughage should be added to the daily ration at a level of 1.5 kg for every 100 kg of body weight. For therapeutic purpose, a dosage of 6.6–11 mg/kg body weight repeated every 6 hr for 24 hr has been suggested by Smith (1979) for goats.

The administration of thiamin to PEM animals generally produces rapid results, sometimes in a matter of hours. Where recognition of the disease has been delayed and irreversible necroses have already developed in the brain, treatment with the vitamin may be useless. The prospects of achieving the satisfactory treatment of animals that are already incapable of standing are very limited. Although treatment improves the condition of such animals, relapses and permanent damage are probable since irreversible changes will have occurred in the central nervous system. Without doubt, PEM is the most important disease arising from a deficiency of thiamin in ruminants, but it is also noteworthy that thiamin can be used effectively as a support in the treatment of the two metabolic disorders of rumen acidosis and ketosis. Even though treatment with thiamin can be therapeutically successful, it does not follow that a deficiency of thiamin contributes to the etiology of these two diseases (Zintzen, 1974).

For horses, parenterally administered thiamin in high doses (1000 mg intramuscularly) has a marked sedative effect that is particularly noticeable in nervous and excited horses (Bräunlich and Zintzen, 1976). Intramuscular or intravenous injections of 1000–2000 mg thiamin also have a digitalis-like effect on the heart. Thiamin is also used in doses of 200–300 mg for the alleviation of muscle cramps (myoglobinuria, melanuric colic, spastic constipation).

Thiamin sources available for addition to feed are the hydrochloride and mononitrate forms. Because of its lower solubility in water, the mononitrate is preferred for addition to premixes. The mononitrate form has somewhat better stability characteristics in dry products than the hydrochloride (Bauernfeind, 1969).

Stability of thiamin in feed premixes can be a problem. More than 50% of the thiamin was destroyed in premixes after 1 month at room temperature (Verbeeck, 1975). When thiamin was observed in premixes without minerals, no losses were encountered when kept at room temperature for 6 months. When the minerals were supplied as sulfates, the losses of thiamin were also tremendously lowered.

For humans in developed countries, for many years a number of foods have been fortified with thiamin (e.g., bread enrichment). Rice enrichment in the Far Eastern countries has dramatically reduced incidence of beriberi. It has even been suggested that addition of thiamin to alcoholic beverages might help to prevent the serious effects of thiamin deficiency in heavy consumers as well as in alcoholics. However, Crane and Price (1983) conclude that the addition of thiamin to beer in the form of alkyl disulfides is not a practical or effective method of enrichment and offers no special advantage.

XI. TOXICITY

Thiamin in large amounts orally is not toxic, and usually the same is true of parenteral doses. Dietary intakes of thiamin up to 1000 times the requirement are apparently safe for most animal species (NRC, 1987). Intolerance to thiamin is relatively rare in humans; daily doses of 500 mg have been administered for as long as a month with no ill effects. For intravenous injection, the lethal doses were 125, 250, 300, and 350 mg/kg for mice, rats, rabbits, and dogs, respectively (Gubler, 1984). Vasodilation, fall in blood pressure, bradycardia, respiratory arrhythmia, and depression result when animals are given thiamin in large doses intravenously. In humans, several reports of toxic reactions occurred following repeated parenteral injections of large doses. These were due to an anaphylactic reaction as a result of sensitization to the vitamin.

7

Riboflavin

I. INTRODUCTION

After isolation of thiamin as the "vitamin B" factor that prevented beriberi and polyneuritis, riboflavin was the first growth factor to be characterized from the remaining B-complex vitamins. Riboflavin in the form of flavin mononucleotide (FMN) and flavin adenine dinucleotide (FAD) functions as a coenzyme in diverse enzymatic reactions. The vitamin is required in the metabolism of all plants and animals and every plant and animal cell contains the vitamin.

Because of microbial ruminal synthesis, adult ruminants apparently do not require dietary riboflavin. Young ruminants, prior to development of the rumen, and other species require dietary sources of riboflavin because of their very limited synthesis by intestinal flora. Riboflavin is one of the vitamins most likely to be deficient in typical swine and poultry diets based on grains and plant protein supplements (i.e., soybean meal). Likewise, human diets low in animal protein products (especially milk and egg products) and leafy vegetables are likely to be deficient in riboflavin.

II. HISTORY

The historical aspects of riboflavin have been reviewed by Wagner-Jauregg (1977), Scott *et al.* (1982) and Loosli (1988). In 1915, it was known that a water-soluble factor or factors promoted growth and prevented beriberi in rats. In 1920, it was found that heating feedstuffs (e.g., yeast) destroyed the beriberi preventive effect more readily than the growth-promoting effect. Therefore, it was determined that this water-soluble B vitamin contained two essential factors, one of which was more stable to heat than the other. The less stable factor was labeled vitamin F (B_1 or thiamin) and the heat-stable factor vitamin G (B_2 or riboflavin).

The biological importance of certain yellow pigments became apparent in 1932 when Warburg and Christain (Germany) isolated an oxidative enzyme ("Old

Yellow Enzyme'') from yeast that was yellow with green fluorescence. They were able to split it into a protein and a nonprotein (pigment) fraction. This was the first identification of a prosthetic or activating group of an enzyme. Thus, riboflavin was found in a coenzyme before it was discovered in free form.

In 1933, Kuhn (Germany) isolated a yellow pigment from egg white that showed green influorescence and had oxidative properties. Kuhn suggested that this growth factor for rats be given the name ''flavin.'' Therefore, the terms ''ovoflavin'' from eggs, ''lactoflavin'' from milk, ''hepatoflavin'' from liver, and ''uroflavin'' from urine were in use. Pure crystalline flavin compounds were found to contain ribose, and thus the name riboflavin became popular.

In 1935, riboflavin was synthesized by two groups, Kuhn's in Germany and Karrer's in Switzerland. The same year, Szent-Györgi proved that the biological activity of the synthetic form was the same as that of the natural vitamin.

III. CHEMICAL STRUCTURE, PROPERTIES, AND ANTAGONISTS

Riboflavin consists of a dimethylisoalloxazine nucleus combined with the alcohol of ribose as a side chain. Riboflavin exists in three forms, as the free riboflavin and as the coenzyme derivatives FMN (riboflavin 5-phosphate) and FAD (Fig. 7.1). The coenzyme derivatives are synthesized sequentially from riboflavin. In the first step, catalyzed by flavokinase, riboflavin reacts with adenosine triphosphate (ATP) to form FMN, then FMN combines with a second molecule of ATP to form FAD in a reaction catalyzed by the enzyme FAD pyrophosphorylase.

Riboflavin is an odorless, bitter, orange-yellow compound that melts at about 280°C. Its empirical formula is $C_{17}H_{20}N_4O_6$, with an elemental analysis of carbon 54.25%, hydrogen 5.36%, and nitrogen 14.89%. Riboflavin is only slightly soluble in water but readily soluble in dilute basic or strong acidic solutions. It is quite stable to heat in neutral and acid, but not alkaline solutions; very little is lost in cooking. Aqueous solutions are unstable to visible and ultraviolet light, instability being increased by heat and alkalinity. When dry it is not affected appreciably by light, but in solution it is quickly destroyed.

Loss in milk during pasteurization and exposure to light is 10–20%. Much larger losses (50–70%) can occur if bottled milk is left standing in bright sunlight for more than 2 hr. Poultry mashes left exposed to direct sunlight for several days and frequently stirred are subject to some loss (Maynard *et al.*, 1979). In alkaline solution, light splits off ribityl residue, rapidly forming lumiflavin (7,8,10-trimethylisoalloxazine). In an acid or neutral solution, light decomposition produces lumichrome (7,8-dimethylalloxazine).

Anti-riboflavin compounds may result from chemical changes in either the

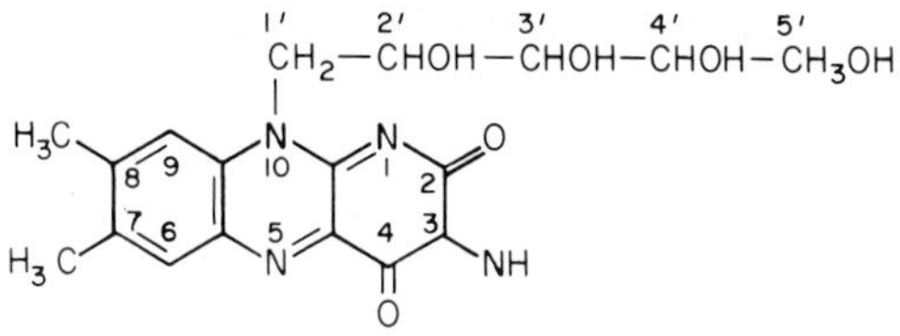

1′ 2′ 3′ 4′ 5′
CH_2—CHOH—CHOH—CHOH—CH_3OH
H_3C N N O
8 9 10 1 2
7 6 5 4 3
H_3C N NH
O

Riboflavin (7,8 dimethyl-10-(D,1′-ribityl)-isoalloxine)

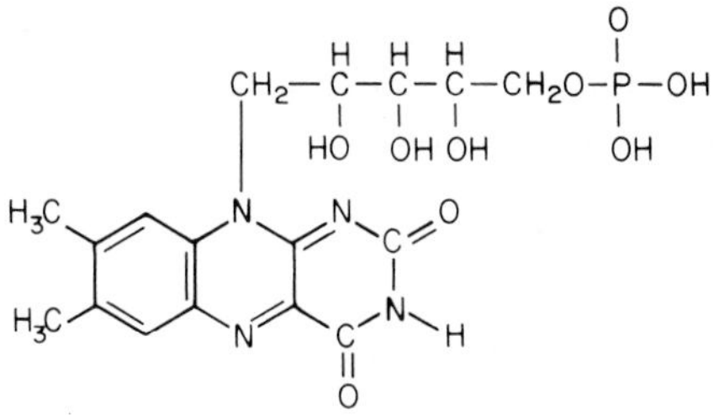

O
H H H
CH_2—C—C—C—CH_2O—P—OH
HO OH OH OH
H_3C N N C=O
H_3C N C N H
O

Flavin mononucleotide (FMN, riboflavin 5-phosphate)

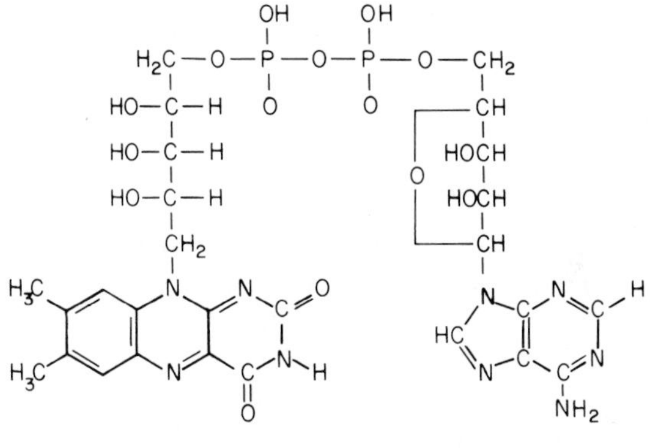

OH OH
H_2C—O—P—O—P—O—CH_2
HO—C—H O O CH
HO—C—H HOCH
HO—C—H O HOCH
CH_2 CH
H_3C N N C=O N C N C H
HC
H_3C N C N H N C C=N
O NH_2

Flavin adenine dinucleotide (FAD)

Fig. 7.1 Structures of riboflavin, flavin mononucleotide (FMN), and flavin adenine dinucleotide (FAD).

isoalloxazine nucleus or in the ribityl side chain. A number of synthetic homologs of riboflavin exist, but none has had experimental or commercial importance. The antagonist D-galactoflavin, 7,8-dimethyl-10-(*d*-1′-dulcityl)isoalloxazine, has been used experimentally in animals and humans to hasten the development of riboflavin deficiency (Cooperman and Lopez, 1984).

IV. ANALYTICAL PROCEDURES

Methods of analyses for riboflavin determination in feeds and biological tissues employ fluorometric or microbiological assays. The first assays measured the biological response in the rat and the chicken, but riboflavin can be assayed more readily by chemical or microbiological methods so animal assays have given way to these. Growth and production of lactic acid by *Lactobacillus casei* are dependent on presence of riboflavin in the medium. Turbidity from the growth of the bacteria can be read in a colorimeter after 16–24 hr of incubation at 37°C.

Riboflavin rarely occurs free in nature, so natural products usually must be treated with acid or enzymes to liberate the riboflavin. If FMN or FAD are to be measured, special nonhydrolytic extraction procedures are required. Unique properties of riboflavin allowing for successful fluorometric analysis are (Scott *et al.*, 1982) (1) riboflavin gives off an intense greenish fluorescence, with intensity proportional to the concentration, (2) it is stable to heat and acid, allowing release from proteins, and (3) it is stable to oxidizing agents (i.e., potassium permanganate) needed to destroy other fluorescing materials. Because riboflavin is readily destroyed by blue or violet light, operations must be done in dim light or amber or red glassware. Also, at a pH of greater than 7.0, dilute solutions may be destroyed by alkali even on clean glassware.

V. METABOLISM

A. Digestion, Absorption, and Transport

Riboflavin is found in feeds as FAD, FMN, and free riboflavin. Riboflavin covalently bound to protein is released by proteolysis digestion. Phosphorylated forms (FAD, FMN) of the riboflavin are hydrolyzed by phosphatases in the upper gastrointestinal tract to free the vitamin for absorption. Free riboflavin enters mucosal cells of the small intestine after apparently being absorbed in all parts of the small intestine. In a comparison between jejunal and ileal cells from guinea pigs, cells from both portions of the small intestine transported riboflavin almost equally (Hegazy and Schwenk, 1983). Cells from deficient animals have a greater maximal absorption uptake of riboflavin (Rose *et al.*, 1986). At low concentrations, riboflavin absorption is an active carrier-mediated process. At high concentrations, however, riboflavin is absorbed by passive diffusion, proportional to concentration.

In mucosal cells, riboflavin is phosphorylated to FMN by the enzyme flavokinase (Cooperman and Lopez, 1984). The FMN then enters the portal system, where it is bound to plasma albumin, and is transported to liver, where it is converted to FAD. A genetically controlled riboflavin binding protein is present in serum and eggs. There is an autosomal recessive disorder in chickens, renal riboflavinuria, in which the riboflavin binding protein is absent. Eggs become riboflavin deficient and embryos generally do not survive beyond the fourteenth day of incubation (Clagett, 1971). Presumably the lack of the specific vitamin transport protein prevents adequate transfer of dietary riboflavin to the developing fetus and riboflavin losses occur via maternal urine. In addition to poultry, specific binding proteins have been found in serum from pregnant cows and rats, human fetal blood, and uterine secretions in the pig. Riboflavin-carrier proteins are induced by estrogens and synthesized in liver (Durgakumari and Adiga, 1986). Rivlin (1984) suggested that there may be physiological mechanisms in pregnancy that facilitate transfer of riboflavin from maternal stores to the fetus in a manner that is fundamentally similar to that in the laying hen.

Hormonal control affects riboflavin metabolism, and specific binding proteins can be greatly increased by estrogen administration (Rivlin, 1984). Thyroid hormones enhance conversion of riboflavin into its coenzyme derivatives and hypothyroidism decreases it. The major site of thyroid hormone control appears to be flavokinase, and physiological doses of thyroxine increase hepatic activity of this enzyme (Rivlin, 1984). Adrenocorticotropic hormone and aldosterone both increase rate of formation of flavin coenzymes from riboflavin.

B. Tissue Distribution, Storage, and Excretion

Animals do not appear to have the ability to store appreciable amounts of riboflavin, with liver, kidney, and heart having the richest concentrations. Liver, the major site of storage, contains about one-third of the total body flavins. A significant amount of free riboflavin exists in retinal tissue, but its function is unclear. Free riboflavin constitutes less than 5% of the stored flavins with 70–90% in the form of FAD. Intakes of riboflavin above current needs are rapidly excreted in urine, primarily as free riboflavin. Minor quantities of absorbed riboflavin are excreted in feces, bile, and sweat.

IV. FUNCTIONS

Riboflavin is required as part of many enzymes essential to utilization of carbohydrates, fat, and protein. FMN and FAD, which contain riboflavin, function as prosthetic groups that combine with specific proteins to form active enzymes, called "flavoproteins." Most flavoproteins contain the FAD form, and a few FMN. Riboflavin in these coenzyme forms acts as an intermediary in the transfer of electrons in biological oxidation–reduction reactions. The enzymes that function aerobically are called oxidases, and those that function anaerobically are called dehydrogenases. The general function is in oxidation of substrate and generation of energy (ATP). By involvement in the hydrogen transport system, flavoproteins function by accepting and passing on hydrogen, undergoing alternate oxidation and reduction.

Flavoproteins may either accept H^+ directly from the substrate (the material being oxidized) or catalyze the oxidation of some other enzyme by accepting hydrogen from it, for example, from the niacin-containing coenzymes, nicotinamide adenine dinucleotide (NAD) and nicotinamide adenine dinucleotide phosphate (NADP). About 40 flavoprotein enzymes may be arbitrarily classified into three groups:

a. $NADH_2$ dehydrogenases—enzymes whose substrate is a reduced pyridine nucleotide and the electron acceptor is either a member of the cytochrome system or some other acceptor besides oxygen.

b. Dehydrogenases—enzymes that accept electrons directly from substrate and can pass them to one of the cytochromes.

c. Oxidases (true)—enzymes that accept electrons from substrate and pass them directly to oxygen; they cannot reduce cytochromes. O_2 is reduced to H_2O_2.

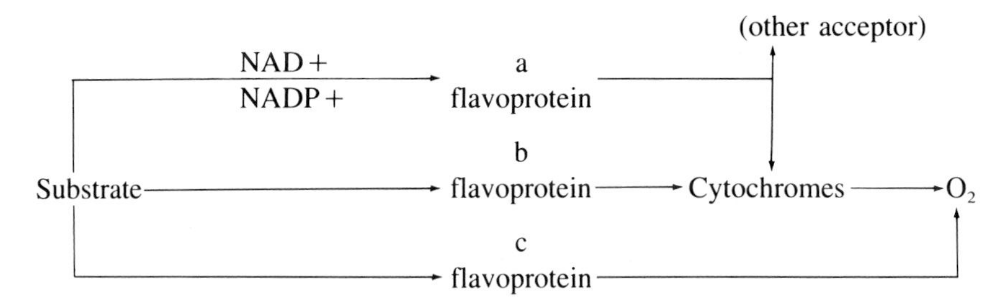

The main function of FMN and FAD in transferring hydrogen between the niacin-containing coenzymes, NAD and NADP, the iron porphyrin cytochromes, and directly from the substrate is illustrated in Fig. 7.2. The sequence of electron acceptors in the early stages of the respiratory chain indicates that coenzyme Q (ubiquinone) acts between flavoprotein and cytochrome *b*. Thus these enzymes are a part of the chain that carries hydrogen from substrates (carbohydrates, amino acids, lipids, etc.) to molecular oxygen, forming water. These flavoprotein pathways are the most important means of electron transport for both mitochondria and microsomes (Scott *et al.*, 1982).

Approximately 40 flavoprotein enzymes participate in electron transfer from metabolites and pyridine nucleotides to molecular oxygen and include:

1. Aerobic dehydrogenases (simple oxidases not containing metals)
   - (a) D and L amino acid oxidases
   - (b) Glycolic acid oxidase
   - (c) Glucose oxidase
   - (d) Warburg's "Old Yellow Enzyme"

1) Succinate dehydrogenase (mitochondrial dehydrogenase)

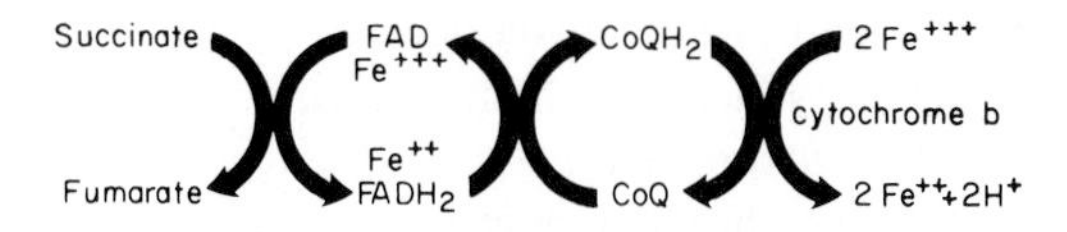

2) Lipoyl dehydrogenase (mitochondrial dehydrogenase)

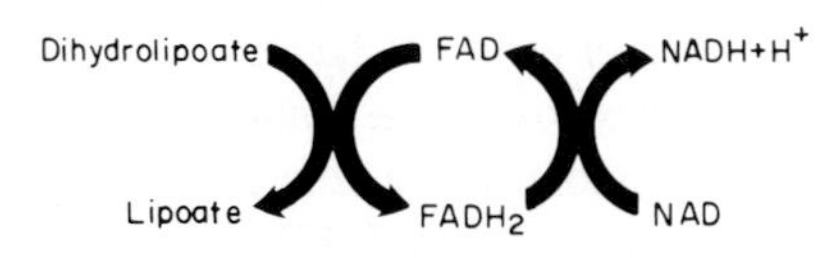

3) NADH dehydrogenase of mitochondria

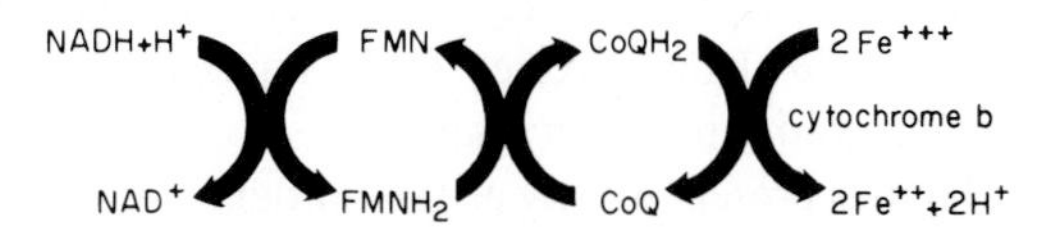

4) NADPH-cytochrome c reductase of microsomes

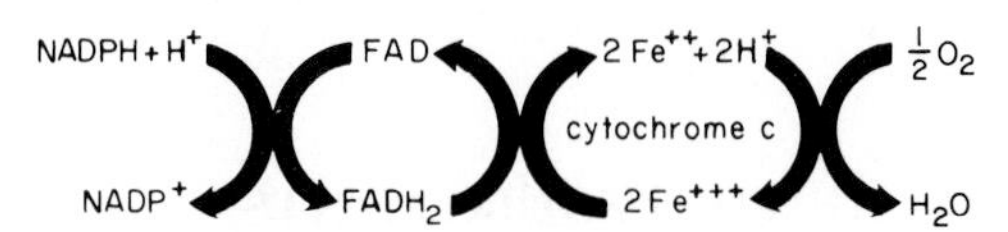

Fig. 7.2 Examples of hydrogen transfer involving flavoproteins. (Modified from Scott *et al.*, 1982.)

2. Oxidases (containing copper, iron, or molybdenum)
   (a) Cuproflavoprotein in butyryl-CoA-dehydrogenase
   (b) Xanthine oxidase (flavoprotein containing iron and molybdenum)
   (c) Molybdoflavoprotein in aldehyde oxidase
   (d) DPNH-cytochrome reductase (iron-containing flavoprotein)
3. Anaerobic dehydrogenases
   (a) Lipoyl dehydrogenase
   (b) Acyl-CoA dehydrogenases and electron-transferring flavoprotein
   (c) Succinic dehydrogenase–fumaric reductase

(d) Others

Choline dehydrogenase
α-Glycerophosphate dehydrogenase
L-Galactone-γ-lactone dehydrogenase
L-Lactate dehydrogenase
D-Lactate cytochrome reductase
Pyridine nucleotide–cytochrome *c* reductases

Riboflavin functions in flavoprotein–enzyme systems to help regulate cellular metabolism, although they are also specifically involved in metabolism of carbohydrates. Riboflavin is also an essential factor in amino acid metabolism as part of amino acid oxidases. These oxidize α amino acids to corresponding amino acids that decompose to give ammonia and a keto acid. There are distinct enzymes oxidizing D amino acids (prosthetic group FAD) and L amino acids (prosthetic group FMN). In addition, riboflavin plays a role in fat metabolism (Copperman and Lopez, 1984) and an FAD flavoprotein is an important link in fatty acid oxidation. This includes the acyl-coenzyme A dehydrogenases, which are necessary for the stepwise degradation of fatty acids. An FMN flavoprotein is required for synthesis of fatty acids from acetate. Thus, flavoproteins are necessary for both degradation and synthesis of fatty acids.

Although riboflavin is present mostly as flavoprotein enzymes FAD and FMN, the retina of the eyes contains free riboflavin in relatively large amounts. The function it fulfills there is still not clear. In avascular tissues such as the cornea it is thought that oxidation takes place by means of a riboflavin-containing enzyme. In deficiency of the vitamin the body attempts oxygenation by vascularization (Marks, 1975).

VII. REQUIREMENTS

Riboflavin requirements vary with heredity, growth, environment, age, activity, health, other dietary components, and synthesis by host. Riboflavin requirements for selected animals and humans are presented in Table 7.1. The majority of species have a requirement between 1 and 4 mg/kg of diet (dry basis). Where sufficient data are available, studies indicate that riboflavin requirements decline with animal maturity and increase for reproductive activity. Chicks receiving diets only partially deficient in riboflavin may recover spontaneously, indicating that requirement rapidly decreased with age (Scott *et al.*, 1982).

Increased dietary fat or protein increases requirements for riboflavin in rats and chickens. It was assumed that high urinary riboflavin excretion during periods of negative nitrogen balance for a number of species was a reflection of impaired riboflavin utilization or retention. However, Turkki and Holtzapple (1982) suggested, in studies with rats, that the effect of protein on riboflavin requirement is related to rate of growth and not to protein intake per se.

TABLE 7.1

Riboflavin Requirements for Various Animals and Humans[a]

Animal	Purpose	Requirement	Reference
Beef cattle	Adult	Microbial synthesis	NRC (1984a)
Dairy cattle	Calf	0.65 mg/kg liquid feed	NRC (1978a)
	Adult	Microbial synthesis	NRC (1978a)
Goat	Adult	Microbial synthesis	NRC (1981b)
Chicken	Leghorn, 0–6 weeks	3.6 mg/kg	NRC (1984b)
	Leghorn, 6–20 weeks	1.8 mg/kg	NRC (1984b)
	Leghorn, laying	2.2 mg/kg	NRC (1984b)
	Leghorn, breeding	3.8 mg/kg	NRC (1984b)
	Broilers, 0–8 weeks	3.6 mg/kg	NRC (1984b)
Turkey	All classes	2.5–4.0 mg/kg	NRC (1984b)
Duck (Peking)	All classes	4.0 mg/kg	NRC (1984b)
Japanese quail	All classes	4.0 mg/kg	NRC (1984a)
Sheep	Adult	Microbial synthesis	NRC (1985b)
Swine	Growing–finishing, 1–20 kg	3.0–4.0 mg/kg	NRC (1988)
	Growing–finishing, 20–110 kg	2.0–2.5 mg/kg	NRC (1988)
	Adult	3.75 mg/kg	NRC (1988)
Horse	Adult (maintenance requirement)	2.2 mg/kg	NRC (1978b)
Mink	Growing	1.5 mg/kg	NRC (1982a)
Fox	Growing	1.25–4.0 mg/kg	NRC (1982a)
Cat	Growing	1 mg/kg	NRC (1986)
Dog	Growing	2–4 mg/kg	NRC (1985a)
Fish	Catfish	9 mg/kg	NRC (1983)
	Trout	3 mg/kg	NRC (1981a)
Rat	All classes	2–4 mg/kg	NRC (1978c)
Mouse	All classes	3.7 mg/kg	NRC (1978c)
Human	Infants	0.4–0.6 mg/day	RDA (1980)
	Children	0.8–1.4 mg/day	RDA (1980)
	Adults	1.2–2.2 mg/day	RDA (1980)

[a]Expressed as per unit animal feed (except for calf) either on an as fed (approximately 90% dry matter) or dry basis (see Appendix Table 1). Human data are expressed as mg/day.

Microbial biosynthesis of riboflavin has been shown to occur in the gastrointestinal tract of a number of animal species and thus affect requirements. However, utilization of this endogenously synthesized riboflavin varies from species to species. Within a single species, utilization depends on diet composition and incidence of coprophagy. Carbohydrates such as starch, cellulose, or lactose are absorbed slowly and increase synthesis. Dextrose, fat, or protein as chief dietary constituents decrease intestinal production, thereby increasing dietary riboflavin requirements. Young rats fed a riboflavin-free, purified diet with sucrose as the only carbohydrate will cease to grow. However, when sucrose is replaced by starch, sorbitol, or lactose, growth is comparable to that of rats supplied with riboflavin (Haenel *et al.*, 1959). Antibiotics, such as tetracycline, penicillin, and streptomycin, reduce the requirements of several animal species for riboflavin or might stimulate microorganisms that synthesize riboflavin. They may inhibit microorganisms in the gut that compete for riboflavin.

Because of microbial ruminal synthesis of riboflavin, ruminants have no dietary requirements. Nevertheless, diet composition influences total microbial synthesis. Data of Miller *et al.* (1983) also suggest a greater riboflavin ruminal synthesis with an increased proportion of dietary concentrates. Buziassy and Tribe (1960) measured increased synthesis of riboflavin with diets containing more protein. These authors also noted that ruminal synthesis was reduced with higher dietary intakes of the vitamin.

Temperature extremes apparently have an effect on riboflavin requirement. According to studies with pigs, riboflavin requirement is substantially higher at a low than at a high environmental temperature (Seymour *et al.*, 1968). At environmental temperatures below 11°C, feed required per unit of gain increased, and rate of gain decreased with decreasing temperature. On the contrary, Onwudike and Adegbola (1984) report riboflavin requirements to be higher for chickens in a tropical environment. A dietary level of 4.1 ppm riboflavin was adequate for egg laying with 5.7 ppm for hatchability, compared to 2.2 and 3.8 ppm, respectively, for the NRC (1984b) requirement. When requirement comparisons between animals residing at different environmental temperatures are calculated, total feed intakes must be considered.

Requirements of humans for riboflavin have been determined from experimental studies with adults and infants. Riboflavin allowance for human infants and adults ranges from 0.4 to 2.2 mg/day (RDA, 1980). Levels are to be increased by 0.3 mg during pregnancy, and by 0.5 mg during lactation and possibly should be related to energy expenditure (Roe *et al.*, 1982). A level of 0.5 mg per 1000 kcal was estimated to be the minimum requirement for adults to maintain a normal urinary excretion. On a similar basis, the infant's minimum daily requirement was estimated as 0.6 mg per 1000 kcal.

VIII. NATURAL SOURCES

Riboflavin is synthesized by green plants, yeasts, fungi, and some bacteria. Rapidly growing, green, leafy vegetables and forages, particularly alfalfa, are a good source, with the vitamin richest in the leaves. Cereals and their by-products have a rather low content, in contrast to their supply of thiamin. Oil meals are fair sources. While grains and protein meals contain some riboflavin, they should not be relied on as the sole source of riboflavin. Riboflavin concentrates obtained from whey and distiller's solubles are important commercial sources, particularly for animal feeds. Although fruits and vegetables are moderately good sources of this vitamin, they are not consumed in sufficient quantities to meet daily requirements (Cooperman and Lopez, 1984). Riboflavin concentration in various foods and foodstuffs is shown in Table 7.2.

For human diets, milk, eggs, liver, heart, kidney, and muscle meat are rich sources. Riboflavin content of milk (i.e., cow or goat) is many times higher than content in their feed because of rumen synthesis. Human milk contains about 0.5 mg riboflavin/liter, while cow's milk is three times as high (i.e., 1.7 mg/liter). For humans in the United States it has been estimated that milk and milk products contribute almost one-half of dietary riboflavin in the diet, with meat, eggs, and legumes contributing about 25%. Fruits, vegetables, and cereal grains contribute about 10% each (Hunt, 1975).

TABLE 7.2

Riboflavin in Foods and Feedstuffs (mg/kg, dry basis)[a]

Food/Feedstuff	mg/kg
Alfalfa hay, sun cured	13.4
Alfalfa leaves, sun cured	23.1
Barley, grain	1.8
Bean, navy (seed)	2.0
Blood meal	2.2
Brewer's grains	1.6
Buttermilk (cattle)	33.1
Chicken, broilers (whole)	15.6
Citrus pulp	2.7
Clover hay, sun cured	17.2
Copra meal (coconut)	3.7
Corn, gluten meal	1.8
Corn, yellow grain	1.4
Cottonseed meal, solvent extracted	5.3
Eggs	3.0
Fish meal, anchovy	8.2
Fish meal, menhaden	5.2
Fish, sardine	5.8
Linseed meal, solvent extracted	3.2
Liver, cattle	92.2
Milk, skimmed, cattle	20.5
Molasses, sugarcane	3.8
Oat, grain	1.7
Peanut meal, solvent extracted	9.8
Rice, bran	2.8
Rice, grain	1.2
Rice, polished	0.6
Rye, grain	1.9
Sorghum, grain	1.4
Soybean meal, solvent extracted	3.2
Soybean seed	3.1
Spleen, cattle	15.3
Timothy hay, sun cured	10.1
Wheat, bran	4.6
Wheat, grain	1.6
Yeast, brewer's	38.1
Yeast, torula	47.6

[a]NRC (1982b).

Riboflavin is one of the more stable vitamins, but can be readily destroyed by ultraviolet rays or sunlight. Appreciable amounts may be lost upon exposure to light; up to one-half the riboflavin content is lost in cooking eggs and pork chops in light and most of the vitamin in milk stored in clear glass bottles. However, processes like pasteurization, evaporation, or condensation have little effect on riboflavin content of milk. Sun drying of fruits and vegetables is likely to lead to substantial losses of vitamin activity. The practice of adding sodium bicarbonate to make vegetables appear fresher accelerates the photodegradation of riboflavin (Rivlin, 1984). Riboflavin in meat is relatively stable during cooking, canning, and dehydration. Milling of rice and wheat results in considerable loss of riboflavin, since most of the vitamin is in the germ and bran, which are removed during this process. About one-half the riboflavin content is lost when rice is milled and whole wheat flour contains about two-thirds more riboflavin than white flour (Cooperman and Lopez, 1984).

IX. DEFICIENCY

A. Effects of Deficiency

Animals and humans are unable to synthesize riboflavin within tissues, therefore requirements are met principally by dietary sources with some intestinal microbial synthesis. Since riboflavin plays many essential roles in the release of food energy and the assimilation of nutrients, it is understandable why a deficiency is reflected in a wide variety of signs that are variable with the species. It is not possible to relate signs, however, to specific biochemical roles that have been established (Maynard *et al.*, 1979). A decreased rate of growth and a lowered feed efficiency are common signs in all species affected. Typical clinical signs often involve the eye, skin, and nervous system. Riboflavin deficiency would not be expected in young nursing animals, as milk is a rich source of the vitamin.

1. Ruminants

Riboflavin is not required in the diet of adult ruminants because ruminal microorganisms synthesize this vitamin in adequate amounts. Confirmation of net synthesis in the rumen can be obtained from work such as that of McElroy and Goss (1940), in which secretion of riboflavin in milk was shown to be equivalent to approximately 10 times dietary intake.

Riboflavin deficiencies have been demonstrated in young ruminants whose rumen flora is not yet established. Failure to provide riboflavin results in redness of the buccal mucosa, lesions in the corner of the mouth, loss of hair, and excessive tear and saliva production. Less specific signs are anorexia, diarrhea, and reduced growth.

2. Swine

Typical swine diets based largely on grains are often borderline to deficient in riboflavin. Signs of riboflavin deficiency in the young growing pig include anorexia, slow growth (Fig. 7.3), rough hair coat, dermatitis, alopecia, abnormal stiffness, unsteady gait, scours, ulcerative colitis, inflammation of anal mucosa, vomiting, cataracts, light sensitivity, and eye lens opacities (Cunha, 1977; NRC, 1988).

In riboflavin-deficient swine, reproduction is impaired (Fig. 7.4). Cunha (1977) summarized the clinical signs for gilts fed a riboflavin-deficient diet during reproduction and lactation as follows: (1) erratic or, at times, complete loss of appetite; (2) poor gains; (3) parturition 4–16 days prematurely; (4) one case of death of fetus in advanced stage with resorption in evidence; (5) all pigs either were dead at birth or died within 48 hr thereafter; (6) enlarged front legs in some pigs, due to gelatinous edema in the connective tissue and generalized edema in many others; and (7) two hairless litters. The longer the period on riboflavin-deficient diets, the more severe the deficiency signs became.

3. Poultry

The most critical requirements for riboflavin are those exhibited by the young chick and the breeder hen. The characteristic sign of riboflavin deficiency in the chick is ''curled-toe'' paralysis. It does not develop, however, when there is absolute deficiency, or when the deficiency is very marked, because the chicks die before it appears. Chicks are first noted to be walking on their hocks with their toes curled inward. Deficient chicks do not move about, except when forced to do so, and their toes are curled inward (Fig. 7.5) both when walking and when resting on their hocks (Scott *et al.*, 1982). Legs become paralyzed, but the birds may otherwise appear normal.

Changes in the sciatic nerves produce the curled-toe paralysis in growing chicks. There is a marked enlargement of sciatic and brachial nerve sheaths with sciatic nerves reaching a diameter four to six times normal size. Histological examinations of affected nerves show definite degenerative changes in myelin sheaths, which when severe may pinch the nerve, producing a permanent stimulus that causes the curled-toe paralysis (Scott *et al.*, 1982). When the curled-toe deformity is long-standing, irreparable damage has occurred in the sciatic nerve and administration of riboflavin no longer cures the condition.

Other signs of riboflavin deficiency are retardation of growth (Figs. 7.6 and 7.7), diarrhea after 8–10 days, and high mortality after about 3 weeks. When chicks are fed a diet deficient in riboflavin, their appetite is fairly good but they grow very slowly and become weak and emaciated. There is no apparent impairment of feather growth; on the contrary, main wing feathers often appear to be disproportionately long.

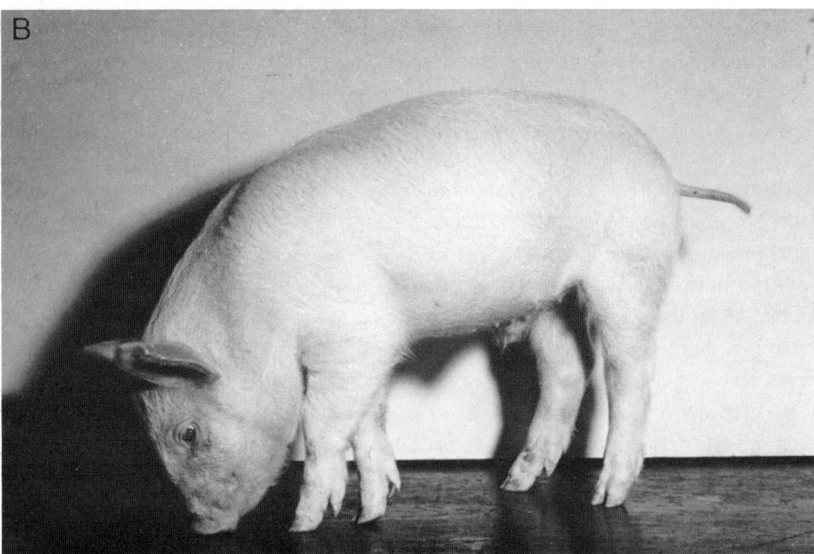

Fig. 7.3 Riboflavin deficiency. The pig in (A) received no dietary riboflavin. Note the rough hair coat, poor growth, and dermatitis. (B) shows pig that received adequate riboflavin. (Courtesy of R. W. Luecke, Michigan Agricultural Experiment Station, East Lansing, Mich., and *J. Nutrition* **52**, 1954, p. 409.)

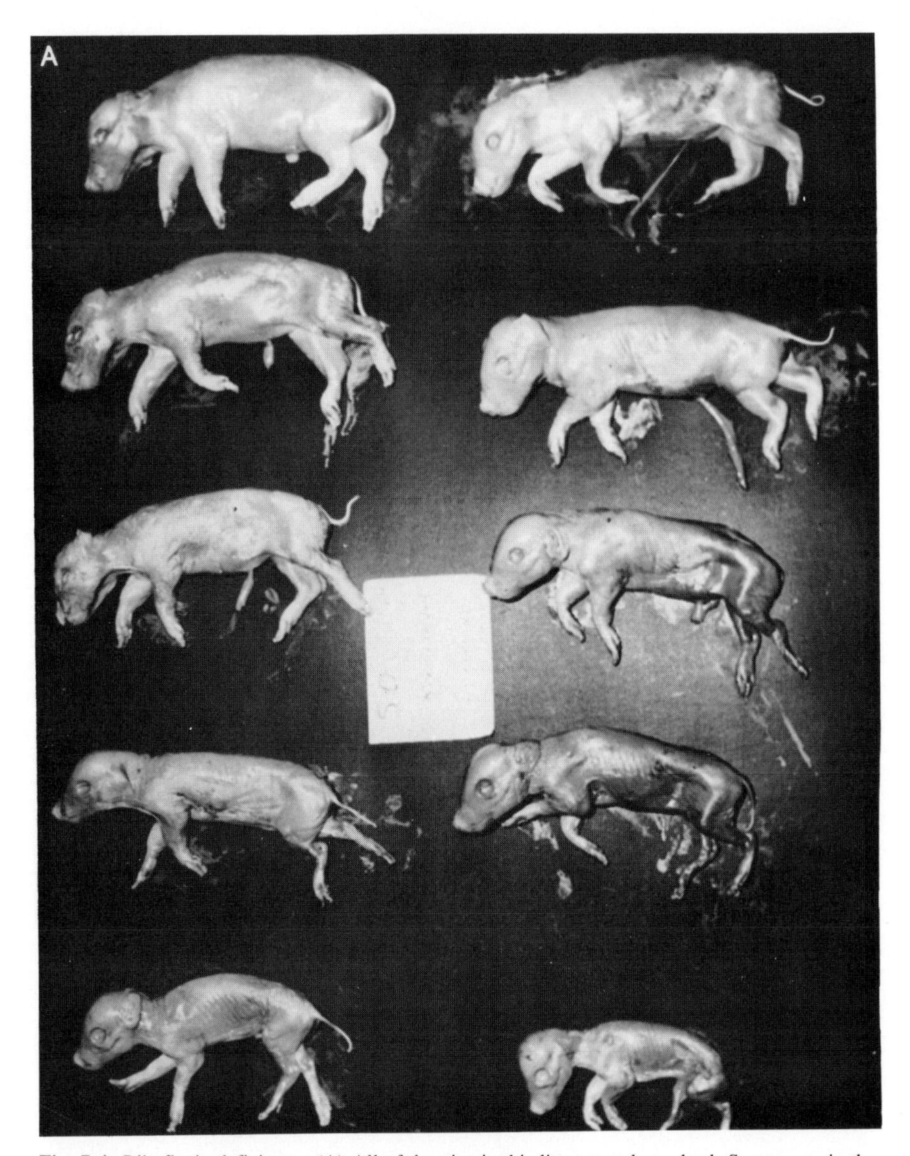

Fig. 7.4 Riboflavin deficiency. (A) All of the pigs in this litter were born dead. Some were in the process of resorption. A few had edema and enlargement of front legs as a result of gelatinous edema. (B) Pigs from a litter in which gelatinous edema was more pronounced. (C) Seven of the ten pigs farrowed were born dead, and the other three were dead within 48 hr. The sow received a riboflavin-deficient diet for a shorter period than the sows farrowing the other two litters. (Courtesy of T. J. Cunha and Washington State University.)

Fig. 7.4 (*Continued*)

Fig. 7.4 (*Continued*)

Signs of riboflavin deficiency in the poult and duckling differ from those in the chick. In the poult, a dermatitis appears in about 8 days, the vent becomes encrusted, inflamed, and excoriated, growth is retarded or completely stopped by about the seventeenth day, and deaths begin to occur about the twenty-first day. In the duckling, there usually are diarrhea and cessation of growth.

In laying poultry, hatchability of incubated eggs is first reduced and subsequently egg production is decreased, roughly in proportion to degree of deficiency. Embryonic mortality has two typical peaks and often a third peak. These are, respectively, on the fourth and twentieth days and on the fourteenth day of incubation. Embryos that fail to hatch from eggs of hens receiving low-riboflavin diets are dwarfed and exhibit pronounced micromelia; some embryos are edematous. The down fails to emerge properly, thus resulting in a typical abnormality termed ''clubbed'' down, which is most common in neck areas and around the vent. The nervous systems of these embryos show degenerative changes much like those described in riboflavin-deficient chicks.

Chicks fed a diet only marginally deficient in riboflavin often recover spon-

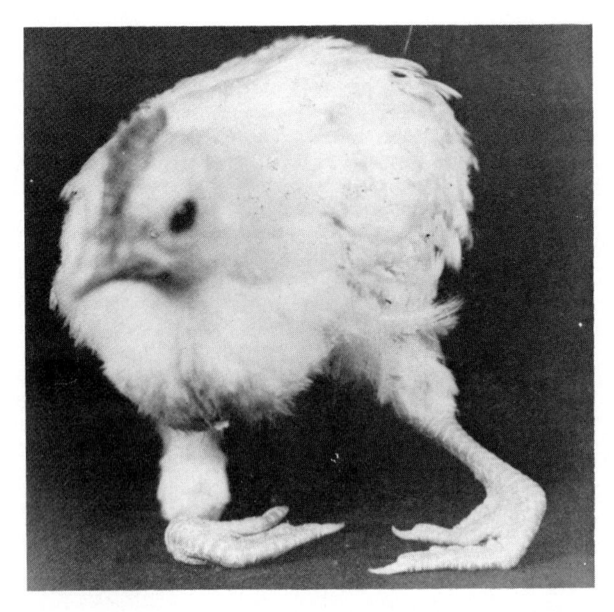

Fig. 7.5 Curled-toe paralysis in a riboflavin-deficient chick. (Courtesy of M. L. Scott, Cornell University.)

taneously. The condition is curable in the early stages, but in its acute stage it is irreversible (NRC, 1984b). There is increasing evidence that vigor and livability of the baby chick are directly tied to amount of riboflavin in the hen's diet (Anonymous, 1969). Ruiz (1987) observed that riboflavin deficiency is more severe in modern strains of chicks and poults than in those used 30–40 years ago.

4. Horses

It is generally felt that riboflavin synthesis in the cecum and colon provides some of the horse's requirement. Horses fed low-riboflavin diets have demonstrated anorexia, sporadic but severe weight loss, general weakness, and poor growth (Pearson *et al.*, 1944). In early studies, riboflavin deficiencies were thought to be a cause of periodic ophthalmia, but other conditions have also been implicated (NRC, 1978b).

5. Other Animal Species

a. Dogs and Cats. Riboflavin-deficient dogs exhibit low growth rates, anemia, and corneal lesions (NRC, 1985a). Cats deficient in the vitamin develop

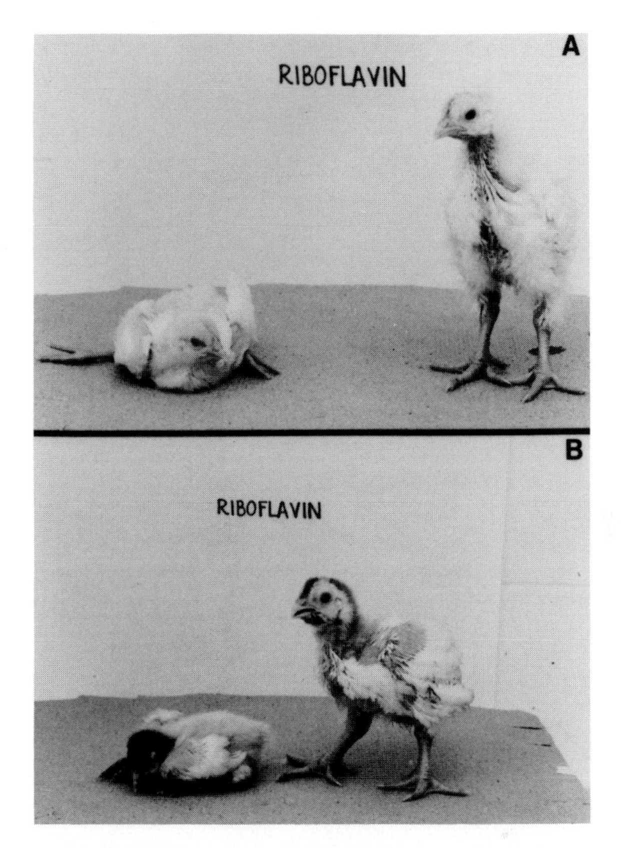

Fig. 7.6 Riboflavin deficiency in the chick. (A) The bird on the left was fed a corn–soybean meal diet without supplemental riboflavin. This chick exhibited the predominant type of paralysis observed at the zero level of riboflavin supplementation. Both birds were female chicks. (B) Same as above but male chicks. (Courtesy of N. Ruiz and R. Harms, University of Florida.)

cataracts, fatty livers, testicular hypoplasia, and alopecia with epidermal atrophy (NRC, 1986).

b. Laboratory Animals. Classical signs of riboflavin deficiency in rats are dermatitis, alopecia, weakness, and decreased growth. Corneal vascularization and ulceration, cataract formation, anemia, myelin degeneration of sciatic nerves and spinal cord, fatty liver, and metabolic abnormalities of hepatocytes may occur (Fig. 7.8) (NRC, 1978c). Riboflavin deficiency in rats brings about an acceleration of the erythrocyte life cycle (Gaetani and D'Aquino, 1987). For the mouse, riboflavin deficiency signs have included poor performance, myelin de-

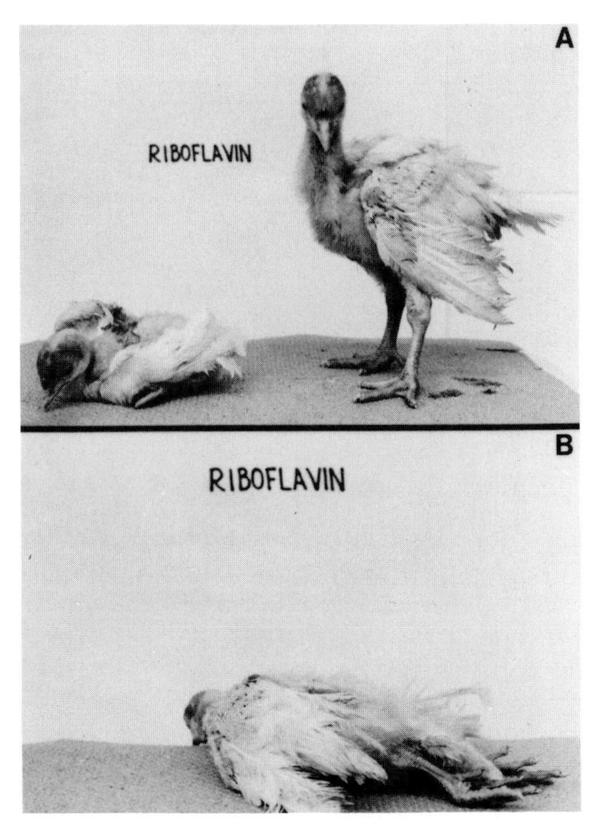

Fig. 7.7 Riboflavin deficiency in the turkey at 21 days of age. (A) The bird on the left was fed a corn–soybean basal diet without supplemental riboflavin. (B) Severe leg paralysis and poor feathering in a turkey poult fed the riboflavin-deficient diet. (Courtesy of N. Ruiz and R. Harms, University of Florida.)

generation in the spinal cord, corneal vascularization with ulceration, and lowered resistance to *Salmonella* infection (NRC, 1978c). Riboflavin-deficient guinea pigs exhibited poor growth, rough hair coats, pale feet, nose, and ears, early death, corneal vascularization, skin atrophy, and myelin degeneration of the spinal cord (NRC, 1978c).

Fig. 7.8 Riboflavin deficiency in the rat. Generalized dermatitis and growth failure in riboflavin-deficient rat. There was marked keratitis of the cornea (A). After 1 month of treatment with riboflavin, the animal showed improvement (B). Growth resumed and ocular and skin lesions practically disappeared. After 2 months of treatment, the rat showed no signs of the original deficiency (C). (Courtesy of Alan T. Forrester, "Scope Manual on Nutrition," The Upjohn Company, Kalamazoo, Mich.)

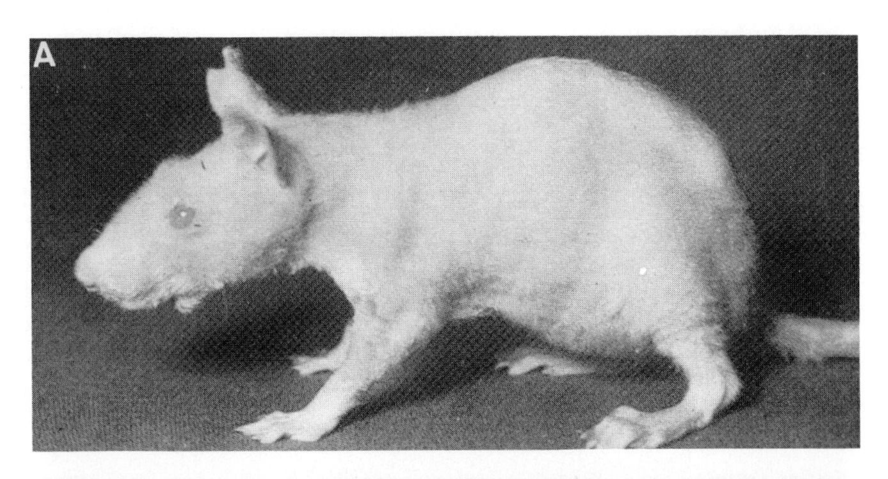
A

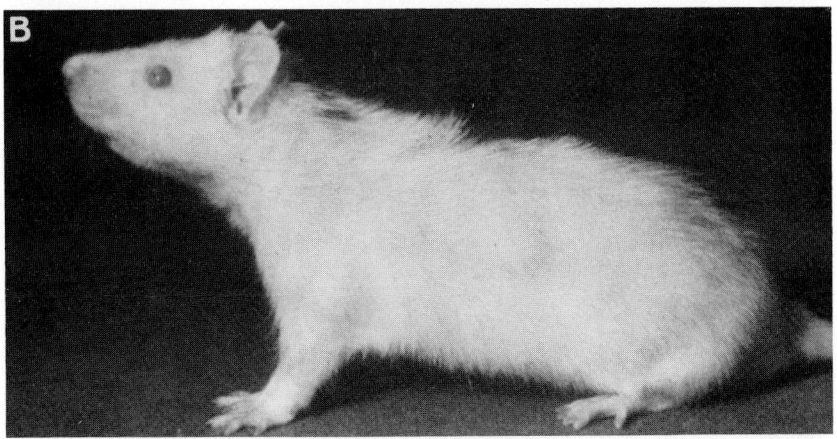
B

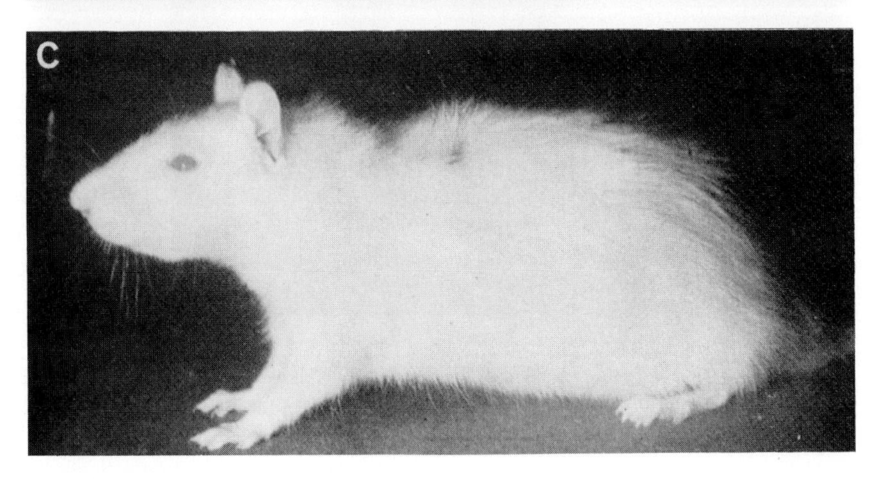
C

c. Nonhuman Primates. For the rhesus monkey, signs of riboflavin deficiency include a ''freckled'' dermatitis, incoordination, faulty grasping reflex, impaired vision, scanty hair coat, and hypochromic, normocytic anemia. Deficiency in cebus monkeys results in weight loss, dermatitis, alopecia, ataxia, and sudden death, but no anemia (NRC, 1978d). Deficiency in the baboon is characterized by apathy, dermatitis, anemia, gingivitis, diarrhea, and adrenal cortical hemorrhage.

d. Foxes and Mink. Riboflavin deficiency in foxes results in a decreased growth rate, signs of muscular weakness, chronic spasms, coma, opacity of the cornea, and a decrease in pigment production in the fur (Fig. 7.9). Anorexia, loss of weight, extreme weakness, and poor breeding results are seen in mink placed on riboflavin-free diets (NRC, 1982a).

Fig. 7.9 Riboflavin dificiency. Right: After 7 weeks on a diet deficient in riboflavin, 12-week-old blue fox showed depigmentation, shedding of fur, and dermatitis. Left: Litter mate was fed the same diet supplemented with riboflavin. (Courtesy of H. Rimeslatten, Agriculture College of Norway, Vollebekk, Norway.)

e. Fish. Channel catfish fed riboflavin-deficient diets developed deficiency signs including anorexia, poor growth, short-body dwarfism, and cataracts (NRC, 1983). Common carp exhibited nervousness, photophobia, and hemorrhages, with the initial deficiency signs of anorexia and nervousness. Signs of deficiency in Japanese eels were poor growth, anorexia, hemorrhage of fins and abdomen, photophobia, and lethargy (NRC, 1983). Signs of riboflavin deficiency in trout and salmon include slow growth, anorexia, inefficient conversion of feed, opaque lens and cornea of the eye, and dark pigmentation of the body (NRC, 1981a).

6. Humans

Riboflavin deficiency signs have been observed in humans consuming nutritionally poor diets and under experimental conditions. As in animals, riboflavin deficiency is believed to retard growth in humans, although no growth experiments have been undertaken. Clinically, riboflavin deficiency in humans is usually observed in conjunction with deficiencies of other B vitamins. The reason for the multiplicity of vitamin deficits is that dietary deficiencies tend to be multiple because the food sources are similar. More recent studies have involved volunteers fed a purified diet deficient only in riboflavin, along with the riboflavin antagonist galactoflavin (Cooperman and Lopez, 1984).

Clinical features of riboflavin deficiency include seborrheic dermatitis around the nose and mouth, soreness and burning of the lips, mouth, and tongue, photophobia, burning and itching of the eyes, superficial vascularization of the cornea, cheilosis, angular stomatitis, glossitis, anemia, and neuropathy (Fig. 7.10) (Rivlin, 1984).

Clinical signs of glossitis initiate with flattening followed by disappearance of the tongue filiform papilla. The fungiform papillae become enlarged, and the tongue color changes to a beefy red color. The tongue of these subjects is sore, and loss of taste sensation develops (Cooperman and Lopez, 1984). Dermatitis due to riboflavin deficiency begins most often in the nasolabial fold and is scaly and oily in character. Similar lesions may also appear around the eyes and on the ears. Dermatitis of the scrotum or vulva is frequently present (Marks, 1975). Corneal vascularization due to riboflavin deficiency always occurs in the entire circumference of the cornea and is nearly always bilateral. Corneal vascularization may be a result of the oxygen requirement of the corneal epithelium. Cells of the cornea have no hemoglobin supply but maintain integrity by intracellular oxidative processes that depend on riboflavin activity. Tears are a rich source of riboflavin. Foy and Mbaya (1977) stated that in vitamin A deficiency the tear ducts are blocked with keratin and vascularization of the cornea follows from lack of tears. Thus, such patients should be given both vitamin A and riboflavin.

One of the first systemic experimental studies in humans resulted in cheilosis in 10 of 18 women consuming a low-riboflavin diet during a 94- to 130-day period (Sebrell and Butler, 1938). The initiation of cheilosis was pallor of the

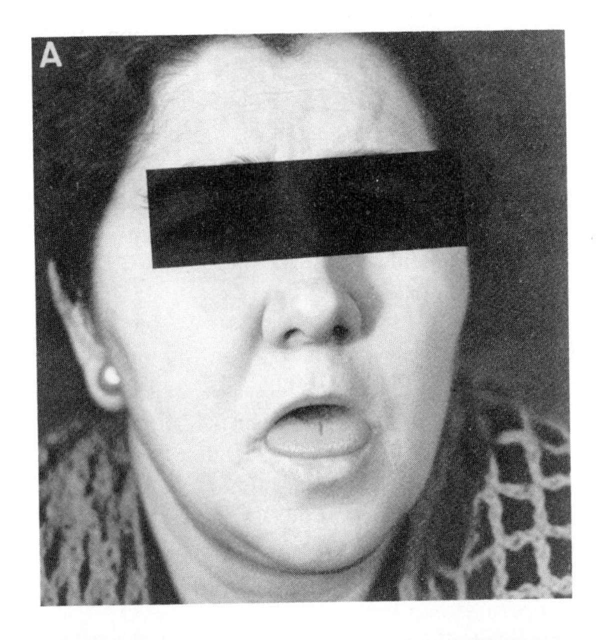

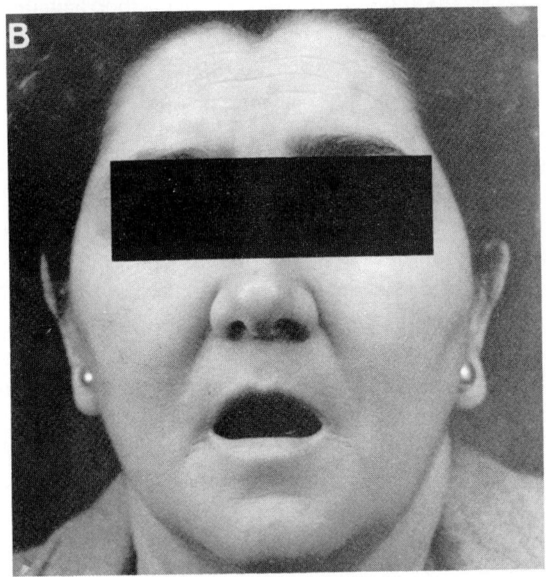

Fig. 7.10 Riboflavin deficiency manifested as fissures at angle of mouth (A). Complete eradication after treatment with riboflavin therapy (B). (Courtesy of Alan T. Forrester, ''Scope Manual on Nutrition,'' The Upjohn Company, Kalamazoo, Mich.)

lips in the angles of the mouth, followed by maceration. Within a few days, superficial transverse fissures appeared in the angles of the mouth.

In developing countries, individuals consuming diets low in green, leafy vegetables and animal products (e.g., especially milk products) are at high risk for riboflavin deficiency. Overt clinical signs of riboflavin deficiency are rarely seen among inhabitants of developed countries. However, the so-called subclinical stage of the deficiency characterized by a change in biochemical indices is common. From Italy, Mobarhan *et al.* (1982) report a marginal deficiency in 18% of people over 65 years of age and 12% of children. The deficiency resulted in a shortening of the life span of erythrocytes. Riboflavin deficiency with biochemical changes, but not necessarily with clinical signs, has been observed in women taking oral contraceptive agents, diabetics, children and adolescents from low socioeconomic backgrounds, children with chronic heart disease, and the aged (Cooperman and Lopez, 1984). In addition to major clinical signs of a deficiency, retarded intellectual development has been observed in children, and even minor degrees of riboflavin deficiency in adults produced marked deterioration of personality (Sterner and Price, 1973).

B. Assessment of Status

Several methods have been used to assess nutritional status of riboflavin. These include clinical signs, blood and urine levels of the vitamin, and measurement of enzymatic coenzyme activity. The most certain way at present of identifying a riboflavin deficiency seems to be to cause a remission of the symptoms in question by feeding riboflavin alone. Measuring riboflavin excretion in the urine is of some value, the limitation being that daily output of riboflavin is usually an estimation of dietary intake. Red blood cell content appears to be a better measure on the basis of human studies. A newer biochemical method has recently been introduced and is based on the change in activity of erythrocyte glutathione reductase (EGR), an FAD-containing enzyme. Experimental studies including those with livestock, laboratory animals, humans, and even fish have shown that EGR activity is a more sensitive and specific indicator of riboflavin deficiency (Frank *et al.*, 1988).

X. SUPPLEMENTATION

Riboflavin is one of the vitamins most likely to be deficient for both nonruminant animals and humans. Before their rumens are developed, young ruminants (up to 2 months of age), if early weaned or dependent on milk replacer, have a dietary need for riboflavin. Cereal grains, though poor sources of riboflavin, are important for people in many developing countries of the world where cereals constitute the major dietary component. Often individuals in developing countries do not have the economic resources to consume the major dietary sources of riboflavin of meat, milk, and dairy products.

Swine and poultry diets based on grains and plant protein sources are likewise often borderline to deficient in riboflavin. Only a few feedstuffs fed to poultry and swine contain enough riboflavin to meet the requirements of growth and reproduction. The trend toward confinement feeding and use of less pasture and/or alfalfa in swine diets has increased need for riboflavin supplementation; both are excellent sources of the vitamin (Cunha, 1977).

Swine and poultry in confinement become more dependent on adequate vitamin (including riboflavin) and trace mineral supplementation as least-cost feed formulation (i.e., using principally corn and soybean meal) limits the number of riboflavin-rich feed ingredients (milk fermentation and fish by-products and dehydrated alfalfa). The greater the variety of feed ingredients, the lower is the chance of vitamin and trace element deficiencies for animals and humans alike.

Riboflavin is commercially available as a crystalline compound produced by chemical synthesis or fermentation. Most of the commercially available riboflavin is made by bacterial synthesis, a cheap and convenient way to produce crystalline riboflavin. It is available to the feed, food, and pharmaceutical industries as a high-potency, USP or feed-grade powder, spray-dried powders, and dry dilutions. The water-soluble, riboflavin 5-phosphate salt is available for liquid oral and parenteral pharmaceuticals. High-potency, USP or feed-grade powders are electrostatic, hygroscopic, and dusty, and thus they do not flow freely and show poor distribution in feeds. In contrast, dilution of riboflavin as a spray-dried powder or dry dilution product or including it in a premix reduces its electrostaticity and hygroscopicity for better flowability and distribution in feeds (Adams, 1978). Diluted riboflavin is added to feeds to increase its mixing properties.

Riboflavin is remarkably stable during heat processing (see Sections III and VIII), however, considerable loss may occur if foods are exposed to light during cooking and some losses occur in feed administered to animals out of doors. Only the portion of the feed exposed to light would be destroyed, therefore this may be of little significance as only the top of concentrate mixtures in automatic feeders would be affected. In dry form, riboflavin is extremely resistant to oxidation—even when heated in air for long periods. While it has been shown that field-cured alfalfa hay exposed to moisture can lose a significant amount of its riboflavin content in a relatively short time, under common circumstances riboflavin has good stability when added to mixed feeds (Anonymous, 1969).

Riboflavin is stable in multivitamin premixes (Frye, 1978). Up to 26% of riboflavin present in pet food is lost during extrusion (Anonymous, 1981a). Storage losses of riboflavin in pelleted feeds are slight, but after pellets are in water for 20 min for fish feeding, about 40% of riboflavin may be lost (Goldblatt *et al.*, 1979).

XI. TOXICITY

A large body of evidence has accumulated that treatment with riboflavin in excess of nutritional requirements has very little toxicity either for experimental animals or for humans (Rivlin, 1978). Most data from rats suggest that dietary levels between 10 and 20 times the requirement (possibly 100 times) can be tolerated safely (NRC, 1987). When massive amounts of riboflavin are administered orally, only a small fraction of the dose is absorbed, the remainder being excreted in the feces. Lack of toxicity is probably because the transport system necessary for the absorption of riboflavin across the gastrointestinal mucosa becomes saturated, limiting riboflavin absorption (Christensen, 1973). Also, capacity of the tissues to store riboflavin and its coenzyme derivatives appears to be limited when excessive amounts are administered.

No case of riboflavin toxicity in humans has been reported and oral doses per kilogram of body weight of 340 mg for mice, 10 g for rats, and 2 g for dogs produced no toxic effects (Cooperman and Lopez, 1984). Only when doses of riboflavin (600 mg/kg body weight) were administered intraperitoneally to rats were ill effects evident (Unna and Greslin, 1942), and then anuria developed and at autopsy crystals were apparent in the collecting tubules and renal pelvis. Rainbow trout, like other animals, are insensitive to excess dietary riboflavin, as levels of 600 mg/kg of diet produced no undesirable effect on growth (Hughes, 1984).

8

Niacin

I. INTRODUCTION

After isolation of thiamin and riboflavin from the "vitamin B complex," niacin was the third B vitamin to be established. Niacin exerts its major physiological effects through its role in the enzyme system for cell respiration. A deficiency of niacin results in the disease pellagra in humans and blacktongue in dogs, and in 1937 scientists discovered that this vitamin would cure these conditions. Soon after, niacin was found to be a dietary essential for pigs, poultry, and other nonruminant animals.

Although the need for niacin for simple-stomached animals has been established, exact requirements are difficult to determine since the vitamin can be synthesized from the amino acid tryptophan. The accepted dogma that there is no dietary requirement in ruminants for B vitamins, including niacin, has recently been changed with supplemental niacin providing substantial benefits under some feeding systems.

II. HISTORY

Niacin has been known to organic chemists since 1867, long before its importance as an essential nutrient was recognized. The history of niacin and its importance as a deficiency have been reported (Darby *et al.*, 1975; Hankes, 1984; Loosli, 1988; Anonymous, 1987). As early as 1911 to 1913 Funk had isolated it from yeast and rice polishings in the course of an attempt to identify the water-soluble antiberiberi vitamin. But interest in niacin was lost when it was found ineffective in curing pigeons of beriberi. Although Funk found that niacin did not cure beriberi, cures were more rapid when it was administered in conjunction with the concentrates containing the antiberiberi vitamin (thiamin). Warburg and co-workers first demonstrated a biochemical function for nicotinic acid when they isolated it from an enzyme in 1935 and showed that it functioned as part of a hydrogen transport system.

Niacin deficiency in humans has been equated with pellagra, the disease considered to be a condition of the corn (maize)-eating population. Pellagra appeared in Europe in the 1730s when corn from the New World became the major staple foodstuff of successive regions around the Mediterranean—in Spain, France, Italy, and eastward. Pellagra was first reported in the United States in 1864 in New York and Massachusetts. Most cases occurred in low-income groups, whose diet was limited to cheap foodstuffs. Diets characteristically associated with the disease were referred to as the three M's—meal (corn), meat (back fat), and molasses—plus poverty.

The name "pellagra" means rough skin, which would be a dermatitis. Other descriptive names for the conditions were "mal del sol" (illness of the sun) and "corn bread fever." After the turn of the century in the United States, particularly the South, it was common for 20,000 deaths to occur annually from pellagra. Even as late as 1941, five years after the cause of pellagra was known, 2000 deaths were reported from the disease. Clinical signs and mortality from pellagra could be referred to as the four D's—dermatitis of areas exposed to the sun, diarrhea, dementia (mental problems), and death. Several mental institutions in the United States, Europe, and Egypt were primarily devoted to care of pellagrins (Darby *et al.*, 1975).

In 1914 Goldberger, a bacteriologist in the United States Public Health Service, was assigned the task of identifying the cause of pellagra. In his studies he observed that the disease was associated with poor diet and poverty, with well-fed persons not contracting the disease. In orphanages, prisons, and mental institutions the therapeutic value of good diets was demonstrated. Goldberger, his wife, and 14 volunteers constituted a "filth squad" who ingested and were injected with various biological materials and/or excreta from pellagrins, thus demonstrating the noninfectious nature of pellagra. At this time researchers and physicians did not want to believe that pellagra was the result of poor nutrition, but rather to result from an infection based on the popularity of the "germ theory" of disease. An important step toward the isolation of the preventative factor for pellagra was discovery of a suitable laboratory animal for testing its potency in various concentrated preparations. It was found that a pellagra-like disease could be produced in dogs (blacktongue).

Following discovery that a crude extract of liver was effective in curing pellagra, and therefore was a source of the preventative factor, Elvehjem and his colleagues (1937) isolated nicotinamide from liver as the factor that would cure blacktongue in dogs. Reports of the dramatic therapeutic effects of niacin in human pellagra quickly followed from several clinics.

In 1945 Krehl and co-workers found that the animo acid tryptophan was as active as niacin in treatment of pellagra. Positive proof that tryptophan was converted to nicotinic acid was published by Heidelberger when he showed that L-[^{14}C]tryptophan was converted to ^{14}C-labeled nicotinic acid in the rat. The conversion of tryptophan to niacin explained why foods rich in animal protein (i.e., milk) prevented and cured pellagra.

III. CHEMICAL STRUCTURE, PROPERTIES, AND ANTAGONISTS

Chemically, niacin is one of the simplest vitamins, $C_6H_5O_2N$. Nicotinic acid and nicotinamide correspond to 3-pyridine carboxylic acid and its amide, respectively (Fig. 8.1). The term niacin is used as a generic descriptor of pyridine 3-carboxylic acid and derivatives exhibiting the same qualitative biological activity of nicotinamide. There are antivitamins or antagonists for niacin. These compounds have the basic pyridine structure, with two of the important antagonists of nicotinic acid being 3-acetyl pyridine and pyridine sulfonic acid. Nicotinic acid and nicotinamide (niacinamide) possess the same vitamin activity; the free acid is converted to the amide in the body. Nicotinamide functions as a component of two coenzymes: nicotinamide adenine dinucleotide (NAD, formerly called DPN) and nicotinamide adenine dinucleotide phosphate (NADP, formerly TPN, Fig. 8.1).

Both nicotinic acid and nicotinamide are white, odorless, crystalline solids soluble in water and alcohol. They are very resistant to heat, air, light, and alkali

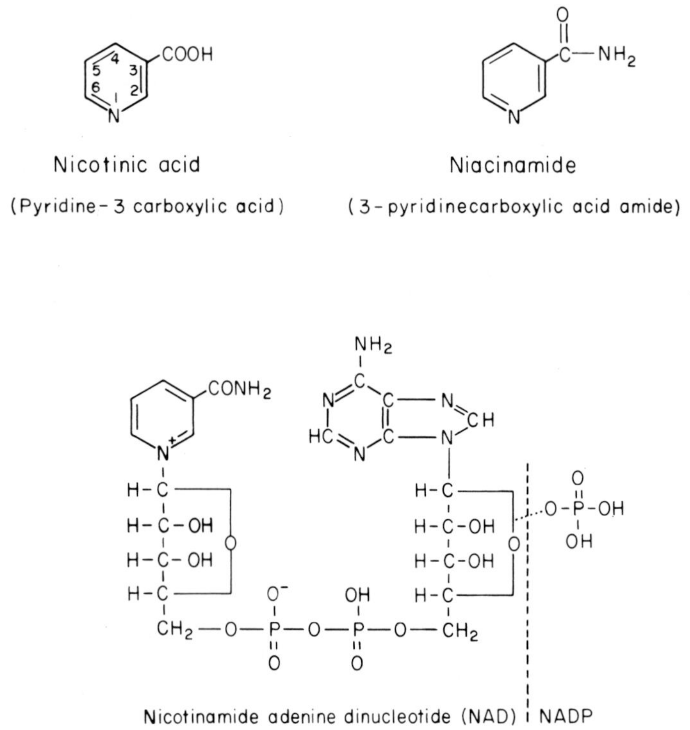

Nicotinic acid

(Pyridine-3 carboxylic acid)

Niacinamide

(3-pyridinecarboxylic acid amide)

Nicotinamide adenine dinucleotide (NAD) | NADP

Fig. 8.1 Chemical structures of nicotinic acid, nicotinamide, nicotinamide adenine dinucleotide (NAD), and nicotinamide adenine dinucleotide phosphate (NADP).

and, thus, are stable in foods. Niacin is also stable in the presence of the usual oxidizing agents. However, it will undergo decarboxylation at a high temperature when in an alkaline medium. Nicotinic acid readily forms salts with such metals as aluminium, calcium, copper, and sodium. When in acid solution, niacin readily forms quaternary ammonium compounds, such as nicotinic acid hydrochloride, which is soluble in water. When in a basic solution the nicotinic acid readily forms carboxylic acid salts.

IV. ANALYTICAL PROCEDURES

The most sensitive method for the determination of niacin and closely related compounds are microbiological. *Lactobacillus plantarum* responds to both forms of the vitamin, whereas *Leuconostoc mesenteroides* measures only nicotinic acid. Niacin must be freed from bound forms before assay. Since niacin is very stable to strong acids, it can be released by acid hydrolysis. Chemical methods of analysis are less sensitive than microbiological procedures and generally require more extensive extraction methods. The cyanogen bromide method of analysis is based on the reaction of pyridine derivatives with cyanogen bromide, forming a colored compound that can be used for quantitative measurement of the vitamin. Bioassay procedures for niacin present two major difficulties in that (1) tryptophan in the diet is converted to niacin in the tissues and (2) niacin is synthesized by intestinal bacteria to varying degrees. Chicks, puppies, and weanling rats have been used for niacin assay.

V. METABOLISM

A. Absorption, Transport, and Storage

Nicotinic acid and its amide are readily and very efficiently absorbed by diffusion at either physiological or pharmacological doses. By employing the gastrointestinal tube technique, niacin was shown to be equally well absorbed from both the stomach and upper small intestine in humans (Bechgaard and Jespersen, 1977). In a steady-state situation, approximately 85% of a 3 g/day dose of niacin was recovered from the urine of humans and absorption was considered almost complete. The mechanism by which nicotinamide nucleotides present in animal foods are absorbed, however, is not known. Whether the coenzyme is hydrolyzed in the lumen of the intestine or after it is taken up by the mucosa is not clear.

Niacin and nicotinamide transport has not been extensively studied. There seems to be a continual transport of the two compounds via the bloodstream such that there is cross-feeding of the tissues and organs for synthesis of pyridine nucleotides. Blood transport of niacin is associated mainly with the red blood cells. Niacin rapidly leaves the bloodstream and enters kidney, liver, and adipose tissues. Evidence indicates that absorbed niacin is actively cycled through the NAD pathway to nicotinamide in the intestinal mucosa (Henderson and Gross, 1979). Absorbed nicotinamide is taken up by tissues and incorporated into its coenzymes.

The tissue content of niacin and its analogs, NAD and NADP, is variable, dependent on the diet and a number of other factors, such as strain, sex, age, and treatment of animals (Hankes, 1984). Although niacin coenzymes are widely distributed in the body, no true storage occurs. The liver is the site of greatest niacin concentration but the amount stored is not great.

B. Excretion

Urine is the primary pathway of excretion of absorbed niacin and its metabolites. At high dosages the half-life of both nicotinic acid and nicotinamide is determined mainly by rate of excretion of the unchanged compound in urine and not by metabolic change. When low dosages were used, both compounds were excreted principally as metabolites rather than as unchanged compounds.

The principal excretory products in humans, dogs, rats, and pigs is the methylated metabolite *N′*-methylnicotinamide or as two oxidation products of this compound, 4-pyridone or 6-pyridone of *N*-methylnicotinamide. On the other hand, in herbivores niacin does not seem to be metabolized by methylation, but large amounts are excreted unchanged. In the chicken, however, nicotinic acid is conjugated with ornithine as either α- or δ-nicotinyl ornithine or dinicotinyl ornithine. The measurement of the excretion of these metabolites is carried out in studies of niacin requirements and of niacin metabolism. Such studies are complicated by the fact that the kinds and relative amounts of these products vary with the species and level of niacin intake (Maynard *et al.*, 1979).

C. Tryptophan–Niacin Conversion

The amino acid tryptophan is a precursor for the synthesis of niacin in the body, and there is considerable evidence that synthesis can take place in the intestine. Investigations indicate that there is a much larger urinary excretion of the metabolite *N′*-methylnicotinamide when tryptophan was administered by stomach tube than when injected parenterally. There is also evidence that synthesis occurs in the developing chick embryo. The extent to which the metabolic requirement for niacin can be met from tryptophan will depend, first, on the amount of tryptophan in the diet and, second, on the efficiency of the conversion of tryptophan to niacin. The pathway of tryptophan conversion to nicotinic acid mononucleotide in the body is shown in Fig. 8.2. Protein, energy, vitamin B_6,

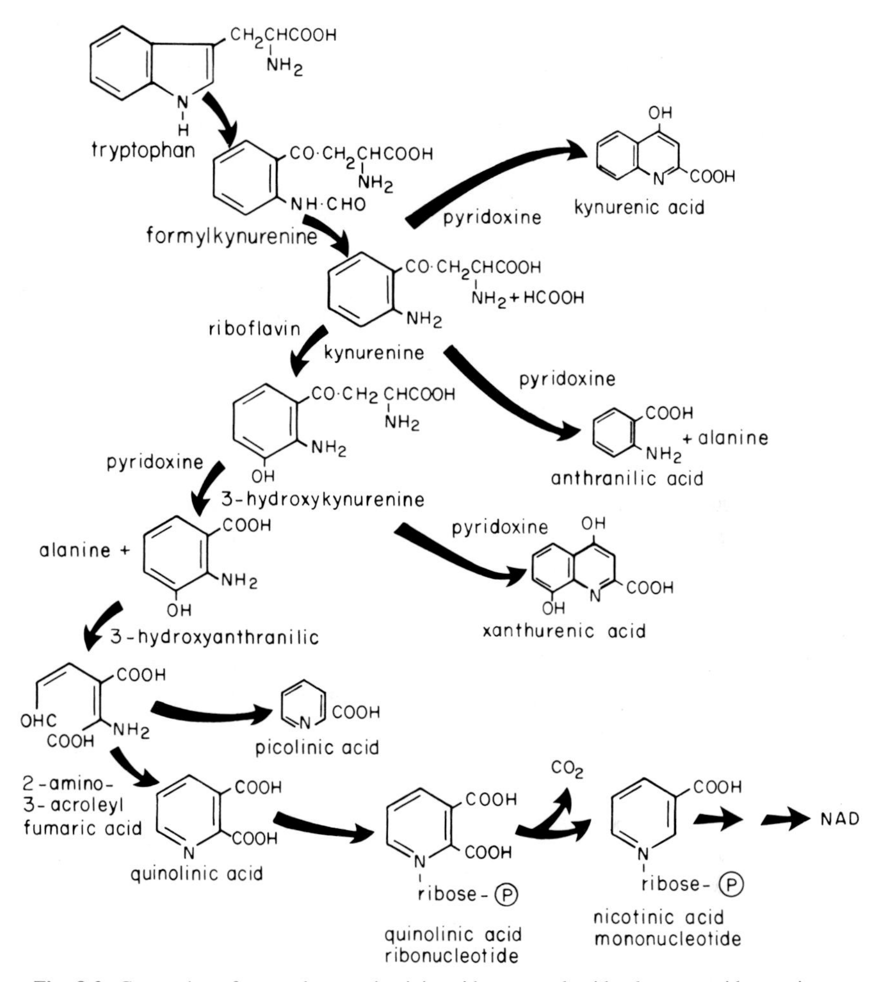

Fig. 8.2 Conversion of tryptophan to nicotinic acid mononucleotide plus some side reactions.

and riboflavin nutritional status and hormones affect one or more steps in the conversion sequence shown in Fig. 8.2, and hence can influence the yield of niacin from tryptophan. It is suggested that pellagra is not simply a disease of niacin deficiency, but more appropriately a disease of tryptophan metabolism (Anonymous, 1987).

At low levels of tryptophan intake, the efficiency of conversion is high. It decreases when niacin and tryptophan levels in the diet are increased. When animals have received deficient levels of tryptophan, increasing amounts of dietary tryptophan are used first to restore nitrogen balance, next to restore blood pyridine nucleotides, and then to be excreted as niacin metabolites. Under starvation or energy restriction, efficiency of conversion increased, while pregnant women can convert tryptophan to niacin more efficiently than can other adults (Narasinga Rao and Gopalan, 1984). Another important factor that affects this conversion ratio is amino acid imbalance due to excess leucine in the diet (see Section IX).

Animal species differ widely in ability to synthesize niacin from tryptophan. From a variety of experiments approximately 60 mg tryptophan is estimated to be equivalent to 1 mg niacin in humans, while the rat is more efficient at a conversion rate of 35–50 mg tryptophan required. Conversion efficiency is probably due to inherent differences in liver levels of picolinic acid carboxylase, the enzyme that diverts one of the intermediates (2-amino-3-acroleylfumaric acid) toward the glutaryl-CoA pathway instead of allowing this compound to condense to quinolinic acid, the immediate precursor of nicotinic acid. Picolinic acid carboxylase in livers of various species has a very close inverse relationship to experimentally determined niacin requirements, that is, the cat has so much of this enzyme that it cannot convert any of its dietary tryptophan to niacin. Thus the cat has an absolute requirement for niacin itself. Conversely, the rat diverts very little of its dietary tryptophan to carbon dioxide and water, and thus is very efficient in converting tryptophan to niacin. The duck has a very high niacin requirement (approximately twice as high as chickens), with considerably higher levels of picolinic acid carboxylase activity (Scott *et al.*, 1982).

VI. FUNCTIONS

The major function of niacin is in the coenzyme forms of nicotinamide, NAD and NADP. Enzymes containing NAD and NADP are important links in a series of reactions associated with carbohydrate, protein, and lipid metabolism. They are especially important in the metabolic reactions that furnish energy to the animal. More than 40 biochemical reactions have been identified and are of paramount importance for normal tissue integrity, particularly for the skin, the gastrointestinal tract, and the nervous system.

The NAD- and NADP-containing enzymes play an important role in biological oxidation–reduction systems by virtue of their capacity to serve as hydrogen-transfer agents. Hydrogen is effectively transferred from the oxidizable substrate to oxygen through a series of graded enzymatic hydrogen transfers. Nicotinamide-containing enzyme systems constitute one such group of hydrogen-transfer agents. In electron transport, oxidation of reduced NAD or NADP is carried out by a second hydrogen-carrying system, the riboflavin coenzymes (see Chapter 7, Fig. 7.2). The nicotinamide moiety of the coenzyme operates in these systems by reversibly alternating between an oxidized quarternary pyridinium ion and a reduced tertiary amine. The transfer of hydrogen is reversible and stereospecific.

NAD is specific for hydrogenases involved in passing electrons on to oxygen via the electron-transport system in the tricarboxylic acid cycle. Here NAD serves as the electron acceptor in three of the four dehydrogenation steps. NADP also participates in dehydrogenation reactions, particularly in the hexose monophosphate (HMP) shunt of glucose metabolism. Reduced NADP has an important role in the synthesis of fats and steroids and is contained in the alcohol–dehydrogenase system, the lactic acid–dehydrogenase system, and others. Both NAD and NADP are involved in degradation and synthesis of amino acids.

Some of these enzymes show a strict specificity for either NAD or NADP, while others utilize these coenzymes equally well. Important metabolic reactions catalyzed by NAD and NADP are summarized below:

1. Carbohydrate metabolism—(a) glycolysis (anaerobic and aerobic oxidation of glucose) and (b) the Krebs cycle.
2. Lipid metabolism—(a) glycerol synthesis and breakdown, (b) fatty acid oxidation and synthesis, and (c) steroid synthesis.
3. Protein metabolism—(a) degradation and synthesis of amino acids and (b) oxidation of carbon chains via the Krebs cycle.
4. Photosynthesis.
5. Rhodopsin synthesis (see Chapter 2, Fig. 2.6).

VII. REQUIREMENTS

Niacin requirements for various animals and humans are presented in Table 8.1. The wide variation in niacin requirements are generally due to (1) niacin is synthesized from tryptophan, thus the niacin requirement depends on dietary tryptophan, and (2) much of dietary niacin is in a bound form unavailable to humans and animals.

Some species, such as the cat, mink, and most fish, apparently lack the ability to convert tryptophan to niacin. For species that have the capacity to synthesize niacin from tryptophan, it is impossible to set the niacin requirement unless the tryptophan level is specified and it is known that the diet is adequate in vitamin B_6, since this vitamin is needed in synthesis of niacin from tryptophan. However, the vitamin B_6 level is adequate in most practical livestock diets.

Wide variations in niacin requirements are reported. Cunha (1982) lists a number of factors that influence niacin requirements:

1. Genetic differences can influence niacin needs, such as selection for meatier, faster-growing animals and increased production levels.

TABLE 8.1

Niacin Requirements for Various Animals and Humans[a]

Animal	Purpose	Requirement	Reference
Beef cattle	Adult	Microbial synthesis	NRC (1984a)
Dairy cattle	Calf	2.6 mg/liter milk	NRC (1978a)
	Adult	Microbial synthesis	NRC (1978a)
Chicken	Leghorn, 0–6 weeks	27.0 mg/kg	NRC (1984b)
	Leghorn, 6–20 weeks	11.0 mg/kg	NRC (1984b)
	Leghorn, laying	10.0 mg/kg	NRC (1984b)
	Leghorn, breeding	10.0 mg/kg	NRC (1984b)
	Broilers, 0–8 weeks	11–27 mg/kg	NRC (1984b)
Turkey	All classes	30–70 mg/kg	NRC (1984b)
Duck (Pekin)	All classes	40–55 mg/kg	NRC (1984b)
Japanese quail	All classes	20–40 mg/kg	NRC (1984b)
Sheep	Adult	Microbial synthesis	NRC (1985b)
Swine	Growing–finishing, 1–5 kg	20 mg/kg	NRC (1988)
	Growing–finishing, 5–110 kg	7–15 mg/kg	NRC (1988)
	Adult	10.0 mg/kg	NRC (1988)
Horse	Adult	Microbial synthesis	NRC (1978b)
Goat	Adult	Microbial synthesis	NRC (1981b)
Fox	Growing	10 mg/kg	NRC (1982a)
Cat	Growing	40 mg/kg	NRC (1986)
Dog	Growing	450 μg/kg body wt.	NRC (1985a)
Fish	Catfish	14 mg/kg	NRC (1983)
	Trout	95 mg/kg	NRC (1981a)
Rat	All classes	20 mg/kg	NRC (1978c)
Mouse	All classes	10 mg/kg	NRC (1978c)
Human	Infants	6–8 mg/day	RDA (1980)
	Children	9–16 mg/day	RDA (1980)
	Adults	13–20 mg/day	RDA (1980)

[a]Expressed as per unit animal feed on an as fed (approximately 90% dry matter) or dry basis (see Appendix Table 1) (except for calf and dog). Human data are expressed as mg/day.

2. The ability to synthesize niacin from tryptophan (see Section V.)

3. Increased stress and subclinical disease level on the farm because of closer and more frequent contact between animals in confinement.

4. Trend toward more intensified operations that may lessen opportunity for coprophagy and will reduce access to pasture.

5. Newer methods of handling and processing feeds that may affect niacin and tryptophan level and availability.

6. Various nutrient interrelationships including amino acid imbalances.

7. Trend toward earlier weaning, which increases need for higher vitamin levels in milk-substitute diets (prestarter and starter feeds).

8. Molds and antimetabolites in feeds that can increase certain nutrient needs.

Niacin requirements of ruminants are unknown with the vitamin normally synthesized in adequate quantities in the rumen. Recently, niacin supplementation has been shown to be beneficial to beef cattle receiving high-concentrate diets and to high-producing dairy cows. In a review, Olentine (1984) has summarized the factors affecting ruminant niacin requirements:

1. Protein balance—excess of leucine, arginine, and glycine increases the requirement.
2. High tryptophan content of feeds—as tryptophan content increases, niacin requirements decrease.
3. Energy content—high-energy rations require more niacin per unit of feed.
4. Antibiotics—depending on the product, niacin requirements are increased.
5. Dietary rancidity—if fat is rancid, niacin requirements are increased.
6. Gastrointestinal synthesis—niacin is synthesized in the gastric and intestinal regions.
7. Availability of niacin in feedstuffs—cereal grains and other feedstuffs have varying degrees of niacin availability.

Human niacin requirements are based on niacin intakes of pellagrins and on human depletion and repletion studies. The minimum adult requirement has been reported to be 4.4 mg niacin equivalents per 1000 kcal. The Food and Agriculture Organization/World Health Organization Expert Group has recommended an intake of 6.6 mg niacin equivalents per 1000 kcal for all age groups (Narasinga Rao and Gopalan, 1984). The RDA (1980) recommends niacin requirements varying from 6 to 19 mg per day (Table 8.1).

VIII. NATURAL SOURCES

Niacin is widely distributed in foods of both plant and animal origin (Table 8.2). Animal and fish by-products, distiller's grains and yeast, various distillation and fermentation solubles, and certain oil meals are good sources. Leafy materials, especially pasture grass, are fair sources. Milk, dairy products, fruits, and eggs are poor sources.

Most species can use the essential amino acid tryptophan and synthesize niacin from it (see Sections V and VII). Because tryptophan can give rise to body niacin, both niacin and tryptophan content should be considered in expressing niacin values of foods. However, since there is a preferential use of tryptophan for protein synthesis before any becomes available for conversion to niacin (Kodicek *et al.*, 1974), it seems unlikely, given the low tryptophan content in many feedstuffs, that tryptophan conversion greatly contributes to the niacin supply.

Niacin is often present in feeds in a bound form that is not available. The niacin in cereal grains and their by-products is in a bound, complex form that

TABLE 8.2

Niacin in Foods and Feedstuffs (mg/kg, Dry Basis)[a]

Food	Niacin	Food	Niacin
Alfalfa hay, sun cured	42	Rice, polished	17
Alfalfa leaves, sun cured	53	Rye, grain	21
Barley, grain	94	Sorghum, grain	43
Bean, navy (seed)	28	Soybean meal, solvent extracted	31
Blood meal	34	Soybean seed	24
Brewer's grains	47	Spleen, cattle	25
Buttermilk (cattle)	9	Timothy hay, sun cured	29
Chicken, broilers (whole)	230	Wheat, bran	268
Citrus pulp	23	Wheat, grain	64
Clover hay, Ladino (sun cured)	11	Whey	11
Copra meal (coconut)	28	Yeast, brewer's	482
Corn, gluten meal	55	Yeast, torula	525
Corn, yellow grain	28		
Cottonseed meal, solvent extracted	48		
Fish meal, anchovy	89		
Fish meal, menhaden	60		
Fish, sardine	81		
Linseed meal, solvent extracted	37		
Liver, cattle	269		
Milk, skimmed, cattle	12		
Molasses, sugarcane	49		
Oat, grain	16		
Pea seeds	36		
Peanut meal, solvent extracted	188		
Potato	37		
Rice, bran	330		
Rice, grain	39		

[a]NRC (1982b).

is virtually unavailable, at least to simple-stomached animals (Luce *et al.*, 1967). It seems that much of this niacin will also be unavailable to rumen microorganisms. Any dietary niacin escaping degradation in the rumen will also likely be unavailable for absorption in the lower gut.

Initial evidence of bound forms of niacin stemmed from observed discrepancy between niacin values in cereals determined colorimetrically before and after hydrolysis with dilute NaOH. Microbiological assays similarly indicated that about 20% more niacin was obtained in dilute alkaline extracts of cereals than in aqueous or acid extracts.

Two types of bound niacin were initially described: (1) a peptide with molecular weight of 12,000 to 13,000, the so-called niacinogens, and (2) a carbohydrate complex with a molecular weight of approximately 2370 (Darby *et al.*, 1975). The name niacytin has been used to designate this latter material from wheat bran. It appears that a number of bound forms of niacin are present in cereals. Ghosh *et al.* (1963), using a microbiological assay, reported that 85 to 90% of the total nicotinic acid in cereals is in a bound form. Mason *et al.* (1973) reported bound niacin to be linked to macromolecules, of which about 60% were polysaccharides and 40% peptides or glycopeptides. Lack of niacin availability may be due to a blocking effect of these molecules, which prevent access of digestive enzymes to the niacin–macromolecule bond, and therefore free niacin would not be available for absorption.

Oilseeds contain about 40% of their total niacin in bound form, while only a small proportion of the niacin in pulses, yeast, crustacea, fish, animal tissue, or milk is bound. By use of a rat assay procedure, Carter and Carpenter (1982) showed that for eight samples of mature cooked cereals (corn, wheat, rice, and milo), only about 35% of the total niacin was available. In the calculation of the niacin content of formulated diets, probably all niacin from cereal grain sources should be ignored or at least given a value no greater than one-third of the total niacin.

Some bound forms of niacin are biologically available but the niacin in corn is particularly unavailable and implicated in the etiology of pellagra among societies that consume large quantities of the grain. Pellagra occurs in Mexico less frequently than one might expect; there the custom is to treat the maize with limewater before making tortillas and in this way the nicotinic acid is liberated. Boiling releases niacin in sweet corn but not mature corn. Roasting of sweet corn, a traditional way of treating corn used by the Hopi Indians, also results in the release of free nicotinic acid (Kodicek *et al.*, 1974). In rat growth assays for available niacin, corn harvested immaturely (''milky stage'') gave values from 74 to 88 μg/g, whereas corn harvested at maturity gave assay values of 16–18 μg/g (Carpenter *et al.*, 1988).

Beverages including tea and coffee also can contribute significantly to niacin intake. Roasted coffee contains more niacin than does raw coffee because the trigonellin present in raw coffee is converted to nicotinic acid when the coffee is roasted. In several Central and South American countries, absence of pellagra among those whose staple is maize is attributed to widespread consumption of coffee.

IX. DEFICIENCY

A. Effects of Deficiency

A deficiency of niacin is characterized by severe metabolic disorders in the skin and digestive organs. The first signs to appear are loss of appetite, retarded growth, weakness, digestive disorders, and diarrhea. The deficiency is found in both human and animal populations that are overly dependent on foods (particularly corn) low in available niacin and its precursor tryptophan.

1. Ruminants

Young ruminants prior to the development of the rumen would be expected to suffer B-vitamin deficiencies, including niacin, when diets contain insufficient quantities of these vitamins. However, early studies with lambs and calves failed to produce a deficiency when fed niacin-free diets. Ability to produce niacin deficiency is dependent on use of a low tryptophan milk diet. From these studies, it may be concluded that a dietary requirement of niacin for ruminants does not exist as long as the level of tryptophan is maintained near 0.2% of the diet. For calves, a diet free of niacin and low in tryptophan produced deficiency signs of sudden anorexia, severe diarrhea, and dehydration followed by sudden death. Supplementation with 2.6 mg of nicotinic acid per liter of milk offered *ad libitum* twice daily prevented the deficiency (Hopper and Johnson, 1955).

Adult ruminants had previously been thought to synthesize adequate niacin by microorganisms and/or conversion from tryptophan. However, the synthesizing ability appears to be low, since production responses from niacin supplementation can be demonstrated in both sheep and cattle. Niacin supplementation is especially beneficial to stressed animals, such as beef cattle being adapted to high-grain diets or lactating cows that have just calved.

Byers (1979) summarized 14 beef cattle studies demonstrating improved gains and feed efficiency by 9.7 and 10.9%, respectively. Growth from all trials was especially beneficial during the adaptation of cattle to feedlot diets (i.e., during the first 40 days). Studies with growing and finishing lambs revealed increased performance with evidence that niacin stimulates rumen microbial protein synthesis (Shields and Perry, 1982). These studies indicate that supplemental niacin (1) increased nitrogen utilization, (2) improved the percentage of absorbed nitrogen retained, (3) reduced urinary nitrogen excretion, and (4) reduced the percentage of nitrogen found as urea nitrogen. All these positive changes point toward improved protein metabolism.

Niacin appears to be effective in enhancing acclimation and adaptation to urea-supplemented diets. However, niacin stimulation of microbial protein production occurs regardless of dietary nitrogen source, but is greatest when natural protein rather than nonprotein nitrogen (urea) was fed (Brent and Bartley, 1984). This may be partly due to the synthesis of niacin from tryptophan. Summaries of beef cattle studies over the total feeding period (73–176 days) indicate that 50 or 100 ppm of niacin is more effective than are higher levels of 150, 250, or 500 ppm, with respect to gain.

Brent and Bartley (1984) found a higher milk production response from niacin to be greater for postpartum cows than for those in midlactation. Niacin response in early postpartum cows might be partially explained via reduction in ketosis.

A report has indicated that about 50% of dairy cows in high-production herds go through borderline ketosis during early lactation (Emery *et al.*, 1964). Fronk and Schultz (1979) indicated that treating ketotic dairy cows with 12-g doses of nicotinic acid daily had a beneficial effect on the reversal of both subclinical and clinical ketosis. More recent studies indicate that 6 g of niacin may be sufficient.

Other workers have reported that supplemental niacin increased microbial protein synthesis. Enhanced production of microbial protein might explain increased milk production, weight gain, and feed efficiency observed when urea-containing rations were supplemented with 250–500 ppm niacin (Cunha, 1982). Daily niacin supplementation of 3–6 g to early lactation dairy cows resulted in slight increases of milk production. Jaster *et al.* (1983) showed only a slight increase in milk fat percentage in six commercial dairy herds supplemented with niacin.

Horner *et al.* (1986) reported that feeding of whole cottonseed and most other dietary fat sources to dairy cows results in a reduction of milk protein percentage and protein yield. Diets supplemented with 0.03% niacin (6 g niacin/20.45 kg dry matter) increased milk protein percentage over diets with 15% whole cottonseed meal. The authors concluded that milk protein depression with whole cottonseed was alleviated by niacin because of stimulation of mammary casein synthesis.

2. Swine

Niacin is one of the B vitamins that would be expected to be deficient for typical swine diets, particularly when corn, which is low in available niacin and tryptophan is fed. Wide variation has been observed in the severity of clinical signs of niacin deficiency in pigs with similar breeding and environmental backgrounds. Occasionally animals appear to thrive with no niacin, and other animals appear to vary in their requirement (Cunha, 1977). During reproduction and lactation, it was not possible to produce niacin deficiency with sows fed a purified diet with either 18 or 26.1% casein (Ensminger *et al.*, 1951). Evidently the diet contained enough tryptophan to supply niacin needs, or the duration of the experiment was not long enough to develop a niacin deficiency.

Signs of niacin deficiency include poor appetite, decreased rate of gain (Fig. 8.3), stomatitis, normocytic anemia and achlorhydria, followed by diarrhea, occasional vomiting, and an exfoliate type of dermatitis and hair loss (Cunha, 1977). Nervous system degenerative changes are reported to occur in the ganglion cells in the posterior root with extensive chromatolysis in the dorsal root (Wintrobe *et al.*, 1945).

Niacin-deficient pigs have inflammatory lesions of the gastrointestinal tract. Ulcerative necrotic lesions of the large intestine are present, which swarm with fusiform bacteria and spirochetes (Fig. 8.4). Diarrhea with foul-smelling feces particularly involves the large intestine, which thickens, is very red, and appears

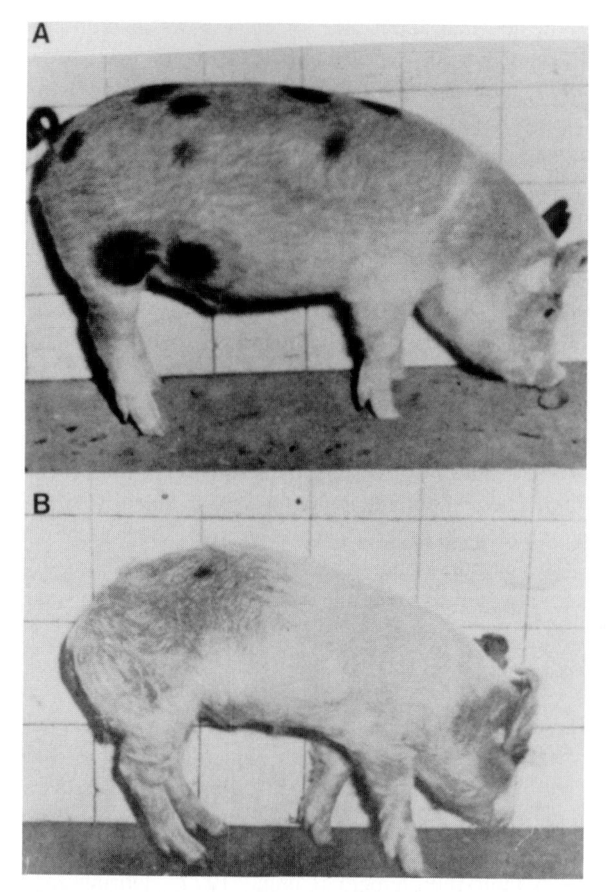

Fig. 8.3 Niacin deficiency. (A) Pig that has received adequate niacin; (B) pig that has not received adequate niacin. The difference in growth and condition is due to the addition of niacin in a diet containing 80% ground yellow corn. (Courtesy of D. E. Becker, Illinois Agriculture Experiment Station.)

weak and "rotten." Enteric conditions may be due to niacin deficiency, bacterial infection, or both. Deficient pigs respond readily to niacin therapy, but infectious enteritis is not benefited. However, adequate dietary niacin probably allows the pig to maintain its resistance to bacterial invasion.

3. Poultry

There is good evidence that chickens—even chick and turkey embryos—are able to synthesize niacin, but that rate of synthesis may be too slow for optimal growth. Ruiz (1987) reports that broilers from 3 to 7 weeks of age do not require

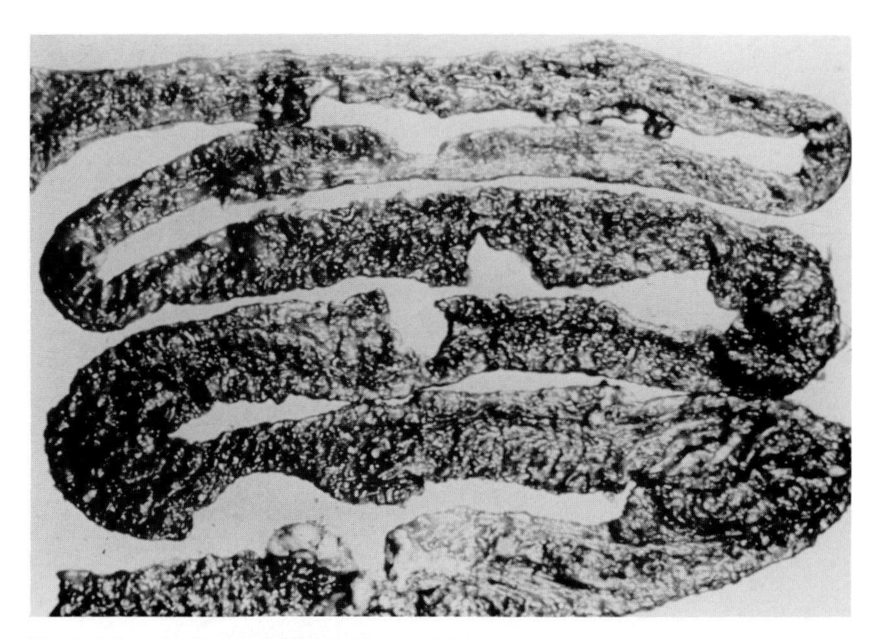

Fig. 8.4 Intestine from deficient pig shows thickened and hemorrhagic mucous membrane, also denuded areas. (Courtesy of R. W. Luecke, Michigan State University.)

supplemental niacin while fed a corn–soybean meal diet, but the vitamin is required from 1 to 21 days of age. It has been claimed that before there can be a marked deficiency of niacin in the chicken, there must first be a deficiency of tryptophan. Experiments using diets containing a limited amount of tryptophan have shown that the chick does require niacin and that a deficiency produces an enlargement of the tibiotarsal joint, a bowing of the legs, poor feathering, and a dermatitis on the feet and head (Scott *et al.*, 1982). The main clinical sign of niacin deficiency in young poultry is an enlargement of the hock joint and bowing of the legs similar to perosis (Figs. 8.5 and 8.6). The main difference between this condition and the perosis of manganese or choline deficiency is that in niacin deficiency the Achilles tendon rarely slips from its condyles.

Niacin deficiency in the chick is characterized by appetite loss and growth failure. The deficiency results in "blacktongue," a condition characterized by inflammation of the tongue and mouth cavity. Beginning at about 2 weeks of age, the entire mouth cavity, as well as the esophagus, becomes distinctly inflamed, growth is retarded, and feed consumption is reduced (NRC, 1984b). There is a loss of weight and both egg production and hatchability are reduced in niacin deficiency of laying hens.

Signs of niacin deficiency in turkeys and ducks, while similar, are much more

Fig. 8.5 Leg disorders in niacin-deficient broiler chicks. Bird on the left was fed a corn–soybean meal diet without supplemental niacin. This bird exhibited bowed legs. (Courtesy of N. Ruiz and R. Harms, University of Florida.)

severe. Compared to the chick, the turkey poult, duckling, pheasant chick, and gosling have higher requirements for niacin. This higher requirement is related to the less efficient conversion of tryptophan to niacin by these species (see Section V). Ducks receiving low-niacin diets show severely bowed legs and ultimately become so crippled and weak that they cannot walk. Niacin deficiency in the turkey is also characterized by a severe bowing of the legs and enlargement of the hock joint. Goslings on purified rations developed perosis and hock deformities that were prevented with nicotinic acid administration (Briggs *et al.*, 1953).

4. Horses

No requirements for niacin have been established for the horse and a deficiency has not been reported (NRC, 1978b). Synthesis of niacin by microflora in the lower digestive tract as well as tissue synthesis from tryptophan should likely preclude niacin deficiency.

5. Other Animal Species

a. Cats. Virtually no niacin is synthesized from tryptophan in the cat (see Section V). Cats deficient in niacin are observed to lose weight and exhibit an-

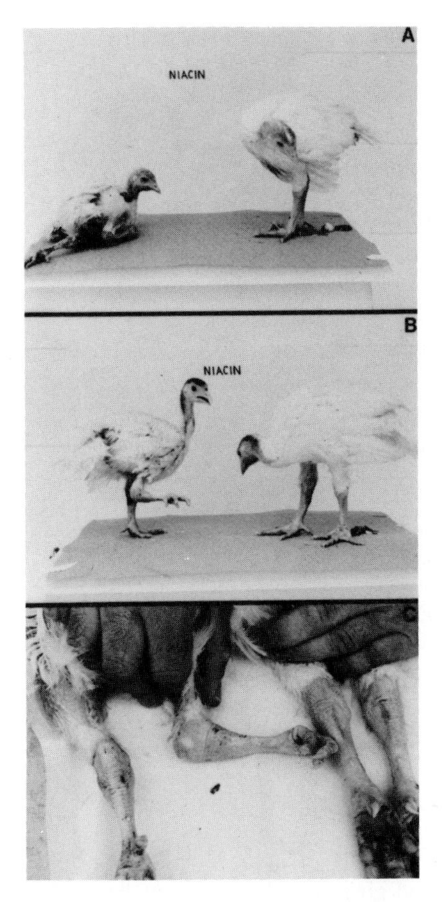

Fig. 8.6 Niacin deficiency in turkey poults. (A, B) The birds on the left side, which were fed a corn–soybean meal diet without supplemental niacin, showed perosis-like signs. (C) Comparison of the legs of the poults in photograph B. (Courtesy of N. Ruiz and R. Harms, University of Florida.)

orexia, weakness, and apathy (NRC, 1986). Thick saliva with a foul odor is drooled from the mouth. The oral cavity is characterized by ulceration of the upper palate, and the tongue is fiery red in color with ulceration and congestion along the anterior border. The fur may be unkempt and diarrhea is present. The deficiency can be associated with respiratory disease, which contributes to an early death.

b. Dogs. The similarity of niacin deficiency signs between the dog and humans was important as the dog became the laboratory animal used to identify the vitamin deficiency. Signs of deficiency in the dog are collectively known as blacktongue. Clinical signs include anorexia, apathy, growth retardation, or loss

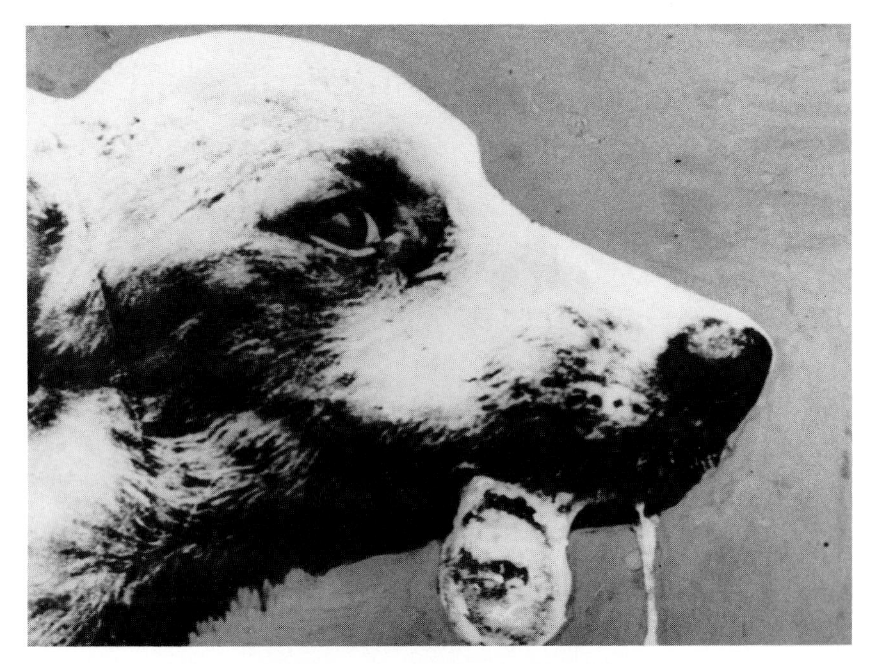

Fig. 8.7 Niacin deficiency. Blacktongue condition in dog exhibiting drooling of thick, ropy saliva. (Courtesy of V. Ramadus Murthy, National Institute of Nutrition, Jamasi-Osmania, Hyderabad, India.)

of body weight and drooling of thick ropy saliva (Fig. 8.7). There is severe cheilosis, glossitis, and gingivitis. Necrotic patches and ulcers may be seen on the oral mucosa and there is a foul odor. The inflammatory changes may extend to the esophagus and eventually to the stomach. There is bloody diarrhea, inflammation and hemorrhagic necrosis of duodenum and jejunum with shortening and clubbing of villi, and inflammation and degeneration of the mucosa of the large intestine. Uncorrected deficiencies lead to dehydration, emaciation, and death (NRC, 1985a).

c. *Laboratory Animals.* Laboratory rodents often do not show characteristic signs of niacin deficiency but exhibit nonspecific signs of appetite loss and growth retardation (NRC, 1978c). Rats fed niacin- and tryptophan-deficient diets developed behavioral changes, convulsions, diarrhea, rough hair coat, and alopecia. Niacin-deficient guinea pigs exhibited poor growth, reduced appetite, pale feet, nose, and ears, drooling, anemia, and a diarrhea tendency. Niacin deficiency has not been reported in the mouse, gerbil, hamster, or vole (NRC, 1978c).

d. Rabbits. In rabbits niacin deficiencies have resulted in pronounced loss of appetite followed by emaciation and diarrhea (NRC, 1977).

e. Foxes and Mink. The mink is similar to the cat concerning niacin metabolism, with very inefficient conversion of tryptophan to niacin. Young mink on a deficient diet displayed nonspecific signs, including weight loss, weak voice, general weakness, and bloody diarrhea (NRC, 1982a). Foxes fed niacin-deficient diets exhibit anorexia, loss of body weight, and typical blacktongue, which is characterized by severe inflammation of the gums and fiery redness of the lips, tongue, and gums (NRC, 1982a).

f. Fish. For fish species studied to date, tryptophan is not an adequate precursor to niacin, with the vitamin deficiency rapidly seen in many fish fed a niacin-deficient diet. Signs of niacin deficiency in trout and salmon include anorexia, reduced growth, poor feed conversion, a photosensitivity or sunburn, intestinal lesions, muscular weakness and spasms, and increased mortality (NRC, 1981a). Channel catfish deficient in niacin show skin and fin lesions, deformed jaws, anemia, exophthalmia, and high mortality (NRC, 1983). Japanese eels exhibit reduced growth, ataxia, skin lesions, and dark coloration. Common carp had skin hemorrhages at 4 weeks and substantial mortality after 5 weeks.

g. Monkeys. Niacin deficiency in the rhesus monkey is characterized by weight loss, apathy, anemia, skin pigmentation, and bloody diarrhea (NRC, 1978d). It is suggested that there is an inefficient conversion of tryptophan to niacin.

6. Humans

Traditionally niacin deficiency in humans has been equated with pellagra. Symptoms of deficiency can be considered under three headings: (1) skin changes, (2) lesions of the mucous membranes of the mouth, tongue, stomach, and intestinal tract, and (3) changes of nervous origin.

The earliest symptom of pellagra is inflammation and soreness of the mouth followed by bilateral symmetrical erythema on all parts of the body exposed to sunlight. Therefore, common sites are the extensor surfaces of the extremities, face, and neck (Fig. 8.8). The lesions are also found at sites of constant irritation, for example, under the breast, scrotum, axilla, and perineum. A characteristic feature of these photosensitive lesions is their clear-cut demarcation from the adjoining, unaffected parts. The appearance of the dermatosis varies with the severity of the disease and from patient to patient.

Gastrointestinal symptoms often include glossitis and stomatitis, the tongue having a characteristic swollen and "beefy-red" appearance. There is also anorexia, abdominal discomfort, and diarrhea. A burning or raw sensation of the

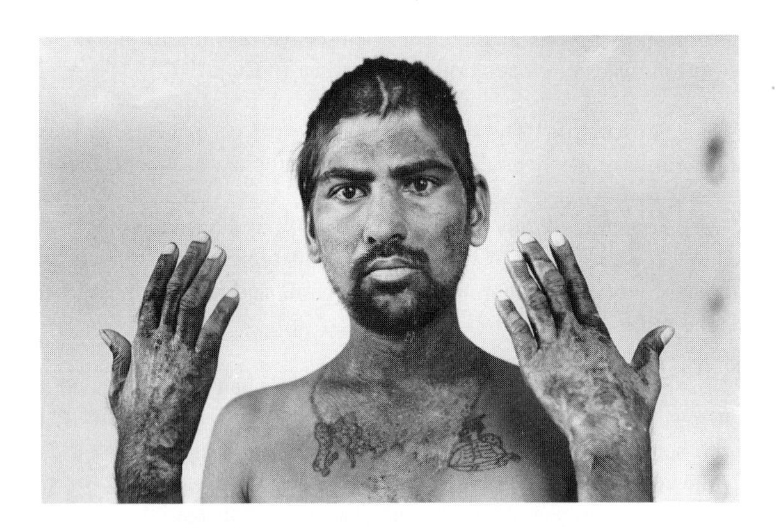

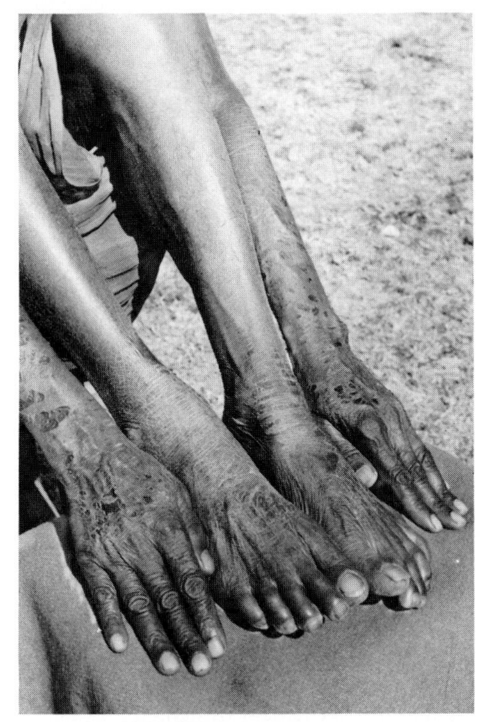

Fig. 8.8 Niacin deficiency. Typical deramal lesions associated with pellagra. Dermatitis is severe when sunlight reacts with exposed skin. Dermal lesions around neck have been referred to as Casal's necklace (upper photo). (Courtesy of V. Ramadus Murthy, National Institute of Nutrition, Jamasi-Osmania, Hyderabad, India.)

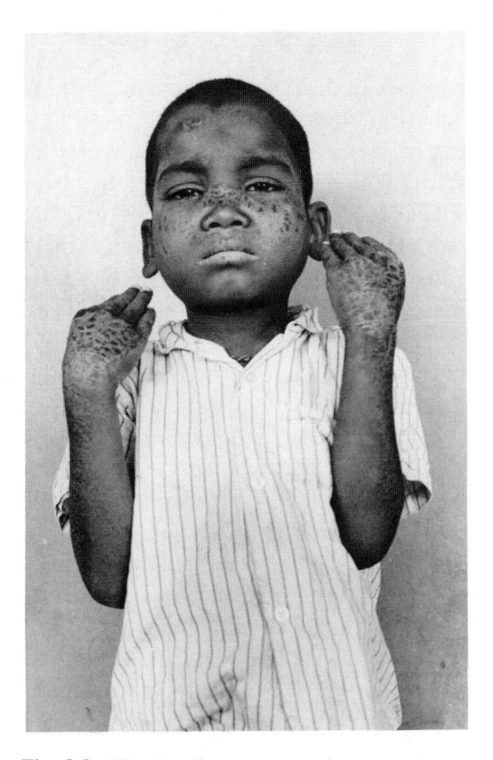

Fig. 8.8 (*Continued*)

mouth, stomach, and rectum has been described by various workers. Angular stomatitis, cheilosis, atrophy of the lingual papillae, and glossitis are commonly seen. However, it is not known whether these are manifestations of niacin deficiency or whether they are due to the associated deficiency of riboflavin and other members of the vitamin B-complex group.

Early mental symptoms include lassitude, lack of ambition, apprehension, depression, morbid fears, and loss of memory, and these may be succeeded by disorientation, confusion, hysteria, and sometimes maniacal outbursts. More general symptoms such as headache, irritability, inability to concentrate, and apathy may also be encountered. Insomnia is a very common complaint. Various psychoses, including manic–depressive syndromes, severe paranoia, hallucinations, etc., may necessitate confinement in a mental asylum. A good percentage of the inmates of such institutions have been reported to be pellagrins, which bears testimony to the serious nature of the mental changes (Gopalan and Rao, 1978).

As is true of deficiencies of the other B vitamins, pellagra often accompanies poverty, chronic alcoholism, dietary peculiarities, fever, hyperthyroidism, pregnancy, and the stress of injury or surgical procedures. Its incidence is much greater during the spring than in any other season. In advanced stages pellagra can be diagnosed by the classic "three D's"—dermatitis, diarrhea, and dementia (death being the "fourth D"). Early stage recognition of the disease must depend on reliable medical and dietary history and careful physical examination.

Pellagra has been endemic in certain areas of Italy, Spain, Rumania, Egypt, and the southeastern part of the United States. Classically, pellagra has been considered to be a disease of the corn-eating population and has practically disappeared from all affluent societies. It is still encountered among the Bantus of South Africa, in the Middle East, in Yugoslavia, and in the corn-eating population of India (Gopalan and Rao, 1978).

It had been accepted that the low tryptophan and unavailable niacin in corn was responsible for pellagra (see Section VIII). However, identification of pellagra in populations subsisting on sorghum, a millet in which niacin is neither so low nor biologically unavailable, raised serious doubts regarding this concept (Gopalan and Srikantia, 1960). Pellagra has been found to be endemic among the poor communities of India where the millet *Sorghum vulgare* is the staple food. The effect of a high intake of leucine and possibly an isoleucine–leucine imbalance has been suggested for pellagra in sorghum-eating as well as corn-eating populations (Gopalan, 1969). Recent studies indicate that concomitant vitamin B_6 deficiency, either primary or brought about by leucine toxicity, may also have an important role (Rao *et al.*, 1975).

B. Assessment of Status

Determination of cellular or blood serum levels of niacin or niacin-dependent enzymes has not proved to be a reliable or acceptable method for evaluating niacin status. To date, a functional biochemical test for assessing body reserves has not been developed (Hankes, 1984). Biochemical assessment of niacin nutritional status is usually based on measurement of urinary metabolites. Measurement of niacin metabolites would be dependent on species, as marked differences in type of metabolites exist among species, (see Section V). For humans and most monogastric species, niacin is excreted largely as methylated products. One of the testing procedures widely used in human population surveys is the estimation of urinary excretion of N'-methylnicotinamide during a period of 4–5 hr after administering a test dose of 50 mg niacinamide (Narasinga Rao and Gopalan, 1984).

X. SUPPLEMENTATION

Niacin supplementation must be considered as a possibility with all classes of animals and in particular with swine and poultry diets. Much of the niacin in common feeds (plant sources) is in a bound form that is not available to animals (see Section VIII). In formulating swine and poultry diets, therefore, niacin values for corn and other cereal grains and their by-product feeds should be almost disregarded. It is best to assume that these feeds provide no available niacin for the pig or chick (Cunha, 1982).

The pig and chick and most species can use the essential amino acid tryptophan to synthesize niacin, but they cannot convert niacin back to tryptophan. Therefore, if a diet contains enough niacin, the tryptophan is not depleted for niacin synthesis. Most monogastric diets do not contain large excesses of tryptophan, particularly diets based on corn. Tryptophan concentrations are not only low in corn, but are largely unavailable. Therefore, one should make sure that monogastric diets are adequate in niacin since it is very low in price and it would be poor economics to satisfy niacin needs by the more expensive tryptophan (Cunha, 1982). Also, some species, such as the cat, mink, and fish, apparently lack the ability to convert tryptophan to niacin.

The most critical time for supplementation is during early growth, when requirements are the highest (Table 8.1). Scott *et al.* (1982) reported that no evidence has been obtained of any need to supplement practical diets of mature poultry with niacin. Niacin requirements as recommended by the National Research Council for swine and poultry may be insufficient. T. J. Cunha for swine and M. L. Scott for poultry have recommended higher niacin requirements as a reasonable safety factor (Cunha, 1982). Higher niacin levels are mainly recommended when subclinical disease level, stress, and higher production rates are encountered. For poultry, recommended levels are generally 12–25% higher than the NRC recommendations. Increased swine requirements are approximately double those of the NRC for growing swine under the more stressful conditions, with breeding animals needing a 4-fold increase (Cunha, 1982). Producers must judge their own livestock operations in relation to stress conditions for animals and other factors that may require higher niacin supplementation levels.

Addition of niacin to a milk replacer is recommended at a level of 2.6 ppm because of lack of synthesis in the undeveloped rumen. Inclusion in diets of growing heifers or lactating dairy cows beyond the first trimester of lactation cannot be justified by current evidence (Schultz, 1983).

Supplementation of niacin for ruminants can be beneficial. If tryptophan is present in large amounts, the microorganisms can probably synthesize considerable niacin. However, in urea diets the microbes do not have enough tryptophan to draw upon and hence use exogenous sources of niacin. Supplementation appears to be of greatest benefit to stressed animals such as beef cattle entering the feedlot, dairy cows immediately postpartum, and cows suffering from subclinical ketosis. Recommended supplemental levels are estimated to be 50–100 ppm niacin for finishing beef cattle and 3–12 g daily for milking cows.

Although general results of field experiments on niacin supplementation for lactating cows are positive for both milk production and fat test, the somewhat erratic results and the occasional lack of statistical significance make it difficult to give an unqualified recommendation to include niacin in diets of all dairy cows in early lactation. However, enough positive evidence is available to recommend it in herds with above average incidence of ketosis and fat cow problems. For individual cows with positive tests indicating subclinical ketosis, a level of 12 g per day is recommended. When used as a general diet additive, a level of 6 g per day is suggested, starting 2 weeks before calving. This recommendation to start before calving is based on the evidence that it inhibits lipid mobilization and thus may prevent abnormal deposition of fat in the liver at calving time. Most evidence would suggest that feeding be continued 8–12 weeks after calving (Schultz, 1983).

Niacin supplementation is important for human populations that depend primarily on corn and sorghum in their diets. Although the condition is largely eliminated in the more technologically advanced countries, it still is found as an undiagnosed complication of malnutrition in today's urban society (Bosco, 1980). Niacin deficiency is usually found in people with chronically bad dietary habits, people who consume low-calorie, low-protein, high-fat, and high-carbohydrate diets, alcoholics, and people with diseases that interfere with digestion and absorption. Also, some individuals with metabolic defects (i.e., Hartnup's disease, deviation of niacin–tryptophan metabolism) require supplemental niacin at levels higher than normally consumed in typical foods.

Pharmacological effects from massive supplemental doses of niacin are reported in humans. Large doses of nicotinic acid (but not the amide) produce vascular dilatation or "flushing" with accompanying sensation of burning of the face and hands. Large doses of nicotinic acid (3 g or more per day) administered orally result in a significant reduction of serum cholesterol and β-lipoprotein cholesterol (Narasinga Rao and Gopalan, 1984). Careful analysis of a large-scale trial in coronary heart disease indicates that it may be slightly beneficial in protecting persons against recurrence of nonfatal myocardial infarction, but not in reducing mortality (Carlson and Rossner, 1975). Niacin is also used in treatment of schizophrenia. Beneficial effects of pharmacological doses of niacin must be weighed against the long-term effects of high doses, including laboratory evidence of diabetes, hepatic injury, and activation of peptic ulcer.

Niacin is available in two forms, niacinamide and nicotinic acid, with both forms providing about the same niacin biological activity. Crystalline products are used in feeds and pharmaceuticals, and dry dilutions in feeds.

Commercially produced niacin is quite stable compared to most other vitamins. Synthetic niacin and the amide were found to be stable in premixes with or without minerals for 3 months (Verbeeck, 1975). Gadient (1986) reports niacin to be insensitive to heat, oxygen, moisture, and light. The retention of niacin activity in pelleted feeds after 3 months at room temperature should be 95–100% as a general rule.

XI. TOXICITY

Harmful effects of nicotinic acid occur at levels far in excess of requirements. Limited research indicates that nicotinic acid and nicotinamide are toxic at dietary intakes greater than about 350 mg/kg of body weight/day (NRC, 1987). Short-term iv administration of niacin at a dose of 2.5 g/kg was needed before 50% of test mice died. Respective values for subcutaneous and oral routes of administration were 2.8 and 4.5 g/kg in mice. Nicotinamide is two to three times more toxic than the free acid (Waterman, 1978).

High levels of nicotinic acid, such as 3 g/day in humans, can cause vasodilation, itching, sensations of heat, nausea, vomiting, headaches, and occasional skin lesions. For dogs, oral administration of 2 g/day of nicotinic acid (133–145 mg/kg of body weight) produced bloody feces in a few dogs, followed by convulsions and death (NRC, 1987).

9

Vitamin B_6

I. INTRODUCTION

Vitamin B_6 refers to a group of three compounds: pyridoxol (pyridoxine), pyridoxal, and pyridoxamine. Their activities are equivalent in animals but not in various microorganisms. Vitamin B_6 acts as a component of many enzymes that are involved in the metabolism of proteins, fats, and carbohydrates. The vitamin is particularly involved in various aspects of protein metabolism.

It has been generally considered that common feed ingredients for typical poultry and swine diets contain adequate amounts of vitamin B_6. Although such conditions are rare, vitamin B_6 can be deficient for poultry and swine in a few situations. Although animal diets are generally adequate in vitamin B_6, there is evidence of low vitamin B_6 status in the human population, especially for young and pregnant or lactating women.

II. HISTORY

The early history of vitamin B_6 can be found in Maynard *et al.* (1979), Scott *et al.* (1982), Driskell (1984), and Loosli (1988). Clinical signs of what was later to be known as vitamin B_6 deficiency were described in 1926 by Goldberger and Lillie in their attempts to produce pellagra in experimental animals. Rats developed a severe dermatitis called acrodynia (painful extremities). In 1934 György first recognized vitamin B_6 as a distinct vitamin and showed that it prevented rat acrodynia and that the condition was not prevented by vitamin B_1 (thiamin), vitamin B_2 (riboflavin), or niacin. Two years later Lepkowsky and associates showed the vitamin to be an essential nutrient for chickens.

Isolation of the vitamin in crystalline form was accomplished independently in five different laboratories in 1938. The structure of the vitamin was first explained by Kuhn and co-workers, who named it "adermin," because it was believed that skin dermatitis was the only clinical sign of the deficiency. The term was discarded when evidence accumulated for nondermal clinical signs for the vitamin deficiency. In view of the chemical structure of the compound (pyridine structure), György proposed pyridoxine as the vitamin name, with this proposal widely adopted. Later, however, from bacterial studies, two other compounds, pyridoxal and pyridoxamine, were identified. Thus, by official action of the Society of Biological Chemists and the American Institute of Nutrition, the original term B_6 became the approved name for this vitamin.

In 1939 chemical synthesis was accomplished by two independent laboratories. Therefore, only five years elapsed between the discovery of the vitamin and its chemical synthesis. The speed of isolation indicates the intense efforts expended in the mid-1930s to identify new vitamins.

III. CHEMICAL STRUCTURE, PROPERTIES, AND ANTAGONISTS

Vitamin B_6 is a relatively simple compound with three substituted pyridine derivatives that differ only in functional group in the 4-position; these are the alcohol (pyridoxine or pyridoxol), the aldehyde pyridoxal, and the amine pyridoxamine (Fig. 9.1). Pyridoxine is the predominant form in plants, whereas pyridoxal and pyridoxamine are the forms generally found in animal products. These three forms have equal activity when administered parenterally to animals but are not equivalent when administered to various microorganisms. Two additional vitamin B_6 forms found in foods are the coenzyme forms of pyridoxal

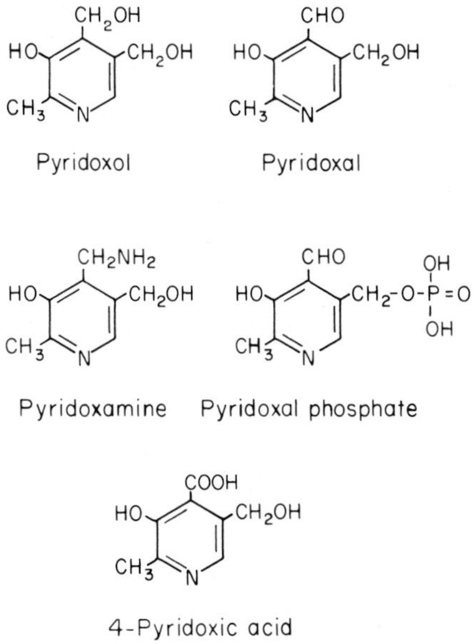

Fig. 9.1 Structural formulas of pyridoxol, pyridoxal, pyridoxamine, pyridoxal phosphate, and 4-pyridoxic acid.

phosphate (PLP) and pyridoxamine phosphate. Various forms of vitamin B_6 found in animal tissues are interconvertible, with vitamin B_6 metabolically active mainly as PLP and to a lesser degree as pyridoxamine phosphate.

Various forms of vitamin B_6 have been shown to be stable to heat, acid, and alkali, although exposure to light, especially in neutral or alkaline media, is highly destructive. Forms of vitamin B_6 are colorless crystals soluble in water and alcohol as both free bases and commonly available hydrochlorides. Commercial preparation of vitamin B_6 is almost exclusively the hydrochloride salt of the alcohol form, pyridoxine hydrochloride.

Several vitamin B_6 antagonists exist that either compete for reactive sites of apoenzymes or react with pyridoxal phosphate to form inactive compounds. Deoxypyridoxine is a powerful antagonist to vitamin B_6 and is commonly employed in experiments to accelerate the vitamin deficiency. Fortunately, this antivitamin does not occur in nature. A particularly important binding compound of vitamin B_6 is isonicotinic acid hydrazide (isoniazid), which is used in tuberculosis treatment. Isoniazid has been extensively used in experiments designed to clarify functions of various enzymes that are dependent on PLP (Price *et al.*, 1957).

The antihypertensive drugs thiosemicarbazide and hydralazine have also been shown to interfere with vitamin B_6 usage. Penicillamine, which is used to remove body copper in copper poisoning and Wilson's disease, is also known to complex PLP. L-Dopa, an antiparkinsonism drug, has been shown to be a vitamin B_6 antagonist. Other drugs also form thiazalidine compounds or inhibit pyridoxine kinase, the enzyme that is needed in formation of PLP (Rothschild, 1982). Oral contraceptives (estrogen component) have been shown to be antagonistic to vitamin B_6, resulting in a deficiency of the vitamin (Salkeld *et al.*, 1973). Presence of a vitamin B_6 antagonist in flax (linseed oil meal) is of particular interest to animal nutritionists. This substance was identified in 1967 as hydrazic acid and was found to have antibiotic properties (Parsons and Klostermann, 1967).

IV. ANALYTICAL PROCEDURES

Methods of analysis for vitamin B_6 must be capable of detecting all forms of the vitamin. In biological systems the vitamin is usually bound to protein and therefore extraction procedures are required. Bioassay using rat or chick growth has the advantage over microbiological procedures in that it measures all biologically active forms and does not require extraction of bound forms of the vitamin. Rat growth assay has been used most widely. Rat and chick bioassay methods are expensive and time-consuming and have been largely replaced by other techniques.

The standard method for quantitation of vitamin B_6 in foods is via microbiological assay. *Saccharomyces uvarum* (formerly *S. carlsbergensis*) is the organism commonly used in the microbiological method. This yeast organism, along with *Streptococcus faecalis* and *Lactobacillus casei,* makes it possible to differentiate between alcohol, amine, and aldehyde vitamin forms. Driskell (1984) lists the following disadvantages in microbiological methods: the procedure is time-consuming, variability exists in growth response of various microorganisms to the vitamin, microorganisms can mutate, microbiologically unavailable complexes of the vitamin may be formed, and microbial growth may be retarded by substances in the food extract.

Gas–liquid chromatography is used to analyze vitamin B_6, but mainly in pure standards. Combinations of chromatographic separation methods with fluorometric determinations may be used successfully for enriched foods and feeds. Vanderslice *et al.* (1980) developed a high-pressure liquid chromatography (HPLC) anion-exchange system that quantitatively separated phosphorylated and unphosphorylated forms of vitamin B_6 in pork, hamburger, carp, and nonfat dry milk. These HPLC techniques may likely become the methods of choice for vitamin B_6 assay.

V. METABOLISM

A. Digestion, Absorption, and Transport

Utilization of dietary vitamin B_6 by animals necessitates digestion and absorption of the five forms known to occur in foods: pyridoxine, pyridoxal, pyridoxamine, pyridoxal phosphate, and pyridoxamine phosphate. Digestion would first involve splitting the vitamin, as it is bound to protein portion of foods. Vitamin B_6 is absorbed mainly in the jejunum, but also in the ileum, by passive diffusion. Absorption from the colon is insignificant, even though colon microflora synthesize the vitamin.

Vitamin B_6 compounds are all absorbed from the diet in the dephosphorylated forms. The small intestine is rich in alkaline phosphatases for the dephosphorylation reaction. After absorption, B_6 compounds rapidly appear in liver, where they are mostly converted into pyridoxal phosphate, considered to be the most active vitamin form in metabolism. Pyridoxamine phosphate is involved only in several metabolic reactions. It is widely accepted that the synthesis of pyridoxal-P and pyridoxamine-P from the three unphosphorylated forms is mediated by pyridoxal kinase and pyridoxamine-P (pyridoxine-P) oxidase. Pyridoxal-P and pyridoxamine-P are also interconvertible through aminotransferases (formerly referred to as the transaminases). Both niacin (as an NADP-dependent enzyme) and riboflavin (as the flavoprotein pyridoxamine phosphate oxidase) are important for conversion of vitamin B_6 forms and phosphorylation reactions.

Although other tissues also contribute to vitamin B_6 metabolism, liver is thought to be responsible for forming pyridoxal phosphate found in plasma. Pyridoxal and pyridoxal-phosphate found in circulation are associated primarily with plasma albumin and red cell hemoglobin (Mehansho and Henderson, 1980). Pyridoxal phosphate accounts for 60% of plasma vitamin B_6. Currently, researchers do not agree on whether pyridoxal or pyridoxal phosphate is the transport form of B_6 (Driskell, 1984).

B. Storage

Only small quantities of vitamin B_6 are stored in the body. The vitamin is widely distributed in various tissues mainly as pyridoxal phosphate or pyridoxamine phosphate. Reports of pyridoxal phosphate content of glycogen phosphorylase suggests that 90% or more of the vitamin B_6 present in muscle might be present in this single enzyme (Henderson, 1984). Because the muscle accounts for about 40% of body weight, and because the muscle contains more vitamin B_6 per gram than do other tissues (except kidney), muscle phosphorylase may account for as much as 70 to 80% of the total body store of the vitamin.

C. Excretion

When B_6 coenzymes are synthesized in excess of the binding capacity of B_6-dependent apoenzymes they are dephosphorylated by alkaline phosphatase with the resultant pyridoxal reutilized or oxidized to pyridoxic acid (Fig. 9.1) by aldehyde oxidase and/or NAD-dependent dehydrogenase. Pyridoxic acid is the major route for elimination of vitamin B_6, which is excreted in urine. Also, small quantities of pyridoxol, pyridoxal, and pyridoxamine, as well as their phosphorylated derivatives, are excreted into the urine (Rabinowitz and Snell, 1949; Henderson, 1984). Biliary excretion and enterohepatic circulation of vitamin B_6 probably play only a minor role in the overall economy of the vitamin (Lui *et al.*, 1983).

VI. FUNCTIONS

Vitamin B_6 in the form of PLP (also named codecarboxylase) and to a lesser degree pyridoxamine phosphate plays an essential role in the interaction of amino acid, carbohydrate, and fatty acid metabolism and the energy-producing citric acid cycle. Over 50 enzymes are already known to depend on vitamin B_6 coenzymes. Pyridoxal phosphate functions in practically all reactions involved in amino acid metabolism, including transamination, decarboxylation, deamination, and desulfhydration, and in the cleavage or synthesis of amino acids.

Aminotransferases are involved in interconversions of a pair of amino acids into their corresponding ketoacids, generally these are α-amino and α-keto acids. For example, amino groups are transferred from glutamate to pyruvate with formation of α-ketoglutarate and alanine, or from aspartate to α-ketoglutarate, forming oxaloacetate and glutamate. Aminotransferases function in both amino acid biosynthesis (nonessential amino acids) and catabolism. Each aminotransferase is specific for a specified pair of amino and keto acids functioning as substrates, but is nonspecific for the other pair. The aminotransferases thus represent an important link between amino acid, carbohydrate, and fatty acid metabolism and the energy-producing citric acid cycle (Fig. 9.2).

Nonoxidative decarboxylation reactions also involve PLP as a coenzyme. Decarboxylases convert amino acids to synthesis of biogenic amines, such as histamine, hydroxytyramine, serotonin, γ-aminobutyric acid, ethanolamine, and taurine, some of which are substances of high physiological activity (regulation of blood vessel diameter, neurohormonal actions, and essential components of phospholipids and bile acids). In accord with their general importance these en-

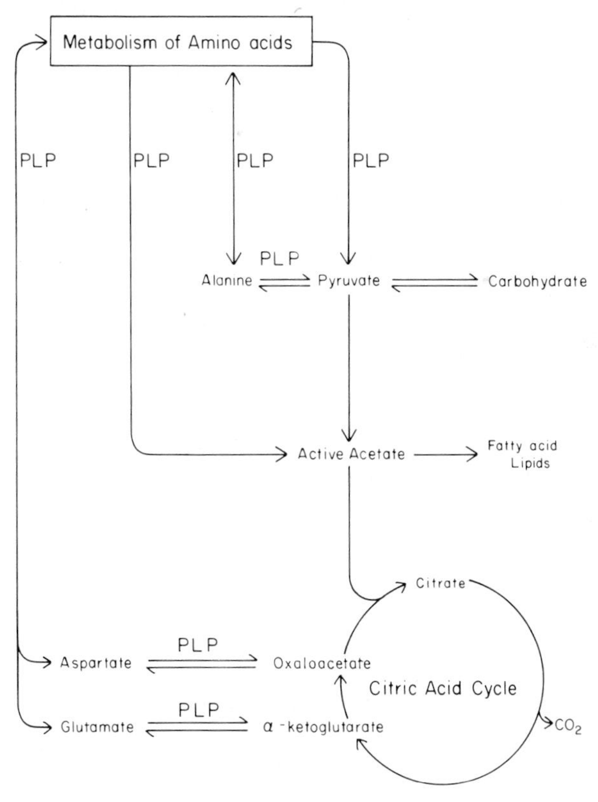

Fig. 9.2 Pyridoxal phosphate (PLP) and amino acid metabolism.

zymatic reactions take place in virtually all organs, most intensively in liver, heart, and brain. Neurological disorders including states of agitation and convulsions result from reduction of B_6 enzymes in the brain, including glutamate decarboxylase and γ-aminobutyric acid transaminase. Maternal restriction of B_6 in rats adversely affected neurogenesis and neuron longevity of the progeny (Groziak and Kirksey, 1987).

Vitamin B_6 is involved in many additional reactions, particularly those involving proteins. The vitamin participates in the following functions (Bräunlich, 1974; Marks, 1975; Driskell, 1984):

1. Deaminases—for serine, threonine, and cystathionine.
2. Desulfhydrases and transulfhyurases—interconversion and metabolism of sulfur-containing amino acids.
3. Synthesis of niacin from tryptophan—hydroxykynurenine is not converted to hydroxyanthranilic acid but rather to xanthurenic acid because of lack of the B_6-dependent enzyme kynureninase (see Chapter 8, Section VI).
4. Formation of δ-aminolevulinic acid from succinyl-CoA and glycine, the first step in porphyrin synthesis.
5. Conversion of linoleic to arachidonic acid in the metabolism of essential fatty acids (this function is controversial).
6. Glycogen phosphorylase catalyzes glycogen breakdown to glucose 1-phosphate. Pyridoxal phosphate does not appear to be a coenzyme for this enzyme but rather affects the enzyme's conformation.
7. Synthesis of epinephrine and norepinephrine from either phenylalanine or tyrosine—both norepinephrine and epinephrine are involved in carbohydrate metabolism as well as in other body reactions.
8. Racemases—pyridoxal phosphate-dependent racemases enable certain microorganisms to utilize D-amino acids. Racemases have not yet been detected in mammalian tissues.
9. Transmethylation by methionine.
10. Incorporation of iron in hemoglobin synthesis.
11. Amino acid transport—all three known amino acid transport systems, (a) neutral amino acids and histidine, (b) basic amino acids, and (c) proline and hydroxyproline, appear to require pyridoxal phosphate.
12. Formation of antibodies—B_6 deficiency results in inhibition of the synthesis of globulins, which carry antibodies.

VII. REQUIREMENTS

Requirement for vitamin B_6 has been found generally to depend on species, age, physiological function, dietary components, the intestinal flora, and other factors that are not yet fully understood. Table 9.1 summarizes the vitamin B_6

TABLE 9.1

Vitamin B_6 Requirements for Various Animals and Humans[a]

Animal	Purpose	Requirement	Reference
Beef cattle	Adult	Microbial synthesis	NRC (1984a)
Dairy cattle	Calf	65 μg/kg body wt	NRC (1978a)
	Adult	Microbial synthesis	NRC (1978a)
Chicken	Leghorn, 0–20 weeks	3.0 mg/kg	NRC (1984b)
	Leghorn, laying	3.0 mg/kg	NRC (1984b)
	Leghorn, breeding	4.5 mg/kg	NRC (1984b)
	Broilers, 0–8 weeks	2.5–3.0 mg/kg	NRC (1984b)
Turkey	All classes	3.0–4.5 mg/kg	NRC (1984b)
Duck (Pekin)	All classes	2.6–3.0 mg/kg	NRC (1984b)
Japanese quail	All classes	3.0 mg/kg	NRC (1984b)
Sheep	Adult	Microbial synthesis	NRC (1985b)
Swine	Growing–finishing, 1–5 kg	2.0 mg/kg	NRC (1988)
	Growing–finishing, 5–110 kg	1.0–1.5 mg/kg	NRC (1988)
	Adult	1.0 mg/kg	NRC (1988)
Horse	Adult	Microbial synthesis	NRC (1978b)
Goat	Adult	Microbial synthesis	NRC (1981b)
Mink	Growing	1.6 mg/kg	NRC (1982a)
Fox	Growing	2.0 mg/kg	NRC (1982)
Cat	Growing	4.0 mg/kg	NRC (1986)
Dog	Growing	60 μg/kg body wt	NRC (1985a)
Fish	Catfish	3.0 mg/kg	NRC (1983)
	Trout	5–15 mg/kg	NRC (1981a)
Rat	All classes	6.0 mg/kg	NRC (1978c)
Mouse	All classes	1.0 mg/kg	NRC (1978c)
Human	Infants	0.3–0.6 mg/day	RDA (1980)
	Children	0.9–1.6 mg/day	RDA (1980)
	Adults	1.8–2.8 mg/day	RDA (1980)

[a]Expressed as per unit of animal feed on an as fed (approximately 90% dry matter) or dry basis (see Appendix Table 1), except for dogs (body weight basis). Human data expressed as mg/day.

requirements for various animal species and humans, with a more complete listing given in Appendix Table 1. Because of microbial synthesis, ruminants have no dietary requirement for vitamin B_6. Young ruminants that do not have a fully developed rumen, however, require a dietary source. Considerable quantities of vitamin B_6 are synthesized in the relatively large intestinal tract of the horse, but whether this is adequately absorbed to meet requirements is controversial. Vitamin B_6 is also produced by microorganisms of the intestinal tracts of other animals and humans, but whether significant quantities are absorbed and utilized is in doubt. Animals practicing coprophagy would obviously be receiving vitamin B_6 from this source.

Breed of animal and environmental temperature have been shown to influence vitamin B_6 requirements. Lucas *et al.* (1946) found that cross-bred chicks (Rhode Island Red × Barred Plymouth Rock) showed a considerably higher requirement for pyridoxine than had previously been found for White Leghorn chicks, and a Japanese strain of chickens was also shown to have a higher B_6 requirement (Scott *et al.*, 1982). Regarding ambient temperature, when rats are housed at 33°C they needed twice as much vitamin B_6 as when they were housed at 19°C (Bräunlich, 1974).

Quantity of dietary protein affects requirement for vitamin B_6 in both animals and humans. Vitamin B_6 requirement is increased when high-protein diets are fed. For example, when feed contained 60% casein instead of 20%, the level of pyridoxine required by mice was three times as high (Miller and Baumann, 1945). Gries and Scott (1972) found a much higher vitamin B_6 requirement in chicks receiving 31% protein than in those receiving a normal 22% protein diet. A number of studies have suggested that amino acid imbalance has an adverse effect on vitamin B_6 status in that weight gain was depressed and survival was decreased when large amounts of a single amino acid were added to rat diets limited in the vitamin. High tryptophan, methionine, and other amino acids increase the need for vitamin B_6 (Scott *et al.*, 1982). Fisher *et al.* (1984) report a consistent deleterious effect of a low-quality protein on vitamin B_6 status in rats. Certain feed antagonists (see Section III), bioavailability of B_6 in feeds (see Section VIII), and nutrients other than protein influence the B_6 requirement. Niacin and riboflavin are needed for interconversions of different forms of vitamin B_6, with an overdose of thiamin reported to produce deficiency of the vitamin in rats (Driskell, 1984).

In humans, vitamin B_6 requirements are increased with higher dietary protein, pregnancy, lactation, oral contraceptives, alcoholic beverages, liver disease, and certain drugs (see Section III), in the elderly population, and for some individuals with inborn errors of metabolism. The daily vitamin B_6 requirement is estimated to be 1.5–2.0 mg for diets containing 100–150 g protein and between 1.0 and 1.5 mg when protein intake is less than 100 g (Driskell, 1984). The estimated requirement is increased to 2.5 mg/day during pregnancy and lactation (RDA, 1980). Low milk vitamin B_6 has been associated with long-term use of oral contraceptives (estrogen component). Both the use of oral contraceptives and alcoholism result in lowered plasma PLP levels. Low plasma PLP levels in alcoholics are believed to result from acetaldehyde, which reduces the protein binding of PLP, thereby promoting its dephosphorylation and oxidation to pyridoxic acid. Some individuals have inborn errors of metabolism that greatly increase their B_6 requirement. Fortunately, the number of such individuals appears to be small but unless they take doses of B_6 starting at birth, they develop convulsions, brain damage, and die (Scriver and Whelan, 1969).

VIII. NATURAL SOURCES

Vitamin B_6 is widely distributed in foods and feeds (Table 9.2). In general, muscle meats, liver, vegetables, whole-grain cereals and their by-products, and nuts are among best sources; few materials are really poor sources with the exception of fruits. The vitamin present in cereal grains is concentrated mainly in bran, the rest containing only small amounts. The richest source is royal jelly produced by bees (5000 μg/g). Most vitamin B_6 in animal products is in the form of pyridoxal and pyridoxamine phosphates. In plants and seeds the usual form is pyridoxol. Many analytical figures for vitamin B_6, especially older ones, were too low because the assays did not measure all the biologically active forms (Scott *et al.*, 1982).

TABLE 9.2

Vitamin B_6 in Foods and Feedstuffs (mg/kg, Dry Basis)[a]

Food	mg/kg	Food	mg/kg
Alfalfa leaves, sun cured	4.4	Sorghum, grain	5.0
Barley, grain	7.3	Soybean meal, solvent extracted	6.7
Bean, navy (seed)	0.3	Spleen, cattle	1.3
Blood meal	4.8	Wheat, bran	9.6
Brewer's grains	0.8	Wheat, grain	5.6
Buttermilk (cattle)	2.6	Whey	3.6
Copra meal (coconut)	4.8	Yeast, brewer's	39.8
Corn, gluten meal	8.8	Yeast, torula	38.9
Corn, yellow grain	5.3		
Cottonseed meal, solvent extracted	6.8		
Crab meal	7.2		
Fish meal, anchovy	5.0		
Fish meal, menhaden	5.1		
Horse meat	0.7		
Linseed meal, mechanically extracted	6.1		
Liver, cattle	18.0		
Milk, skimmed, cattle	4.5		
Molasses, sugarcane	5.7		
Oat, grain	2.8		
Oat, grouts	1.2		
Pea seeds	1.7		
Peanut meal, solvent extracted	6.9		
Potato	15.5		
Rice, grain	5.0		
Rice, polished	0.4		
Rye, grain	2.9		

[a]NRC (1982b).

The level of vitamin B_6 contained in all feeds is affected by processing and subsequent storage. Vitamin B_6 loss during cooking, processing, refining, and storage has been reported to be as high as 70% (Shideler, 1983) or in the range of 0–40% (Birdsall, 1975). Losses may be caused by heat, light, and various agents that can promote oxidation. Two-hour sunlight exposure may destroy half the vitamins in milk.

In the early 1950s several infants in the United States were diagnosed as deficient in vitamin B_6 as a result of consuming a commercially sterilized, liquid, milk formula. Later studies showed that the vitamin underwent rapid destruction during the autoclaving procedure used in the production of formulas. Also, human milk often does not contain sufficient vitamin B_6 for the newborn (see Section IX), however, human milk could be an adequate source of vitamin B_6 because vitamin levels respond rapidly to dietary intake (Kirksey, 1981).

Data on vitamin B_6 content of feeds are generally insufficient and information on the vitamin's bioavailability is lacking. Bioavailability of B_6 was found to be greater from beef than from corn meal, spinach, or potatoes (Nguyen and Gregory, 1983). LeKlem *et al.* (1980) reported that in young adult men, vitamin B_6 bioavailability in whole-wheat bread is lower than that in white bread enriched with pure pyridoxine. Vitamin B_6 in whole-wheat bread and peanut butter was 75 and 63%, respectively, as available as that from canned tuna (Kabir *et al.*, 1983).

IX. DEFICIENCY

A. Effects of Deficiency

Characteristics of vitamin B_6 deficiency are retarded growth, dermatitis, epileptic-like convulsions, anemia, and a partial alopecia. Because of the predominant function of the vitamin in protein metabolism, in vitamin B_6 deficiency a fall in nitrogen retention is observed, feed protein is not well utilized, nitrogen excretion is excessive, and impaired tryptophan metabolism may result.

1. Ruminants

Amounts of vitamin B_6 that are normally synthesized by microorganisms in the digestive tract of ruminants, when these are fully developed, are enough to meet needs of these animals. McElroy and Goss (1939) reported 6–10 mg/kg of vitamin B_6 in dried rumen contents of mature sheep, even though the diet contained only 1.5 mg/kg. Therefore, no deficiency signs of vitamin B_6 have yet been observed in mature ruminants.

Vitamin B_6 has been shown to be essential for the young calf when selected experimental diets are used. Some calves showed clinical signs of anorexia, scouring, convulsive seizures, and death in 3–4 weeks, while others grew poorly to 4 months without showing these extreme deficiency signs (Johnson *et al.*, 1947). Pathological studies indicated some demyelination of peripheral nerves and hemorrhages in the epicardium.

2. Swine

In growing pigs, clinical signs of vitamin B_6 deficiency include a poor appetite, slow growth (Fig. 9.3), microcytic hypochromic anemia, epileptic-like fits or convulsions (Fig. 9.4), fatty infiltration of the liver, diarrhea, rough hair coats, scaly skin, a brown exudate around the eyes, demyelination of peripheral nerves, and subcutaneous edema (Bauernfeind, 1974; Bräunlich, 1974; Cunha, 1977). The first and most conspicuous sign in baby pigs that vitamin B_6 is insufficient is a loss of appetite that, if the deficiency is severe, may appear in less than 2 weeks. This is accompanied by reduced growth, vomiting, diarrhea, and a peculiar compulsion to lick.

When deficiency of vitamin B_6 reaches an advanced stage (probably due to degeneration of the peripheral nerves), disordered movement and ataxia appear. Finally, convulsions develop at irregular intervals, but are, apparently, stimulated by excitement, because they are most often observed at feeding time. Between

Fig. 9.3 A 6-week-old B_6-deficient pig weighing only 3.6 kg. (Courtesy of R. W. Luecke and E. R. Miller, Michigan Agriculture Experiment Station, and *J. Nutrition* **62** (1957), 405.)

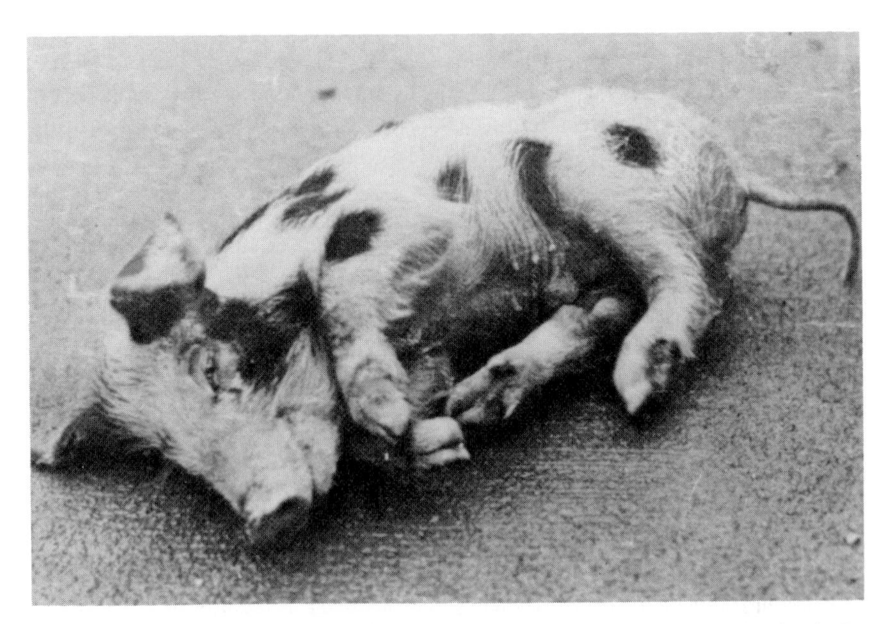

Fig. 9.4 This pig is having an epileptic-like seizure while receiving a diet low in vitamin B_6. (Courtesy of E. H. Hughes and H. Heitman, California Agricultural Experiment Station.)

these convulsions, pigs lie down and are apathetic and unresponsive (Bräunlich, 1974). Bräunlich (1974) suggested that a vitamin B_6 deficiency may go unnoticed in swine because of a lack of visible signs associated specifically with the deficiency. Metabolic disorders may be revealed only by poor appetite, slow growth, and inefficient feed utilization. In some experiments with vitamin B_6, protein retention by pigs deficient in the vitamin was reduced to less than half of that shown in animals receiving sufficient amounts of the vitamin.

During reproduction and lactation, sows fed a corn–sorghum–soybean meal diet responded to vitamin B_6 supplementation at a level of 4.4 mg/kg of feed (Adams *et al.*, 1967). Vitamin B_6 supplementation of 11 mg/kg of feed produced a slightly superior daily gain in weight, more piglets born alive, and a smaller number of resorbed fetuses as compared with control sows that received only 1 mg/kg of vitamin B_6 (Ritchie *et al.*, 1960).

3. Poultry

Chicks fed a vitamin B_6-deficient diet have little appetite and grow slowly, with plumage failing to fully develop. Chicks maintained on a B_6-deficient diet exhibited general weakness after a few days of deprivation. The birds squat in a characteristic posture, with wings slightly spread and head resting on the ground (Bräunlich, 1974).

A more specific sign of B_6 deficiency is the nature of the nervous condition that develops. Deficient chicks are abnormally excitable. As deprivation continues, nervous disorders become increasingly severe (Bräunlich, 1974). There is a trembling and vibration of the tip of the tail, with movement stiff and jerky. Chicks will run aimlessly about with lowered head and drooping wings. Finally, convulsions develop, during which chicks fall on their side or back, with the legs scrabbling. Violent convulsions cause complete exhaustion and may lead to death. These clinical signs may be distinguished from those of encephalomalacia by the relatively greater intensity of activity during a seizure (Scott *et al.*, 1982).

Similar signs of a vitamin B_6 deficiency have been observed in turkey poults: loss of appetite, poor growth (Fig. 9.5), oversensitivity and cramps, and finally death. Ducklings not receiving enough vitamin B_6 grow slowly and development of plumage is poor, but cramps and other nervous disorders appear only after a long period of deficiency.

Signs for B_6 deficiency in chicks will appear very rapidly after introduction of a B_6-deficient feed. Fuller and Kifer (1959) reported that signs of a deficiency appeared on the eighth day. Chronic, borderline B_6 deficiency produces perosis. Usually one leg is severely crippled and one or both of the middle toes may be bent inward at the first joint (Gries and Scott, 1972). A marked increase in

Fig. 9.5 Vitamin B_6-deficient poult (about 4 weeks old) on left and a normal poult on right. (Courtesy of T. W. Sullivan, University of Nebraska.)

gizzard erosion was found in vitamin B_6-deficient chicks (Daghir and Haddad, 1981).

For adult poultry, vitamin B_6 deficiency results in reduced egg production and hatchability as well as decreased feed consumption, weight loss, and death. A severe deficiency (levels of vitamin B_6 below 0.5 mg/kg of diet) causes rapid involution of the ovary, oviduct, comb, and wattles in mature laying hens. Involution of testes, comb, and wattle occurs in vitamin B_6-deficient adult cockerels (Scott *et al.*, 1982).

4. Horses

No experimental information is available on vitamin B_6 requirements or deficiency in the horse. It has been shown that the horse synthesizes vitamin B_6 in the intestinal tract, with the assumption that adequate quantities are absorbed. However, some researchers believe that racehorses need vitamin B_6 supplementation because when these horses undergo intensive training they need a high proportion of protein in their diets that would increase B_6 requirements (Bräunlich, 1974).

5. Other Animal Species

a. Dogs and Cats. A typical consequence of a B_6 deficiency in both young and old dogs is a microcytic, hypochromic anemia. In addition to decreased appetite and body weight loss, pathological changes include ataxia, cardiac dilatation and hypertrophy, congestion of various tissues, and demyelination of peripheral nerves (NRC, 1985a). In kittens deficient in B_6, there is growth depression, a mild microcytic, hypochromic anemia, convulsive seizures, and irreversible kidney lesions consisting of areas of tubular atrophy and dilatation (NRC, 1986).

b. Laboratory Animals. Rat fed diets deficient in vitamin B_6 develop symmetrical scaling dermatitis on the tail, paws, face, and ears, microcytic anemia, hyperexcitability, convulsions, and reduced reproductive performance in both sexes (NRC, 1978c). For mice, deficiency signs include poor growth, hyperirritability, posterior paralysis, necrotic degeneration of the tail, and alopecia. Loss of appetite is an early and constant result of B_6 deficiency in hamsters. Although acrodynia is characteristic of vitamin B_6 deficiency in rats, it is not found in hamsters. The fur of B_6-deprived hamsters has an unkempt appearance, and crusted lesions are occasionally observed on lips and mouth (NRC, 1978c).

c. Rabbits. In rabbits, vitamin B_6 deficiency signs include a scaly thickening of the skin of ears and inflammation around eyes and nose. In a few cases there is a pronounced acrodynia with severe encrustation and inflammation of nose, eyes, and paws. Alopecia develops most often on forelegs and is accompanied by skin inflammation and desquamation (Bräunlich, 1974). Neurological signs observed include convulsions, partial paralysis, and contracture of extremities. In the advanced stage, some rabbits are unable to move their hind legs at all, the muscles of the hindleg having become severely atrophied.

d. Nonhuman Primates. Vitamin B_6 deficiency signs in rhesus monkeys consist of weight loss, hypochromic microcytic anemia, apathy, and ataxia. There is widespread arteriosclerosis, fatty metamorphosis of liver, hepatic necrosis and cirrhosis, dental carries, oxaluria, and neuronal degeneration of the cerebral cortex (NRC, 1978d). With cebus monkeys, deficiency consisted of weight loss, a profound hypochromic microcytic anemia, hair loss, and dermatitis, especially about the hands and toes.

e. Foxes and Mink. Signs of deficiency in growing kits appeared after about 2 weeks on a B_6-deficient diet and included reduced feed intake, weight loss, diarrhea, brown exudate around the nose, excessive lacrimation, swelling and puffiness around the nose and face region, apathy, muscular incoordination, convulsions, and finally death (NRC, 1982a). Reproduction is also affected in mink with testes degeneration and resorption of embryos. Vitamin B_6 deficiency in the fox results in anorexia, growth cessation, and a decrease in hemoglobin.

f. Fish. Most fish are sensitive to vitamin B_6 deficiency and show neurological disorders in 3–8 weeks (Lovell, 1987). Clinical signs for catfish include hyperirritability, erratic swimming, anorexia, tetany, and greenish-blue coloration. For cold-water fish, signs of a deficiency include general nervous disorders, epileptic-like seizures, hyperirritability, lowered resistance to handling, erratic and spiral swimming, rapid breathing, flexing of opercles, and rapid onset of rigor mortis after death (NRC, 1981a).

6. Humans

A high proportion of the human population receives inadequate dietary vitamin B_6, particularly young and pregnant or lactating women. Typical vitamin B_6 deficiency symptoms in humans include hypochromic microcytic anemia, loss of weight, abdominal distress, vomiting, hyperirritability, epileptic-type convulsions in infants, depression and confusion followed by convulsions in adults, and electroencephalographic abnormalities (Driskell, 1984). Seborrhea-like lesions developed about the eyes, nose, and mouth and some subjects showed cheilosis and glossitis. A depression of the lymphocyte count was a fairly constant finding.

Convincing evidence of the essential nature of vitamin B_6 in human nutrition was first provided by reports in the United States of infants who developed irritability and convulsive seizures (and sometimes died) when fed autoclaved infant formulas. This autoclaving destroyed much of the vitamin B_6, and infants did not receive adequate quantities of the vitamin. Infants receiving human milk are also sometimes at risk as generally this is a poor source of the vitamin. Other conditions responsive to B_6 first seen in infants are several inborn errors of metabolism resulting in vitamin B_6 dependency.

Both pregnancy and use of oral contraceptives increased the vitamin B_6 requirement (Miller, 1986). Contractor and Shane (1970) discovered that pregnant women had lower blood levels of pyridoxal phosphate than did nonpregnant controls. Brin (1971) reported that pregnant women had increased urinary metabolites of tryptophan. He also found fetal blood levels of B_6 two to three times greater than the maternal level, indicating a depletion of the mother's B_6 stores. According to Heller *et al.* (1973), vitamin B_6 supplementation is necessary in approximately 50% of pregnant women to maintain normal B_6 coenzyme saturation of aspartate aminotransferase.

Alcohol and certain drugs (see Section III) interfere with the metabolic functions of vitamin B_6. Vitamin B_6 deficiency is viewed as an important nutritional complication in alcoholism, and it has been established that alcoholics absorb less vitamin B_6 than do control subjects. Alcoholics have a decreased ability to liberate vitamin B_6 from its bound form and several studies suggest that alcohol interferes with conversion of pyridoxine to pyridoxal phosphate (Bonjour, 1980). Certain drugs, including the antituberculosis drug isonicotinic hydrazide (INH), have been shown to increase need of vitamin B_6. INH forms hydrazone complexes with pyridoxal phosphate, which inactivates the coenzyme (Roe, 1973). Studies suggest that vitamin B_6 availability is limited in hyperthyroidism and consequently B_6 should be added to antithyroid medication for hyperthyroid patients (Wohl *et al.*, 1960).

Vitamin B_6 is affected by uremia and urinary calculi and has been associated with inadequacy of the vitamin. Many symptoms of uremia are quite similar to vitamin B_6 deficiency. Convulsions, peripheral neuritis, and depression of the immune response are seen in both conditions. Dobbelstein and co-workers (1974) suggested the uremic toxins exert an inhibitory effect on pyridoxal phosphate. The increase in oxalic as well as xanthurenic acid in the urine of B_6-deficient humans has been suggested as a cause for urinary calculi. In a study by Lilum *et al.* (1981), 85% of 100 rats produced calcium oxalate stones within 6 weeks of being placed on a vitamin B_6-free diet.

B. Assessment of Status

Several biochemical procedures have been developed and utilized for assessment of vitamin B_6 status of humans and animals. Deficiency leads to an impairment of protein utilization even before visible signs occur. The most widely used approach to assess vitamin B_6 status in humans is measuring urinary metabolites of tryptophan following a tryptophan load test. In a B_6 deficiency, tryptophan is not converted to niacin and is metabolized to other intermediate products at a greater rate than in a nondeficient state. Metabolites most often measured are xanthurenic acid and kynurenic acid. Although the tryptophan load test has been widely used in assessment of vitamin B_6 status, results obtained via this method should be interpreted with care, as several other metabolic and hormonal factors are known to be involved in tryptophan metabolism. The tryptophan load test is a good parameter for the assessment of the vitamin B_6 status of population groups.

Direct measurement of one or more forms of vitamin B_6 in plasma or urine has not been a useful approach largely because of analytical limitations. In recent years, however, use of apoenzymes such as tyrosine aminotransferase, tyrosine decarboxylase, tryptophanase, or aspartate aminotransferase has made possible the accurate determination of plasma pyridoxal phosphate. Also, one of the more common methods for assessing vitamin B_6 status is through measurement of aminotransferases in red blood cells (Cinnamon and Beaton, 1970). Measurements are made before and after addition of pyridoxal phosphate. An increase in aminotransferases after addition of pyridoxal phosphate suggests a deficiency. Driskell (1984) concludes that the best vitamin B_6 assessment parameter for vitamin B_6 status in clinical cases is measurement of either the coenzyme stimulation of erythrocyte alanine aminotransferase activity or the plasma pyridoxal phosphate level.

X. SUPPLEMENTATION

Vitamin B_6 is one of the B vitamins that is least likely to be deficient in livestock diets. Only before the rumen is developed should B_6 supplementation for ruminants be considered. Bräunlich (1974) suggested that the best possible concentration of vitamin B_6 in "milk substitutes" for calves is probably higher than 2.4 mg/kg. Because of its wide dietary distribution, vitamin B_6 is seldom deficient in typical swine and poultry diets. Evidence to date indicates that corn, soybean meal, and other ingredients used to supply energy and protein in practical diets for these species usually also provide sufficient vitamin B_6. Consequently, it is not often added in supplemental form.

Under certain conditions vitamin B_6 supplementation is warranted for practical growing and breeding diets for poultry and other monogastric or simple-stomached animals. Fuller *et al.* (1961) believed that although corn–soybean meal practical poultry breeder diets probably contain sufficient vitamin B_6 to support hatchability, there is little margin of safety. The amount of supplemental vitamin B_6 recommended for monogastric species varies from 1 to 10 mg/kg of diet depending on species, age, activity, stress of performance, and field use experience (Bauernfeind, 1974). Reasons for needed supplementation of vitamin B_6 include the following (Perry, 1978): (1) great variations in amounts of B_6 in individual ingredients, (2) variable bioavailability of this vitamin in ingredients, (3) losses reported during processing of ingredients, (4) discrepancies between activity for test organisms versus those for animals, and (5) a higher pyridoxine requirement due to a marginal level of methionine in the diet.

Variability of vitamin B_6 in feeds depends on the sample origin, conditions of growth, climate, weather conditions, and other local factors (see Section VIII). For products of animal origin, concentration depends on level of nutrition and environment of original animals. Yen *et al.* (1976) determined available vitamin B_6 in corn and soybean meal using a chick growth assay. Corn was found to be 38–45% available and B_6 in soybean meal 58–62% available. There was little difference in availability between corn samples not heated and heated to 120°C. However, corn heated at 160°C contained significantly less available B_6. Level of vitamin B_6 contained in feedstuffs is also affected by processing and subsequent storage. In one report a loss of 30% of B_6 contained in alfalfa meal during the coarse-milling and pelleting processes was observed (Bräunlich, 1974). Ink and Henderson (1984) reported that bioavailability can be as low as 40–50% after heat processing of foods.

Predominant losses of vitamin B_6 activity in food occur in the pyridoxal and pyridoxamine forms, with pyridoxine being the more stable form. Supplemental vitamin B_6 is reported to have a higher bioavailability and stability than the naturally occurring vitamin, which in retorted milk products exhibited only 50% of the bioavailability of synthetic B_6 or B_6 in formulas that were fortified with the vitamin prior to thermal processing (Tomarelli *et al.*, 1955).

Commercially, vitamin B_6 is available as crystalline pyridoxine hydrochloride and various dilutions. Pyridoxine hydrochloride contains 82.3% vitamin B_6 activity. The dry dilution is used in feeds, while the crystalline product is used in parenteral and oral pharmaceuticals as well as in feeds. For human nutrition, pyridoxine hydrochloride is used by pharmaceutical companies in the preparation of capsules, tablets, and ampoules. These tablets used for prophylactic purposes usually contain 2 mg per daily dose. For therapeutic purposes, 10- to 150-mg tablets are taken one to three times daily.

The recovery of vitamin B_6 as pyridoxine hydrochloride in a multivitamin premix not containing trace minerals was 100%, even after 3 months' storage at 37°C. However, stability in a premix containing trace minerals was poor with only 45% recovery after 3 months at 37°C (Adams, 1982). Verbeeck (1975) found vitamin B_6 to be stable in premixes with minerals as sulfates. However, if minerals in the form of carbonates and oxides are used, 25% of the vitamin can be lost over a 3-month period. Stress agents such as choline chloride help catalyze this destruction. Gadient (1986) considers pyridoxine to be very sensitive to heat, slightly sensitive to moisture and light, and insensitive to oxygen. Retention of B_6 activity in pelleted feeds after 3 months at room temperature should be 80–100% as a general rule. The retention of pyridoxine in an extruded fish feed after 1 month at room temperature was found to be 56%.

In recent years data have accumulated for the need of vitamin B_6 supplementation for humans, especially for young and pregnant or lactating women. Requirements for B_6 are higher for individuals during pregnancy, use of oral contraceptives, certain drug therapy, radiation sickness, overuse of alcohol, hyperthyroidism, uremia, urinary calculi, and in errors of metabolism (see Section IX). Vitamin B_6 is one of several vitamins currently popular in "orthomolecular" megadose therapy. It has been used for a variety of conditions including premenstrual syndrome and behavioral disorders. Pyridoxine tablets are available over the counter in dosages of 30–500 mg.

XI. TOXICITY

In common with other members of the vitamin B complex, vitamin B_6 has a very low toxicity. Prolonged feeding of pyridoxine·HCl to rats (2.5 mg per day), puppies (20 mg per kilogram body weight per day), and monkeys (10 mg per kilogram body weight per day) produced no toxic signs (Unna and Antopol, 1940). Rats, rabbits, and dogs tolerated 1 g/kg body weight without ill effects, however, larger doses produced adverse manifestations in all three species. Marked impairment of coordination and/or righting reflexes were observed within about 3 days, followed by severe convulsions, paralysis, and death. Pyridoxal is apparently twice as toxic as pyridoxine or pyridoxamine. Evidence from dog and rat studies suggests that more than 1000 times the requirement would be needed in diets to produce signs of toxicity (NRC, 1987).

Humans are reported to tolerate daily doses of 20–1000 mg per day of pyridoxine·HCl for prolonged periods without deleterious effects. These doses are 10–500 times greater than the recommended daily dietary allowance of vitamin B_6 for adults (Haskell, 1978). Vitamin B_6 is one of several vitamins currently popular in megadose therapy (see Section X). Humans have developed a sensory neuropathy as a direct result of taking 2000–6000 mg of pyridoxine daily (Schaumburg *et al.*, 1983). Subjects developed ataxia and severe sensory nervous system dysfunctions. Substantial improvement occurred with pyridoxine withdrawal.

10

Pantothenic Acid

I. INTRODUCTION

After the discovery of thiamin, riboflavin, and niacin as B-complex vitamins, researchers realized that at least one other unidentified factor remained. Two additional vitamins, pyridoxine (vitamin B_6) and pantothenic acid, were found in the late 1930s as fractions of yeast and liver. Tissue extracts from a variety of biological materials provided a growth factor for yeast that was identified as "pantothenic acid," derived from the Greek word "pantos," meaning "found everywhere."

Pantothenic acid is found in two enzymes, coenzyme A and acyl carrier protein, which are involved in many reactions in carbohydrate, fat, and protein metabolism. Although this vitamin occurs in practically all feedstuffs, the quantity present is generally insufficient for optimum performance of poultry and swine and other monogastric species. There are no reports of deficiency in adult ruminants because of microbial synthesis. Pantothenic acid deficiency occurs only rarely in humans.

II. HISTORY

Historical aspects of pantothenic acid have been reviewed (Scott *et al.*, 1982; Fox, 1984; Loosli, 1988). During the 1930s several independent research programs concentrated on either a growth factor for microorganisms or a chick antidermatitis factor. Research to identify pantothenic acid was also closely associated with studies of vitamin B_6 (pyridoxine). Both vitamins were fractions of the "vitamin B_2" complex and were found associated in biological materials. Since pyridoxine was adsorbed on Fuller's earth, from which it could be eluted, and pantothenic acid was not, pantothenic acid was sometimes referred to as "filtrate factor" and pyridoxine as "eluate factor." Pantothenic acid deficiency was first described in the chick as a pellagra-like dermatitis by Norris and Ringrose in 1930. Since the "filtrate factor" cured dermatitis in chicks, but not rats, one early name of the vitamin was "chick antidermatitis factor."

In independent studies in 1933, R. J. Williams and associates fractionated "bios," a growth factor for yeast, particularly *Saccharomyces cerevisiae*. They concentrated the factor, determined many of its properties, and provided the name pantothenic acid. Snell and associates independently studied the nature of an essential factor for lactic acid bacteria. It was soon found that the filtrate factor, the chick antidermatitis factor, and the unknown factor required by yeast and lactic acid bacteria were all pantothenic acid. In 1940 the structure was determined and in the same year the vitamin was synthesized independently in three different laboratories.

III. CHEMICAL STRUCTURE, PROPERTIES, AND ANTAGONISTS

Pantothenic acid is an amide consisting of pantoic acid (α, γ-dihydroxy-β,β′-dimethylbutyric acid) joined to β-alanine. Pantothenic acid and the biologically active coenzyme A, which contains the vitamin as an essential component, are shown in Fig. 10.1. The vitamin is derivatized at its carboxyl end by β-mer-

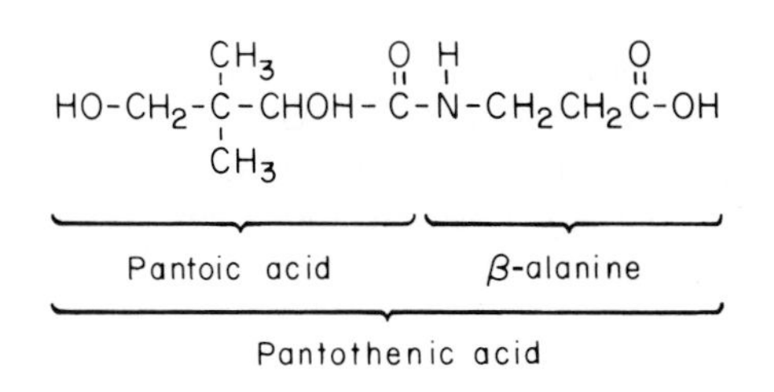

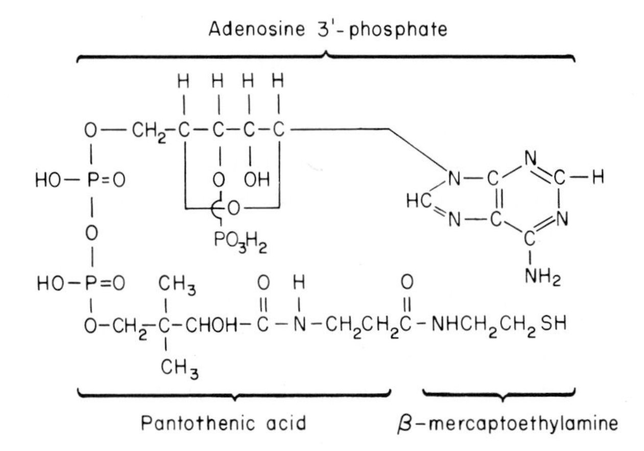

Fig. 10.1 Structures of pantothenic acid and coenzyme A.

captoethylamine and at its alcoholic end by phosphate to form a pseudodinucleotide containing phosphoadenylic acid. Therefore coenzyme A contains the vitamin combined with adenosine 3′-phosphate, pyrophosphate, and β-mercaptoethylamine. Another metabolically active form of pantothenic acid is acyl carrier protein. The alcohol equivalent to pantothenic acid, panthenol, is more easily absorbed and is converted to the acid *in vivo* (Marks, 1975).

Analogs of pantothenic acid that have replaced the β-alanine with other amino acids such as α-alanine, β-aminobutyric acid, aspartic acid, leucine, or lysine are inactive. The most common anatagonist of panthothenic acid is ω-methylpantothenic acid, which has been used to produce a deficiency of the vitamin in humans (Hodges *et al.*, 1958). Other antivitamins include pantoyltaurine, phenylpantothenate, and antimetabolites of the vitamin containing alkyl or aryl ureido and carbamate components in the amide part of the molecule (Fox, 1984).

The free acid of the vitamin is a viscous, pale yellow oil readily soluble in water and ethyl acetate. The oil is extremely hygroscopic and is easily destroyed by acids, bases, and heat. Maximum heat stability occurs at pH 5.5–7.0.

Calcium pantothenate is the pure form of the vitamin used in commerce. It crystallizes as white needles from methanol and is reasonably stable to light and air. Pantothenic acid is optically active (characteristic of rotating a polarized light). It may be prepared as either the pure dextrorotatory (*d*-) form or the *dl*-form; the racemic form has approximately one-half the biological activity of *d*-calcium pantothenate. Only the dextrorotatory form, *d*-pantothenic acid, is effective as a vitamin. The *l*-pantothenic acid is inactive for organisms requiring the intact vitamin.

IV. ANALYTICAL PROCEDURES

Pantothenic acid content of various substances has been analyzed by microbiological assay, animal bioassay, radioimmunoassay, and chemical methods (Fox, 1984). The complex nature of natural products has hindered development of chemical methods for assay of the vitamin. Chick growth bioassays have been used extensively and measure bound forms as well as free pantothenic acid. Radioimmunoassay (Wyse and Hansen, 1977) and an automated fluorometric assay technique (Roy and Buccafuri, 1978) have been reported to be successful for pantothenic acid determination.

Microbiological procedures for determination of pantothenic acid are widely employed but require that the vitamin be freed from the coenzyme form. Values for pantothenic acid content of feedstuffs that were obtained by microbiological procedures performed before proper methods were devised for liberating the vitamin from its bound form are much too low. Procedures for cleavage of bound forms to obtain free pantothenic acid can include use of (1) intestinal phosphatase to cleave the phosphate linkage and (2) pigeon or chick liver enzyme preparation to break the linkage between mercaptoethylamine and pantothenic acid. Bound forms cannot be hydrolyzed as this will destroy the vitamin. A number of organisms can be used in the assay method, with *Lactobacillus plantarum* (formerly *Lactobacillus arabinosus*) widely used.

V. METABOLISM

Pantothenic acid is found in feeds in both bound (largely as coenzyme A) and free forms. It is necessary to liberate the pantothenic acid from the bound forms in the digestive process prior to absorption. Little information is available on digestion, absorption, and transport of the vitamin. Pantothenic acid, its salt, and the alcohol are absorbed from the intestinal tract, probably by diffusion (Marks, 1975). Within tissues pantothenic acid is converted to coenzyme A and other compounds in which the vitamin is a functional group (Sauberlich, 1985). Free pantothenate appears to be efficiently absorbed; in the dog between 81 and 94% of an oral dose of sodium [^{14}C]pantothenate was absorbed (Taylor *et al.*, 1974). Measurement of pantothenic acid bioavailability in adult men consuming a typical United States diet ranged from 40 to 61% with an average of 50% (Sauberlich, 1985). Urinary excretion represents the major route of body loss of absorbed pantothenic acid, with prompt excretion when taken in excess. Most pantothenic acid is excreted as the free vitamin, but some species (e.g., dog) excrete it as β-glucuronide (Taylor *et al.*, 1972).

Animals and humans do not appear to have the ability to store appreciable amounts of pantothenic acid, with organs such as liver and kidney having the highest concentrations. The majority of pantothenic acid in blood exists in red blood cells as coenzyme A; serum contains no coenzyme A but does contain free pantothenic acid. A number of human studies suggest that stores of pantothenic acid are not great as tissue levels dropped dramatically and/or clinical signs developed anywhere from 7 to 12 weeks on diets low in the vitamin (Marks, 1975; Fry *et al.*, 1976). Urinary pantothenic acid concentrations dropped from 3 to 0.8 mg in just 63 days in human subjects fed low quantities of the vitamin (Fry *et al.*, 1976).

VI. FUNCTIONS

Pantothenic acid is a constituent of two important coenzymes, coenzyme A and acyl carrier protein (ACP). Coenzyme A is found in all tissues and is one of the most important coenzymes for tissue metabolism. The important role of coenzyme A is summarized in Fig. 10.2, with biochemical reactions catalyzed by coenzyme A presented in Table 10.1.

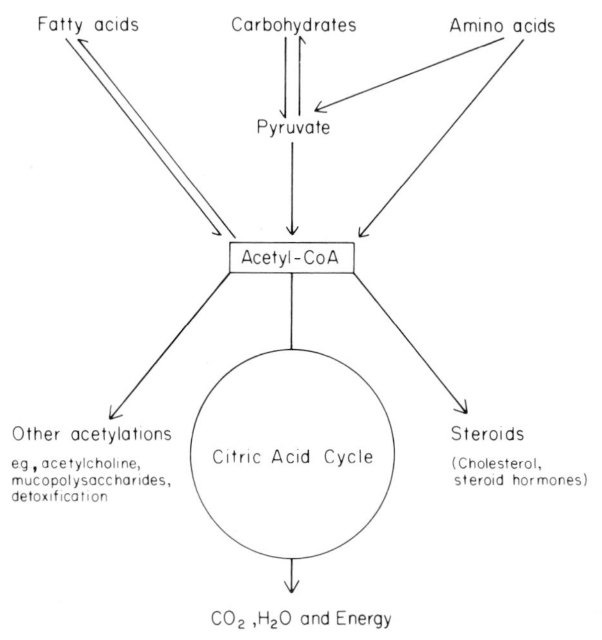

Fig. 10.2 The important role of coenzyme A in metabolism. (Modified from Marks, 1975.)

The most important function of coenzyme A is to act as a carrier mechanism for carboxylic acids (Marks, 1975). Such acids when bound to coenzyme A have a high potential for transfer to other groups and are normally then referred to as "active." The most important of these reactions is the combination of coenzyme A with acetate to form "active acetate" with a high-energy bond that renders acetate capable of further chemical interactions. As an example, it is utilized directly by combination with oxaloacetic acid to form citric acid, which enters the Krebs citric acid cycle. Its combination with two-carbon fragments from fats, carbohydrates, and certain amino acids to form acetyl coenzyme A is an essential step in their complete metabolism, because the coenzyme enables these fragments to enter the citric acid cycle.

Coenzyme A functions as a carrier of acyl groups in enzymatic reactions involved in synthesis of fatty acids, cholesterol, and sterols, in the oxidation of fatty acids, pyruvate, and α-ketoglutarate, and in biological acetylations. In the form of active acetate, acetic acid can also combine with choline to form acetylcholine, the chemical transmitter at the nerve synapse, and can be used for detoxification of drugs including sulfonamides.

Coenzyme A has an essential function in lipid metabolism, fatty acids are activated by formation of the coenzyme derivative, and degradation by removal

TABLE 10.1

Selected Biochemical Reactions Catalyzed by Coenzyme A[a]

Enzyme	Pantothenate derivative	Reactant	Product	Site
Pyruvic dehydrogenase	CoA	Pyruvate	Acetyl-CoA	Mitochondria
α-Ketoglutarate dehydrogenase	CoA	α-Ketoglutarate	Succinyl-CoA	Mitochondria
Fatty acid oxidase	CoA	Palmitate	Acetyl-CoA	Mitochondria
Fatty acid synthetase	Acyl carrier protein	Acetyl-CoA, Malonyl-CoA	Palmitate	Microsomes
Propionyl-CoA carboxylase	CoA	Propionyl-CoA, carbon dioxide	Methylmalonyl-CoA	Microsomes
Acyl-CoA synthetase	Phosphopantetheine	Succinyl-CoA, GDP + P_i	Succinate, GTP + CoA	Mitochondria

[a]Modified from Olson (1984).

of acetate fragments in beta oxidation also uses another molecule of coenzyme A. These active acetate fragments may directly enter the citric acid cycle or combine to form ketone bodies. A pantothenic acid-dependent enzyme, ACP, that replaces coenzyme A during building of the carbon chain in synthesis of fatty acids was recognized in 1965 (Pugh and Wakil, 1965). ACP is a protein with a sulfhydryl group covalently attached to acetyl, malonyl, and intermediate-chain acyl groups. The sulfhydryl group at the binding site of the ACP was identified as a cysteine residue, thioethanolamine. Discovery of thioethanolamine and β-alanine residues suggested that both ACP and coenzyme A have similar acyl-binding sites.

Decarboxylation of ketoglutaric acid in the citric acid cycle yields succinic acid, which is then converted to the "active" form by linkage with coenzyme A. Active succinate and glycine are together involved in the first step of heme biosynthesis. Pantothenic acid also stimulates synthesis of antibodies, which increase resistance of animals to pathogens. It appears that when pantothenic acid is deficient, the incorporation of amino acids into the blood albumin fraction is inhibited, which would explain why there is a reduction in the titer of antibodies (Axelrod, 1971).

VII. REQUIREMENTS

Pantothenic acid requirements for selected animals and humans are presented in Table 10.2. For growth and reproduction, the majority of species have a dietary requirement between 5 and 15 mg/kg. For egg production by chickens the pantothenic acid requirement is very low (2.2 mg/kg) compared to a requirement

TABLE 10.2

Pantothenic Acid Requirements for Various Animals and Humans[a]

Animal	Purpose	Requirement	Reference
Beef cattle	Adult	Microbial synthesis	NRC (1984a)
Dairy cattle	Calf	130 μg/kg body wt	NRC (1978a)
	Adult	Microbial synthesis	NRC (1978a)
Chicken	Leghorn, 0–20 weeks	10.0 mg/kg	NRC (1984b)
	Leghorn, laying	2.2 mg/kg	NRC (1984b)
	Leghorn, breeding	10.0 mg/kg	NRC (1984b)
	Broilers, 0–8 weeks	10.0 mg/kg	NRC (1984b)
Turkey	4–8 weeks	11.0 mg/kg	NRC (1984b)
	8–24 weeks	9.0 mg/kg	NRC (1984b)
Duck (Pekin)	0–7 weeks	11.0 mg/kg	NRC (1984b)
Japanese quail	All classes	10.0–15.0 mg/kg	NRC (1984b)
Sheep	Adult	Microbial synthesis	NRC (1985b)
Swine	Growing–finishing, 1–5 kg	12.0 mg/kg	NRC (1988)
	Growing–finishing, 5–110 kg	7.0–10.0 mg/kg	NRC (1988)
	Adult	12.0 mg/kg	NRC (1988)
Horse	Adult	Microbial synthesis	NRC (1978b)
Goat	Adult	Microbial synthesis	NRC (1981b)
Fox	Growing	8.0 mg/kg	NRC (1982a)
Cat	Growing	5.0 mg/kg	NRC (1986)
Dog	Growing	400 μg/kg body wt	NRC (1985a)
Fish	Catfish	10–20 mg/kg	NRC (1983)
	Trout	10–20 mg/kg	NRC (1981a)
Rat	All classes	8.0 mg/kg	NRC (1978c)
Mouse	All classes	10.0 mg/kg	NRC (1978c)
Human	Children	4–7 mg/day	RDA (1980)
	Adults	4–7 mg/day	RDA (1980)

[a]Expressed as per unit of animal feed either on an as fed (approximately 90% dry matter) or dry basis (see Appendix Table 1) (except for calf and dog). Human data expressed as mg/day.

of 10 mg/kg for growth and reproduction. Requirements are based on typical consumption levels. When energy density of diets is increased intake is reduced so that higher dietary concentrations of pantothenic acid and other vitamins are required. When the level of energy in the rations of broilers was raised from 2870 to 3505 kcal/kg, intake of pantothenic acid fell by 19.1%, because appetite is mainly controlled by intake of energy (Friesecke, 1975).

Apparently there is a wide variation in pantothenic acid requirements among breeds and among animals within the same breed. Data from Michigan show that for one-half of growing pigs 9.13 mg/kg of pantothenic acid was sufficient for growth, whereas the remaining half required more than 9.13 but less than 13.5 mg/kg (Luecke *et al.*, 1953).

Antibiotics have a sparing effect on the pantothenic acid requirement. A dietary level of 22 mg/kg aureomycin for weanling pigs (McKigney *et al.*, 1957) and 10 mg/kg of procaine penicillin to turkey poults (Slinger and Pepper, 1954) reduced the pantothenic acid requirement for these species.

If the rumen is functioning normally, ruminal microflora will synthesize enough pantothenic acid to satisfy ruminant needs, however, biosynthesis will depend on the composition of feed. Vitamin synthesis is reduced with diets high in cellulose but increases with higher quantities of easily soluble carbohydrates (Virtanen, 1966). Certain amounts of B-complex (including pantothenic acid) vitamins are synthesized in the large intestine of monogastric animals. It is doubtful, however, whether much benefit is derived as only limited pantothenic acid absorption occurs in the large intestine, with greatest benefit found in animals that practice coprophagy (Friesecke, 1975).

Interrelationships with other vitamins on pantothenic acid requirements are known, for example, that among pantothenic acid and vitamin B_{12}, ascorbic acid, and biotin (Scott *et al.*, 1982). Pantothenic acid requirement of chicks from B_{12}-depleted hens was found to be greater than that of chicks from normal hens. A 5-fold increase in coenzyme A content of liver was found in B_{12}-deficient chicks and rats. There is no evidence of sparing action of B_{12} on pantothenic acid needs for swine as has been observed in poultry (Luecke *et al.*, 1953). There is some evidence that pantothenic acid is involved in ascorbic acid synthesis in plants and animals, with ascorbic acid sparing the pantothenic acid requirement in rats (Barboriak and Krehl, 1957). Also, there have been suggestions of a possible interrelationship between folacin and biotin with pantothenate. Both vitamins were found necessary for pantothenic acid utilization in the rat (Wright and Welch, 1943). The inclusion of biotin in the diet of a pantothenic acid-deficient pig was effective in prolonging the life of the pig, but caused the pantothenic acid deficiency symptoms to appear in half the time (Colby *et al.*, 1948).

Dietary fat and protein influence the pantothentic acid requirement. Pigs fed a diet deficient in pantothenic acid and high in fat failed to gain weight, exhibited a lower feed efficiency, and developed deficiency signs more quickly than those fed diets low in fat (Sewell *et al.*, 1962). Nelson and Evans (1945) found that rats deficient in pantothenic acid fed a high-protein diet excreted more pantothenic acid and had accelerated growth and survival rates in comparison with rats fed a low-protein diet. The superiority of the high-protein diet may be due to the decreased level of dietary carbohydrate, which would presumably require coenzyme A for metabolism.

The Food and Nutrition Board of the National Research Council (RDA, 1980) has not established a recommended human dietary allowance for pantothenic acid but states that 4 to 7 mg per day is a safe and adequate dosage for children and adults. Women who are pregnant or lactating may have an increased requirement of about 30% (Plaut *et al.*, 1974).

VIII. NATURAL SOURCES

Pantothenic acid is widely distributed in foods of animal and plant origin. Alfalfa hay, peanut meal, cane molasses, yeast, rice bran, green leafy plants, wheat bran, brewer's yeast, fish solubles, and rich polishings are good sources for animals. For human foods, in general, liver, kidney, yeast, egg yolk, and fresh vegetables are the best sources, followed by milk and meat, with lowest levels in grains, fruits, and nuts. Royal jelly of bees is a rich source (510 ppm), with ovaries of the cod containing over 2000 ppm. The pantothenic acid content of foods and feedstuffs is shown in Table 10.3.

Many swine and poultry diets are borderline in supplying pantothenic acid requirements and many are deficient in this vitamin. Corn- and soybean meal-based diets are apt to be deficient in pantothenic acid. Milling by-products such

TABLE 10.3

Pantothenic Acid in Foods and Feedstuffs (mg/kg, Dry Basis)[a]

Food/feedstuff	mg/kg	Food/feedstuff	mg/kg
Alfalfa hay, sun cured	28.6	Rice, bran	25.2
Alfalfa leaves, sun cured	32.4	Rice, grain	9.1
Barley, grain	9.1	Rice, polished	3.9
Bean, navy (seed)	2.3	Rye, grain	9.1
Blood meal	2.6	Sorghum, grain	12.5
Brewer's grains	8.9	Soybean meal, solvent extracted	18.2
Buttermilk (cattle)	40.1	Soybean seed	17.3
Citrus pulp	14.3	Spleen, cattle	8.2
Clover hay, ladino (sun cured)	1.1	Timothy hay, sun cured	7.9
Copra meal (coconut)	6.9	Wheat, bran	33.5
Corn, gluten meal	11.2	Wheat, grain	11.4
Corn, yellow grain	6.6	Whey	49.6
Cottonseed meal, solvent extracted	15.4	Yeast, brewer's	118.4
Eggs	27.0	Yeast, torula	100.6
Fish meal, anchovy	10.9		
Fish meal, menhaden	9.4		
Fish, sardine	11.8		
Linseed meal, solvent extracted	16.3		
Liver, cattle	164.9		
Milk, skimmed, cattle	38.6		
Molasses, sugarcane	50.3		
Oat, grain	8.8		
Pea seeds	21.0		
Peanut meal, solvent extracted	50.7		
Potato	22.0		

[a]NRC (1982b).

as rice bran and wheat bran are good sources, being two to three times higher than the respective grains.

Pantothenic acid is reported to be fairly stable in feedstuffs during long periods of storage (Scott *et al.*, 1982). The authors indicate that heating during processing may cause considerable losses, especially if temperatures attain 100°–150°C for long periods of time and pH values are above 7 or below 5.

Gadient (1986) considers pantothenic acid to be slightly sensitive to heat, very sensitive to moisture, and not very sensitive to oxygen or light. Pelleting was reported to cause only small losses of the vitamin. As a general guideline, pantothenic acid activity in normal pelleted feed over a period of 3 months at room temperature should be 80–100%.

IX. DEFICIENCY

A. Effects of Deficiency

On the basis of observations on pantothenic acid-deficient animals and studies in human volunteers, deficiency of the vitamin is shown in the following signs and symptoms:

1. Reduced growth and feed conversion efficiency.
2. Lesions of skin and its appendages.
3. Disorders of the nervous system.
4. Gastrointestinal disturbances.
5. Inhibition of antibody formation and thus decreased resistance to infection.
6. Impairment of adrenal function.

Clinical signs of pantothenic acid deficiency take many forms and differ from one animal species to another. For humans, additional emotional and neurological symptoms include hyperventilation, irritability, insomnia, depression, headache, and dizziness. Since pantothenic acid is so widely distributed in nature, clearcut deficiency symptoms in humans are rarely found in practice. Typical pantothenic acid deficiency does, however, occur under certain feeding regimes with animals.

1. Ruminants

Pantothenic acid is not required in the diet of adult ruminants, because ruminal microorganisms synthesize this vitamin in adequate amounts. Sheppard and Johnson (1957) have experimentally produced pantothenic acid deficiencies in calves. Major clinical signs were anorexia, reduced growth, rough hair coat, dermatitis, diarrhea, and eventual death. The most characteristic pantothenic acid deficiency sign in the calf is scaly dermatitis around the eyes and muzzle. Anorexia and diarrhea follow after 11–20 weeks on a deficient diet, and calves become weak and unable to stand and may develop convulsions. They are susceptible to mucosal infection, especially in the respiratory tract. Postmortem studies have revealed moderate sciatic and peripheral nerve demyelination. If deficient calves were treated with calcium pantothenate, they responded with increased appetite and weight gains and subsequent reversal of dermatitis.

2. Swine

Many swine diets are borderline in supplying pantothenic acid and many are deficient in the vitamin (Cunha, 1977). Nonspecific signs of pantothenic acid deficiency in swine include reduced growth rate, bloody diarrhea, loss of appetite, poor general condition, lowered blood pantothenic acid, and reduced feed conversion.

A characteristic sign of a pantothenic acid deficiency in the pig is locomotor disorder (especially of hindquarters), which has been described by Goodwin (1962). In the early stages of such a deficiency, the movement of back legs becomes stiff and jerky. Standing animals show a slight tremor of the hindquarters. When deficiency persists, this particular action of the back legs grows more exaggerated and resembles a characteristic military gait, termed "goose step"

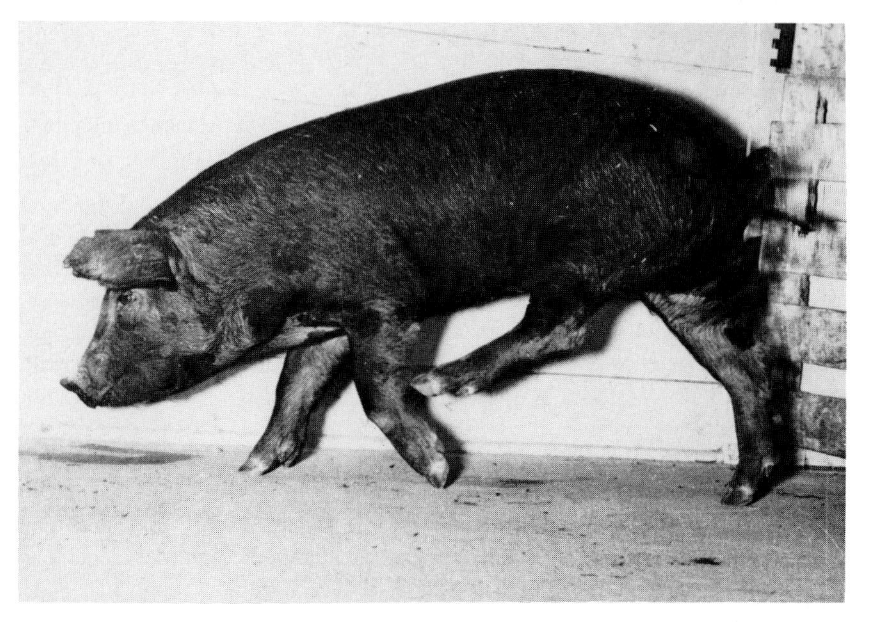

Fig. 10.3 Goose-stepping in pantothenic acid deficiency. (Courtesy of R. W. Luecke, Michigan Agriculture Experiment Station.)

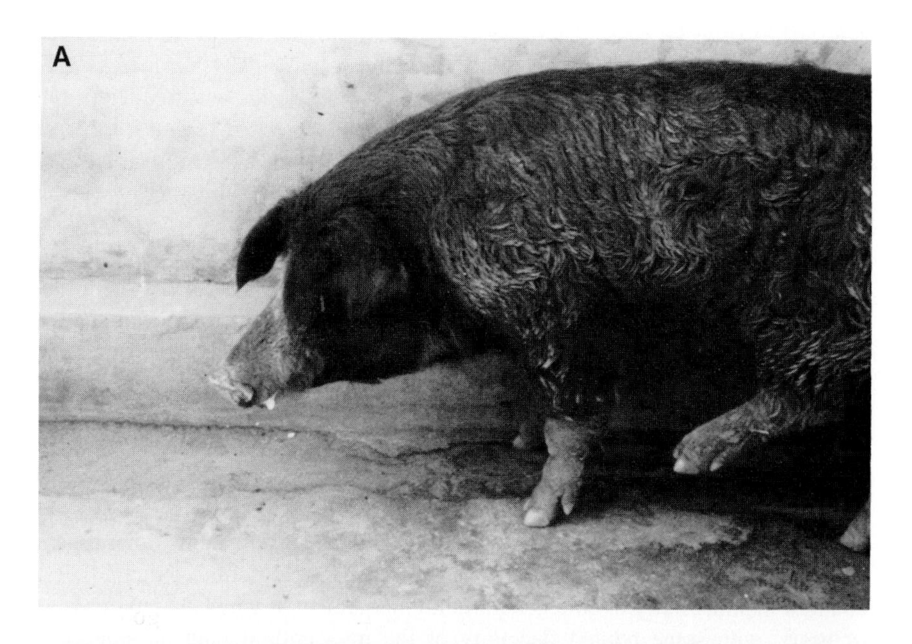

Fig. 10.4 Pantothenic acid deficiency. Locomotor incoordination (goose-stepping) (A). Affected pig often will fall sideways or, with its back legs spread apart, will assume a posture that resembles that of a sitting dog (B). (Courtesy of L. R. McDowell, University of Florida.)

(Fig. 10.3) (Luecke *et al.*, 1953). The condition may be so severe that, as the pig moves forward, the back legs will touch the belly. Finally, an increasingly severe paralysis of the hindquarters develops. Affected pigs will frequently fall sideways or with their back legs spread apart, a posture that resembles that of a sitting dog (Fig. 10.4). The chief microscopic lesion is a chromatolysis of isolated cells of the dorsal root ganglia, followed by a demyelinating process in brachial and sciatic nerves.

Pigs suffering from pantothenic acid deficiency have scurfy skin and thin hair and a brownish secretion around the eyes. The dermatosis associated with deficiency appears principally on the shoulders and behind the ears; the skin appears dirty and scaly. Skin becomes reddened, and the bristles on the rump and along the spine loosen and fall. The dermatosis extends to the mucosa, where it becomes manifest as necrotic enteritis, ulceration, and hemorrhages in the large intestine (Ullrey *et al.*, 1955). As a consequence, the feces contain blood. Goodwin (1962) observed various degrees of gastritis and, occasionally, peritonitis and intestinal fissures.

Ullrey *et al.* (1955) observed pathological changes in some of the other organs of sows and baby pigs maintained on pantothenic acid-deficient diets. These changes included fatty liver degeneration, enlargement of adrenals, enlargement of heart, with some related flaccidity of the myocardium, and intramuscular hemorrhages. Histopathological studies showed degenerative changes and necroses of the tissue cells.

Pantothenic acid is particularly important in sow fertility, with insufficient quantities of the vitamin resulting in complete reproductive failure. Female hogs fed low-pantothenic acid diets developed fatty livers, enlarged adrenal glands, intramuscular hemorrhage, heart dilation, diminution of ovaries, and improper uterus development (Ullrey *et al.*, 1955). Ensminger *et al.* (1951) reported that although gilts on a low-pantothenic acid diet became pregnant, they did not farrow or show any signs of pregnancy. Necropsy of gilts revealed macerating feti in the uterine horns in all cases. Minimal pantothenic acid sufficient to result in normal farrowing may still result in an abnormal locomotion in suckling pigs from sows that had received diets low in the vitamin (Teague *et al.*, 1971).

3. Poultry

The major lesions of pantothenic acid deficiency for poultry appear to involve the nervous system, adrenal cortex, and skin (Scott *et al.*, 1982). Pantothenic acid deficiency reduces normal egg production and hatchability. Subcutaneous hemorrhage and severe edema are signs of pantothenic acid deficiency in the developing chick embryo. In chickens, there is first a decline of growth, followed by decline in feed conversion and retardation of feather growth. Plumage becomes rough and ruffled, and feathers become brittle and may fall; next a dermatosis rapidly develops in chicks. Corners of the beak and the area below the beak are

Fig. 10.5 Pantothenic acid deficiency. Note dermatitis around beak. (Courtesy of G. F. Combs, University of Maryland.)

the worst affected, but the disorder is also observed in feet. Outer layers of skin between toes and on bottoms of feet peel off and small cracks and fissures appear. In some cases skin layers of feet thicken and cornify and wart-like protuberances develop on balls of the feet. The foot problem is usually exacerbated by bacterial invasion of the lesions. Within 12–14 days of receiving a deficient diet, the margins of the eyelids are sealed closed by a viscous discharge (Fig. 10.5 illustrates the typical deficiency syndrome in the chick).

Pantothenic acid concentrations in liver are reduced during deficiency. Liver is hypertrophied and varies in color from a faint yellow to a dirty yellow. Nerves and fibers of the spinal cord show myelin degeneration, with these degenerating fibers occurring in all segments of the cord down to the lumbar region (Scott *et al.*, 1982).

In young chicks clinical signs of pantothenic acid deficiency are difficult to differentiate from those of biotin deficiency—both cause severe dermatitis, broken feathers, perosis, poor growth, and mortality (See Chapter 11, Section IX). In pantothenic acid deficiency, dermatitis of the feet is evident over the toes, in contrast to biotin deficiency, which primarily affects the foot pads and is often more severe than deficiency of pantothenic acid.

Signs of pantothenic acid deficiency in young turkeys are similar to those in

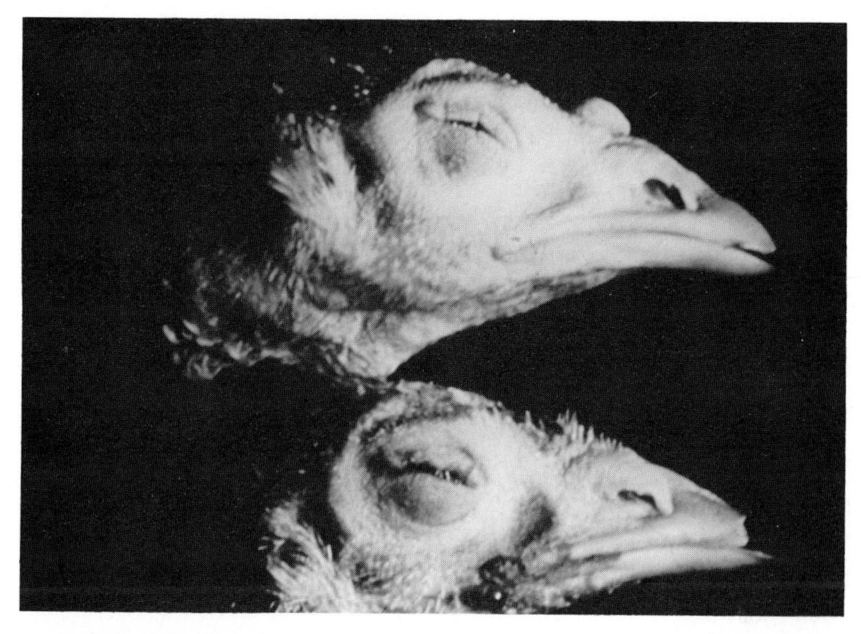

Fig. 10.6 Pantothenic acid deficiency. BSW turkey with dermatitis on lower beak and at angle of mouth. Sticky exudate formed on eyelid resulted in encrustation and caused swollen eyelids to remain stuck together. Control hen above. (Courtesy of T. M. Ferguson (deceased) and J. R. Couch, Texas A&M University.)

young chickens and include general weakness, dermatitis, and sticking together of eyelids (Fig. 10.6). Young ducks do not show the usual signs seen in chickens and turkeys, except retarded growth; however, their mortality rate is very high.

A pantothenic acid deficiency does not normally affect egg production but severely depresses hatchability and chicks that hatch may be too weak to survive. Embryonic mortality in pantothenic acid deficiency occurs usually during the last few days of incubation. A direct linear relationship exists between diet pantothenic acid and hatchability. Beer *et al.* (1963) fed a purified diet to White Leghorn hens that contained 0.9 mg of pantothenic acid/kg of diet. They found that the hens required addition of 1.0 mg of pantothenic acid/kg of diet for optimum egg production, at least 4.0 mg/kg for maximum hatchability, and 8.0 mg/kg for optimum hatchability and viability of offspring. Dawson *et al.* (1962) reported that turkey breeder hens fed a diet deficient in pantothenic acid demonstrated a high embryonic mortality during the first week of development. After 17 days, the surviving embryos were small, poorly feathered, and showed signs of edema, hemorrhaging, fatty livers, and pale dilated hearts.

4. Horses

A pantothenic acid deficiency has not been produced with the horse. Likewise, no requirement for the vitamin has been reported. Pearson and Schmidt (1948) reported that urinary excretion of pantothenic acid varied with intake, but no difference was noted in growth rates.

5. Other Animal Species

Experimental pantothenic acid deficiencies have been induced in most animal species studied. However, the deficiency is not present in animals with sufficient microbial fermentation (e.g., ruminants and horses) or animals that practice sufficient coprophagy (e.g., rabbit).

a. Dogs and Cats. Pantothenic acid-deficient dogs exhibit erratic appetites, reduced rates of growth, lowered antibody response, and reduced blood concentrations of cholesterol and total lipids. Deficient dogs have reduced pantothenate concentrations in urine, blood, liver, muscle, and brain. In terminal stages of deficiency, dogs exhibit spasticity of the hindquarters, sudden prostration, or coma, usually accompanied by rapid respiratory and cardial rates and possibly convulsions (NRC, 1985a). For kittens, pantothenic acid deficiency is characterized chiefly by emaciation (NRC, 1986). Moderate to marked fatty metamorphosis of the liver occurs, with both fine and coarse vacuolar formation.

b. Laboratory Animals. A deficiency retards the development of weight in all laboratory animals; weight gains are reduced, or weight may even be lost if deficiency is severe. Feed consumption is reduced because of appetite loss. Animals become apathetic and noticeably less active (Friesecke, 1975).

In rats, lack of pantothenic acid results in a premature graying of hair (achromotrichia). For this reason pantothenic acid has been called the "anti-gray-hair factor." Another term applied to rats on a deficient diet is "bloody whiskers." Porphyrin from the lachrymal glands of the rat accumulates on the whiskers in pantothenic acid deficiency and gives the appearance of blood. The same condition can be produced when water intake is reduced to 50% of normal, so it may be due to adrenal degeneration. Pantothenic acid deficiency also induces exfoliative dermatitis, oral hyperkeratosis, necrosis, and ulceration of the gastrointestinal tract. Focal or generalized hemorrhagic necrosis of the adrenals may occur, and death results after 4–6 weeks of deficiency (NRC, 1978c). Deficient rats have impaired antibody synthesis, decreased serum globulins, and decreased antibody-forming cells in response to antigen. Reproduction is affected with litter size and birth weight reduced. Newborn rats exhibit neurological defects in the form of locomotor disturbances, incoordination of movements, and motor spasms.

Pantothenic acid deficiency signs in growing mice are reported to be weight loss, hair loss, particularly on the ventral surface, flanks, and legs, dermatosis, partial posterior paralysis, other neurological abnormalities, and achromotrichia. Young guinea pigs fed a purified, pantothenic acid-deficient diet developed signs of deficiency such as decreased growth rate, anorexia, weight loss, rough coat, diarrhea, weakness, and death. Adult guinea pigs fed a pantothenic acid-deficient diet died within 10–41 days. Many of them had adrenal and gastrointestinal hemorrhages (NRC, 1978c).

c. Nonhuman Primates. Pantothenic acid deficiency in rhesus monkeys resulted in reduced growth, anemia, hair loss, and ataxia (NRC, 1978d).

d. Foxes and Mink. Deficiency signs of a pantothenic acid-deprived mink include severe emaciation, enlarged adrenals, and general gastric ulceration with petechial (minute) hemorrhages in the jejunum (NRC, 1982a). No definitive pantothenic acid studies in foxes have been completed.

e. Fish. Signs of pantothenic acid-deficient trout are manifested as a condition called dietary gill disease, including clubbed, exudate-covered gill lamellae, swollen opercles, fused gill filaments, abnormal swimming near the surface of the water, anorexia, poor feed conversion, loss of weight, and high mortality within 8–10 weeks (NRC, 1981a). Clubbed, exudate-covered, fused gill filaments (lamellae) are the characteristic result of deficiency in channel catfish (Lovell, 1987).

6. HUMANS

In humans, pantothenic acid deficiency has not been observed under natural conditions except associated with severe malnutrition. The main reason is that pantothenic acid deficiency is rare because of the vitamin's widespread distribution in foods. Deficiency has been produced in volunteers by feeding a pantothenic acid-deficient diet, but more rapidly by also including the antagonist ω-methylpantothenic acid (see Section III) in the diet. The deficient diet alone may require about 12 weeks to produce recognizable symptoms (Hodges *et al.*, 1958). Symptoms most commonly present were persistent and annoying fatigue, headache, muscle weakness, and cramps. Some subjects demonstrated impaired motor coordination and peculiar gait. Other deficiency manifestations are gastrointestinal disturbances, apathy, depression, cardiovascular instability, increased susceptibility to infections, and impaired adrenal function.

A deficiency of the vitamin was reported to cause "burning feet syndrome" during World War II among prisoners in Japan and the Philippines (Gopalan, 1946). Burning feet is a syndrome occasionally described for malnourished people in developing countries. Symptoms include abnormal skin sensations of the feet and lower legs, which were increased by warmth and lessened by cold. Administration of large doses of calcium pantothenate improved the ability of the subjects to withstand stress.

B. Assessment of Status

Urinary excretion has been used in humans to estimate status of pantothenic acid as well as the vitamin content of the diet. Urinary excretions of less than 1 mg per day are considered to be in the deficiency zone corresponding to intakes of possibly less than 4 mg per day (Olson, 1984). Like urine, the pantothenic acid content of human milk reflects the pantothenic acid content of the diet. Johnston *et al.* (1981) reported a correlation of 0.62 between dietary and milk content of pantothenic acid. Total pantothenic acid levels in whole blood below 80 μg per deciliter would be suggestive of the vitamin deficiency (Olson, 1984). Studies of pantothenic acid deficiency in various animal species indicate lowered tissue levels and decreased urinary excretion of the vitamin and decreased tissue coenzyme A (Nelson, 1978). However, present data are insufficient to establish pantothenic acid status for animals based on critical concentrations. There is a close relationship between dietary pantothenic acid in hen diets, egg production, and newly hatched chicks (Pearson *et al.*, 1945), indicating potential for status determination.

X. SUPPLEMENTATION

Monogastric animal diets based on grains, particularly corn, are routinely supplemented with pantothenic acid. Scott *et al.* (1982) concludes that practical diets usually contain sufficient pantothenic acid for all classes of chickens but a number of factors may influence the requirement for this vitamin. Because of this, supplemental calcium pantothenate is usually added to diets for chicks and breeding hens. Scott (1966) indicated that the pantothenic acid requirement for poultry may have to be increased 60 to 80% because of a lack of availability from bound forms in feeds. Clinical pantothenic acid deficiency signs appear to be completely reversible, if not too far advanced, by oral treatment or injection with the vitamin followed by restoration of an adequate level of pantothenic acid in the diet.

Pantothenic acid is available as a commercially synthesized product for addition to feed. It is available as *d*- or *dl*-calcium pantothenate. One gram of *d*-calcium pantothenate is equivalent to 0.92 g of *d*-pantothenic acid activity, while the combination of 1 g of the *dl* form has 0.46 g of *d*-pantothenic acid activity. A racemic mixture (equal parts of *d*- and *l*-calcium pantothenate) is generally sold to the feed industry. Because livestock and poultry can biologically utilize only the *d* isomer of pantothenic acid, nutrient requirements for the vitamin are routinely expressed in the *d* form.

Feed-grade pantothenic acid products are available in a number of potencies. Products that are sold on the basis of racemic mixture content can be misleading and confusing to a buyer who is not fully aware of the biological activity supplied by *d*-pantothenic acid. To avoid confusion, the label should clearly state the grams of *d*-calcium pantothenate or its equivalent per unit weight and the grams of *d*-pantothenic acid.

A straight racemic mixture (90%) is available to the feed industry but its hygroscopic and electrostatic properties contribute to handling problems. Because it readily picks up moisture, it sticks to bags, cans, and scoops and can become hard after prolonged exposure to air. Its electrostatic properties cause it to cling to metallic and other objects and losses can be significant. Through complexing procedures, several companies now market free-flowing and essentially nonhygroscopic and nonelectrostatic products.

Verbeeck (1975) reported calcium pantothenate to be stable in premixes with or without minerals and regardless of the mineral form. Losses of calcium pantothenate may occur, however, in premixes containing niacin or other substances of acidic nature. If a calcium pantothenate–calcium chloride complex is used instead of the plain calcium pantothenate, this problem should be alleviated.

In the United States, estimates of pantothenic acid dietary intakes in the human adult population range between 5 and 20 mg/day (Chung *et al.*, 1961). These levels are higher than estimated requirements of 4 to 7 mg/day for children and adults (RDA, 1980). Therefore, supplementation would not normally be recommended unless associated with general malnutrition. Humans at greatest risk and needing supplementation include alcoholics and diabetics, with adrenal and digestive dysfunctions also influencing need for the vitamin (Fox, 1984).

XI. TOXICITY

Pantothenic acid has only limited pharmacological effects when administered to animals or humans. Dexpanthenol, the alcohol synthetically derived from *d*-pantothenic acid, has been administered to increase gastrointestinal peristalsis and applied topically to alleviate itching and improve minor skin irritations (Fox, 1984). Pantothenic acid has been reported to have a protective effect against radiation sickness (Egarova and Perepelkin, 1979). Pantothenic acid toxicity has not been reported in humans. The LD_{50} in mg per kg body weight for mice and rats has been reported to range from 0.83 to 10.0 for calcium pantothenate (Unna and Greslin, 1941). For the rat, 100 times the dietary requirement resulted in nonfatal liver damage (NRC, 1987).

11

Biotin

I. INTRODUCTION

For many years it was believed that supplemental biotin was not required in swine and poultry diets because of its wide distribution in feedstuffs and because of known biotin synthesis by the intestinal microflora. However, in the mid-1970s, field cases in these species were found under modern production systems. These conditions, characterized by specific clinical signs, responded to supplemental biotin. On the basis of these findings nutritionists have had to reexamine the role of biotin in livestock and poultry diets.

Incidence of biotin deficiency has been found occasionally when humans and animals have consumed excessive quantities of raw eggs, which contain a biotin complexing factor (avidin). Likewise biotin deficiency is reported in children with inborn errors of metabolism when there is an insufficiency of biotin-dependent enzymes. Such cases in children respond dramatically to high-level dietary supplementations with biotin.

II. HISTORY

Discovery of the physiological significance of biotin is the history of the merging of different lines of investigation that appeared to be unrelated (Maynard *et al.*, 1979; Scott *et al.*, 1982; Bonjour, 1984). Biotin was the name given to a substance isolated from egg yolk by Kögl and Tönnis in 1936 that was necessary for yeast growth. This substance was discovered to be identical with a growth factor named "coenzyme R" that was required for legume nodule bacteria.

The toxic properties of feeding raw egg white to animals were first observed by Bateman in 1916. Clinical signs of dermatitis and hair loss due to "egg-white injury" were prevented by several researchers by the feeding of certain foods, notably liver and kidney. Szent-György studied the chemistry of this "protective factor" in certain foods, which he named "factor H" in 1937. The term vitamin H was chosen because the factor protected the "Haut," the German word for skin.

In 1940 Szent-György and associates found that biotin, vitamin H, and coenzyme R were the same substance. Other names given to this factor were protective factor X, egg-white injury protection factor, factor S, and factor W (or vitamin B_w). Egg-white injury resulted from its rendering dietary biotin unavailable because of a specific constituent known as avidin. Heat was found to destroy avidin, therefore revealing that this biotin antagonistic factor was heat labile. Avidin is a protein that is a secretory product of the mucosa of the oviduct and therefore is found in the albuminous part of eggs.

The structure and properties of biotin were established by United States and European investigators between 1940 and 1943. The first chemical synthesis was completed by Harris and associates of the Merck Company in 1945.

III. CHEMICAL STRUCTURE, PROPERTIES, AND ANTAGONISTS

The chemical structure of biotin in metabolism includes a sulfur atom in its ring (like thiamin) and a transverse bond across the ring (Fig. 11.1). As determined by isolation and synthesis, biotin is 2-keto-3, 4-imadazilido-2-tetrahydrothiophenevaleric acid. It is a monocarboxylic acid with sulfur as a thioether linkage. Biotin with its rather unique structure contains three asymmetric carbonations and therefore eight different isomers are possible. Of these isomers only one contains vitamin activity, *d*-biotin. The stereoisomer *l*-biotin is inactive.

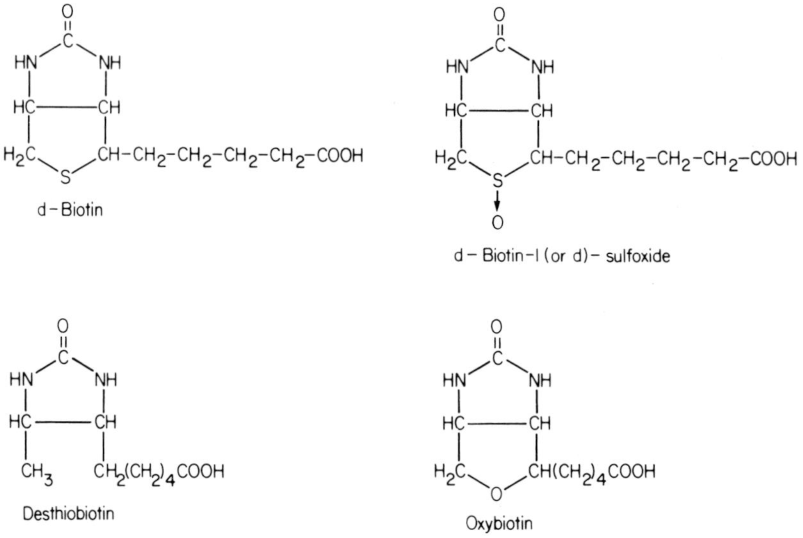

d-Biotin

d-Biotin-l (or d)-sulfoxide

Desthiobiotin

Oxybiotin

Fig. 11.1 Structures of biotin and some derivatives.

Biotin crystallizes from water solution as long, white needles. Its melting point is 232–233°C. Free biotin is soluble in dilute alkali and hot water and practically insoluble in fats and organic solvents. Biotin is quite stable under ordinary conditions. It is destroyed by nitrous acid, other strong acids, strong bases, and formaldehyde and is also inactivated by rancid fats and choline (Scott *et al.*, 1982). It is gradually destroyed by ultraviolet radiation.

Structurally related analogs of biotin (Fig. 11.1) can vary from no activity to partial replacement value and to antibiotin activity. Mild oxidation converts biotin to the sulfoxide and strong oxidation converts it to sulfone. Strong agents result in sulfur replacement by oxygen, resulting in oxybiotin and desthiobiotin. Oxybiotin has some biotin activity for chicks (one-third) but less for rats (one-tenth). Both desthiobiotin and biotin sulfone are active for yeast but are inhibitory to bacteria. Oleic acid and related compounds, in the presence of aspartic acid, satisfy the biotin requirement of many microorganisms, but are inactive for some and actually toxic for others (McDowell, 1985b).

IV. ANALYTICAL PROCEDURES

A number of methods are available for estimating biotin content of feeds and supplements. Bioassay procedures with animals (rats and chicks) and microorganisms are employed because of the difficulty of establishing chemical or physical methods suitable for natural materials. However, Achuta Murthy and Mistry (1977) reported progress in the use of colorimetric, gas chromatographic, and polarographic methods of analyses. Nevertheless, such analytical methods are not capable of identifying amounts of this vitamin that are biologically available to the animal.

In the biological estimation, rats and chicks are made biotin deficient by the use of special diets. Rats are made deficient by resorting to the use of sulfonamides, prevention of coprophagy, feeding egg white (avidin), or combinations of these techniques. However, a deficiency in the chick can be produced by a diet low in biotin alone. For both species, biotin content of the test sample can be calculated from growth–response curves. Most studies of biotin content of feedstuffs have been conducted by microbiological assay. *Lactobacillus arabinosus* is often employed with other suitable organisms including *L. casei* and *Saccharomyces cerevisiae*. The procedure with microorganisms includes heating with H_2SO_4 to liberate bound forms and removing unsaturated fatty acids that interfere with the assay. Recently a simple, rapid, and economic method for assessing biotin availability of feedstuffs was developed using egg yolk and plasma biotin concentrations after hens were fed specific feeds (Buenrostro and Kratzer, 1984). An additional test to evaluate feeds is based on the pyruvate carboxylase activity in chicken blood (Whitehead *et al.*, 1982).

V. METABOLISM

Biotin exists in natural materials in both bound and free forms with much of the bound biotin apparently not available to animal species. For poultry often less than one-half of the microbiologically determined biotin in a feedstuff is biologically available (Scott, 1981). Naturally occurring biotin is found partly in the free state (fruit, milk, vegetables) and partly in a form bound to protein in animal tissues, plant seeds, and yeast. Naturally occurring biotin is often bound to the amino acid lysine or to protein.

The few studies conducted in animals on biotin metabolism revealed that biotin is absorbed as the intact molecule in the first third to half of the small intestine (Bonjour, 1984). Information is very limited on biotin transport, tissue deposition, and storage in animals and humans. McCormick and Olson (1984) report that biotin is transported as a free water-soluble component of plasma, is taken up by cells via active transport, and is attached to its apoenzymes.

Disappearance of an intravenous dose of radioactive biotin from blood of biotin-deficient rats was more rapid than for controls (Petrelli *et al.*, 1979). Also rate and extent of deposition into deficient liver, particularly mitochondrial and cytosolic fractions, were favored. This research supports the concept of homeostatis mechanisms responding to provide biotin in relation to needs. All cells contain some biotin, with larger quantities in liver and kidneys. Intracellular distribution of biotin corresponds to known localization of biotin-dependent enzymes (carboxylases).

Investigations into biotin metabolism in animals are difficult to interpret as biotin-producing microorganisms exist in the intestinal tract distal to the cecum. Often the amount of biotin excreted in urine and feces together exceeds total dietary intake, whereas urinary biotin excretion is usually less than intake. ^{14}C-labeled biotin showed the major portion of intraperitoneally injected radioactivity to be excreted in the urine and none in the feces or expired as CO_2 (Lee *et al.*, 1973).

VI. FUNCTIONS

Biotin is an essential coenzyme in carbohydrate, fat, and protein metabolism. It is involved in conversion of carbohydrate to protein and vice versa as well as conversion of protein and carbohydrate to fat. Biotin also plays an important role in maintaining normal blood glucose levels from metabolism of protein and fat when the dietary intake of carbohydrate is low. As a component of several carboxylating enzymes it has the capacity to transport carboxyl units and to fix carbon dioxide (as bicarbonate) in tissue. Biotin serves as a prosthetic group in a number of enzymes in which the biotin moiety functions as a mobile carboxyl carrier. The biotin prosthetic group is linked covalently to the ε-amino group of a lysyl residue of the biotin-dependent enzyme.

In carbohydrate metabolism, biotin functions in both carbon dioxide fixation and decarboxylation, with the energy-producing citric acid cycle dependent on the presence of this vitamin. Specific biotin-dependent reactions in carbohydrate metabolism are:

—Carboxylation of pyruvic acid to oxaloacetic acid.
—Conversion of malic acid to pyruvic acid.
—Interconversion of succinic acid and propionic acid.
—Conversion of oxalosuccinic acid to α-ketoglutaric acid.

In protein metabolism, biotin enzymes are important in protein synthesis, amino acid deamination, purine synthesis, and nucleic acid metabolism. Biotin is required for transcarboxylation in degradation of various amino acids. Deficiency of the vitamin in mammals hinders the normal conversion of the deaminated chain of leucine to oxaloacetate. Ability of liver homogenates to incorporate labeled CO_2 was directly related to biotin level of the animal (Terroine, 1960). Likewise, ability to synthesize citrulline from ornithine is reduced in liver homogenates from biotin-deficient rats.

Acetyl-CoA carboxylase catalyzes addition of CO_2 to acetyl-CoA to form malonyl-CoA. This is the first reaction in the synthesis of fatty acids. A cytosolic multienzyme complex, fatty acid synthetase, then accomplishes synthesis of palmitate from malonyl-CoA. Biotin is required for normal long-chain unsaturated fatty acid synthesis and is important for essential fatty acid metabolism (Kramer *et al.*, 1984). Deficiency in rats inhibited arachidonic acid (20 : 4 ω-6) synthesis from linoleic acid (18 : 2 ω-6) while increasing linolenic acid (18 : 3 ω-3) and its metabolite (22 : 6 ω-3).

VII. REQUIREMENTS

Estimates of biotin requirements for various animals and humans are presented in Table 11.1. Biotin requirements are difficult to establish because of variability in feed content and bioavailability. Likewise it is difficult to obtain a quantitative requirement for biotin because the vitamin is synthesized by many different microorganisms and certain fungi. These microorganisms are found in the lower part of the intestinal tract, a region in which absorption of nutrients is generally reduced. However, it is believed that intestinal microflora make a significant contribution to the body pool of available biotin. In general, combined urinary and fecal excretion of biotin exceeds the dietary intake. What is not known for

TABLE 11.1

Biotin Requirements for Various Animals and Humans[a]

Animal	Purpose	Requirement	Reference
Beef cattle	Adult	Microbial synthesis	NRC (1984a)
Dairy cattle	Calf	10 μg/kg body wt	NRC (1978a)
	Adult	Microbial synthesis	NRC (1978a)
Chicken	Leghorn, 0–6 weeks	0.15 mg/kg	NRC (1984b)
	Leghorn, 6–20 weeks	0.10 mg/kg	NRC (1984b)
	Leghorn, laying	0.10 mg/kg	NRC (1984b)
	Leghorn, breeding	0.15 mg/kg	NRC (1984b)
	Broilers	0.15 mg/kg	NRC (1984b)
Turkey	0–8 weeks	0.20 mg/kg	NRC (1984b)
	8–24 weeks	0.10–0.15 mg/kg	NRC (1984b)
	Breeding hens	0.15 mg/kg	NRC (1984b)
Japanese quail	Starting and growing	0.3 mg/kg	NRC (1984b)
	Breeding	0.15 mg/kg	NRC (1984b)
Sheep	Adult	Microbial synthesis	NRC (1985)
Swine	Growing–finishing	0.05 mg/kg	NRC (1988)
	Breeding–lactating	0.20 mg/kg	NRC (1988)
Horse	Adult	Microbial synthesis(?)	NRC (1978b)
Goat	Adult	Microbial synthesis	NRC (1981b)
Cat	All classes	0.07 mg/kg	NRC (1986)
Mink	Growing	0.12 mg/kg	NRC (1982a)
Hamster	All classes	0.6 mg/kg	NRC (1978c)
Trout	Growing	0.05–0.25 mg/kg	NRC (1981a)
Human	Infants	35–50 μg/day	RDA (1980)
	Children	65–120 μg/day	RDA (1980)
	Adults	100–200 μg/day	RDA (1980)

[a]Expressed as per unit of animal feed either on an as fed (approximately 90% dry matter) or dry basis (see Appendix Table 1) (except for calf). Human requirements expressed as μg/day. Often requirements are unknown, with the best recommendation used.

the various species is the extent of microbial synthesis or the biotin availability to the host.

It is concluded that microorganisms contribute to animal and human requirements because the use of some sulfa drugs, such as sulfathalidine, can induce deficiency under some circumstances. Microorganisms that provide significant quantities of biotin to most species apparently supply a variable and undependable amount of biotin for poultry (Scott, 1981). Biotin deficiency was less severe in cecectomized as compared with normal chickens, indicating that cecal microorganisms do not supply chickens with significant amounts of biotin but instead compete with the host animal for dietary biotin, thus increasing the requirement (Sunde *et al.*, 1950).

Rate and extent of biotin synthesis may be dependent on the level of other dietary components. In poultry and rats it has been shown that polyunsaturated fats, ascorbic acid, and other B vitamins may influence the demand for biotin. Addition of polyunsaturated fatty acids to fat-free, biotin-deficient diets increased severity of dermal lesions (Roland and Edwards, 1971). Biotin is rapidly destroyed as feeds become rancid. Pure biotin was inactivated to the extent of 96% in 12 hr when linoleic acid of a high peroxide number was added to the diet (Pavcek and Shull, 1942). In the presence of α-tocopherol, this destruction amounted to only 40% after 48 hr.

VIII. NATURAL SOURCES

Biotin is present in many foods and feedstuffs (Table 11.2). The richest sources of biotin are royal jelly, liver, kidney, yeast, blackstrap molasses, peanuts, and eggs. Most fresh vegetables and some fruits are fairly good sources. Corn, wheat, other cereals, meat, and fish are relatively poor sources.

Less than one-half of the biotin in various feedstuffs is biologically available, as determined by microbiological assay (Frigg, 1976, 1984). Thus, it is important to know the chemical form of biotin (e.g., bound or unbound) as well as its overall content in feed. Oilseed meals, alfalfa meals, and dried yeast are excellent sources of biologically available biotin. Meat meal and fish meals contain biotin of relatively poor biological availability. Grains are poor sources of biotin, with corn and oats more available sources than wheat, hull-less barley, or regular barley. Frigg (1976) reported that the biotin content of wheat and barley may be totally unavailable. Sorghum biotin availability is 10–20% compared to a value of 75–100% for corn (Buenrostro and Kratzer, 1984).

The biotin content of cereal grains appears to be influenced by variety, season, and yield (in particular the endosperm/pericarp ratio) (Brooks, 1982). Harvest conditions, postharvest treatment, and storage conditions all appear to play a part in determining biotin content and may also affect the availability. Biotin in feedstuffs may be destroyed by heat curing, solvent extraction, and improper storage conditions, while pelleting has little effect on biotin content of feed (McGinnis, 1986a,b). There is considerable variation in biotin content within individual sources. For example, 65 samples of corn analyzed for biotin varied between 0.012 and 0.072 mg/kg and 20 samples of meat meal between 0.008 and 0.20 mg/kg (Brooks, 1982). Milling of wheat or corn and canning of corn, carrots, spinach, or tomatoes reduced biotin concentrations (Bonjour, 1984). After a 6-month storage of baby foods, an approximate 15% reduction in biotin content was found (Karlin and Foisy, 1972).

TABLE 11.2

Typical Biotin Contents and Bioavailability in Various Foods and Feedstuffs (As Fed Basis)[a]

Food or feedstuff	Biotin levels (μg/g)	Biotin availability (%)
Alfalfa meal, dehydrated	0.33	75–100
Bacon	0.07	—
Barley	0.13–0.17	20–50
Beef, steak	0.04	—
Cabbage	0.02	—
Carrots	0.03	—
Chicken	0.10	—
Corn, yellow	0.06–0.10	75–100
Corn gluten meal	0.11–0.19	62
Cottonseed meal, solvent extracted	0.08–0.47	100
Distiller's solubles, dried	0.44–1.1	90
Eggs, whole	0.25	—
Fish meal	0.20–0.55	30
Liver, beef	1.00	—
Meat and bone meal	0.07–0.31	85
Milk, cow's	0.05	—
Molasses, blackstrap	0.7	—
Oats	0.11–0.39	32
Peanut meal	1.63	53
Rice bran	0.42	—
Rice polishings	0.38	23
Skim milk, dried	0.33	—
Sorghum	0.13–0.29	10–60
Soybean meal	0.18–0.50	100
Wheat	0.06–0.18	0
Wheat bran	0.09	18

[a]Modified from NRC (1982b).

IX. DEFICIENCY

A. Effects of Deficiency

Biotin is important for normal function of the thyroid and adrenal glands, the reproductive tract, and the nervous system. Its effect on the cutaneous system is most dramatic since a severe dermatitis is the major obvious clinical sign of biotin deficiency in livestock and poultry.

1. Ruminants

Biotin deficiency, as identified by hindquarter paralysis, decreased urinary excretion of biotin, and correction of the problem with biotin injections, has been reported in calves (Wiese *et al.*, 1946). However, no evidence for a biotin deficiency has been produced in animals with functional rumens. Need for biotin supplementation even in diets of preruminant calves and lambs is unlikely (Church, 1979).

2. Swine

Biotin deficiency was produced in 1946 in swine by feeding a purified diet containing sulfathalidine or raw egg white (Lindley and Cunha, 1946; Cunha *et al.*, 1946). For many years it was concluded that biotin supplementation was not needed since it is synthesized by intestinal microorganisms and is widely distributed in feed. Nevertheless, deficiency clinical signs were observed by feed company personnel and scientists under field conditions. However, it was the 1970s before a greater awareness to the problem of biotin field deficiencies became apparent (Cunha, 1984a).

Biotin deficiency results in reduced growth rate and impaired feed conversion as well as producing a wide variety of clinical signs. Clinical signs associated with biotin deficiency include alopecia (hair loss), a dermatitis characterized by dryness, roughness, and a brownish exudate, ulceration of the skin, inflammation of the mouth mucosa, hindleg spasticity, and transverse cracking of the soles and tops of hooves (Cunha, 1977). Figures 11.2 and 11.3 show clinical signs of biotin deficiency in swine.

Growth depression may become evident in biotin-deficient swine before clinical signs are seen. The first clinical signs are generally excessive hair loss and dermatitis, with complete hair loss in severe cases. Dermatitis first appears as scaly skin, often starting on the ears, neck, shoulder, and tail and eventually spreading over the entire body. In later stages, crust and cracks appear on the face and extremities.

After 5–7 weeks on a biotin-deficient diet, swine may show claw defects. In a biotin deficiency the claw horn becomes soft and rubbery and poorly resistant to abrasions. Depending on flooring type on which the animal is kept this may have little effect or may lead to the development of cracks and necrotic lesions, resulting in extreme lameness (Glättli, 1975). Secondary infections may gain entry through claw cracks and result in infected joints, which may lead to premature removal from the herd. Feeding and breeding activities are also adversely affected; in particular, the sow becomes unable to support the weight of the boar. Also, because the hog's ability to eat may be impaired, these problems will obviously lead to economic losses.

Tagwerker (1983) noted that foot claw lesions were responsible for 4–8% of all sows culled in Europe. Also, he noted a Denmark study that showed 8.5% of claw lesions in biotin-supplemented sows, compared to 25% for controls. In Holland after biotin supplementation, culling rate due to lameness was decreased from 25 to 14% (de Jong and Sytsema, 1983).

Fig. 11.2 The two pigs in the middle are biotin deficient. Note the hair loss and dermatitis. (Courtesy of T. J. Cunha and Washington State University.)

Cunha (1984a) noted that in most of the 40 countries he visited during the past 30 years, biotin deficiency signs were observed in swine operations. These deficiency signs observed under field conditions occurred in only 10–20% of sows or less. Baby pigs nursing these sows usually showed no biotin deficiency signs but responded to biotin supplementation. Unfortunately many swine producers are of the opinion that it is natural for a swine herd to have a few animals with hair loss, dermatitis, and cracked feet and, therefore, are not overly concerned when a small percentage of sows exhibit these clinical signs (Cunha, 1984a).

In a field study, sows had severe lameness and impaired reproduction (Anonymous, 1981a). After these sows received supplemental biotin, normal foot health and normal reproductive performance were restored. Recently researchers found that sows housed in total confinement showed a positive response to conception rate and weaning to first estrus and a trend to larger litters and first parity when supplemented with biotin (Bryant *et al.*, 1985). In an earlier study (Anonymous, 1981b), sows fed supplemental biotin had more pigs born alive (9.8 vs. 8.1), more pigs weaned (7.8 vs. 6.8), increased litter weight at weaning (71.0 vs. 64.5 kg), and reduced time interval from weaning to first estrus after weaning (6.2 vs. 15.3 days) compared to unsupplemented controls.

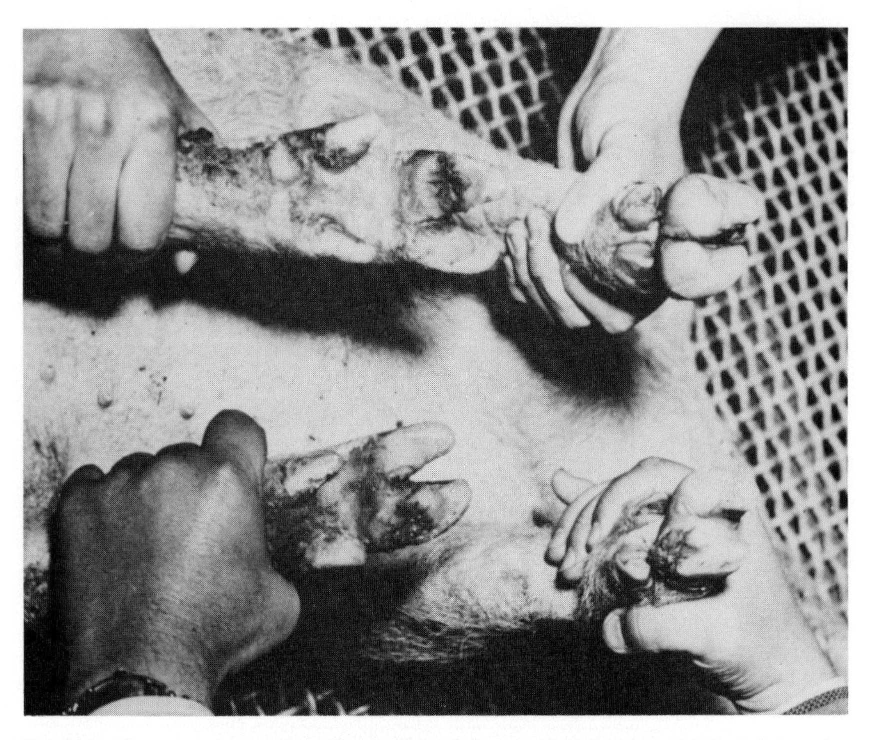

Fig. 11.3 Note transverse cracking of the soles and the tops of the hooves of biotin-deficient pigs. (Courtesy of T. J. Cunha and Washington State University.)

3. Poultry

Biotin requirement in the turkey is higher than that of the chick so more field problems with biotin deficiency have arisen in turkeys. Biotin deficiency in chicks and poults results in a wide range of clinical signs (Fig. 11.4–11.7) with considerable variation in time of appearance of individual signs (NRC, 1984b). The principal effects in both species are reduced growth rate and feed efficiency, disturbed and broken feathering, dermatitis, and leg and beak deformities. First signs of a deficiency are usually growth depression and loose feathering; signs of dermatitis then appear and, finally, disorders of the leg (perosis) and beak become apparent. With dermal lesions, bottoms of feet become rough and calloused and contain deep fissures that show some hemorrhaging. Feet problems are usually exacerbated by bacterial invasion of lesions. Toes may become necrotic and slough off. Tops of feet and legs usually show only a dry scaliness. Lesions appear in the corner of the mouth and slowly spread to the whole area around the beak. Eyelids eventually swell and stick together.

Dermal lesions have a characteristic order of appearance, although speed of

Fig. 11.4 Perosis (bone deformities) in chicks as a result of biotin deficiency. Chicks showed perosis as early as 17 days, with rigid limb joints resulting in a stilted walk. (Courtesy of Hoffmann–La Roche Inc., Nutley, N. J.)

onset depends on deficiency severity. For chicks fed severely deficient diets, dryness and flakiness of the feet first become noticeable at about 14 days of age, and slight encrustations and superficial fissures develop on the undersurfaces of the feet at about 18 days (Whitehead, 1978). These increase in severity until, by about 25 days, the fissures become hemorrhagic. Between 3 and 4 weeks, dermatitis may also appear on the eyelids and, as this develops, the bird becomes unable to keep the lids apart and they eventually become stuck together.

Dermal lesions are similar to those of pantothenic acid deficiency (see Chapter 10). However, with biotin deficiency, lesions occur first in feet and later around beak and eyes, whereas in pantothenic acid deficiency the signs occur first in corners of mouth and eyes and only in prolonged cases appear in the feet. Because of the difficulty of a differential diagnosis between the two vitamins, it is often necessary to examine the diet composition and decide which is most likely to be deficient.

Biotin deficiency is a cause of hock disorders in both poults and chicks. The major deficiency sign affecting market turkeys is severe leg weakness. Lesions caused by biotin deficiency are brought about by chondrodystrophy, a condition in which bone mineralization is normal but linear growth of long bones is impaired. Chondrodystrophy caused by biotin deficiency can result in shortening

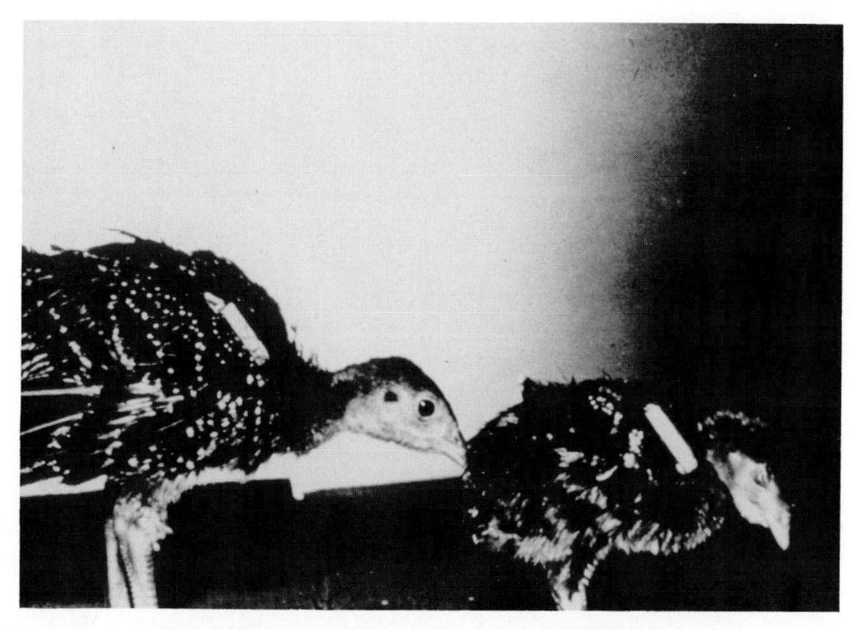

Fig. 11.5 Comparison of normal and biotin-deficient Broad Breasted Bronze male turkeys at 3 weeks of age. (Courtesy of D. C. Dobson, Utah State University.)

of metatarsal bones and perosis. Perosis occurs when irregular bone development results in enlargement and deformity of the hock joint. Crippling in turkeys can occur as early as 3–4 weeks of age, and often it seems to disappear at 6–7 weeks. Then, it reappears with great severity between 13 and 16 weeks (Scott, 1981). At this stage, the birds are unable to walk and, thus, can be trampled by other turkeys. Perosis can occur at any stage in the growth of the turkey from 3 to 24 weeks of age. In general, young chickens are less susceptible than poults to leg disorders, although biotin deficiency does cause problems of the same type in chicks as in poults (Whitehead, 1978). Once the deformities of perosis occur, biotin administration is not curative.

For turkeys, dry and brittle feathers usually accompany the other signs of clinical biotin deficiency. Bronze poults can exhibit white barring of the feathers, usually affecting just tom turkeys. Likewise, deficient chicks have rough and broken feathering with head and breast feathers often having a spiky, matted appearance.

Poor egg production and hatchability will result from clinical biotin deficiency. For breeder chickens, biotin deficiency will reduce hatchability but is less likely to affect egg production. Clinical signs and conditions associated with biotin deficiency in chick embryos and/or newly hatched chicks include bone deformities

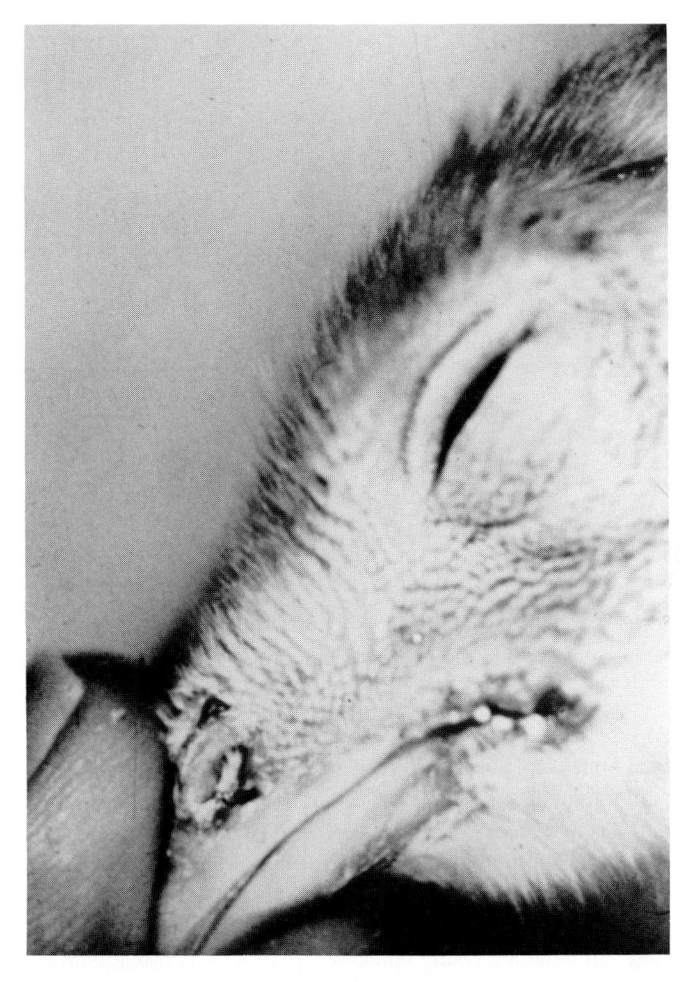

Fig. 11.6 Biotin deficiency in poult. Note lesion at apex of mouth. (Courtesy of L. S. Jensen and Washington State University.)

(perosis), impaired muscular coordination (ataxia), skeletal deformities (e.g., crooked legs), extensive foot webbing, abnormal cartilage development (chondrodystrophy), embryonic mortality, twisted, malformed beak ("parrot beak"), and reduced size. For turkeys, Ferguson *et al.* (1961) reported that biotin deficiency resulted in a marked decrease in hatchability and a high rate of embryonic mortality during the first week of incubation. At the end of the second week, hatchability decreased from 83 to 14%. At the end of the third week hatchability was zero. Feeding the biotin-deficient diet for a long period of time resulted in egg production abruptly decreasing after 13 weeks.

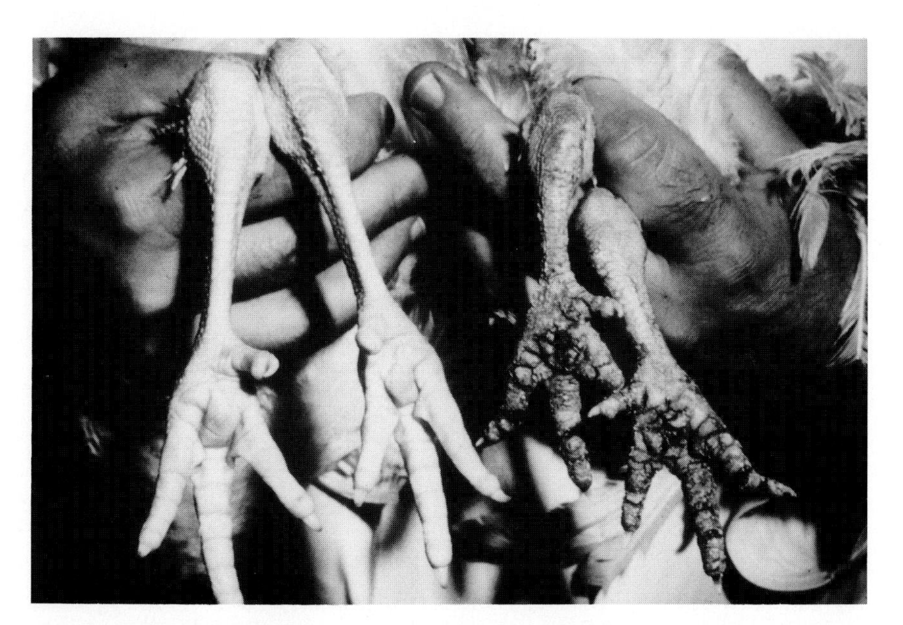

Fig. 11.7 Severe foot-pad lesions in growing turkey as a result of biotin deficiency. Less severe lesions are more common. (Courtesy of Richard Miles, University of Florida.)

4. Horses

Biotin is synthesized in the lower digestive tract of the horse, but information on dietary requirement is lacking (NRC, 1978b). Hoof integrity for a number of cases has been reported to improve as a result of biotin supplementation (Fig. 11.8). Such cases were characterized by horn that tended to crumble at the lower edges of the hoof walls and that was generally prone to poor conformation and damage to the walls, soles, and the white line junction (Comben *et al.*, 1984).

5. Other Animal Species

For a number of species, including guinea pigs, rats, hamsters, dogs, cats, and rabbits, no deficiency signs have been observed when animals are fed purified diets and/or natural diets containing low biotin concentrations. However, with these same species, feeding a biotin-deficient diet containing raw egg white (Fig. 11.9) produced typical deficiency signs often including weight loss and alopecia.

a. Mice. Biotin supplementation improved reproduction and lactation for mice when added to a purified diet devoid of egg white.

b. Monkey. Biotin deficiency was produced in adolescent rhesus monkeys with thinning of hair and loss of hair color as a result of a low-biotin diet for

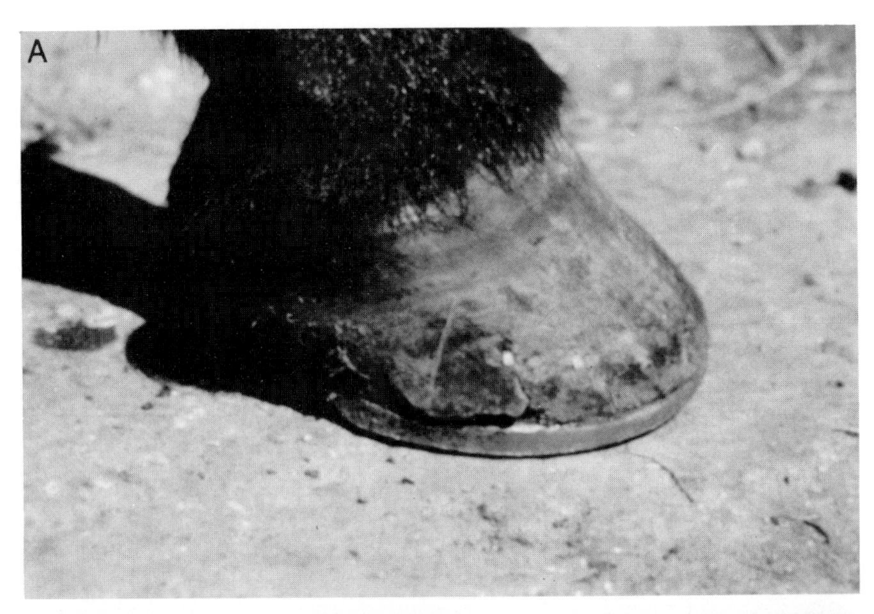

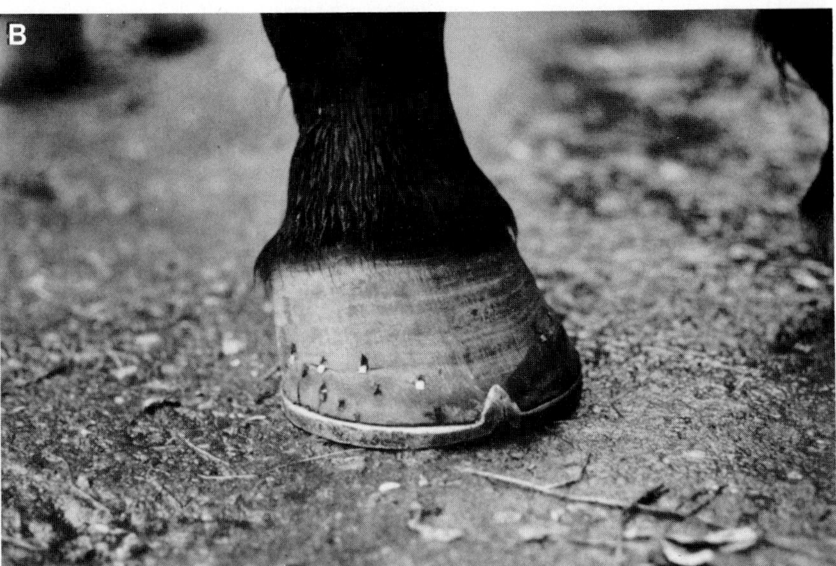

Fig. 11.8 Hoof condition of a 7-year-old, 18-hands heavyweight show hunter resulting from biotin supplementation. The horse (A) had a history of tender feet. The walls of the hooves were very weak, crumbling at their lower edges, with large areas breaking out and detaching when shoes were nailed. Five months after 15 mg biotin/day supplementation (B), walls of the hooves were thicker and harder so that nailing was achieved. (Courtesy of N. Comben, R. J. Clark, and D. J. B. Sutherland and permission of *Veterinary Record* **115** (1984), 642–645.)

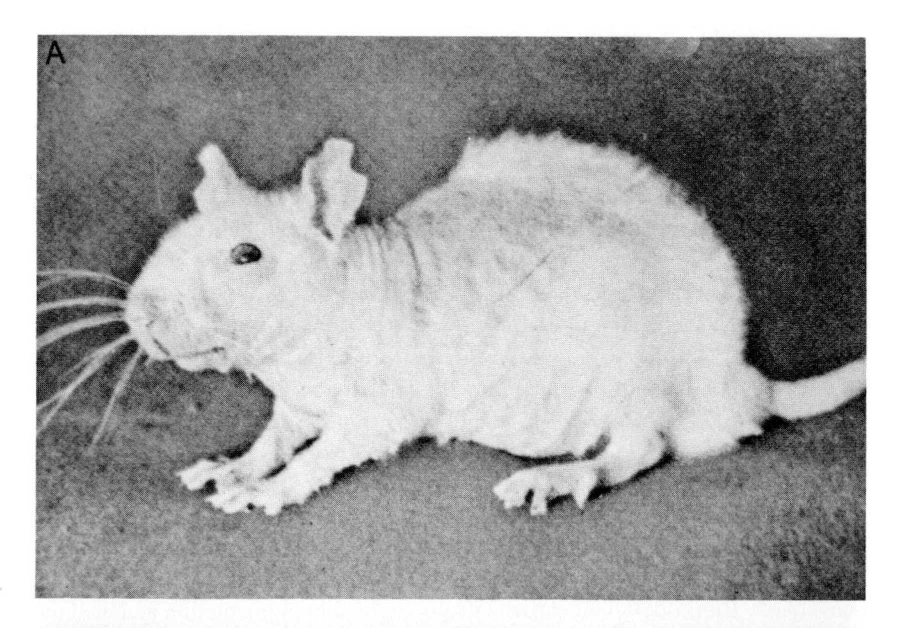

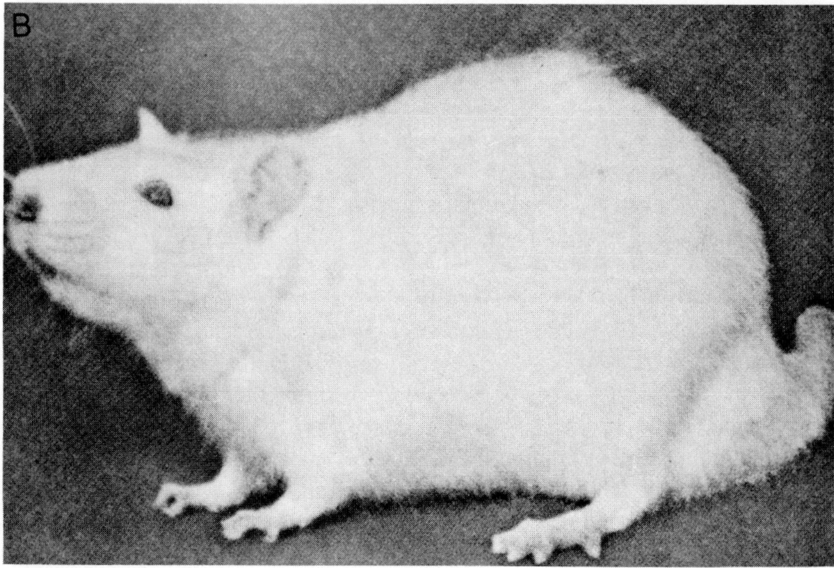

Fig. 11.9 Egg-white injury as a result of feeding a rat raw egg white, which contains a biotin antagonist, avidin (A). The resulting dermatitis has progressed in the rat to generalized alopecia. After 3 months of treatment, the animal is normal (B). (Courtesy of Alan T. Forrester, ''Scope Manual on Nutrition,'' The Upjohn Company, Kalamazoo, Mich.)

long periods (15–28 months) Waisman *et al.*, 1945). Use of diets with egg white and/or sulfonamides greatly hastened and intensified biotin deficiency signs.

c. Foxes and Mink. Biotin deficiencies have been produced by feeding purified diets or diets containing egg products to growing mink. Apparently the short intestinal tract of the mink and rapid feed passage make it impossible for the animal to obtain adequate biotin from that synthesized by its intestinal flora. Stout *et al.* (1966) reported biotin deficiency in mink when diets contained 40% or more offal from breeder hen turkeys. Presence of raw eggs in the offal was presumed responsible for the deficiency. Later studies in Oregon have shown that inclusion of spray-dried egg in the diet can have a similar effect (Wehr *et al.*, 1980). For both mink and foxes biotin deficiency resulted in graying or discoloration of hair (Fig. 11.10) and in extreme cases hair loss (NRC, 1982a).

d. Fish. Biotin is synthesized in many species of fish by intestinal microflora, but supplemental biotin may be required for maximum growth and to prevent clinical deficiency signs (NRC, 1981a, 1983). Signs of biotin deficiency in trout and salmon include anorexia, slow growth, increased mortality, and poor feed conversion. Biotin-deficient catfish develop light skin pigmentation and anemia. Japanese eels fed a biotin-deficient diet exhibited reduced growth, abnormal swimming behavior, and dark coloration.

6. HUMANS

Except in infants, there is no evidence of spontaneous biotin deficiency in humans. This is probably due to the ubiquitous nature of the vitamin in diets plus benefits derived from microbial synthesis. Biotin deficiency occurs in individuals consuming a large number of raw eggs daily because of the antagonistic effect of the egg-white protein avidin (see Section II). Adult volunteers fed 200 g dehydrated egg white daily for 5 weeks displayed symptoms of mild depression, lassitude, somnolence, hallucination, anxiety with muscle pain, and hyperesthesia (Sydenstricker *et al.*, 1942). After 8 weeks, anorexia and a striking grayish pallor with dermatitis and desquamation occurred. Termination of the experiment resulted because of a marked fall in food intake.

Some children suffer from an inborn error of metabolism resulting from lack of biotin-dependent carboxylases. Two conditions in infants, seborrheic dermatitis and desquamative erythroderma (Leiner's disease), are apparently connected with biotin deficiency since children suffering from these diseases have subnormal biotin in blood and urine and respond to supplementation with the vitamin.

Biotin deficiency can result when insufficient biotin is added to infant milk formulations or to total parenteral nutrition solutions. Although the evidence is circumstantial, some researchers suggest that sudden infant death syndrome may be related to low biotin intakes of bottle-fed infants (Bonjour, 1984). Cases of

Fig. 11.10 The newborn fox pup on the left is from a biotin-deficient dam that received a diet containing raw egg white. Thin, gray pelt and deformed legs are apparent. A control diet containing cooked egg white was fed to the dam of the pup on the right which is also newborn. (Courtesy of A. Helgebostad, Veterinary College of Norway, Heggedal.)

low circulating plasma biotin have also been reported for alcoholics, the elderly, epileptics, and burn victims.

B. Assessment of Status

Detection of subclinical or marginal biotin deficiency, which is most likely to occur under field conditions, is difficult. Field observations are important, but it is essential to realize that there may be no clinical signs in spite of a deficiency. If clinical signs are seen, they may not all be present and only a small percentage of animals may exhibit such clinical signs. Chemical, chick, or microbiological assays often give different results for feed biotin concentrations. Likewise, there is a great variability in feed biotin concentrations and availability in various feeds (See Section VIII).

At present, blood biotin concentrations lack dependability as a good indicator of biotin status of pig (Cunha, 1984a). This is particularly true with borderline deficiencies, in which plasma biotin varies with dietary intake and shows considerable individual variation among pigs. In chickens, several reports have indicated that plasma biotin concentration below 100 ng/100 ml is indicative of biotin deficiency (Scott *et al.*, 1982). In three experiments using chickens and turkeys, plasma pyruvic carboxylase activity was positively related to supplemental biotin in both species (Whitehead and Bannister, 1978). Unfortunately, unlike in poultry, pig erythrocytes contain no mitochondria and as a result enzyme activity levels in plasma are too low to be meaningful. Robinson and Lovell (1978) report that biotin status in fish may be assessed by measuring pyruvate carboxylase activity in the liver. Biotin concentration in egg yolk is an additional method of evaluating the status of the vitamin (see Section IV). In both poultry and swine the ratio of the blood fatty acids is likewise indicative of biotin status, with palmitoleic acid particularly increasing in relation to stearic acid (Edwards, 1974).

Evaluation of human biotin status employs blood and urinary concentrations of the vitamin. As with animals, there are considerable differences and variations among humans that are difficult to explain. From the available data, a biotin level in urine of approximately 70 nmol/liter and a circulating level in plasma of around 1500 nmol/liter seem to indicate an adequate supply of biotin for humans (Bonjour, 1984).

X. SUPPLEMENTATION

Of all vitamins considered for supplementation, biotin is currently the most expensive. In view of the considerable cost of synthetically produced biotin, the question of biotin supplementation of practical diets is economically significant.

Although clinical signs were found for poultry and swine under experimental conditions in the 1940s, for many years it was believed that supplemental biotin was not needed in swine and poultry diets because of the production of biotin by the animal's intestinal microflora. In 1967 for poultry and in the mid-1970s for swine, interest was rekindled when more field cases occurred than in the past. Interference with the biosynthesis of biotin by intestinal bacteria can individually or collectively lead to a biotin deficiency. These interferences can be in the form of therapeutic administration of antibacterial agents and modern housing systems limiting animals' access to feces. Additionally, biotin deficiencies are now more prevalent because of limited bioavailability of biotin found in some grains (e.g., wheat, barley, sorghum) and in some animal protein sources (e.g., meat meal, poultry by-product meal), biotin antagonists including molds, feed rancidity, and improved genetic characteristics for greater production. Cunha (1984a) summarized the possible reasons for more frequent occurrence of biotin deficiencies in swine (Table 11.3) with the majority of these reasons likewise applicable to the poultry industry. Biotin deficiency in fur-bearing animals caused by feeding of egg products is economically disastrous, because of interference with fur coloration, and supplementation is mandated.

Some reports conclude that foot lesions in adult sows healed in a matter of weeks (Tagwerker, 1974) when sows were supplemented with high levels of biotin, while Brooks *et al.* (1977) found biotin supplementation to result in a 28% reduction in lesions after 6 months. Pigs housed on badly designed floors have little opportunity for recovery as traumatic injury exceeds capacity of the hoof for growth and repair. Brooks (1982) summarized data from Great Britain, Denmark, and Switzerland and concluded that where foot lesions already existed, dietary supplements of 2000–3000 μg of biotin per kilogram of diet were beneficial.

For many species receiving typical diets, supplementation is not needed because of adequate sources in feed ingredients and/or intestinal microflora synthesis. As an example of dietary adequacy, practical diets for channel catfish made from the commonly used ingredients soybean meal, corn, and menhaden fish meal did not need supplemental biotin (Lovell and Buston, 1984). For some species, including equines, need for supplemental biotin is controversial. If hoof problems exist, biotin supplementation has been suggested (Comben *et al.*, 1984) in mg biotin/animal/day as follows: donkeys and ponies, 5–10 mg; riding horses, 15 mg; and heavy horses, up to 30 mg. The recommendation is to continue supplementation as increased hoof horn strength may be anticipated within 3–5 months.

Supplementation of human diets is generally considered unnecessary because of adequate dietary intakes as well as intestinal microbial synthesis. Exceptions would be for individuals consuming raw eggs and infants with inborn errors of metabolism. Eggs are a good source of biotin; however, the biotin contained in

TABLE 11.3

Possible Explanations Why Biotin Deficiencies Have Become More Prevalent in Swine and Poultry Operations in Recent Years[a]

Increased use of confinement, which lessens opportunity for coprophagy (feces eating); feces containing biotin synthesized in the intestines.
Increased use of grain–soybean meal diets, with less utilization of biotin-rich feeds including whey, fermentation by-products, yeast, dehydrated alfalfa, and pasture.
Biotin antagonists in feeds such as streptavidin, certain antimicrobial drugs, and dieldrin (a pesticide). *Streptomyces* are molds affecting biotin availability that are found in soil, moldy feeds, manure, and litter.
Rancidity in feeds causes biotin to be readily destroyed. Biotin in feedstuffs may be destroyed by heat curing, solvent extraction, pelleting, and improper storage conditions. Length of storage, temperature, and humidity result in biotin loss.
Reduced intestinal synthesis and/or absorption of biotin may result from diseases and other conditions affecting the gastrointestinal tract.
Improved genetic characteristics (breed, type, and strain) and intensified production for faster weight gains and better feed conversion. Increased reproductive performance with more farrowings per sow yearly or increased egg production in poultry operations.
Decreased level of biotin or its availability in feeds because of new plant varieties, new feed production practices, and processing methods.
For swine, reduced feed consumption (thus reduced biotin intake) by sows to avoid excess weight, which is detrimental to reproduction. Sows now receive 1.4–2.3 kg feed/day during gestation while previously had received 2.7–3.2 kg.
Interrelationships between biotin and other nutrients affect requirements. Dietary fats, pantothenic acid, vitamin B_6, vitamin B_{12}, folacin, thiamin, riboflavin, *myo*-inositol, and ascorbic acid are related to biotin requirements and metabolism.
Lower levels and availability of biotin in feeds. Different batches of the same feed may vary considerably in biotin. Biotin in feeds exist in both free and bound forms. Certain feeds have a low availability (i.e., wheat, barley, sorghum). A shift from use of corn, soybean meal, and cottonseed meal often results in diets with less available biotin.

[a]Adapted from Cunha (1984a).

egg yolk is inadequate to counteract the deficiency of the vitamin caused by the avidin in egg while, so that unheated dried whole egg is deficient in this vitamin (Kratzer *et al.*, 1988). In addition, since the advent of total parenteral nutrition, biotin deficiency has resulted in some cases in which biotin was omitted from intravenous fluids (Bonjour, 1984).

Biotin is commercially available as a 100% crystalline product or as dilutions (45.5 mg/kg and as a 1% premix). The form containing vitamin activity is *d*-biotin, which occurs in nature and is the commercially available form. A 2% spray-dried biotin product is also available for use in either feed or drinking water. An example of supplemental biotin provided in water for biotin-deficient young turkeys is 0.25 mg biotin per gallon of drinking water for 3–4 weeks and for older birds 0.50 mg per gallon (Bauernfeind, 1969).

Losses of biotin during storage can be considerable. Biotin is readily destroyed by feed rancidity (Pavcek and Shull, 1942). Buying fresh feeds, storing them only short periods of time, and keeping them dry and in a well-ventilated storage area will minimize rancidity problems. Also, the diet should be low in feeds high in prooxidants and/or properly protected by an effective antioxidant to avoid destruction of biotin, vitamin E, selenium, and other nutrients. Biotin is relatively stable in multivitamin premixes as well as natural sources in feeds, but losses of the vitamin during processing do occur. Processing losses of biotin in extruded pet food have been reported at about 15% (Anonymous, 1981a).

XI. TOXICITY

For both humans and animals biotin is apparently not toxic even in large doses. However, Paul *et al.* (1973) reported that an acute dose of biotin (5 mg/100 g body weight) caused irregularities of the estrus cycle with heavy infiltration of leukocytes in the vagina of the rat up to 14 days after treatment. In studies with rats, a dose of biotin at least 5000 or 10,000 times the daily requirement had no deleterious effects (Mittelholzer, 1976). Studies with poultry and swine indicate a safety tolerance of 4–10 times their requirement, with maximum tolerable level probably much higher (NRC, 1987). Likewise in humans, no adverse effects have resulted from oral or intravenous administration of high doses over prolonged periods of time.

12

Folacin

I. INTRODUCTION

Folacin is a generic term used to describe folic acid and related compounds that exhibit the biological activity of folic acid. A number of researchers in both developed and developing countries have reported a high incidence of folacin deficiency in pregnant women. It has been estimated that up to one-third of all pregnant women in the world may experience a folacin deficiency of varying severity (Rothman, 1970). Because of their rapid growth rate, cancer cells have an exceptionally high folacin requirement. Therefore, drugs that inhibit folacin-requiring enzymes are widely used in medicine for cancer chemotherapy. Megaloblastic anemia of pregnancy, resulting from low folacin intakes, is associated with poverty and poor diet selection. While folacin deficiency is extremely common in women 16 to 40 years of age because of the effects of pregnancy and lactation, it is rare in men less than 60 years of age. After age 60, folacin deficiency is equally high in both men and women.

For animals, folacin needs are met principally by dietary sources and to some extent by intestinal bacterial synthesis. Different species vary in ability to utilize microbial intestinal synthesis as a source of folacin. Folacin is a feed additive of general use in poultry diets. Swine diets often are not provided with additional folacin, however, supplementation practices are being reevaluated and the vitamin may be required for young swine and sows during pregnancy.

II. HISTORY

In the 1930s and early 1940s, a number of active substances were described that were effective against nutritional deficiencies in humans, certain animals, and bacteria. Factors that later were found to be folacin-related included vitamin M, factor U, vitamin Bc, Bc conjugate, *Lactobacillus casei* factor, SLR (*Streptococcus lactis* R.) factor, folic acid, citrovorum factor, and others (Maynard *et al.*, 1979). The early history of folacin is discussed by Blakley and Benkovic (1984), Scott *et al.* (1982), and Loosli (1988). Willis in 1931 demonstrated a factor from yeast that was active in treating a tropical macrocytic anemia seen in women of India. An anemia preventive factor for monkeys was found in yeast or liver extracts and designated vitamin M in 1935 by Day and associates. In 1939 Hogan and Parrot prevented anemia in chicks with a factor in liver called Bc. In the late 1930s, factors needed for growth and anemia prevention in poultry were referred to as Bc, factor U, or factor R. In 1940 a growth factor for *Lactobacillus casei* was found and in the same year a growth factor for *Streptococcus lacti* was found in spinach and named folic acid. Folic acid was isolated from the liver with the structure and synthesis accomplished by the Lederle group, who named it pteroylglutamic acid on the basis of chemical structure (Maynard *et al.*, 1979). Active forms were found to contain a formyl group or methyl group attached to the number 5 nitrogen of the pteridine nucleus.

Confusion existed in the 1940s concerning the identity of these various factors because both *L. casei* factor and folic acid were active for both microorganisms and animals, whereas vitamin M, factor R, and vitamin Bc were active for monkeys and chicks but not microorganisms. This was resolved when studies showed that folacin exists in nature in both free and bound (additional glutamic acid molecules) forms. Incubating vitamin M with rat or chick liver enzymes markedly increased activity for *L. casei* and *S. faecalis*. Folacin conjugases were also shown to occur in hog kidney and chick pancreas. These studies, therefore, showed that most species except microorganisms are able to utilize bound forms.

III. CHEMICAL STRUCTURE, PROPERTIES, AND ANTAGONISTS

Folacin is the group name to distinguish naturally occurring compounds of this class, the pure substance being designated pteroylmonoglutamic acid. The chemical structure of folacin (pteroylglutamic acid) is shown in Fig. 12.1. Its chemical structure contains three distinct parts. Reading from right to left this compound consists of glutamic acid, *p*-aminobenzoic acid (PABA), and a pteridine nucleus, the last two making up pteroic acid. Thus the name pteroylglutamic acid was suggested. The PABA portion of the vitamin structure was once thought to be a vitamin. If the folacin requirement is met there is no need to add PABA to the diet (see Chapter 16).

Much of the folacin in natural feedstuffs is conjugated with varying numbers of extra glutamic acid molecules. Folacin as pteroyloligo-γ-L-glutamates ($PteGlu_n$) is generally from one to nine glutamates long, with n indicating the number of glutamyl residues. Polyglutamate forms, usually of 3–7 glutamyl residues, of folacin are the natural coenzymes that are most abundant in every tissue examined (Wagner, 1984). The conjugated forms with two or more glutamic acid residues are joined by γ-glutamyl linkages to the single glutamic acid moiety of the vitamin. Synthetic folacin, however, is in the monoglutamate form.

It has been concluded that there are more biologically active forms of folacin

pteridine nucleus | p-aminobenzoic acid (PABA) | glutamic acid

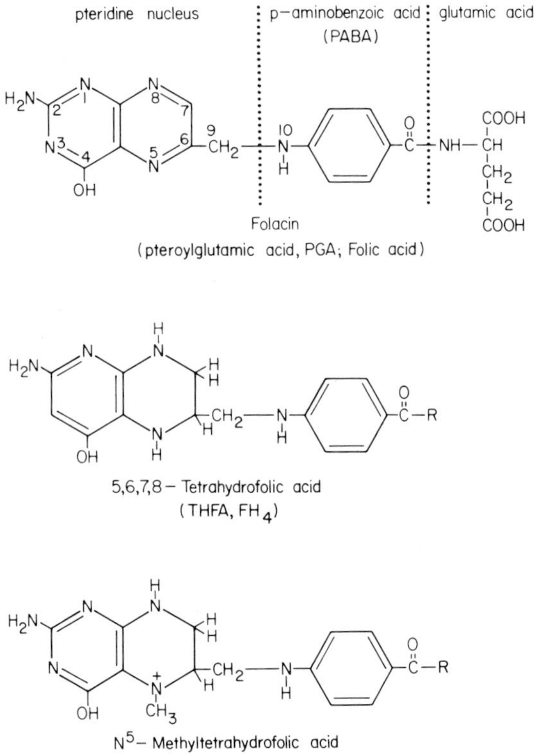

Folacin
(pteroylglutamic acid, PGA; Folic acid)

5,6,7,8-Tetrahydrofolic acid
(THFA, FH_4)

N^5-Methyltetrahydrofolic acid

Fig. 12.1 Structures of folacin compounds (R is one or more glutamic acid molecules).

than any other known vitamin. Naturally occurring pteroylpolyglutamates constitute a large family of closely related compounds arising from modifications of the three parts of the parent compound pteroylglutamic acid. Changes in the state of reduction of the pteridine moiety, addition of various kinds of one-carbon substituents, and addition of glutamic acid residues lead to a wide array of compounds. Baugh and Krumdieck (1971), on the basis of the three known states of reduction of the pyrazine ring, the six different one-carbon substituents that may occur at N-5 and/or N-10, and assuming that the polyglutamyl chain would have no more than seven glutamyl residues, calculated that the theoretical number of folacins approached 150. However, this figure includes compounds that have never been identified in natural materials. Since it is clear now that the polyglutamyl chain reaches at least 8 or 9 residues in animal tissues (and as many as 12 in bacterial cells), the number of folacin compounds that might be expected to occur in animal tissues still approaches 100 compounds. The active forms of folacin contain a formyl group or a methyl group attached to the number 5 or number 10 nitrogens of the compound, or a methylene group between nitrogens 5 and 10. Tetrahydrofolic acid is the principal coenzyme form while the main storage form is 5-methyltetrahydrofolic acid (Fig. 12.1).

Folacin is a yellowish-orange crystalline powder, tasteless and odorless, and insoluble in alcohol, ether, and other organic solvents. It is slightly soluble in hot water in the acid form but quite soluble in the salt form. It is fairly stable to air and heat in neutral and alkaline solution, but unstable in acid solution. From 70 to 100% of folacin activity is destroyed on autoclaving at pH 1 (O'Dell and Hogan, 1943). It is readily degraded by light and ultraviolet radiation. Cooking can reduce folacin food content considerably (see Section VIII).

A great variety of folacin analogs have been prepared, principally for anticancer and antimicrobial therapy (Brody *et al.*, 1984). The folate analogs 4-desoxy-4-amino[N-10]methylpteroylglutamic acid (amethopterin and methotrexate) and 5-fluorouracil are drugs found to be effective in cancer chemotherapy. Since folacin deficiency is more detrimental to cells that are rapidly growing, these antagonists are used as potent antibacterial and antitumoral agents. Folacin antagonists can act by (1) blocking conversion of pteroylmonoglutamic acid to tetrahydrofolic acid by binding to dihydrofolic acid reductase or (2) blocking the transfer of single-carbon units from tetrahydrofolic acid to acceptors such as in synthesis of methionine or purines (Scott *et al.*, 1982). Sulfonamides, though not folacin analogs, are analogs of the folacin biosynthetic intermediate PABA and are widely used as antibacterial agents (Brown, 1962). By competing with PABA, sulfonamides prevent folacin synthesis so the microorganisms cannot multiply, with the result that an important source of folacin to the animal is reduced or eliminated.

IV. ANALYTICAL PROCEDURES

Determination of folacin in biological materials is a difficult analytical problem because of the existence of a number of folacin complexes that exhibit the vitamin activity. High sensitivity is essential in methods of analysis because of the low concentration of folacin in foods and feedstuffs. Both microbiological and chemical methods have been employed to identify folacin and its derivatives. Because folacin in natural feedstuffs is conjugated with varying numbers of extra glutamic acid molecules, the bound forms must be freed to be active for the assay microorganisms. Conjugase enzymes, widely distributed in animal tissues, are capable of releasing free folacin. Typically conjugases from pancreas, liver, or kidney are used to reduce the conjugate to the monoglutamate form. The microbiological assay can use *Lactobacillus casei,* which responds to all the common monoglutamate forms of folacin, and a differential assay using *S. faecalis* and *L. citrovorum*. A radiometric microbiological method has also been reported

(Chen *et al.*, 1983). For a biological assay of feeds, the chick is the preferred animal.

Various methods have been developed for separation of folacins by high-pressure liquid chromatography (HPLC). Separations of monoglutamyl folacins have been accomplished by anion-exchange, paired-ion reverse-phase, and conventional reverse-phase HPLC methods (Gregory *et al.*, 1984). Quantitation of the major folacin compounds in biological materials has been reported by Gregory *et al.* (1984) to combine fluorometric determination with HPLC.

V. METABOLISM

A. Digestion, Absorption, and Transport

Polyglutamate forms are digested via hydrolysis to pteroylmonoglutamate prior to transport across the intestinal mucosa. The enzyme responsible for the hydrolysis of pteroylpolyglutamate is a γ-carboxypeptidase known as folate conjugase (Baugh and Krumdieck, 1971). Most likely several conjugase enzymes are responsible for hydrolysis of the long-chain folate polyglutamates to the monoglutamates, which are then taken up by the mucosal cell (Rosenberg and Newmann, 1974). Pteroylmonoglutamate is absorbed predominantly in the duodenum and jejunum, apparently by an active process involving sodium. The sequence of intestinal absorption of conjugated folate would appear to be mucosal uptake followed by hydrolysis to simple folate and subsequent exit of this compound from the mucosal cell. However, Kesavan and Noronha (1983) suggested from results with rats that luminal conjugase is a secretion of pancreatic origin and that the hydrolysis of polyglutamate forms occurred in the lumen rather than at the mucosal surface or within the mucosal cell.

The bulk of information currently available on folacin transport is consistent with a specific active transport process. Dietary folates, after hydrolysis and absorption from the intestine, are transported in plasma as monoglutamate derivatives, predominantly as 5-methyltetrahydrofolate. The monoglutamate derivatives are then taken up by cells in tissues by specific transport systems. There the pteroylpolyglutamates, the major folacin form in cells, are built up again in stepwise fashion by the enzyme folate polyglutamate synthetase.

Wagonfeld *et al.* (1975) showed in *in vitro* studies that the rate-limiting step in folacin absorption is transport of the monoglutamyl folacin into the mesenteric circulation rather than hydrolysis of polyglutamyl folacins, which is very rapid. However, in cases where the conjugase is strongly inhibited, absorption of polyglutamyl folacins may be reduced. Low availability in orange juice is due to the inhibition of conjugase activity by the low pH produced by large amounts of citric acid. Therefore, drugs that drastically alter intraluminal pH and/or inhibit conjugase activity will also decrease folacin absorption.

In humans, a zinc deficiency resulted in a decreased intestinal hydrolysis of pteroylpolyglutamate, suggesting that intestinal conjugase is a zinc-dependent enzyme (Tamura *et al.*, 1978). From a different aspect of a zinc–folacin relationship, Ghishan *et al.* (1986) showed that zinc transport is significantly decreased when folacin is present in the intestinal lumen and folacin transport is likewise decreased with the presence of zinc.

Specific folate-binding proteins (FBPs) that bind folacin mono- and polyglutamates are known to exist in many tissues and body fluids, including liver, kidney, small intestinal brush border membranes, leukemic granulocytes, blood serum, and milk (Tani and Iwai, 1984). The physiological roles of these FBPs are unknown, although they have been suggested to play a role in folacin transport analogous to the intrinsic factor in the absorption of vitamin B_{12}. The FBPs may play a role in tissue storage of folacin by protecting the polyglutamate derivatives from the action of degradative or hydrolytic enzymes (Brody *et al.*, 1984). Tani and Iwai (1984) reported that FBP of bovine milk affected bioavailability of folacin *in vivo* by resulting in a more uniform absorption throughout the small intestine and reduced loss of 5-methyltetrahydrofolacin in the urine.

Studies showed that about 79–88% of labeled folacin is absorbed, and that absorption is rapid since serum concentrations usually peak about 2 hr after ingestion. The mean availability of folacin in seven separate food items was found to be close to 50%, ranging from 37 to 72% (Babu and Skrikantia, 1976). Folacin from brewer's yeast was only 10% available. Availability of synthetic monoglutamates and polyglutamates was about the same, indicating that conjugase hydrolysis of the polyglutamates was not limiting.

B. Storage

Folacin is widely distributed in tissues largely in the conjugated polyglutamate forms. Normal body stores in humans have been estimated at 5–10 mg, with approximately half in the liver (Brody *et al.*, 1984). Well-nourished adults may have stores that can meet normal body requirements for up to 4 or 5 months (Baker and DeMaeyer, 1979). Body folacin stores are small at birth and are rapidly depleted, particularly in small premature infants. Under a condition of vitamin B_{12} deficiency, there are defects in the conversion of pteroylmonoglutamates to polyglutamate forms that lead to a decreased tissue ability to retain intracellular folacin. Vitamin B_{12} deficiency has been shown to lead to functional folacin deficiency even when folacin intake and absorption are normal. Giugliani *et al.* (1985) reported a significantly higher serum folacin concentration for patients receiving vitamin B_{12} supplementation during pregnancy.

C. Excretion

Urinary excretion of folacin represents a small fraction of total excretion. Fecal folacin concentrations are quite high, often higher than intake, representing not only undigested folacin but more importantly the considerable bacterial synthesis of the vitamin in the intestine. In tracer studies for the first 24 hr after the dose, much of the dose in urine is as intact folacins. After longer time periods, all the excreted label was in degradative products, primarily pteridines and acetamidobenzoylglutamate (Murphy *et al.*, 1976). Over half of labeled dietary folacin is excreted in feces. This is expected as bile contains high levels of folacin due to enterohepatic circulation, with most of bile folacin reabsorbed in the intestine.

VI. FUNCTIONS

Folacin, in the form 5,6,7,8-tetrahydrofolic acid, is indispensable in transfer of single-carbon units in various reactions, a role analogous to that of pantothenic acid in the transfer of two-carbon units. The one-carbon units can be formyl, forminino, methylene, or methyl groups. Some biosynthetic relationships of one-carbon units are shown in Fig. 12.2. These one-carbon units are generated primarily during amino acid metabolism and are used in the metabolic intercon-

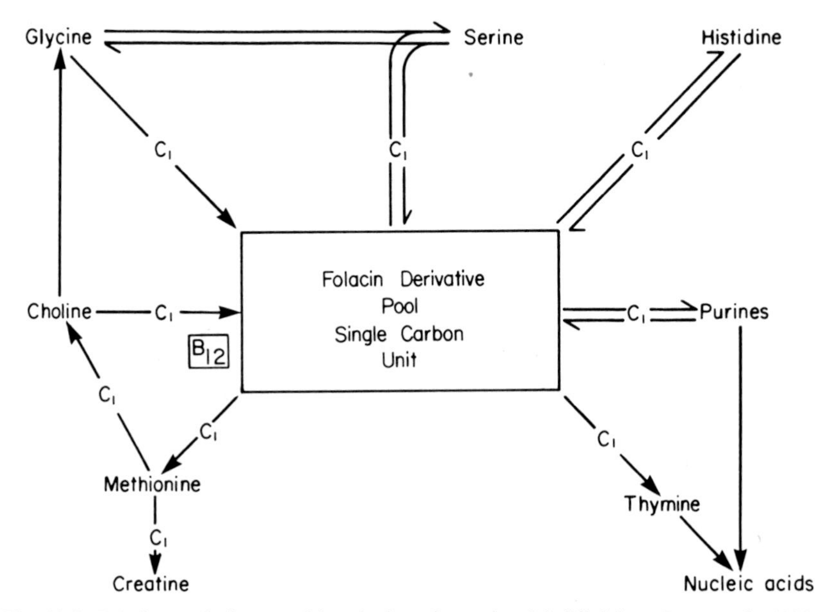

Fig. 12.2 Folacin metabolism requiring single-carbon units. (Modified from Scott *et al.*, 1982.)

versions of amino acids and in the biosynthesis of the purine and pyrimidine components of nucleic acids that are needed for cell division. The important physiological function of tetrahydrofolic acid (THF) consists of binding the C_1 units to the vitamin molecule and thus transforming them to "active formic acid" or "active formaldehyde" so that these are interconvertible by reduction or oxidation and transferable to appropriate acceptors.

Ability of the vitamin as a coenzyme is governed by three factors: (1) nature of the carbon-containing substituent attached at either the N-5 or N-10 position, or bridging them, (2) state of oxidation or reduction of the pyrazine ring, and (3) number of glutamic acid units attached in gamma linkage to the glutamate of PGA. Certain polyglutamates may serve as cofactors for one enzyme while inhibiting another (White *et al.*, 1976).

Folacin polyglutamates work at least as well or better than the corresponding monoglutamate forms in every enzyme system examined (Wagner, 1984). It is now accepted that the pteroylpolyglutamates are the acceptors and donors of one-carbon units in amino acid and nucleotide metabolism, while the monoglutamate is merely a transport form.

Glutamate chain length of folacin polyglutamate may affect metabolism of one-carbon units (Foo and Shane, 1982). In hamster ovary cells with normal extracellular methionine, polyglutamate chain lengths were longer with high levels of octaglutamates, nonaglutamates, and even decaglutamates occurring, while with suboptimal levels of methionine, shorter chain lengths were found. This observed phenomenon of a profound effect of extracellular methionine concentration on glutamate chain elongation may be interpreted to mean that there is a regulatory action of polyglutamate chain length on one-carbon metabolism.

Specific reactions involving single-carbon transfer by folacin compounds are (1) purine and pyrimidine synthesis, (2) interconversion of serine and glycine, (3) glycine-α-carbon as a source of C_1 units for many syntheses, (4) histidine degradation, and (5) synthesis of methyl groups for such compounds as methionine, choline, and thymine.

Purine bases (adenine and guanine) as well as thymine are constituents of nucleic acids and with a folacin deficiency there is a reduction in the biosynthesis of nucleic acids essential for cell formation and function. Hence, deficiency of the vitamin leads to impaired cell division and alterations of protein synthesis; these effects are most noticeable in rapidly growing tissues. In absence of adequate nucleoproteins, normal maturation of primordial red blood cells does not take place and hematopoiesis is inhibited at the megaloblast stage. As a result of this megaloblastic arrest of normal red blood cell maturation in bone marrow, a typical peripheral blood picture results that is characterized by macrocytic anemia. White blood cell formation is also affected, resulting in thrombopenia, leukopenia, and old, multilobed neutrophils.

In folacin deficiency, formiminoglutamic acid (FIGLU), formed as an intermediate in degradation of histidine, can no longer be transformed completely into glutamate and formiminotetrahydrofolic acid, and is therefore excreted in urine. This excretion is suitable as a biochemical criterion for diagnosis of folacin deficiency, appearing at an early stage of deficiency.

Vitamin B_{12} is also closely associated with the progress of the folacin-dependent reactions of intermediary metabolism. Vitamin B_{12} has two main effects in facilitating folacin: (1) vitamin B_{12} regulates the proportion of methyl to nonmethyl tetrahydrofolates according to the methyl trap theory and (2) vitamin B_{12} is necessary for transport of methyl-THF across the cell membrane and promotes folacin retention by tissues. According to the methyl trap concept (Herbert and Zalusky, 1962), vitamin B_{12} deficiency decreases the formation of methionine from homocysteine and methyl-THF by the B_{12}-dependent methionine synthetase. This results in an increase in methyl-THF and a decrease in THF, which is the active coenzyme form that functions in the degradation of FIGLU and formate.

Vitamin B_{12} is necessary in the reduction of one-carbon compounds of the oxidation stage of formate and formaldehyde, and in this way it participates, with folacin, in biosynthesis of labile methyl groups. Folacin is also essentially involved in all these reactions of labile methyl groups. The metabolism of labile methyl groups plays an important role for the body in the biosynthesis of methionine from homocysteine and of choline from ethanolamine. Folacin has a sparing effect on requirements of choline, with the importance of both folacin and vitamin B_{12} on synthesis of choline discussed in Chapter 14.

Folacin is needed to maintain the immune system. The functioning of the immune system was severely inhibited by folacin deficiency in rats (Kumar and Axelrod, 1978), which is probably mediated through a reduction in DNA synthesis, resulting in impaired nuclear division.

VII. REQUIREMENTS

Various animal species differ markedly in their requirements for folacin. Because of microbial synthesis in their digestive tracts, ruminants have no dietary requirement for folacin. Only young ruminants that do not have a fully developed rumen would be expected to require a dietary source. Folacin requirements for monogastric species would be dependent on degree of intestinal folacin synthesis and utilization by the animal. Animals that practice coprophagy would also have a lower dietary need for folacin, as feces is a rich source of the vitamin (Abad and Gregory, 1987). The majority of species apparently do not require dietary folacin because of their ability to utilize microbial intestinal synthesis. However, poultry, guinea pigs, and primates (including humans) develop deficiencies on low dietary folacin. Even though deficiencies can be produced with special diets, corn, soybean meal, and other common feedstuffs in a practical poultry diet should provide ample folacin under most conditions (Scott *et al.*, 1982).

Self-synthesis of folacin is dependent on dietary composition. For poultry, some research has indicated higher folacin requirements for very high protein diets, or when sucrose was the only source of carbohydrates (Scott *et al.*, 1982). Keagy and Oace (1984) reported that dietary fiber had an effect on folacin utilization; xylan, wheat bran, and beans stimulated folacin synthesis in the rat, reflected as higher fecal and liver folacin.

The levels of antibacterials added to the feed will affect microbial synthesis of folacin. Sulfa drugs, which are commonly added to livestock diets, are folacin antagonists (see Section IX). Even in the chicken, sulfa drugs have been shown to increase the requirement (Scott *et al.*, 1982). Moldy feeds (e.g., aflatoxins)

TABLE 12.1

Folacin Requirements for Various Animals and Humans[a]

Animal	Purpose	Requirement	Reference
Beef cattle	Adult	Microbial synthesis	NRC (1984a)
Dairy cattle	Adult	Microbial synthesis	NRC (1978a)
Chicken	Leghorn, 0–6 weeks	0.55 mg/kg	NRC (1984b)
	Leghorn, 6–14 weeks	0.25 mg/kg	NRC (1984b)
	Laying	0.25 mg/kg	NRC (1984b)
	Breeding	0.35 mg/kg	NRC (1984b)
	Broilers, 0–6 weeks	0.55 mg/kg	NRC (1984b)
	Broilers, 6–8 weeks	0.25 mg/kg	NRC (1984b)
Turkey	Growing, 0–8 weeks	1.0 mg/kg	NRC (1984b)
	Growing, 8–16 weeks	0.8 mg/kg	NRC (1984b)
	Growing, 16–24 weeks	0.7 mg/kg	NRC (1984b)
	Breeding hens	1.0 mg/kg	NRC (1984b)
Japanese quail	All classes	1.0 mg/kg	NRC (1984b)
Sheep	Adult	Microbial synthesis	NRC (1985b)
Swine	All classes	0.3 mg/kg	NRC (1988)
Horse	Adult	20 mg/day[b]	NRC (1978b)
Goat	Adult	Microbial synthesis	NRC (1981b)
Fox	Growing	0.2 mg/kg	NRC (1982a)
Mink	Growing	0.5 mg/kg	NRC (1982a)
Cat	Adult	0.8 mg/kg	NRC (1986)
Dog	Growing	0.2 mg/kg	NRC (1985a)
Fish	Trout and salmon	1.0–5.0 mg/kg	NRC (1981a)
Rat	All classes	1.0 mg/kg	NRC (1978c)
Human	Infants	30–45 μg/day	RDA (1980)
	Children	100–300 μg/day	RDA (1980)
	Adults	400–800 μg/day	RDA (1980)

[a]Expressed as per unit of animal feed either on an as fed (approximately 90% dry matter) or dry basis (see Appendix Table 1). Requirements established for some species, while only suggested for others. Human requirements expressed as μg/day.

[b]Horses responded to level, but no establishment of requirement.

have also been shown to contain antagonists that inhibit microbial intestinal synthesis in swine (Purser, 1981).

Folacin requirements are dependent on the form in which it is fed and concentrations and interrelationships of other nutrients. Deficiencies of choline, vitamin B_{12}, iron, and vitamin C all have an effect on folacin needs. Although most folacin in poultry feedstuffs is present in conjugated form, the young chick is fully capable of utilizing it. On the contrary, Baker *et al.* (1978) reported that human patients over 60 years of age utilized conjugated forms of folacin much less efficiently than monoglutamates.

Folacin requirements are related to type and level of production. The more rapid the growth or production rates, the greater is the need for folacin because of its role in DNA synthesis. In poultry the requirement for egg hatchability is higher than that for production (NRC, 1984b). Table 12.1 summarizes the folacin requirements for various livestock species and humans with a more complete listing given in Appendix Table 1.

For humans the burden of lactation on maternal folacin reserves is estimated to be 20–50 μg per day, varying with the folacin content and volume of milk (Matoth *et al.*, 1965). This estimate, based on production of 850 cc of milk of average folacin content, should be doubled to meet the needs of mothers producing milk with high folate content. An additional 100 μg of dietary folacin should provide for the absorption of the excess need. The recommended daily allowance (RDA, 1980) for folacin is necessarily in excess of the minimum daily requirement to provide a margin of safety and allow for losses in preparation of foods as well as the decreased availability of food folacin as compared to folacin in the monoglutamate form.

VIII. NATURAL SOURCES

Folacin is widely distributed in nature, almost exclusively as THF acid derivatives, the stable ones having a methyl or formyl group in the 5-γ position, and generally possessing three or more glutamic acid residues in glutamyl linkages. Only limited amounts of free folacin occur in natural products, with most feed sources containing predominantly polyglutamyl folacin. However, in seeds or fruit, which presumably store the vitamin, a considerable amount is present as a monoglutamate. A high proportion of monoglutamate forms of folacin are found in milk and soybeans. Much of the folacin in milk is available in the monoglutamate form, which is necessary for absorption by the newborn (Wagner, 1984). The predominant form of folacin in such vegetables as spinach, asparagus, broccoli, lettuce, yeast, rice, and peas is the N-10 formyl derivative. Folates in tissues such as liver, kidney, and red cells are predominantly pentaglutamates.

Folacin is abundant in green leafy materials and organ meats. Soybeans, other beans, nuts, some animal products, and citrus fruits are good sources. Cereal grains, milk, and eggs are generally poor sources of the vitamin. Any animal or human diet without green leafy materials or animal protein, especially organ meats, is likely to be low in folacin. Human diets may be particularly low during

TABLE 12.2

Folacin in Foods and Feedstuffs (mg/kg)[a]

Alfalfa meal	5.5
Asparagus, fresh	1.4
Barley, grain	0.6
Blood meal	0.1
Brewer's grains	7.7
Broccoli, fresh	1.69
Broccoli (boiled), fresh	0.65
Cabbage, fresh	0.3
Cabbage (boiled), fresh	0.16
Carrot, roots	1.2
Coconut meal	1.5
Corn (Maiz), grain	0.3
Corn (Maiz), gluten meal	0.3
Cottonseed meal, solvent extracted	2.8
Fish meal, anchovy	0.2
Fish meal, menhaden	0.2
Linseed meal, solvent extracted	1.4
Liver, cattle	8.4
Milk, skimmed, cow's	0.7
Orange juice, fresh	1.4
Peanut meal, solvent extracted	0.7
Rice, bran	2.4
Rice, grain	0.4
Rice, polished	0.2
Rye, grain	0.7
Sorghum, grain	0.2
Soybean meal	0.7
Soybean, seeds	3.9
Spinach, fresh	1.93
Spinach (boiled), fresh	0.91
Sugarcane molasses	0.1
Timothy hay, sun cured	2.3
Wheat, bran	1.6
Wheat, grain	0.5
Whey, cattle	0.9
Yeast, brewer's	10.3

[a]Dry basis (unless otherwise specified). Concentrations from NRC (1982a) and Brody *et al.* (1984).

the winter season when green vegetables and citrus fruits are less plentiful. The folacin content of typical foods and feedstuffs is shown in Table 12.2.

A considerable loss of folacin (50–90%) occurs during cooking or processing of foods. Folacin is sensitive to light and heating, particularly in acid solution. In aerobic conditions, destruction of most folacin forms is significant with heating, with reduced folacins more stable in foods because of the relatively anerobic conditions and because folacin is protected from light (Brody *et al.*, 1984). Fresh cabbage and broccoli are good sources of folacin and contain on a wet weight basis 0.30 and 1.69 ppm, respectively. However, when these vegetables are boiled and the water is discarded, losses are considerable, reducing concentrations to 0.16 and 0.65 ppm, respectively (Leichter *et al.*, 1978).

IX. DEFICIENCY

Folacin deficiency has been produced experimentally in many animal species, with macrocytic anemia (megaloblastic anemia) and leukopenia (a reduced number of white cells) being consistent findings. Tissues that have a rapid rate of cell growth or tissue regeneration, such as epithelial lining of the gastrointestinal tract, the epidermis, and bone marrow, are principally affected (Hoffbrand, 1978).

For some animals such as the chick, guinea pig, and monkey, the presence of adequate amounts of folacin in the diet is essential, and deficiency signs can readily be induced by feeding a diet deficient in the vitamin. In other animals, for example, the rat, dog, and pig, folacin produced by the intestinal microflora is usually adequate to meet requirements. Consequently, deficiency signs do not develop unless an intestinal antiseptic is also included in the diet to depress bacterial growth.

A. Effects of Deficiency

1. Ruminants

Folacin synthesis occurs in the rumen. However, young animals that do not have a fully developed rumen would be expected to suffer from a folacin deficiency. A folacin deficiency has not been demonstrated in the calf but Draper and Johnson (1952) have reported a deficiency in lambs fed synthetic diets. The disease was characterized by leukopenia followed by diarrhea, pneumonia, and death. Folacin therapy promoted regeneration of white cells and 0.39 mg per liter of milk in control animal diets prevented the deficiency. There was no indication of folacin deficiency in calves fed a synthetic milk containing 52 mg of folacin per kilogram of liquid feed fed at 10% of liveweight (Wiese *et al.*, 1947).

2. Swine

Until recently folacin deficiency in swine had only been produced by the simultaneous feeding of sulfa drugs, indicating that intestinal synthesis was adequate to meet needs. Deficiencies were not observed when young pigs were fed only purified diets or low natural diets alone (Johnson *et al.*, 1948b). Feeding a purified diet containing 2% sulfasuxidine to weanling pigs resulted in reduced gains and alopecia (Cartwright and Wintrobe, 1949). The pigs also developed a mild normochromic, normocytic anemia; in bone marrow there was a decrease in the ratio of leukocytes to erythrocytes and an increase in the number of immature nucleated red cells. Positive response was obtained after supplementation with folacin. Cunha *et al.* (1948) found that folacin was needed for normal hematopoiesis with 8-week-old pigs fed a purified diet for 21 weeks with sulfasuxidine. A normocytic anemia resulted that was prevented by folacin, whereas a more severe anemia was produced by using a crude folacin antagonist. A combination of folacin and biotin was more effective than folacin in counteracting the anemia. Lindemann and Kornegay (1986) reported that combination of the antibiotic mixture ASP 250 (includes chlortetracycline, sulfamethazine, and penicillin) and folacin to a corn–soybean meal diet increased gains and feed consumption, with no effect of either alone.

More severe deficiency signs that responded to folacin supplementation were induced by feeding diets containing a sulfonamide and a folacin antagonist (Welch *et al.*, 1947). Under such circumstances, pigs became listless, had a reduced growth rate, and developed diarrhea. Hematological manifestations were severe macrocytic anemia, leukopenia with a more marked reduction in the number of polymorphonucleocytes, and mild thrombocytopenia. Cartwright *et al.* (1952) reported a combined folacin and vitamin B_{12} deficiency for pigs receiving a purified soybean protein diet that included a folacin antagonist. Growth rate was reduced and macrocytic anemia, leukopenia, and neutropenia developed with erythroid hyperplasia of the bone marrow. Folacin supplementation immediately resulted in a normal blood and bone marrow picture but growth was decreased and the blood picture subsequently relapsed.

In addition to sulfa drugs and other folate antagonists, moldy feeds can increase the need for the vitamin. In feeding trials involving use of corn with mold infestation, additional folacin increased growth rate up to 15% and improved feed efficiency up to 9% (Purser, 1981). Folacin supplementation was of no value when normal corn was fed.

Recently, inadequate folacin has been associated with suboptimal reproductive performance of sows. A dramatic decrease in serum folacin concentrations was observed during early and mid-gestation that may be associated in part with embryonic mortality (Matte *et al.*, 1984a). In a separate trial, folacin was administered intramuscularly according to a schedule that maintained serum folacin at approximately the same level between weanling and 60 days of gestation (Matte *et al.*, 1984b). Average live litter size was 12 piglets per litter for sows receiving folacin and flushing treatments as compared with 10.5 for sows without any treatment. More research is required to determine if supplemental dietary folate will reduce embryonic death loss under differing management systems.

3. Poultry

Poultry are more susceptible to lack of folacin than other farm livestock, as a deficiency can readily be produced by feeding a folacin-deficient diet. Megaloblastic arrest of erythrocyte formation in bone marrow causes a severe macrocytic anemia as one of the first signs. Folacin deficiency in chicks is also characterized by poor growth, very poor feathering, an anemic appearance, and perosis (Figs. 12.3 and 12.4). The chicks become lethargic and feed intake declines. As anemia develops, the comb becomes waxy white and mucous membrane of the mouth becomes pale (Siddons, 1978). Turkey poults fed a folacin-deficient diet show reduced growth rate and increased mortality (Fig. 12.5). The birds develop a spastic type of cervical paralysis in which the neck is stiff and extended but with only a moderate degree of anemia. Poults with cervical paralysis will die within 2 days after the onset of these signs unless folacin is administered

Fig. 12.3 Note depigmentation and reduction in growth of the folacin-deficient bird on left. (Courtesy of G. F. Combs, Department of Poultry Science, University of Maryland.)

Fig. 12.4 Folacin-deficient chick at 5 weeks of age. Note the weakened condition of legs and the way the bird holds the left wing. The chicken is suffering from cervical paralysis. Deficient bird will shake the end of the wing, and the whole bird will quiver at times. (Courtesy of M. L. Sunde, University of Wisconsin.)

immediately (Scott *et al.*, 1982). Erythrocytes of deficient birds tend to be larger in diameter, and their nuclei are less dense than those of birds receiving supplementary folacin (Schweigert *et al.*, 1948).

Folacin deficiency also results in poor feather development for chicks and turkeys with the shafts weak and brittle. Folacin along with lysine and iron is required for feather pigmentation, as depigmentation occurs in colored feathers during a deficiency of the vitamin.

It appears that egg production is less affected by folacin deficiency than the development of the chick or poult. An inadequate intake of folacin by breeding hens results in poor hatchability and a marked increase in embryonic mortality (Fig. 12.6), which occurs during the last days of incubation. A deformed beak and bending of the tibiotarsus are signs of the embryonic deficiency. Folacin requirement for egg production is less than that for hatchability. Taylor (1947) reported that 0.12 mg folacin/kg diet was satisfactory for egg production, but higher levels were required for good hatchability.

Folacin deficiency has sometimes been associated with perosis, or ''slipped tendon.'' Pollard and Creek (1964) demonstrated histologically that the lesions

Fig. 12.5 Folacin deficiency in turkeys. Poult was hatched from hen fed diet low in folacin. (Courtesy of M. L. Sunde, University of Wisconsin.)

of folacin-deficient bones and cartilage are different from those produced by choline or manganese deficiencies. Abnormal structure of the hyaline cartilage is found in folacin-deficient chicks and ossification is retarded. These disorders are not found in chicks deficient in choline or manganese, although bone deformities and slipped tendons are found in both types of disorders. However, Bechtel (1964) claimed that choline is only effective in preventing perosis when sufficient folacin is present in the diet. Dietary choline content has been shown to affect the chicks' requirement for folacin. When the diet contained adequate choline, the folacin requirement was 0.46 mg/kg diet but this increased to 0.96 mg/kg diet when the diet was choline deficient (Young *et al.*, 1955). Increasing the protein content of the diet has also been shown to increase the incidence and severity of perosis in chicks receiving low levels of dietary folacin. It is suggested that this increased requirement of folacin in high protein diets for poultry is a consequence of greater demand for folacin in uric acid formation (Creek and Vasaitis, 1963).

Folacin appears to be necessary for cell mitosis. In the absence of folacin, oviduct growth is not increased in estrogen-treated chicks. The production of water-soluble proteins (particularly the albumin fraction) in the hormone-stimulated oviduct is also greatly reduced, and there is an alteration in the amino

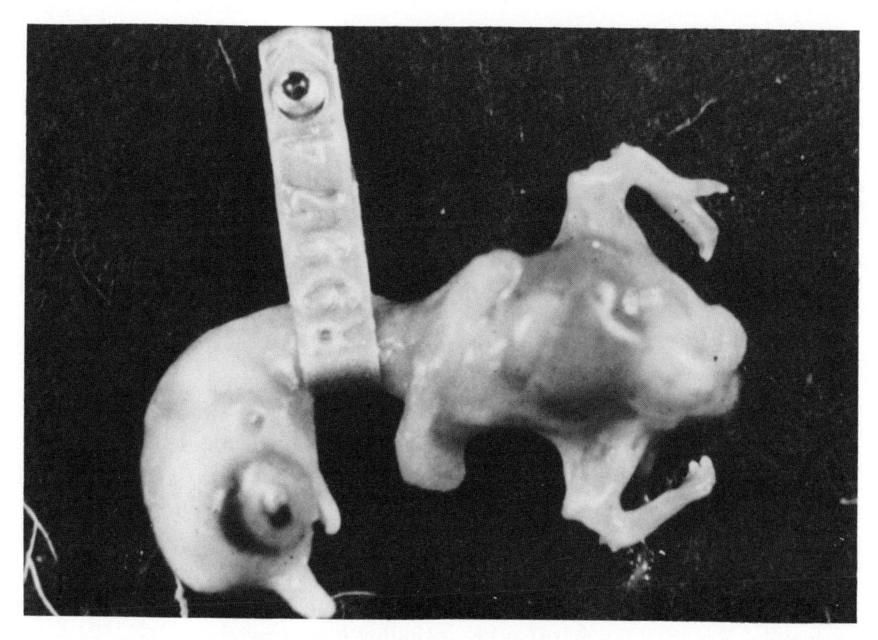

Fig. 12.6 Folic acid deficiency. Abnormal embryo from an egg laid by a hen on a low-folacin diet. (Courtesy of M. L. Sunde, University of Wisconsin.)

acid composition of these proteins. The percentages of arginine, leucine, serine, and tryptophan are decreased and those of glycine and methionine increased (Siddons, 1978).

4. Horses

It has been shown that the horse synthesizes folacin in the intestinal tract. This synthesis may not be sufficient as Seckington *et al.* (1967) reported a case of folacin deficiency in a 7-year-old gelding that had been receiving a diet lacking in fresh grass for many months. The deficient animal had poor performance associated with low serum folacin. Administration of 20 mg of folacin dramatically increased performance as well as elevated blood folacin.

5. Other Animal Species

Folacin deficiency has been reported using only purified feed ingredients or only diets naturally low in the vitamin for a number of species, including the dog, fox, guinea pig, hamster, mink, monkey, salmon, and trout. For other species, including the rat, cat, carp, and catfish, folate deficiencies are difficult to establish unless use is made of an intestinal antiseptic (sulfa drug) or folate antagonist. For mice and certain other species, care must be taken to prevent coprophagy if a folacin deficiency is to be produced on a diet without the aid of an antagonist or sulfa drug.

a. Cats. Folacin deficiency was characterized by weight loss, anemia (macrocytic tendencies), and leukopenia. Blood-clotting time was increased and plasma iron concentrations were elevated (Carvalho da Silva *et al.*, 1955).

b. Dogs. Folacin deficiency results in erratic appetite, decreased gain, water exudate from eyes, glossitis, leukopenia, hypochromic anemia, and decreased antibody response to infectious canine hepatitis and canine distemper virus (NRC, 1985a). A positive response was obtained by subcutaneous injections of folacin.

c. Laboratory Animals. Rodent species are generally similar in relation to folacin deficiency signs by exhibiting reduced growth, anemia, leukopenia, evidence of reduced protein synthesis, reduced folacin tissue levels, and an impaired antibody response (NRC, 1978c; Siddons, 1978). An increase in urinary excretion of FIGLU occurs with folacin deficiency in the rat and mouse. A deficiency of folacin in the pregnant mouse has an adverse effect on reproduction and lactation. Folacin deficiency has been shown to have a teratogenic effect on the rat, with congenital abnormalities in offspring including hydrocephalus.

d. Monkeys. After 3–9 months, rhesus monkeys on a low-folacin diet lost weight, became inactive, and developed a megaloblastic macrocytic anemia between 3 and 18 months (NRC, 1978d; Siddons, 1978). A variety of clinical and hematological signs arise as a result of the deficiency, including weight loss, anorexia, listlessness, mucoid to bloody diarrhea, gingivitis and oral ulceration, increased susceptibility to infection, especially dysentery, macrocytic anemia, leukopenia, and an increased excretion of FIGLU. Leukopenia is the most characteristic hematological sign, and in some monkeys it may prove fatal without the development of anemia.

e. Foxes and Mink. Both adult and growing foxes receiving a folacin-deficient diet develop anorexia, body weight loss, and a decrease in hemoglobin, erythrocyte, and leukocyte concentrations (NRC, 1982b). Folacin deficiency in mink results in anorexia, growth depression, diarrhea, and ulcerative hemorrhagic gastritis (NRC, 1982b; Siddons, 1978).

f. Fish. Folacin deficiency signs have not been demonstrated in experiments with catfish, carp, and red sea bream, but for eel a deficiency results in anorexia, poor growth, and dark coloration (NRC, 1983). Signs of folacin deficiency in trout and salmon include anorexia, reduced growth, poor feed conversion, and macrocytic, normochromic, megaloblastic anemia characterized by pale gills

(NRC, 1981a). Combined dietary deficiencies of folacin and vitamin B_{12} accelerated development of a more pronounced anemia in these fish.

6. Humans

For humans, folacin deficiency is probably the most common vitamin deficiency in the world. Infants, adolescents, elderly persons, and pregnant women seem particularly vulnerable. Studies involving the World Health Organization in various countries suggest that up to one-third of all pregnant women in the world have folacin deficiency (Herbert, 1981). In a clinic in New York City, 16% of pregnant women were deficient with erythrocyte folacin below 150 ng/ml. A further 14% had erythrocyte folacin concentrations suggestive of a deficiency. In a study in Paris, 18% of immigrant pregnant women were deficient in folacin (Hercberg *et al.*, 1987). Megaloblastic anemia of pregnancy was of high incidence among black South African women, and 40% of apparently otherwise healthy pregnant women had morphological and biochemical evidence of folate deficiency (Colman, 1982).

Adolescent girls have a greater nutritional requirement for folacin in relation to body size than do adult women (Heald, 1975). Additional folacin demand of pregnancy and poor dietary habits may compromise their growth potential and increase risk in pregnancy. Folacin deficiency in the older population is due in part to an impairment of dietary folate utilization (Baker *et al.*, 1978). Elderly subjects (greater than 60 years) were unable to utilize polyglutamate forms of folacin effectively compared to younger subjects, indicating a lack of conjugase enzymes. Infants have a high requirement for folacin, with preterm infants reported to have significantly lower erythrocyte folacin than those born at term (EK, 1980). These data would indicate that there is a heightened maternofetal transfer of folacin during the last few weeks of pregnancy.

Typical reaction to folacin deficiency in the human is a megaloblastic red cell maturation in the bone marrow with a resulting macrocytic anemia (Fig. 12.7), which is accompanied by leukopenia. The macrocytic anemia that occurs resembles pernicious anemia without the nervous system involvement. Glossitis, gastrointestinal lesions, diarrhea, and intestinal malabsorption may accompany macrocytic anemia. Likewise, clinical findings of folacin deficiency include pallor, weakness, forgetfulness, sleeplessness, and bouts of euphoria (Herbert, 1962). The most rapidly proliferating tissues of the body, such as the bone marrow for blood cell production, have the greatest requirement for DNA synthesis and thus are principally affected when severe deficiency occurs.

The sequence of signs in the development of human folacin deficiency was reported by Herbert (1967) as he examined biochemical and hematological changes in himself as he consumed a low-folacin diet over time (Table 12.3). After only 3 weeks of dietary folacin deprivation his serum folacin dropped from 7 to less than 3 ng/ml. However, low erythrocyte folacin did not appear until 4 months

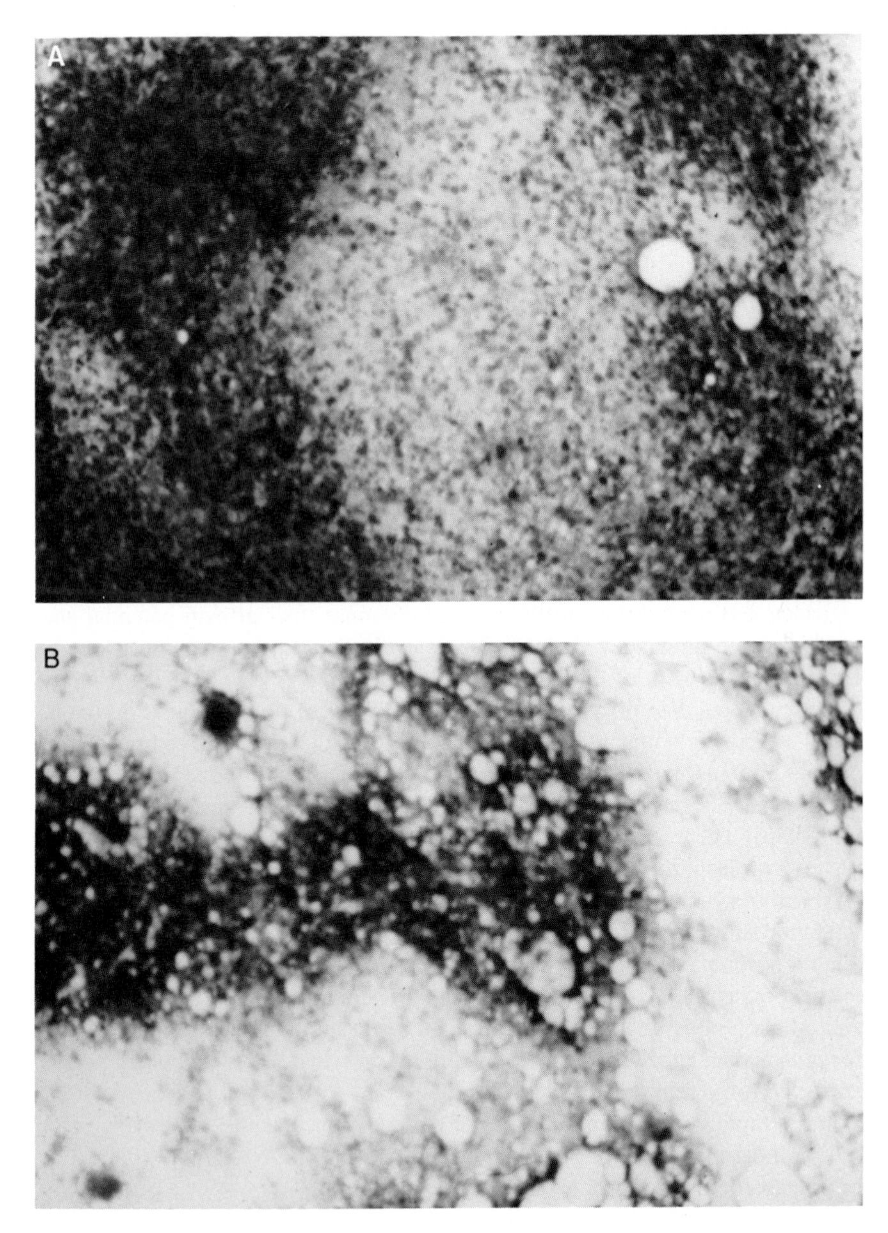

Fig. 12.7 Pernicious anemia characterized by marked hyperplasia in bone marrow (A) compared to normal (B). Deficiency of either folacin or vitamin B_{12} results in ineffective erythropoiesis. (Courtesy of R. R. Streiff, Veterans Administration, University of Florida.)

TABLE 12.3

Progressive Folacin Deficiency Sequence of Events in a 35-Year-Old Human Male[a]

Time (weeks)	Biochemical and hematological events
3	Low serum folacin (<3 ng/ml)
7	Hypersegmentation of neutrophils (leukocytes)
13	High urine formiminoglutamic acid (FIGLU)
17	Low RBC folacin (<20 ng/ml)
19–20	Megaloblastic anemia

[a]Modified from Herbert (1967).

after initiation of folacin deprivation. At about 4.5 months, bone marrow became megaloblastic and anemia occurred.

Apart from features due to megaloblastic anemia or to abnormalities of the epithelial cell surfaces, sterility and gastrointestinal abnormalities are established effects of folacin deficiency (Hoffbrand, 1978). Sterility in both men and women and, in a small percentage of cases, widespread reversible melanin pigmentation of the skin, mainly affecting the skin creases and nail-beds, are reported (Fleming and Dawson, 1972). Folacin deficiency may be related to abnormalities of pregnancy, including postpartum hemorrhage, antepartum hemorrhage, congenital malformation, and prematurity. Clinical signs related to the gastrointestinal tract include sore tongue, angular cheilosis, loss of appetite, and diarrhea (Rose, 1971). The tongue may appear red and shiny or smooth, pale, and atrophic. Some researchers have shown structural and functional jejunal changes in nutritional folacin deficiency that can be reversed by supplemental folacin (Hoffbrand, 1978). There is a complicated relationship between tropical sprue and folacin deficiency. Nutritional folate deficiency may predispose to the disease and that deficiency due to reduced diet and to malabsorption may aggravate the small intestinal lesion.

Folacin deficiency is often associated with chronic alcoholism. Between 40 and 87% of alcoholics admitted to municipal hospitals in the United States have low serum folacin and between 40 and 61% have megaloblastic anemia (Halsted and Tamura, 1979). Folacin deficiency appears to be a result of a complex interaction between nutritional deprivation and chronic alcohol ingestion. Alcohol specifically inhibits hematopoiesis and Halsted *et al.* (1973) reported that the jejunal uptake of folacin was significantly greater in subjects that remained sober.

B. Assessment of Status

Assessment of nutritional status of folacin can involve dietary evaluation, clinical signs, response to supplementation, and laboratory analysis. For dietary history, humans or animals (particularly poultry) that have not received green leafy plant sources or organ meats would suggest a reduced folacin intake. Because liver contains a high percentage of stored folacin, concentration in this organ would serve as a folacin status indicator. On low-folacin diets, liver concentrations are depleted in a few months. Clinical signs of folacin deficiency are extremely variable and are less precise than laboratory analysis to confirm a deficiency.

In humans, a positive diagnosis of folacin deficiency is usually made by the finding of subnormal serum and erythrocyte levels. Cutoff values of <3.0 ng/ml for serum folacin and <140 ng/m for erythrocyte folacin are the basis for estimation of the prevalence of low values of the vitamin (Senti and Pilch, 1985). Patients with vitamin B_{12} deficiency exhibit all the clinical, hematological, and many of the biochemical features of folacin deficiency. Thus, erythrocyte folacin assay is better used in conjunction with the values for serum vitamin B_{12}.

Measurement of tissue content of several forms of folacin and urinary excretion of FIGLU, a histidine catabolite that cannot be normally metabolized in folacin deficiency, is utilized to assess folacin nutrition. It has been shown that the amount of FIGLU excreted in urine roughly parallels erythrocyte folate level and the hepatic folacin level in both anemic and nonanemic folacin-deficient patients and thus, like the erythrocyte and hepatic folacin levels, appears to be a satisfactory index of tissue folate stores (Herbert, 1967).

Deficiencies of iron and vitamin C can be related to folacin status, but it has not yet been proven that folacin requirement is increased in either iron deficiency anemia or scurvy. The frequent occurrence of combined iron and folacin deficiencies has led to the suggestion that iron deficiency may be responsible for the development of a secondary folacin deficiency. Iron deficiency has been reported to influence folacin metabolism in pregnant women. The stress of lactation in rats superimposed on iron deficiency was found to alter milk folacin concentration, resulting in folacin depletion in rat pups (Kochanowski *et al.*, 1983). Both iron and folacin are required for normal hematopoiesis, with anemia resulting from lack of either nutrient. Iron deficiency may mask the changes in the developing erythroblasts but does not affect the white cell abnormalities. Anemia in humans is often assumed to be the result of iron deficiency, however, Bailey *et al.* (1980) reported that folacin deficiency was much more prevalent than iron deficiency in a lower-income pregnant Florida population. Whether the association of scurvy and megaloblastic anemia is due to increased folacin requirements secondary to vitamin C lack or is merely the result of a double nutritional deficiency is unresolved. However, it is interesting to speculate that many cases of vitamin C deficiency could likewise be deficiencies of folacin since both vitamins are rich in green plants and citrus fruits and lacking in most other food sources.

In most human patients with folacin deficiency, a combination of factors leads to negative folacin balance. Poor diet is usually the major cause, since few diseases cause such malabsorption or poor folacin utilization to such a severity that a good intake of the vitamin cannot overcome losses. It is likely that severe folacin deficiency may occur in tropical sprue, and congenital specific malabsorption of folacin, despite a normal dietary folacin content (Hoffbrand, 1978).

X. SUPPLEMENTATION

Folacin needs for livestock are often met by good practical diets (Maynard *et al.*, 1979), and for most species, substantial quantities of folacin are provided through microbial synthesis. Nevertheless, field observations have been made on folacin-insufficient diets. Green forage is an excellent source of folacin. Supplementation of folacin would be most needed when animals are in confinement without access to green grazing or preserved green forages. The successful treatment of field cases of folacin deficiency with supplemental folacin has demonstrated that commercial feeds do not always supply adequate quantities of the vitamin to poultry.

Of farm livestock, poultry would most likely need supplemental folacin and then only under certain conditions. Newly weaned pigs may also need additional folacin in their diets for optimum growth and feed efficiency. Folacin may be of little benefit when pigs receive only low levels of sulfa drugs and consume grains relatively free of toxin-producing molds. However, since a large percentage of the U.S. corn crop contains some mold contamination, folacin supplementation should have a positive effect in many commercial hog operations as well as for other livestock enterprises (Purser, 1981). This would likely be an even more important consideration in developing tropical countries where conditions favoring mold growth are optimized. Individual responses to folacin supplementation to counteract mold effect will obviously vary with the class of livestock being fed, species of mold present, and the levels of toxin encountered (Bhavanishankar *et al.*, 1986). Purser (1981) fed supplemental folacin to pigs weaned at either 3,4, or 5 weeks of age. At the end of the 4-week feeding period, the younger pigs showed the greatest response to diets providing the supplemental vitamin.

Crystalline folacin, produced by chemical synthesis, is available for feeds, foods, and pharmaceuticals. Although folacin is only sparingly soluble in water, the sodium salt is quite soluble and is used in injections as well as feed supplements (McGinnis, 1986b; Tremblay *et al.*, 1986).

Gadient (1986) considers folacin to be very sensitive to heat and light, slightly sensitive to moisture, and insenitive to oxygen. Folacin can be lost during storage of premixes, particularly at elevated temperatures (Frye, 1978). After 3 months of room-temperature storage, 43% of the original folacin activity was lost. Verbeeck (1975) found folacin to be stable in premixes without minerals but there may be as much as 50% loss in a premix with minerals kept at room temperature for 3 months. Adams (1982) reported only 38% retention of folacin activity in a premix without minerals after 3 weeks at 45°C. However, he reported 57% retention of activity after 3 months at room temperature. Slinger *et al.* (1979) reported processing and storage losses of folacin in fish feeds of 5 to 10% for steam-pelleted crumbles and 3–7% for extruded crumbles, depending on dietary level. Scott (1966) indicated that an adjustment of 10–20% in the folacin level in poultry feed may be necessary because of pelleting losses.

Folacin supplements (monoglutamate form) for humans ranging from 100 to 1000 μg per day have been recommended by different investigators. During an observation period of 6 weeks, a daily oral supplement of 100 μg of PGA maintained the serum folacin in the normal range, 50 μg maintained it in a range that was diagnostically indeterminate, and 25 μg did not prevent a fall in serum folacin to deficient levels (RDA, 1980). Oral supplementation appears to be desirable to maintain maternal stores and to keep pace with the increased folacin turnover that is seen in rapidly growing tissue.

Supplementation with large doses of folacin will result in curing the macrocytic anemia in human patients with the vitamin B_{12} deficiency, pernicious anemia. However, folacin will not prevent the often irreversible neurological lesions of vitamin B_{12} deficiency or pernicious anemia. Therefore, large quantities of supplemental folacin should not be taken indiscriminately as they may obscure a diagnosis of vitamin B_{12} deficiency.

XI. TOXICITY

Folacin generally has been regarded as a nontoxic vitamin (NRC, 1987). Acute intravenous toxicity is very low, with the LD_{50} in mg per kg body weight being: mice, 600; rat, 500; rabbit, 410; guinea pig, 120. In rats, most of the deaths occurred within 30 min of injection (Anonymous, 1961). Rabbits given 50 mg/kg/day intraperitoneally for 10 weeks were possibly retarded in growth and did not differ in blood picture, number of deaths, or general appearance, but did show signs of renal injury at autopsy. Folacin has a low acute and chronic toxicity for humans. In adults, no adverse effects were noted after 400 mg/day for 5 months and after 10 mg/day for 5 years (Brody *et al.*, 1984).

13

Vitamin B_{12}

I. INTRODUCTION

Vitamin B_{12} was the last vitamin to be discovered (1948) and the most potent of the vitamins, with the lowest concentrations required to meet daily requirements. Vitamin B_{12} is unique in that it is synthesized in nature only by microorganisms; therefore, it is usually not found in plant feedstuffs. Consequently, humans and other monogastric species who subsist entirely on plant foods would be susceptible to vitamin B_{12} deficiency. It is also unique in that the trace element cobalt is an integral part of the molecule.

The discovery of this vitamin was dramatic and made possible by the combined efforts of microbiologists, biochemists, nutrition scientists, and physicians working in various laboratories. Three seemingly unrelated conditions attributed to lack of the vitamin or its precursor were identified: (1) a fatal anemia in humans, (2) a potent growth factor for monogastric species, and (3) a relationship to cobalt, the lack of which resulted in wasting diseases in ruminants.

II. HISTORY

The history of vitamin B_{12} in human and animal nutrition is both exciting and stimulating and has been reviewed (Sebrell and Harris, 1968; Folkers, 1982; Loosli, 1988). In 1824 Combe described a fatal anemia, pernicious anemia, and suggested that it could be related to a disorder of the digestive tract. The existence of an unknown factor in liver, effective in treatment of pernicious anemia, was recognized in 1926 when Minot and Murphy showed that large amounts (120–240 g/day) of raw liver given by mouth daily would alleviate this previously fatal disease.

During the next 20 years, research resulted in concentrating the activity of 400 g of liver to 1 mg of active substance. From 1929 onward, Castle postulated that pernicious anemia was due to the interaction of a dietary (extrinsic) factor and an intrinsic factor produced by the stomach. Mixing beef muscle and gastric juice prevented anemia, thus a factor in gastric juice was the intrinsic factor while a different substance in beef muscle was the extrinsic factor. Castle took the extraordinary step of using his own stomach to process food for his anemic patients, then regurgitated his partially digested meals to supplement their diets.

For many years following the discovery that liver contained a substance that could cause a remission of pernicious anemia, scientists strove unsuccessfully to isolate from liver the antipernicious anemia (APA) factor. Progress was slow because no experimental animal for laboratory trials exhibited this condition and thus human patients with pernicious anemia were required to evaluate results of fractionation studies. In 1947 Shorb of the University of Maryland reported that a factor (LLD factor) in liver extract required by the bacterium *Lactobacillus lactis* was in concentrations bearing an almost linear relationship to the APA activity of the extract. Making use of this organism, Rickes and co-workers in the United States isolated, in crystalline form, a factor from the liver that cured pernicious anemia. They named this factor vitamin B_{12}. At the same time Smith in England isolated the APA factor. West (1948) confirmed the clinical activity of the vitamin, which prevented pernicious anemia with a single dose of 3–6 μg. Although isolated by the two laboratories in 1948, it was not until 1956 that its very complicated structure was ascertained. In 1961 Lenhert and Hodgkin reported the structure of the enzyme form of vitamin B_{12}.

Attempts to raise pigs and chickens by feeding all vegetable diets resulted in poor performance. In 1926 it was recognized that liver extracts and other concentrates of animal origin stimulated growth of rats, chicks, and pigs. Because the true nature of the active principle of such animal products was unknown, it was called "animal protein factor" (APF). It was also referred to as the "chick growth factor" and the same factor was likewise found essential for hatchability. Manure from cattle was also found to contain the factor.

Vitamin B_{12} became available for animal experiments in 1948 as the crystalline vitamin B_{12} was isolated from liver and crude fermentation products, which contained considerable concentrations of the vitamin. To the surprise of researchers, the growth effect of pure vitamin B_{12} in animals was not quite as good as that observed from APF supplements. Eventually it was found that crude vitamin B_{12} concentrates contained more than one active principle, namely, vitamin B_{12}, some essential amino acids, and small concentrations of compounds with antibiotic activity. The growth-promoting effects of antibiotics were discovered practically as a by-product of research on vitamin B_{12} in animal nutrition. These discoveries were important considerations to providing complete confinement for certain classes of livestock.

The significance of vitamin B_{12} for ruminants was discovered to be the requirement of cobalt by rumen microorganisms in order to synthesize the vitamin (McDowell, 1985a). Cobalt, the central ion in vitamin B_{12}, was shown to be a dietary essential for sheep in 1935 by Underwood and others in Australia (Underwood, 1977). The Australians showed the deficiency to cause debilitating diseases of sheep known as "coast disease" and "wasting disease." In Florida in 1937, cobalt deficiency was reported by Becker and co-workers (1965) to be responsible in part for ''salt sick'' cattle, a severe wasting disease. In 1951, Smith and co-workers at Cornell discovered that injections of vitamin B_{12} prevented all signs of a cobalt deficiency.

III. CHEMICAL STRUCTURE, PROPERTIES, AND METHODS OF ANALYSIS

Vitamin B_{12} is now considered by nutritionists as the generic name for a group of compounds having B_{12} activity. These compounds have very complex structures (Ellenbogen, 1984). The empirical formula of B_{12} is $C_{63}H_{88}O_{14}N_{14}PCo$, and among its unusual features is the content of 4.5% cobalt. The structure of one B_{12} compound, cyanocobalamin, is shown in Fig. 13.1. Vitamin B_{12} resembles a porphyrin structure consisting of four pyrrole nuclei coupled directly to each other, with the inner nitrogen atom of each pyrrole coordinated with a single atom of cobalt. The basic tetrapyrrole structure is the corrin nucleus, which positionally is a planar structure coupled below to the nucleotide 5,6-dimethylbenzimidazole and above to cyanide or some other derivative. The large ring formed by the four reduced rings is called ''corrin'' because it is the core of the vitamin. The vitamin belongs to the corrinoid group of compounds that have a corrin nucleus, however, numerous other corrinoids do not possess vitamin B_{12} activity. The name cobalamins is used for compounds in which the cobalt atom is in the center of the corrin nucleus.

In vitamin B_{12} the base is coupled directly to the cobalt atom, and an ester linkage from the phosphate group of the nucleotide to the propionic acid group of the D ring of the corrin nucleus adds further stability to the molecule. Cyanide, which lies above the planar ring, is attached to the cobalt atom, and thus the name cyanocobalamin. The cyanide can be replaced by other groups including OH (hydroxycobalamin), H_2O (aquacobalamin), NO_2 (nitrocobalamin), and CH_3 (methylcobalamin). All these compounds are referred to as cobalamins and have activity. In addition, several other compounds, referred to as ''pseudo'' vitamin B_{12} complexes or vitamin B_{12}-like factors that have some activity, have been isolated or synthesized. Their structure differs regarding the nucleotide moiety. These pseudovitamins are probably intermediates of the biosynthesis of vitamin B_{12} and are found in sewage, manure, rumen contents, and residues from fermentation. Some of these pseudovitamin B_{12} compounds are analogs without a nucleotide, while others contain a nucleotide other than 5,6-dimethylbenzimidazole.

The isolation of coenzyme forms of vitamin B_{12} led to the recognition that cyanocobalamin is not the naturally occurring form of the vitamin but is rather an artifact that arises from the original isolation procedure. Adenosylcobalamin, hydroxocobalamin, methylcobalamin, cyanocobalamin, and sulfitocobalamin have been determined in feedstuffs with the first three being the most common. Deox-

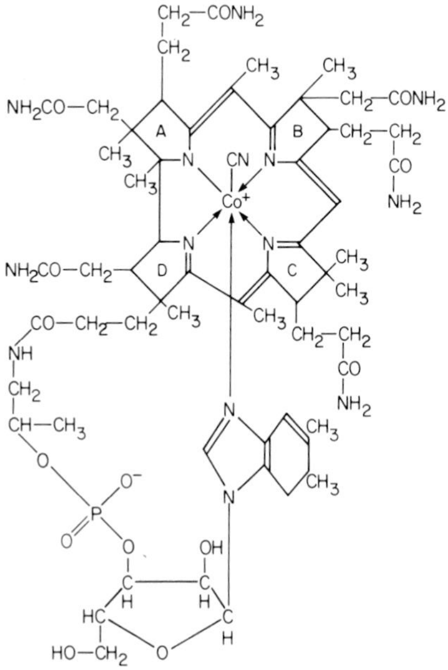

Fig. 13.1 Structure of vitamin B_{12} (cyanocobalamin).

yadenosylcobalamin, hydroxocobalamin, and methylcobalamin are predominant forms in human tissue (Farquharson and Adams, 1976). Cyanocobalamin, however, is the most widely used form of cobalamin in clinical practice because of its relative availability and stability. Most metabolic studies utilize cyanocobalamin.

Vitamin B_{12} is a dark-red, crystalline, hygroscopic substance, freely soluble in water and alcohol but insoluble in acetone, chloroform, or ether. Cyanocobalamin has a molecular weight of 1354 and is the most complex structure and heaviest compound of all the vitamins. Oxidizing and reducing agents and exposure to sunlight tend to destroy its activity. Losses of vitamin B_{12} during cooking are usually not excessive because it is stable at temperatures lower than 250°C.

IV. ANALYTICAL PROCEDURES

Several techniques are employed for analysis of vitamin B_{12}. Chemical assays, including spectrophotometric and colorimetric procedures, have been developed for pharmaceutical preparations but are not sensitive enough for determination of the vitamin in natural materials (Scott *et al.*, 1982). Colorimetric procedures rely on measurement of cyanide released or a color complex with 5,6-dimethylbenzimidazole.

Microbiological assays for vitamin B_{12} are sensitive and can be applied to crude materials. Microbiological methods using *Lactobacillus leichmannii* can determine quantities less than 0.01 μg of the vitamin per milliliter of assay solution. The organism responds, however, to deoxyribonucleosides and to several B_{12} pseudovitamins. Treatment of the sample with alkali destroys vitamin B_{12}, leaving the deoxyribonucleosides intact; thus the vitamin B_{12} plus pseudo forms can be determined by difference. The protozoans *Euglena gracilis* and *Ochromonas malhamensis* are also successfully used for vitamin B_{12} determination.

Vitamin B_{12} assays involving higher animals are somewhat more difficult and time-consuming than microbiological assays. Large stores of vitamin B_{12} found in young, growing animals reared from normal mothers present the biggest problem (Ellenbogen, 1984). A biological assay for vitamin B_{12} uses growth of chicks hatched from eggs of vitamin B_{12}-deficient hens, and assays have been conducted with young rats born from depleted dams. Thyroid-stimulating material often is added to assay diets to increase the B_{12} requirement in the young animal.

The isotope dilution methods for the assay of cobalamins have been replacing the microbiological methods in recent years, and at present they are used more widely than most other methods. These assays measure the extent to which cobalamin, after first being liberated from bound materials, competes with radioactively labeled cyanocobalamin for binding sites on a protein (Ellenbogen, 1984).

V. METABOLISM

A. Digestion, Absorption, and Transport

Passage of vitamin B_{12} through the intestinal wall requires intervention of certain carrier compounds able to bind the vitamin molecule. Vitamin B_{12} in the diet is bound to food proteins. In the stomach, the combined effect of low pH and peptic digestion releases the vitamin, which is then bound to a nonintrinsic factor–cobalamin complex (Toskes *et al.*, 1973). Vitamin B_{12} is bound preferentially to nonintrinsic protein factor in the acid medium of the stomach rather than to intrinsic factor. The B_{12} remains bound to nonintrinsic protein in the slightly alkaline environment of intestine until pancreatic proteases (i.e., trypsin) partially degrade the nonintrinsic factor proteins and thereby enable B_{12} to become bound exclusively to intrinsic factor. Therefore, patients with pancreatic insufficiency absorb B_{12} poorly, and this malabsorption is completely corrected by administration of pancreatic enzymes or purified trypsin.

A prerequisite for intestinal absorption of physiological amounts of cobalamin is binding to intrinsic factor. Intrinsic factor is a glycoprotein (mucoprotein) synthesized and secreted by parietal cells of the gastric mucosa. Atrophy of the fundus, where intrinsic factor is produced, and lack of free HCl (achlorhydria) are usually associated with pernicious anemia. The formation of this intrinsic factor complex serves to protect the vitamin from bacterial utilization and/or degradation as it traverses the lumen of the small intestine to the terminal ileum, where absorption occurs (Ellenbogen and Highley, 1970). The intrinsic factor–B_{12} complex is transiently attached to an ileal receptor. The proximal small intestine does not have the ability to enhance absorption of the vitamin, only the ileum has this property. In the ileum the intrinsic factor moiety of the intrinsic factor–B_{12} complex binds to a specific receptor protein on the microvillus membrane of brush borders of intestinal epithelial cells. Next there is transport of vitamin B_{12} from the receptor–intrinsic factor–B_{12} complex through the epithelial cell to portal blood.

When B_{12} enters the portal blood it is no longer bound to intrinsic factor but to specific transport proteins called transcobalamins. Three binding proteins have been identified in normal human serum and are designated as transcobalamin I, II, and III. The transcobalamins are probably synthesized, at least in part, by liver and have been shown to deliver B_{12} to various tissues such as liver, kidney, spleen, heart, lung, and small intestine (Rothenberg and Cotter, 1978). Transcobalamin II appears to be primarily concerned with transport of vitamin B_{12}, whereas transcobalamin I is involved in storage of the vitamin. Less information is available on transcobalamin III, but this binding protein has more similar characteristics to transcobalamin I than II. Transcobalamin III may also scavenge undesired analogs of B_{12} from serum for excretion via bile (Kanazawa *et al.*, 1983).

To summarize the B_{12} absorption for most species studied, the following are required: (1) adequate quantities of dietary B_{12}, (2) normal stomach for breakdown of food proteins for release of B_{12}, (3) normal stomach for production of intrinsic factor for absorption of B_{12} through the ileum, (4) normal pancrease (trypsin) required for release of bound B_{12} prior to combining the vitamin with the intrinsic factor, and (5) normal ileum with receptor and absorption sites. Additional factors that diminish vitamin B_{12} absorption include deficiencies of protein, iron, and vitamin B_6, thyroid removal, and dietary tannic acid (Anonymous, 1984a).

Intrinsic factor concentrates prepared from one animal's stomach do not in all cases increase B_{12} absorption in other species or in humans. There are structural differences in the B_{12} intrinsic factor among species. Likewise, species differences exist for B_{12} transport proteins (Polak *et al.*, 1979). Intrinsic factor has been demonstrated in humans, monkey, pig, rat, cow, ferret, rabbit, hamster, fox, lion, tiger, and leopard. It has not, to the present, been detected in dog, guinea pig, horse, sheep, chicken, and a number of other species.

There is evidence that absorption of vitamin B_{12} does not completely depend on active intervention of the intrinsic factor. Both active and passive mechanisms exist for absorption of B_{12} (Herbert, 1968). The passive mechanism, probably diffusion, is operative throughout the digestive tract and becomes practically important only in the presence of large quantities of the vitamin, in excess of those present in most foods.

About 3% of ingested cobalt is converted to vitamin B_{12} in the rumen. Of the vitamin B_{12} produced, only 1–3% is absorbed. As in most species, the absorptive site for ruminants is the lower portion of the small intestine. Substantial amounts of B_{12} are secreted into the duodenum and then reabsorbed in the ileum.

B. Tissue Distribution and Storage

In normal human subjects, vitamin B_{12} is found principally in the liver; the average amount is 1.5 mg. Kidneys, heart, spleen, and brain each contain about 20–30 μg (Ellenbogen, 1984). Vitamin B_{12} is stored in the liver in the largest quantities for most animals that have been studied, but it is stored in the kidney of the bat.

Henderickx *et al.* (1964) reported a total retention of 20–23% of an oral dose of vitamin B_{12} in pigs, about two-thirds of which was present in liver. Even though vitamin B_{12} is a water-soluble vitamin, Kominato (1971) reported a tissue half-life of 32 days, indicating a considerable degree of tissue storage.

To become metabolically active, vitamin B_{12} must be converted into one of its various coenzyme forms. This transformation takes place mainly in liver but also in kidneys. Most of the cobalamins in humans occur as two coenzymatically active forms, adenosylcobalamin and methylcobalamin. Methylcobalamin constitutes 60–80% of total plasma cobalamin, while adenosylcobalamin is the major cobalamin in all cellular tissues, constituting about 60–70% in the liver and about 50% in other organs (Ellenbogen, 1984). Cyanocobalamin is converted within cells to either methylcobalamin, a coenzyme for methyltransferase, or adenosylcobalamin, the coenzyme for mutase.

C. Excretion

The main excretion of absorbed vitamin B_{12} is via urinary, biliary, and fecal routes. Total body loss ranges from 2 to 5 μg daily in humans (Shinton, 1972). Urinary excretion of the intact vitamin B_{12} by kidney glomerular filtration is minimal. Biliary excretion via feces is the major excretory route. Approximately 0.5–5 μg of cobalamin is secreted into the alimentary tract daily, mainly in bile (Ellenbogen, 1984). The majority of cobalamin excreted in bile is reabsorbed; at least 65–75% is reabsorbed in ileum by means of the intrinsic factor mechanism.

VI. FUNCTIONS

Vitamin B_{12} is an essential part of several enzyme systems that carry out a number of very basic metabolic functions. Specific biochemical reactions in which cobalamin coenzymes participate are of two types: (1) those that contain 5′-deoxyadenosine linked covalently to the cobalt atom (adenosylcobalamin) and

(2) those that have a methyl group attached to the central cobalt atom (methylcobalamin). Most reactions involve transfer or synthesis of one-carbon units, for example, methyl groups.

Vitamin B_{12} is metabolically related to other essential nutrients such as choline, methionine, and folacin. Interrelationships of these nutrients with vitamin B_{12} and in particular to transmethylation and biosynthesis of labile methyl groups are discussed in Chapters 12 and 14. Though the most important tasks of vitamin B_{12} concern metabolism of nucleic acids and proteins, it also functions in metabolism of fats and carbohydrates. A summary of B_{12} functions would be: (1) purine and pyrimidine synthesis, (2) transfer of methyl groups, (3) formation of proteins from amino acids, and (4) carbohydrate and fat metabolism. A general function of B_{12} is to promote red blood cell synthesis and to maintain nervous system integrity, which are functions noticeably affected in the deficient state.

Vitamin B_{12} is necessary in reduction of one-carbon compounds of formate and formaldehyde, and in this way it participates with folacin in biosynthesis of labile methyl groups. Formation of labile methyl groups is necessary for biosynthesis of purine and pyrimidine bases, which represent essential constituents of nucleic acids. Disorders of nucleic acid synthesis in vitamin B_{12} deficiency are connected with this. The purine bases (adenine and guanine) as well as thymine are constituents of nucleic acids and with a folacin deficiency there is a reduction in biosynthesis of nucleic acids essential for cell formation and function. Hence, deficiency of either folacin or B_{12} leads to impaired cell division and alterations of protein synthesis; these effects are most noticeable in rapidly growing tissues.

Deficiency of B_{12} will induce a folacin deficiency by blocking utilization of folacin derivatives. A vitamin B_{12}-containing enzyme removes the methyl group from methylfolate, thereby regenerating tetrahydrofolate (THF), from which is made the 5,20-methylene-THF required for thymidylate synthesis. Because methylfolate returns to the body's folacin pool only via the vitamin B_{12}-dependent step, vitamin B_{12} deficiency results in folacin being "trapped" as methylfolate, and thus becoming metabolically useless. The "folate trap" concept explains why hematological damage of vitamin B_{12} deficiency is indistinguishable from that of folacin deficiency by alleging that in both instances the defective synthesis of DNA results from the same final common pathway defect, namely, an inadequate quantity of 5,10-methylene-THF to participate adequately in DNA synthesis (Herbert and Zalusky, 1962).

Metabolism of labile methyl groups plays a significant part in biosynthesis of methionine from homocysteine. A vitamin B_{12}-requiring enzyme, 5-methyltetrahydrofolate-homocysteine methyltransferase, catalyzes reformation of methionine from homocysteine according to the reaction

$$\text{5-Methyltetrahydrofolate} + \text{homocysteine} \rightleftarrows \text{methionine} + \text{tetrahydrofolate}$$

The mechanism of converting homocysteine to methionine has utilized a methyl group from folacin, the mechanism for maintaining folacin in a reduced form.

Activity of this enzyme is depressed in liver of vitamin B_{12}-deficient sheep (MacPherson, 1982), which could lead to a deficiency of available methionine that may account for impairment of nitrogen metabolism in vitamin B_{12}-deficient sheep.

Overall synthesis of protein is impaired in vitamin B_{12}-deficient animals. Wagle *et al.* (1958) demonstrated that rats and baby pigs deprived of vitamin B_{12} were less able to incorporate serine, methionine, phenylalanine, and glucose into liver proteins. There is good reason to believe that impairment of protein synthesis is the principal reason for the growth depression that is frequently observed in animals deficient in vitamin B_{12} (Friesecke, 1980).

In the metabolism of animals, propionate of dietary or metabolic origin is converted into succinate, which then enters the tricarboxylic acid (Krebs) cycle. Because propionate is a three-carbon and succinate a four-carbon compound, this process requires the introduction of a one-carbon unit. Methylmalonyl-CoA isomerase (mutase) is a vitamin B_{12}-requiring enzyme (5′-deoxyadenosylcobalamin) that catalyzes the conversion of methylmalonyl-CoA to succinyl-CoA. Flavin and Ochoa (1957) established that for succinate production the following steps are involved:

$$\text{Propionate} + \text{ATP} + \text{CoA} \rightleftarrows \text{propionyl-CoA}$$

$$\text{Propionyl-CoA} + CO_2 + \text{ATP} \rightleftarrows \text{methylmalonyl-CoA (a)}$$

$$\text{Methylmalonyl-CoA (a)} \rightleftarrows \text{methylmalonyl-CoA (b)}$$

$$\text{Methylmalonyl-CoA (b)} \rightleftarrows \text{succinyl-CoA}$$

Methylmalonyl-CoA (a) is an inactive isomer. Its active form (b) is converted into succinyl-CoA by a methylmalonyl isomerase, or methylmalonyl mutase (fourth reaction).

Vitamin B_{12} is a metabolic essential for all animal species studied and vitamin B_{12} deficiency can be induced with the addition of high dietary levels of propionic acid. However, metabolism of propionic acid is of special interest in ruminant nutrition because large quantities are produced during carbohydrate fermentation in the rumen. Propionate production proceeds normally but in cobalt or vitamin B_{12} deficiency its rate of clearance from blood is depressed and methylmalonyl-CoA accumulates. This results in an increased urinary excretion of methylmalonic acid and also loss of appetite because impaired propionate metabolism leads to higher blood propionate levels, which are inversely correlated to voluntary feed intake (MacPherson, 1982). Injection of cobalt-deficient animals with vitamin B_{12} produces an overnight improvement in appetite, whereas oral dosing with cobalt takes from 7 to 10 days to produce the same effect.

A further important function of vitamin B_{12} in intermediary metabolism consists of maintaining glutathione and sulfhydryl groups of enzymes in the reduced state (Marks, 1975). The reduced activity of glyceraldehyde-3-phosphate dehydrogenase, which needs glutathione as a coenzyme, is possibly responsible for carbohydrate metabolism being impaired in a vitamin B_{12} deficiency. Vitamin B_{12} also influences lipid metabolism via its effect on the thiols.

VII. REQUIREMENTS

Vitamin B_{12} requirements are exceedingly small; an adequate allowance is only a few micrograms per kilogram of feed, making the most potent of vitamins. Estimated requirements of vitamin B_{12} for various animals and humans are presented in Table 13.1.

The vitamin B_{12} requirements of various species depend on the levels of several other nutrients in the diet. Excess protein increases the need for B_{12} as does performance level. The B_{12} requirement seems to depend on the levels of choline, methionine, and folacin in the diet and B_{12} is interrelated with ascorbic acid metabolism (Scott *et al.*, 1982). Sewell *et al.*, (1952) showed that B_{12} has a sparing effect on the methionine needs of the pig. A reciprocal relationship occurs between B_{12} and pantothenic acid in chick nutrition, with pantothenic acid sparing the B_{12} requirement. Dietary ingredients may also affect the requirement, as wheat bran has been shown to reduce availability of vitamin B_{12} in humans (Lewis *et al.*, 1986).

Dietary need depends on intestinal synthesis and tissue reserves at birth. Intestinal synthesis probably explains frequent failures to produce a B_{12} deficiency in pigs and rats on diets designed to be B_{12} free. The deficiency can be readily produced in rats, however, when coprophagy is completely prevented (Barnes and Fiala, 1958). Coprophagous animals and poultry on deep litter receive excellent supplies of B_{12} from microbial fermentation. Poultry obtain some vitamin B_{12} by direct absorption of the vitamin produced by bacterial synthesis in the intestine (NRC, 1984b), however, the amount from this source is not reliable.

Requirement for vitamin B_{12} in ruminant diets is closely associated with their requirement for cobalt since this trace mineral is a component of the B_{12} molecule. Ruminant animals have the ability to synthesize vitamin B_{12} provided they are supplied with an adequate dietary supply of cobalt (0.07–0.2 ppm) and have a normally functioning rumen. Under typical conditions a rumen would be functional for synthesis of all B vitamins at 6–8 weeks of age. Therefore, only young ruminants that do not have a fully developed rumen would be expected to require a dietary source of B_{12}.

Cobalt content of the diet is the primary limiting factor for synthesis of vitamin B_{12} by ruminal microflora. However, studies indicate that synthesis of vitamin B_{12} could be restricted even when the diet is adequate in cobalt, as several factors can influence synthesis and perhaps utilization. On high-concentrate diets there is a decrease in vitamin B_{12} synthesis and more analogs are produced than the

TABLE 13.1

Vitamin B_{12} Requirements for Various Animals and Humans[a]

Animal	Purpose	Requirement	Reference
Beef cattle	Adult	Microbial synthesis[b]	NRC (1984a)
Dairy cattle	Calf	0.34–0.68 μg/kg body wt	NRC (1978a)
	Adult	Microbial synthesis[b]	NRC (1978a)
Chicken	Leghorn, 0–6 weeks	9 μg/kg	NRC (1984b)
	Leghorn, 6–20 weeks	3 μg/kg	NRC (1984b)
	Leghorn, laying–breeding	4 μg/kg	NRC (1984b)
	Broilers, 0–6 weeks	9 μg/kg	NRC (1984b)
	Broilers, 6–8 weeks	3 μg/kg	NRC (1984b)
Turkey	All classes	3 μg/kg	NRC (1984b)
Japanese quail	All classes	3 μg/kg	NRC (1984b)
Sheep	Adult	Microbial synthesis[b]	NRC (1985b)
Swine	Growing–finishing	5–20 μg/kg	NRC (1988)
	Breeding–lactating	15 μg/kg	NRC (1988)
Horse	Adult	Microbial synthesis[b]	NRC (1978b)
Goat	Adult	Microbial synthesis[b]	NRC (1981b)
Dog	Growing	26 μg/kg	NRC (1985a)
Cat	All classes	20 μg/kg	NRC (1986)
Mink	All classes	30 μg/kg	NRC (1982b)
Rabbit	All classes	Microbial synthesis	NRC (1977)
Salmon	Growing	2–3 μg/kg	NRC (1981a)
Trout	Growing	2–3 μg/kg	NRC (1981a)
Rat	Growing	50 μg/kg	NRC (1978c)
Hamster	Growing	10 μg/kg	NRC (1978c)
Human	Infants	0.5–1.5 μg/day	RDA (1980)
	Children	2–3 μg/day	RDA (1980)
	Adults	3 μg/day	RDA (1980)
	Pregnancy–lactation	4 μg/day	RDA (1980)

[a]Expressed as per unit of animal feed either on an as fed (approximately 90% dry matter) or dry basis (see Appendix Table 1) (except for dairy calf). Human data expressed as μg/day.

[b]Only young ruminants have a dietary need for B_{12} prior to ruminal development. Practical vitamin B_{12} deficiency is a secondary result of cobalt deficiency. Suggested cobalt requirements for ruminants range from 0.07 to 0.2 mg/kg of diet.

vitamin itself (Sutton and Elliot, 1972). These natural analogs have little or no vitamin B_{12} activity.

A ruminant may require more vitamin B_{12} than a monogastric animal of comparable size (Brent, 1985). Vitamin B_{12} is essential as a cofactor for methylmalonyl-CoA isomerase, an enzyme necessary for propionic acid utilization that is produced in much greater quantities in ruminants.

Vitamin B_{12} requirements for humans have been estimated from three different types of studies (Ellenbogen, 1984): (1) the amount necessary to treat megaloblastic anemia from vitamin B_{12} deficiency, (2) comparison of blood and liver concentration in normal and cobalamin-deficient subjects, and (3) body stores and turnover rates of the vitamin. Obviously, the requirement for B_{12} will be substantially higher for humans lacking intrinsic factor or other conditions that affect absorption and metabolism of the vitamin (see Section V).

VIII. NATURAL SOURCES

The origin of vitamin B_{12} in nature appears to be microbial synthesis. It is synthesized by many bacteria but apparently not by yeasts or by most fungi. There is no convincing evidence that the vitamin is produced in tissues of higher plants or animals. Synthesis of this vitamin in the alimentary tract is of considerable importance for animals; if sufficient cobalt is available, ruminants are independent of external sources of vitamin B_{12}.

Foods of animal origin are reasonably good sources—meat, liver, kidney, milk, eggs, and fish (Table 13.2). Kidney and liver are excellent sources, and these organs are richer in vitamin B_{12} from ruminants than from most nonruminants. Vitamin B_{12} presence in tissues of animals is due to the ingestion of

TABLE 13.2

Vitamin B_{12} Concentrations of Various Foods and Feedstuffs (ppm, Dry Basis)[a]

Blood meal	49.0
Crab meal	475
Distiller's solubles	3
Fish solubles	1007
Fish meal, anchovy	233
Fish meal, herring	467
Fish meal, menhaden	133
Fish meal, tuna	324
Horse meat	142
Liver meal	542
Meat meal	72
Milk, skim, cow's	54
Poultry by-product meal	322
Spleen, cow	247
Wheat, grain	1
Whey, cow	20
Yeast	1

[a]NRC (1982a).

vitamin B_{12} in animal foods or from intestinal or ruminal synthesis. Among the richest sources are fermentation residues, activated sewage sludge, and manure.

Plant products are practically devoid of vitamin B_{12}. The vitamin B_{12} reported in higher plants in small amounts may result from synthesis by soil microorganisms and excretion of the vitamin into soil, with subsequent absorption by the plant. Root nodules of certain legumes contain small quantities of B_{12}. Certain species of seaweed (algae) have been reported to contain appreciable quantities of vitamin B_{12} (up to 1 μg/g of solids). Seaweed does not synthesize vitamin B_{12}, but it is synthesized by the bacteria associated with seaweed and then concentrated by the seaweed (Scott *et al.*, 1982).

IX. DEFICIENCY

The result of vitamin B_{12} deficiency in humans is a megaloblastic anemia (pernicious anemia) and neurological lesions. A deficiency of vitamin B_{12} in humans usually is conditioned by a deficiency of intrinsic factor necessary for its absorption or is found in humans consuming strict vegetarian diets. For animals, pernicious anemia, or in fact any anemia, is not characteristic of a vitamin B_{12} shortage. In rats, guinea pigs, swine, and poultry, vitamin B_{12} functions as a growth factor, although a mild anemia does occur in a small percentage of deficient swine. In ruminants, vitamin B_{12} deficiency is closely associated with their requirement for cobalt, since the trace mineral is a component of the B_{12} molecule. For all species, as a vitamin B_{12} deficiency progresses, there is a concurrent depletion of vitamin B_{12} (and cobalt) in serum and tissue reserves.

A. Effects of Deficiency

1. Ruminants

A vitamin B_{12} deficiency can occur in young ruminants as long as the microflora of the forestomachs is not yet far enough developed and hence unable to furnish sufficient amounts of the vitamin. Lassiter *et al.* (1953) demonstrated vitamin B_{12} deficiency in calves less than 6 weeks old that received no dietary animal protein. Clinical signs characterizing the deficiency included poor appetite and growth, muscular weakness, demyelination of peripheral nerves, and poor general condition. Young lambs (up to 2 months of age), if weaned early, likewise have a need for dietary vitamin B_{12} (NRC, 1985b). In vitamin B_{12}-deficient lambs, there is a sharp decrease of vitamin B_{12} concentrations in blood and liver before signs like anorexia, loss of body weight, and a decrease in hemoglobin concentration are observed.

As cobalt is required for biosynthesis of vitamin B_{12}, a lack of cobalt may cause a deficiency of the vitamin in adult ruminants. Cobalt-deficient soils occur in large areas of many countries and therefore grazing ruminants may be particularly affected by the deficiency. With the exception of phosphorus and copper, cobalt deficiency is the most extensive mineral limitation to grazing livestock in tropical countries (McDowell *et al.*, 1984).

Deficiency signs for cobalt are not specific, and it is often difficult to distinguish between an animal having a cobalt deficiency and malnutrition due to low intake of energy and protein and an animal that is diseased or parasitized. Acute clinical signs of cobalt deficiency include lack of appetite, rough hair coat, thickening of the skin, anemia (normocytic and normochromic), wasting away (Fig. 13.2), and eventually death, if the animals are not moved to "healthy" pastures or if cobalt supplements are not made available. These clinical signs are identical to those of simple starvation and may indicate that the effect of a lack of cobalt may be simply the effect on appetite, rather than a direct effect of the mineral on the body itself. In a severe deficiency, mucous membranes become blanched, skin turns pale, a fatty liver develops, and the body becomes almost devoid of fat.

Prior to recognition of cobalt deficiency in livestock in many parts of the world, cattle could be maintained on deficient pastures only if they periodically moved to "healthy" ground. Cobalt deficiency can be prevented by moving animals for a few months every year to a "healthy" region, preferably during the rainy season. An example of the necessity of periodically moving animals was illustrated in a disease condition known as "toque" in Espirito Santo, Brazil (Tokarnia *et al.*, 1971). The disease was observed when animals stayed for a period longer than 60 to 180 days on certain pastures. Sick animals isolated themselves from the rest of the herd, were apathetic, showed loss of appetite, rough hair coat, and dry feces, and lost body condition. If the animals were not moved from the pasture, they died, but if they were taken to a pasture where the disease did not occur, the animals recovered quickly.

Cobalt subclinical deficiencies or borderline states are extremely common and are characterized by low production rates unaccompanied by clinical manifestations or visible signs (McDowell, 1985a). Subclinical deficiencies often go unnoticed, thereby resulting in great economic losses to the livestock industry. No estimate can be made of the effect of cobalt subdeficiencies on animal performance in general, but, in many areas of the world, it is one of the major causes of poor production.

2. SWINE

The general signs of a vitamin B_{12} deficiency in pigs are comparable to those observed in other species, principally a loss of appetite, variable feed intake, and a dramatic growth decline (Fig. 13.3). In addition, sometimes there is rough skin and hair coat, vomiting and diarrhea, voice failure, and a slight anemia

Fig. 13.2 Cobalt deficiency in Florida, U.S.A. (A) shows a cobalt-deficient heifer that had access to a Fe–Cu salt supplement. Note the severe emaciation, resulting from failure to synthesize B_{12}. Her blood contained 6.6 g of hemoglobin per 100 ml on February 25, 1937. (B) is the same heifer fully recovered with a Fe–Cu–Co salt supplement while on the same pasture. (Courtesy of R.B. Becker, University of Florida.)

(Catron *et al.*, 1952). A microcytic to normocytic anemia is typical, however, observations on anemia are not unanimous and are sometimes contradictory. Nervous disorders occur in the pig, including increased excitability, unsteady gait, and posterior incoordination. The thymus and spleen become atrophied, while liver and tongue are frequently enlarged as a result of proliferation of granulomatous tissue.

In the reproducing animal, litter size and pig survival are reduced. Abortions, small litters and birth weights, some deformities, and inability to rear young

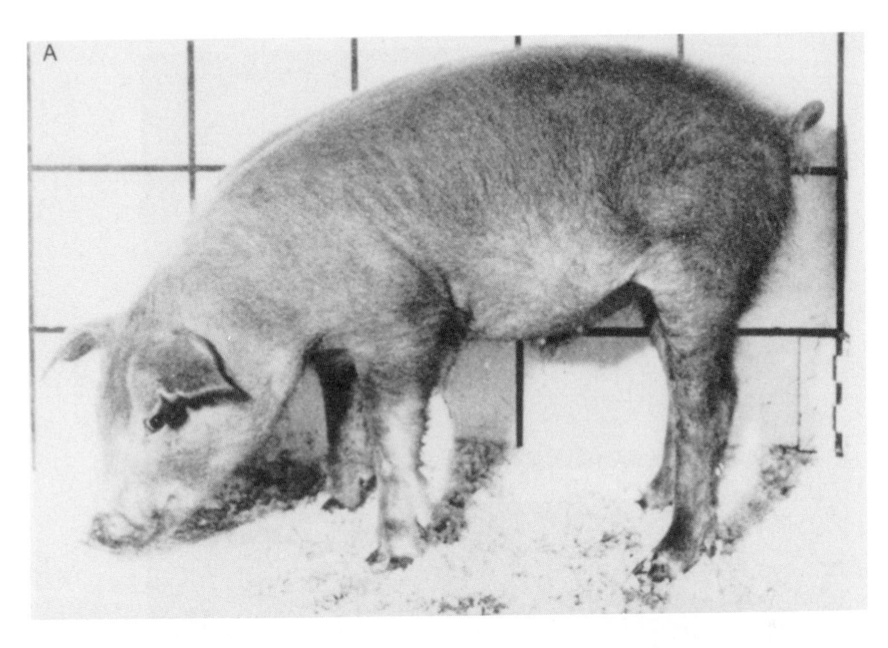

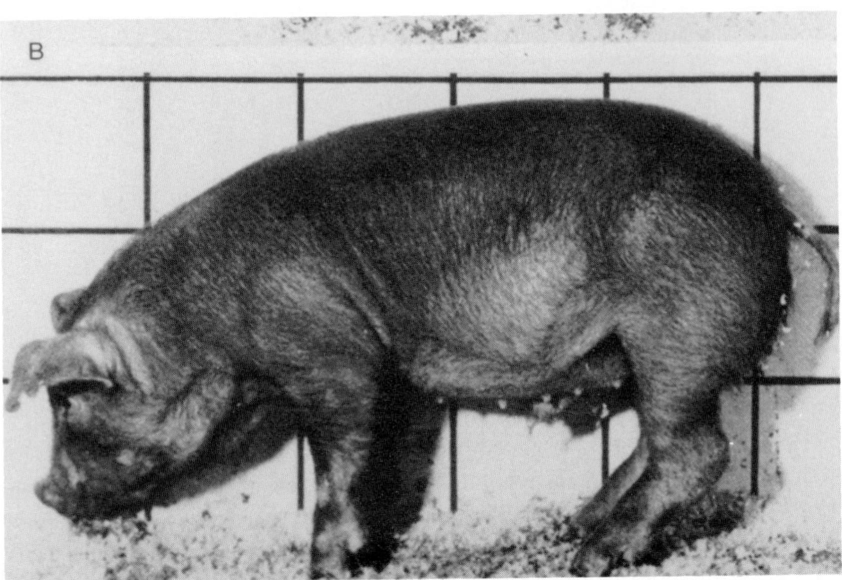

Fig. 13.3 Vitamin B_{12} deficiency. (A) pig deficient in vitamin B_{12}. Note rough hair coat and dermatitis. (B) control pig. (Courtesy of the late D. V. Catron and Iowa State University.)

occur in breeding sows. Later estrus, fewer corpora lutea, and fewer embryos are produced in B_{12}-deficient animals. During reproduction and lactation, vitamin B_{12} supplementation has been shown to increase birth weights and survival of young pigs (Vestal *et al.*, 1950). Under some conditions reproductive performance of sows has been improved by inclusion of higher than recommended levels of dietary vitamin B_{12} (Cunha, 1977). The response is evidenced by an increase in the number and birth weight of pigs.

3. Poultry

In growing chicks, turkey poults, and quail, vitamin B_{12} deficiency reduces body weight gain, feed intake, and feed conversion. Vitamin B_{12} deficiency in growing chicks and turkeys may result in a nervous disorder and defective feathering. It has also been related to leg weakness and perosis, however, this appears to be a secondary effect. Perosis may occur in vitamin B_{12}-deficient chicks or poults when the diet lacks choline, methionine, or betaine as sources of methyl groups. Addition of B_{12} may prevent perosis under these conditions because of its effect on synthesis of methyl groups. Additional clinical signs in B_{12} deficiency include anemia, gizzard erosion, and fattiness of heart, liver, and kidneys. In hens, body weight and egg production are maintained despite a deficiency, but B_{12} has an important influence on egg size (Scott *et al.*, 1982).

Hatchability of incubated eggs may be severely reduced if the breeder's diet contained inadequate vitamin B_{12}. Two to five months may be needed to deplete hens of vitamin B_{12} stores to such an extent that progeny will hatch with low vitamin B_{12} reserves. The rate of depletion is most rapid when hens are fed high-protein diets (Scott *et al.*, 1982).

Vitamin B_{12}-deficient embryos die about the seventeenth day and show leg myoatrophy, malposition of the head, multiple hemorrhages, enlarged hearts and thyroids, and fatty livers. The most obvious change in B_{12}-deficient embryos is myoatrophy of the leg, a condition characterized by atrophy of thigh muscles (Olcese *et al.*, 1950). Chicks that do hatch without adequate carryover of vitamin B_{12} from the dam have a high rate of mortality.

4. Horses

It has been concluded that for mature horses, supplemental vitamin B_{12} is not necessary, as the vitamin can be synthesized in the large intestine, where it is absorbed (NRC, 1978b). Reports that supplemental vitamin B_{12} increased vitamin B_{12} concentration in blood of weaned foals suggest that foals might benefit from receiving supplementation of vitamin B_{12} or cobalt.

5. Other Animal Species

a. Cats. Keesling and Morris (1975) found that kitten growth was slowed and excretion of methylmalonic acid was elevated with vitamin B_{12} deficiency.

b. Dogs. Uncomplicated vitamin B_{12} deficiency has not been described in the dog (NRC, 1985a). However, earlier reports noted reduced growth (Arnrich *et al.*, 1952) and impaired reproduction (Campbell and Phillips, 1952) with B_{12} deficiency.

c. Fish. Salmon and trout fed low dietary B_{12} showed a high variability in numbers of fragmented erythrocytes and in hemoglobin values, with a tendency for microcytic, hypochromic anemia (NRC, 1981a). Channel catfish fed a vitamin B_{12}-deficient diet exhibited reduced growth rates and lower hematocrit values (NRC, 1983). Intestinal microbial synthesis of vitamin B_{12} has been shown to be sufficient to meet the requirements for carp and tilapia.

d. Laboratory Animals. In practically all laboratory animals, growth is retarded in vitamin B_{12} deficiency (NRC, 1978c), and there are also changes in relative weights of certain organs. In B_{12}-deficient rats, there are kidney lesions and mucoid structures in the urinary bladders, with a reduced muscle mass and a fibrotic degeneration of heart. In rats, a deficiency in the diet of the mother can result in hydrocephalus, eye defects, and bone defects in the newborn, and mothers may eat the offspring. In mice deficient in B_{12}, there is death of young, retarded growth, and renal atrophy. Under practical conditions, vitamin B_{12} deficiency would not be expected for most laboratory animals because of intestinal synthesis and the significant amounts of the vitamin obtained by coprophagy.

e. Mink. Mink kits deficient in vitamin B_{12} exhibit anorexia, loss of body weight, and severe fatty degeneration of liver (NRC, 1982b). Such deficiency is unlikely in these animals, which are commonly fed diets rich in animal protein.

f. Monkeys. Megaloblastic anemia reminiscent of pernicious anemia in humans has not been reported in vitamin B_{12}-deficient monkeys. However, abnormalities of the central and peripheral nervous systems in several species of monkeys have been attributed to B_{12} deficiency as a result of vegetarian diets (NRC, 1978d).

g. Rabbits. High urinary and fecal excretion rates of B_{12} have been reported in rabbits receiving diets practically devoid of the vitamin (NRC, 1977). As a result of coprophagy, the rabbit should not be deficient in vitamin B_{12}, assuming adequate cobalt is available.

6. Humans

In humans, pernicious anemia, a fatal megaloblastic anemia with neurological involvement, is the result of vitamin B_{12} deficiency. This megaloblastic anemia is not found in animals. This finding by itself does not elucidate whether the condition is due to a B_{12} or folate deficiency. Either vitamin B_{12} or folacin supplementation will cure the megaloblastic anemia, however, folacin is ineffective in preventing degenerative changes in the nervous system. In macrocytic anemia, erythrocytes are larger than normal and show great variation in size and a normal hemoglobin saturation. The bone marrow shows a megaloblastic pattern of red cell maturation as opposed to the usual normoblast pattern.

In addition to megaloblastic anemia, the most prominent signs and symptoms of vitamin B_{12} deficiency are weakness, tiredness, pale and smooth tongue with inflammation (Fig. 13.4), dyspnea, splenomegaly, leukopenia, thrombocytopenia, achlorhydria, paresthesia, neurological changes, loss of appetite, loss of weight, and low serum cobalamin levels. The condition results in stiffness of limbs, progressive paralysis, mental disorders, diarrhea, and finally death. Neurological B_{12} deficiency results in axon degeneration of nerves in the spinal cord. The deficiency produces patchy, diffuse, and progressive demyelination. The clinical picture of the diffuse, uneven demyelination is one of an insidiously progressive neuropathy, often beginning in the peripheral nerves and progressing centrally to involve the posterior and lateral columns of the spinal cord (Herbert, 1984). In addition to subacute combined degeneration of the cord, hypovitaminosis B_{12} may give rise to a severe psychosis with extensive mental deterioration.

The function of B_{12} and folacin in DNA synthesis accounts for some of the pathological findings when a deficiency of either of these vitamins occurs. The most rapidly proliferating tissues of the body, such as bone marrow for blood cell production, have the greatest requirement for DNA synthesis and thus are principally affected by megaloblastic anemia when severe deficiency of either B_{12} or folacin occurs. Disturbed division and nuclear maturation of proliferating epithelial cells can be observed in buccal mucosal scrapings and intestinal biopsies of B_{12}-deficient patients. The proliferating epithelial cells, as is true for bone marrow, have a high B_{12} and folacin requirement for DNA synthesis.

Vitamin B_{12} deficiency in humans is influenced by one or more of the following considerations:

1. Dietary intake—Inadequate vitamin B_{12} intake can occasionally be seen in geriatric patients and in vegetarians. A completely vegetarian diet, one devoid of meat, eggs, and dairy products, can produce B_{12} deficiency if consumed for several years. In some areas of India, religion and poverty result in a basic diet of polished rice, cereals, some vegetables, and fruit, with an inadequate intake of milk and eggs (Rothenberg and Cotter, 1978). There have also been reports of cobalamin deficiency in infants breast-fed by strictly vegetarian mothers (Higginbottom *et al.*, 1978).

2. Failure of absorption or transport—Pernicious anemia is most commonly acquired because of a failure to secrete intrinsic factor. This may be due to deranged stomach activity or following total gastrectomy. Total gastrectomy in

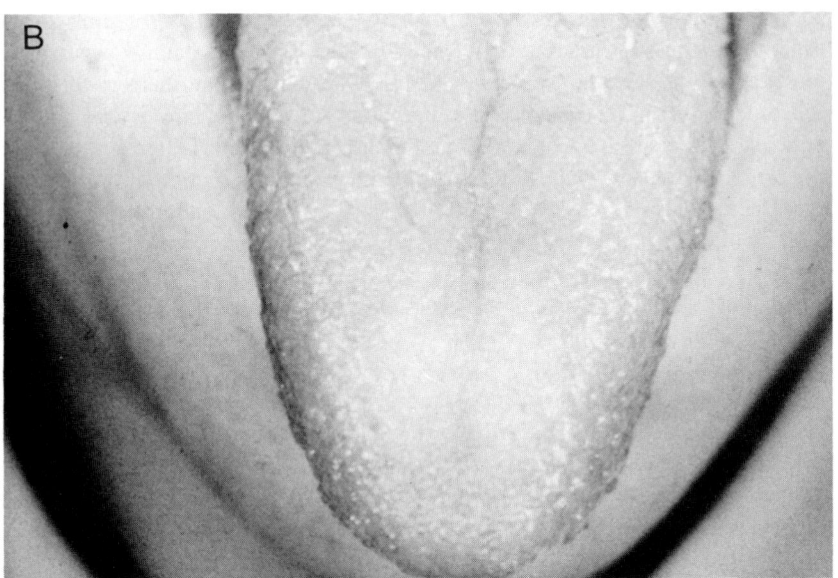

Fig. 13.4 Pernicious anemia. In addition to megaloblastic anemia, an additional sign with vitamin B_{12} deficiency is a pale, smooth tongue with inflammation (A). The normal tongue in (B) has papilla. The smooth tongue is found in one-third to one-half of pernicious anemia patients. (Courtesy of R. R. Streiff, Veterans Administration, University of Florida.)

humans always produces cobalamin deficiency, since it completely removes the site and source of intrinsic factor secretion.

Patients with certain small intestine defects have failure of B_{12} absorption. Impaired absorption of cobalamin is a regular manifestation of tropical sprue. Patients with lesions such as blind loops or small bowel diverticula have demonstrated that an inappropriate bacterial overgrowth in stagnant areas may introduce into the intestinal stream sufficient organisms to absorb all or adsorb much of the dietary cobalamin. Fish tapeworm (*Diphyllobothrium latum*) infestation is a well-recognized cause of impaired B_{12} absorption, as the worm sequesters the vitamin as it progresses through the small intestine. Other conditions that result in reduced B_{12} absorption or transport include excessive intakes of alcohol or certain drugs, chronic pancreatitis, diseased B_{12} ileal receptors, and abnormalities related to B_{12} transport proteins.

3. Storage—Vitamin B_{12} stores in the human body exceed the daily requirement by about 1000-fold (Ellenbogen, 1984), which helps explain why clinical B_{12} deficiency due to dietary insufficiency is not more common. It may take as long as 5–7 years after cessation of intrinsic factor secretion before any outward signs of pernicious anemia are evident. Moreover, the small intestine contains microflora that can synthesize significant amounts of cobalamin.

4. Heredity—Pernicious anemia, inherited as an autosomal dominant trait, chiefly affects persons past middle age and results in lack of intrinsic factor production. The incidence of pernicious anemia in the general population is 1–2 per 1000 and about 25 per 1000 among relatives of pernicious anemia patients (Ellenbogen, 1984).

B. Assessment of Status

A positive diagnosis of vitamin B_{12} deficiency is usually made by the finding of subnormal serum and tissue B_{12} concentrations. Low serum levels are associated with low body content of the vitamin. Microbiological assay methods or radioisotopic dilution techniques may be used for this purpose. The normal range for vitamin B_{12} in human serum (or plasma) is 200–900 pg/ml. Values between 150 and 200 pg/ml strongly suggest B_{12} deficiency (Rothenberg and Cotter, 1978). Analysis of serum levels of both B_{12} and folate can differentiate which nutrient is deficient for a given megaloblastic anemia (see Chapter 12).

Different methods involving various radioactive isotopes of cobalt are available for determining the absorption of an oral dose of radioactive B_{12} in humans. The Schilling urinary excretion test is one of the simplest and most generally used. The basis of this test is giving large does of unlabeled B_{12} to saturate blood binding protein so any absorbed radioactive B_{12} will be excreted by the kidney. Other procedures include direct determination of fecal radioactivity after a test dose of B_{12}-labeled cobalt (i.e., ^{58}Co or ^{56}Co) or measurement of hepatic uptake of B_{12}.

Biochemical detection of B_{12} deficiency for both humans and animals includes increased urinary excretion of formiminoglutamic acid (FIGLU) following loading with histidine and greater percentage of serum folate as 5-methyl-THF rather than THF. Since vitamin B_{12} participates in demethylation of methylfolate, patients with pernicious anemia show an accumulation of the methylfolate and an increased excretion of FIGLU. For deamination of histidine, THF is required rather than methyl-THF.

Methylmalonic acid is often excreted in the urine in increased quantities in vitamin B_{12} deficiency. A vitamin B_{12} enzyme is required for conversion of propionate to succinate and with the vitamin deficiency, methylmalonic acid is excreted. Methylmalonic acid concentrations would be higher in all species with a B_{12} deficiency, but particularly high in ruminants because of the large quantities of propionic acid metabolized.

For ruminants, the best indicators of a cobalt deficiency are low levels of cobalt and B_{12} in tissues, loss of appetite, elevated blood pyruvate, and elevated urinary methylmalonic acid. The levels of cobalt in the livers of sheep and cattle are sufficiently responsive to changes in cobalt intake to have value in the detection of cobalt deficiency, with liver vitamin B_{12} an even more reliable criterion. Values of 0.10 μg vitamin B_{12}/g wet weight or less are "clearly diagnostic of cobalt deficiency disease" (Underwood, 1979). Liver cobalt concentrations in the range of 0.05 to 0.07 ppm (dry basis) or below are critical levels indicating deficiency (McDowell, 1985b). While herbage and tissue analyses are helpful in diagnosing the deficiency, the definite proof is the prompt improvement in feed intake following the feeding of cobalt or injection of vitamin B_{12}.

X. SUPPLEMENTATION

Vitamin B_{12} is produced by fermentation and is available commercially as cyanocobalamin for addition to feed. Vitamin B_{12} is only slightly sensitive to heat, oxygen, moisture, and light (Gadient, 1986). Verbeeck (1975) reported vitamin B_{12} to have good stability in premixes with or without minerals regardless of source of the minerals. Scott (1966) indicated that there is apparently little effect of pelleting on vitamin B_{12} content of feed.

Results of a large number of animal experiments are about equally divided between those reporting a positive response to dietary cyanocobalamin and those reporting little or no response. Variable responses may be due to several factors: initial body stores, environmental sources of the vitamin (such as molds, soil, and animal excreta), microbial synthesis in the intestinal tract, and adequacy or deficiency of other nutrients that influence B_{12} requirement.

Vitamin B_{12} is normally added to diets of all classes of swine and poultry. Swine and poultry raised in confinement, in management systems where there is less access to feces for coprophagy, should have a greater dietary requirement for the vitamin. Although dietary supplements would be recommended, injections of vitamin B_{12} are often given to animals with a poor health appearance. Animals coming into a feedlot are sometimes given B_{12} injections, along with other vitamins, as insurance against animals not quickly adapting to new feeding regimes. This use of B_{12} may be warranted under certain conditions where stress, disease, or parasites lower feed intake, impair ruminal function, and/or reduce intestinal absorption.

Available cobalt must be present in the animal diet for ruminal and intestinal synthesis to occur. Cobalt deficiency in ruminants can be cured or prevented through treatment of soils or pastures with cobalt-containing fertilizers or by direct oral administration of cobalt to the animal through free-choice mineral supplements (McDowell, 1985a). Cobalt-deficient livestock respond quickly to cobalt treatment and recover appetite, vigor, and weight; this serves as an easy, practical test to determine whether a cobalt deficiency exists. Large and frequent injections of vitamin B_{12} can effectively prevent or cure cobalt deficiency, but are much more expensive. Oral dosing or drenching with dilute cobalt solutions is likewise satisfactory if the doses are regular and frequent. An additional method of providing cobalt developed in Australia uses an orally administered, heavy pellet (''bullet'') made of cobalt oxide plus finely divided iron that remains in the reticulorumen for an extended period of time.

The need for continual cobalt supplementation is difficult to assess, as the incidence of cobalt deficiency can vary greatly from year to year, from an undetectable mild deficiency to an acute stage. Lee (1963) illustrated this variation in a 14-year experiment with sheep in southern Australia. Half the ewes, replacements, and progeny were dosed with cobalt and remained healthy. The undosed half had the following performance for the 14 years: in 2 years, lambs were unthrifty, but there were no deaths; in 3 years, growth rate of the lambs was slightly retarded; in 4 years, 30–100% of the lamb crop was lost; in 5 years, the performance of the remaining stock was as good as that of dosed animals.

For human therapy, during the stage of relapse of pernicious anemia, intramuscular doses of vitamin B_{12} at the rate of 15–30 μg/day should be given (Marks, 1975). A single injection of 100 μg or more will produce complete remission in any patient whose vitamin B_{12} deficiency is not complicated by unrelated systemic disease or other factors. When pernicious anemia is due to inadequate absorption, 1 μg of the vitamin by injection daily is adequate therapy. Remission is sustained for life by monthly injections of 100 μg of vitamin B_{12} (Herbert, 1984). In patients treated successfully with intramuscular vitamin B_{12}, the first response occurs within about 2 days of the start of treatment and consists of an intense feeling of well-being and increase in appetite.

Single, large oral doses (1000 μg) of vitamin B_{12} without intrinsic factor have proven effective in treatment of pernicious anemia. Absorption of a small amount of B_{12} from massive doses is independent of the action of intrinsic factor and is believed to occur by a "mass-action" effect, resulting in diffusion of some of the vitamin. An oral dose of at least 150 μg a day is deemed necessary to maintain the pernicious anemia patient. Single weekly oral doses of 1000 μg satisfactorily maintain some pernicious anemia patients (Ellenbogen, 1984). For vegetarian patients who have a normal secretion of intrinsic factor, 1.0 to 1.5 μg daily of vitamin B_{12} orally is sufficient to prevent pernicious anemia.

Supplementation with large doses of vitamin B_{12} has falsely been advocated for various disorders in humans. The red color of the vitamin along with its almost total lack of known toxicity makes it an almost ideal placebo. It is unfortunately used in large quantities to defraud unsuspecting buyers.

XI. TOXICITY

Addition of vitamin B_{12} to food in amounts far in excess of need or absorbability appears to be without hazard. Dietary levels of at least several hundred times the requirement are safe (NRC, 1987). However, the maximum tolerable dietary cobalt for ruminants is estimated at 5 ppm. Cobalt toxicosis in cattle is characterized by a mild polycythemia, excessive urination, defecation, and salivation, shortness of breath, and increased hemoglobin, red cell count, and packed cell volume. Although most researchers consider the margin between safe and toxic doses to be wide enough to make a toxicity under natural conditions to be unlikely, several reports have appeared in the literature concerning cobalt toxicity. These reports almost invariably involve management mistakes in formulating mineral mixtures.

14

Choline

I. INTRODUCTION

Choline is considered essential to the animal organism and is utilized both as a building unit and as an essential component in regulation of certain metabolic processes. Choline is tentatively classified as one of the B-complex vitamins even though it does not entirely satisfy the strict definition of a vitamin. Choline, unlike B vitamins, can be synthesized in the liver, is required in the body in greater amounts, and apparently functions as a structural constituent rather than as a coenzyme. Also, existence of choline in essential body constituents was recognized long before the first vitamin was discovered. Regardless of classification, choline is an essential nutrient for all animals and a required dietary supplement for some species (e.g., poultry and swine).

II. HISTORY

Choline was isolated by Streker from the bile of pigs in 1849 and by Von Balb and Hirschbrunn from an alkaloid of white mustard seed *(Sinapis alba)* in 1852 (Griffith and Nyc, 1971). Streker isolated the compound from lecithin, to which he gave the name ''choline.'' The chemical structure of choline was established in 1867 by Bayer (Scott *et al.*, 1982). Choline's acceptance as a biologically essential compound resulted from studies in 1929 in which acetylcholine was isolated from the spleen of a horse (Dale and Dudley, 1929). In 1932 choline was discovered to be the active component of pure lecithin previously shown to prevent fatty livers in rats (Best *et al.*, 1934). Betaine, considered to be only a methyl donor, was found to have a similar lipotropic effect in rats and dogs. Later studies have shown that choline is required for both growth and perosis prevention in poultry, for prevention of a ''spraddled-leg'' condition in swine, and is effective for various human conditions involving mobilization of liver lipids.

III. CHEMICAL STRUCTURE AND PROPERTIES

Choline is a β-hydroxyethyltrimethylammonium hydroxide and is depicted in Fig. 14.1. Pure choline is a colorless, viscid, strongly alkaline liquid that is notably hygroscopic. Choline is soluble in water, formaldehyde, and alcohol and has no definite melting or boiling point. Choline chloride is produced by chemical synthesis for use in the feed industry, although other forms exist for this purpose. Choline chloride exists as deliquescent white crystals that are very soluble in water and alcohols. Aqueous solutions are almost pH neutral.

Choline is widely distributed in nature as free choline, acetylcholine (Fig. 14.1), and more complex phospholipids and their metabolic intermediates. It is an integral part of the lecithins (Fig. 14.1), which account for its occurrence, in combination at least, in all plant and animal cells (Griffith and Nyc, 1971).

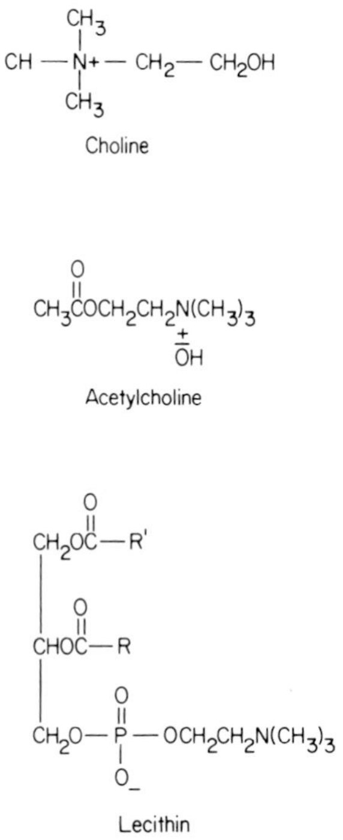

$CH-N^{+}(CH_3)(CH_3)-CH_2-CH_2OH$

Choline

$CH_3C(=O)OCH_2CH_2N(CH_3)_3^{+}\ \overline{O}H$

Acetylcholine

$CH_2OC(=O)-R'$

$CHOC(=O)-R$

$CH_2O-P(=O)(O^{-})-OCH_2CH_2N(CH_3)_3$

Lecithin

Fig. 14.1 Structural formulas for free choline, acetylcholine, and lecithin. R′ refers to any fatty acid.

IV. ANALYTICAL PROCEDURES

Determination of choline is complicated by its various forms in biological materials. Free choline can be extracted with water or alcohol, but more precise procedures must be used for extraction of total choline. Precautions have to be taken for the preparation of biological tissues, which contain enzymes that rapidly hydrolyze phosphate esters of choline in nervous tissues when animals are killed (Chan, 1984). Much of the evidence for presence of free choline in biological materials is unreliable owing to delay in the preparation of extracts, with resulting release of choline by autolysis (Griffith and Nyc, 1971). For example, dog liver contained 0–43 mg choline if extracted immediately after death of the animal and 136–164 mg choline/kg if extracts were made 5 hr postmortem.

The classic and most widely used procedure for quantitative determination of choline is the colorimetric reineckate method (AOAC, 1965). Other methods of analysis include enzymatic, fluorometric, gas chromatographic, photometric, and polarographic techniques. Enzymatic radioisotopic assay and gas chromatography provide high sensitivity and specificity and have been widely used in recent years (Chan, 1984). Microbiological assay for choline generally employs *Neurospora crassa,* while biological assay methods make use of conversion of choline to acetylcholine in isolated tissues, as well as growth response of chicks. Growth response may be an unreliable measure of choline concentration as it is influenced by other constituents (e.g., methionine) in choline-limiting diets (Pesti *et al.*, 1981).

V. METABOLISM

Choline is present in the diet mainly in the form of lecithin, with less than 10% present as either the free base or sphingomyelin. Choline is released from lecithin and sphingomyelin by digestive enzymes of the gastrointestinal tract, although 50% of ingested lecithin enters the thoracic duct intact (Chan, 1984). Choline is absorbed from the jejunum and ileum mainly by an energy- and sodium-dependent carrier mechanism. In the guinea pig, choline was taken up by ileal cells about three times faster than by jejunal cells (Hegazy and Schwenk, 1984). Preferential location of the transport system in ileal cells supports earlier findings in the rat and hamster (Sanford and Smyth, 1971). These results are expected since choline is released from lecithin in the proximal and middle small intestine.

Only one-third of ingested choline appears to be absorbed intact. The remaining two-thirds of the choline is metabolized by intestinal microorganisms to trimethylamine, which is excreted in the urine between 6 and 12 hr after consumption (De La Huerga and Popper, 1952). In contrast, when an equivalent amount of choline is consumed as lecithin, less urinary trimethylamine is excreted, with the majority of the metabolite appearing in urine 12–24 hr after consumption (De La Huerga and Popper, 1952). Dietary choline is the principal factor governing excretion, with presence or absence of other sources of protein and of fat having relatively little effect.

VI. FUNCTIONS

Choline functions in four broad categories in the animal body:

1. Choline is a metabolic essential for building and maintaining cell structure. As a phospholipid it is a structural part of lecithin (phosphatidylcholine), certain plasmologens (phosphatidylcholine), and the sphingomyelins. Choline is incorporated into phospholipid by being converted to phosphoryl choline, then to cytidine diphosphate choline, and finally reacting with phosphatidic acid to lecithin. Lecithin is a part of animal cell membranes and lipid transport moieties in cell plasma membranes. In the prevention of perosis, choline is required as

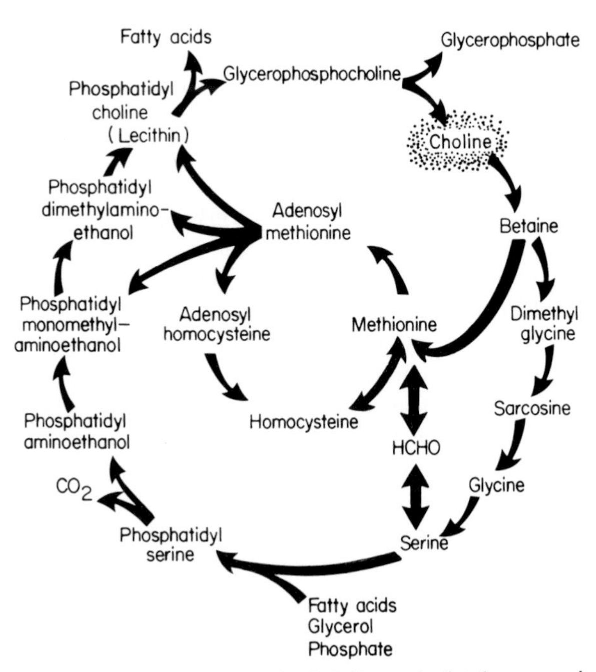

Fig. 14.2 Metabolic pathway for the synthesis of choline and related compounds. In the HCHO step, folacin and vitamin B_{12} are required for synthesis of methyl groups and metabolism of the one-carbon unit. (Adapted from Scott *et al.*, 1982.)

a constituent of the phospholipids needed for normal maturation of the cartilage matrix of the bone. Various metabolic functions and synthesis of choline are depicted in Fig. 14.2.

2. Choline plays an essential role in fat metabolism in the liver. It prevents abnormal accumulation of fat (fatty livers) by promoting its transport as lecithin or by increasing the utilization of fatty acids in the liver itself. Choline is thus referred to as a "lipotropic" factor because of its function of acting on fat metabolism by hastening removal or decreasing deposition of fat in liver.

3. Choline is essential for the formation of acetylcholine, the agent released at the termination of the parasympathetic nerves. It makes possible the transmission of nerve impulses from presynaptic to postsynaptic fibers of the sympathetic and parasympathetic nervous systems. For example, acetylcholine released by the stimulated vagus nerve causes a slowing of heartbeat, and also oviduct contraction results from the action of acetylcholine.

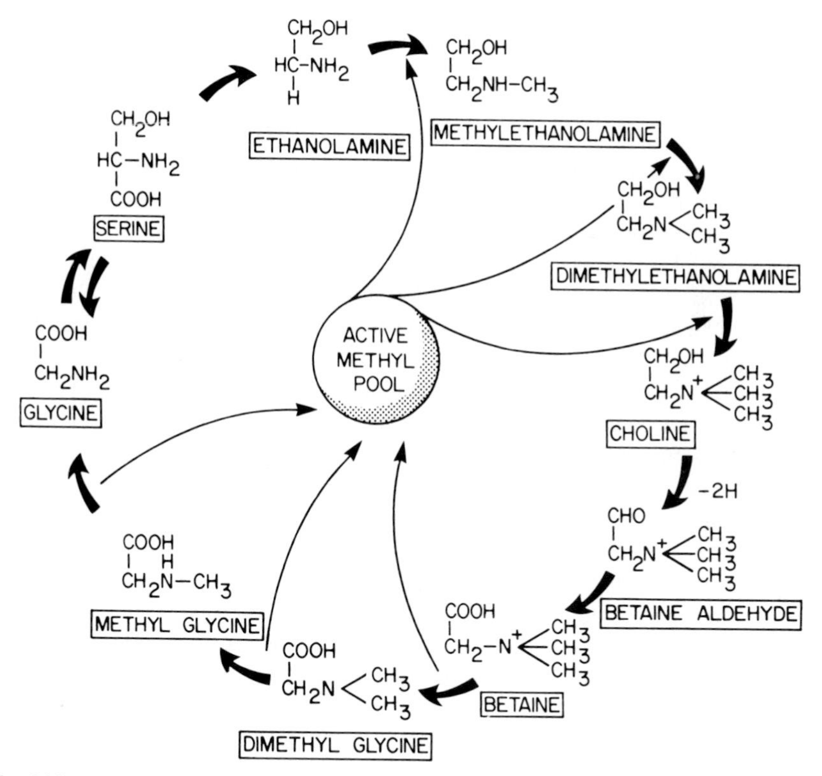

Fig. 14.3 Choline cycle showing metabolic generation and use of active methyl groups. (Adapted from Umbrett, 1960.)

4. A fourth function of choline is as a source of labile methyl groups for formation of methionine from homocystine and of creatine from guanidoacetic acid. However, practical significance of the choline–homocystine interrelationship is of no real importance to feeding animals since natural proteins contain very little of the metabolic intermediate homocystine (Ruiz *et al.*, 1983). To be a source of methyl groups, choline must be converted to betaine, which has been shown to perform methylation functions as well as choline in some cases. However, betaine fails to prevent fatty livers and hemorrhagic kidneys. A diagram of the choline cycle, including related compounds and donations to the body's methyl pool, is shown in Fig. 14.3.

VII. REQUIREMENTS

Choline, unlike most vitamins, can be synthesized by most species, although in many cases not in sufficient amounts or rapidly enough to satisfy all the animal's needs. Dietary factors such as methionine, betaine, *myo*-inositol, folacin, and vitamin B_{12} or the combination of different levels and composition of fat, carbohydrate, and protein in the diet, as well as the age, sex, caloric intake, and growth rate of animals, all have influence on the lipotropic action of choline and thereby requirement of this nutrient (Mookerjea, 1971).

Studies have shown that vitamin B_{12} and folacin reduce requirement for choline in chicks and rats (Welch and Couch, 1955). Folacin and vitamin B_{12} are required for the synthesis of methyl groups and metabolism of the one-carbon unit. Biosynthesis of labile methyl from a formate carbon requires folacin, while B_{12} plays a role in regulated transfer of the methyl group to tetrahydrofolic acid. Therefore, marked increases in choline requirement have been observed under conditions of folacin and/or vitamin B_{12} deficiency.

The two principal methyl donors functioning in animal metabolism are choline and methionine, which contain "biologically labile methyl groups" that can be transferred within the body. This phenomenon is called transmethylation. Methionine furnishes a methyl group that can combine with ethanolamine to form choline and, in reverse, methyl groups from choline (via betaine) can unite with homocystine to form methionine (du Vigneaud *et al.*, 1939). Therefore, dietary adequacy of both methionine and choline directly affect requirements of each other. Other than exogenous sources of methyl groups from choline and methionine, methyl group formation from *de novo* synthesis of formate carbons is reduced with folate and/or vitamin B_{12} deficiencies.

Most animals can synthesize sufficient choline for their needs provided enough methyl groups are supplied. As an example, methionine in the pig can completely replace that portion of the choline needed for transmethylation. Thus, at methionine levels in excess of the physiological requirement, 4.3 mg (or 3.69 with

OH group considered in molecular weight of choline) methionine provides the same methylating capacity as 1 mg choline (NRC, 1979). Young poultry, on the contrary, are unable to benefit from methionine or betaine as a dietary replacement for choline unless methylaminoethanol or dimethylaminoethanol is in the diet, as they appear unable to methylate aminoethanol when fed a purified diet (Jukes, 1947). Later studies showed that the chick can synthesize microsomal methylaminoethanol and choline from *S*-andenosylmethionine, but, unlike the pig, at an insufficient rate to cover its needs (Norvell and Nesheim, 1969).

The metabolic needs for choline can be supplied in two ways: either by dietary choline or by choline synthesis in the body, which makes use of labile methyl groups. For selected species, body synthesis sometimes cannot take place fast enough to meet choline needs for rapid growth and thus clinical signs of deficiency result. Since choline functions in the prevention of fatty livers, hemorrhagic kidneys, and perosis it does not act as a true vitamin since it is incorporated into phospholipids (via cytidine diphosphocholine). Therefore, unlike a typical B vitamin, the choline molecule becomes an integral part of the structural component of liver, kidney, or cartilage cells (Scott *et al.*, 1982).

In general, males are more sensitive to choline deficiency than females (Wilson, 1978). Growth hormone seemed to increase the choline requirement in rats independent of its ability to promote growth and increase food intake (Hall and Bieri, 1953). Cortisone and hydrocortisone have been reported to decrease severity of renal necrosis and hydrocortisone reduced the amount of hepatic lipid in choline-deficient rats (Olson, 1959).

Excess dietary protein increases the young chick's choline requirement. Ketola and Nesheim (1974) observed that over three times as much choline was needed for maximum growth of chicks when fed a diet containing 64% protein than when fed 13%. Diets high in fat aggravate choline deficiency and thus increase requirement. Fatty liver is generally enhanced by fats containing a high proportion of long-chain saturated fatty acids (Hartroft, 1955). Choline deficiency develops to a greater degree in rapidly growing animals with deficiency lesions more severe in these animals.

Use of antibiotics or prevention of coprophagy in rats decreases dietary choline requirement (Barnes and Kwong, 1967). Likewise, dietary requirement of choline is reduced in germ-free rats. Intestinal flora convert choline to trimethylamine, a lipotropically inactive compound that is excreted in urine. Suppression of intestinal flora by antibiotics would result in choline being more available than as inactive trimethylamine.

Recommended dietary allowances for choline or amounts typically fed for a number of animal species are listed in Table 14.1. Requirements for choline have generally been determined through the use of purified diets and recommendations often do not take into account bioavailability from feedstuffs, individual animal variation, or effects of other dietary factors. Choline deficiency

TABLE 14.1

Choline Requirements for Various Animals and Humans[a]

Animal	Purpose	Requirement	Reference
Beef cattle	Adult	Microbial synthesis	NRC (1984a)
Dairy cattle	Calf milk replacer	260 mg/kg liter	NRC (1978a)
	Adult	Microbial synthesis	
Chicken	Leghorn, 0–6 weeks	1300 mg/kg	NRC (1984b)
	Leghorn, 6–14 weeks	900 mg/kg	NRC (1984b)
	Laying–breeding	Unknown	NRC (1984b)
	Broilers, 0–3 weeks	1300 mg/kg	NRC (1984b)
	Broilers, 6–8 weeks	500 mg/kg	NRC (1984b)
Turkey	Growing, 0–4 weeks	1900 mg/kg	NRC (1984b)
	Growing, 16–20 weeks	950 mg/kg	NRC (1984b)
Japanese quail	Starting and growing	2000 mg/kg	NRC (1984b)
	Breeding	1500 mg/kg	NRC (1984b)
Sheep	Adult	Microbial synthesis	NRC (1985b)
Swine	Growing, 1–5 kg	0.6 g/kg	NRC (1988)
	Growing, 5–110 kg	0.3–0.5 g/kg	NRC (1988)
	Adult	1.00–1.25 g/kg	NRC (1988)
Horse	Adult	Microbial synthesis	NRC (1978b)
Goat	Adult	Microbial synthesis	NRC (1981b)
Cat	Adult	2400 mg/kg	NRC (1986)
Dog	Growing	1250 mg/kg	NRC (1985a)
Rabbit	All classes	0.12% choline chloride	NRC (1977)
Catfish	All classes	4000 mg/kg	NRC (1983)
Trout	All classes	1000 mg/kg	NRC (1981a)
Human	All classes	Unknown	RDA (1980)

[a]Expressed as per unit of animal feed either on an as fed (approximately 90% dry matter) or dry basis (see Appendix Table 1) (except for dairy calf).

has not been specifically demonstrated in humans and thus no requirements have been established.

VIII. NATURAL SOURCES

All naturally occurring fats contain some choline and thus it is supplied by all feeds that contain fat. Egg yolk (1.7%), glandular meats (0.6%), brain and fish (0.2%) are the richest animal sources and germ of cereals (0.1%), legumes (0.2–0.35%), and oilseed meals are the best plant sources (SYNTEX, 1979). Choline is largely absent in most fruits and vegetables. However, other foods consumed by animals and humans can be good sources. Infant formulas based on soy protein have much less choline than human or cow's milk (Zeisel *et al.*, 1986). Choline content of typical foods and feedstuffs is shown in Table 14.2.

TABLE 14.2

Typical Choline Concentrations in Various Foods and Feedstuffs (ppm, Dry Basis)[a]

Food/Feedstuff	ppm
Alfalfa meal, dehydrated	1,370
Bakery waste, dehydrated	1,005
Barley	1,177
Blood meal, dehydrated	848
Brewer's grains, dehydrated	1,757
Buttermilk (cattle), dehydrated	1,891
Chicken	6,288
Citrus pulp, dehydrated	867
Copra (coconut) meal	1,036
Corn, yellow	567
Cottonseed meal	2,965
Crab meal	2,179
Fish meal, anchovy	4,036
Liver meal	12,281
Milk, skimmed, dehydrated	1,480
Molasses, sugarcane	1,012
Oats	1,116
Peanut meal	2,120
Potato, dehydrated	2,879
Rice	1,076
Rice bran	1,357
Sesame	1,655
Sorghum	737
Soybean meal	2,916
Wheat	1,053
Wheat bran	1,797

[a]Selected from United States–Canadian Tables of Feed Composition (NRC, 1982).

Corn is low in choline, with wheat, barley, and oats containing approximately twice as much choline as corn.

Little is known of the biological availability of choline in natural feedstuffs. Using a chick assay method, soybean meal was found to contain a high proportion of biologically available choline. Dehulled regular soybean meal and whole soybeans were tested and appeared to range in availability from 60 to 75% (Molitoris and Baker, 1976).

IX. DEFICIENCY

The most common signs of choline deficiency include poor growth, fatty livers, perosis, hemorrhagic tissue (particularly in kidneys and certain joints), and hypertension. In addition, choline deficiency in poultry can result in reduced egg

TABLE 14.3

Clinical Signs and Lesions in Choline-Deficient Animals[a,b]

Signs and lesions	Rat	Mouse	Hamster	Guinea Pig	Rabbit	Monkey	Dog	Cat	Calf	Pig	Chicken	Duck	Turkey
Fatty Liver	+	+	+[c]	+	+	+	+	+[c]	+	+	+	+	+
Cirrhosis	+	+[c]			+[c]	+	+						
Hemorrhagic kidney	+	+[c]			+[d]				+[d]	+[d]			
Growth failure	+	+	+[c]	+	+	+	+	+	+	+	+	+	+
Hypolipemia	+						+						
Perosis											+	+	+
Impaired reproduction	+	+	+							+			
Impaired lactation	+		+						+				
Reduced egg production											+		
Muscle defects	+				+				+	+			
Hemorrhagic lesions	+	+											
Arterial sclerosis	+	+					+				+		
Myocarditis and necrosis	+	+											
Bradycardia	+												
Hypertension	+												
Anemia	+				+		+			+			
Edema, ascites	+												

[a]Modified from Wilson (1978).
[b]In many cases lesions have not been studied or reported in various species.
[c]Produced with difficulty; see text.
[d]Renal degeneration only.

production and increased chick mortality. In general, severity of clinical signs in animal species is influenced by other dietary factors, including methionine, vitamin B_{12}, folacin, and dietary fat (see Section VII). When feed intake and consequently growth are depressed by choline deficiency, severity of choline deficiency is then reduced. Clinical signs and lesions in choline-deficient animals are summarized in Table 14.3.

A. Effects of Deficiency

1. Ruminants

The ability of ruminants to synthesize the B vitamins including choline is well known. However, it was found that ruminal synthesis and/or the unsupplemented diet do not always supply enough choline to meet feedlot cattle demands. Improved performance of feedlot cattle has been related to use of dietary choline, although not consistently in all experiments. Several reports from Washington (Swingle and Dyer, 1970) and Maryland (Rumsey, 1975) have shown increased gains by as much as 6–7% and improved feed efficiency by 2.5–8% for finishing cattle when supplemented with 500–750 ppm dietary choline. Thus choline, under certain conditions of high-concentrate feeding, may be limiting in the diet.

Researchers are showing increasing interest in choline in view of its possible lipotropic effect in high-producing dairy cows. Atkins *et al.* (1983) and Erdman *et al.* (1984) studied the effects of supplementing choline on roughage adequate diets for dairy cattle and concluded that added choline had little effect on milk production, although slight increases were seen in feed intake and milk fat percentage.

An apparent choline deficiency syndrome was produced with a synthetic milk diet containing 15% casein (Johnson *et al.*, 1951). Within 6–8 days, calves developed extreme weakness with labored breathing and were unable to stand. Supplementation with 260 mg of choline per liter of milk replacer prevented these deficiency signs.

2. Swine

Choline deficiency in the young pig results in unthriftiness, poor conformation (short-legged and pot-bellied), lack of coordination in movements, a characteristic lack of proper rigidity in joints (particularly the shoulders), fatty infiltration of liver, characteristic renal glomerular occlusion, and some tubular epithelial necrosis (Cunha, 1977). These clinical signs resulted from low-methionine diets (0.8%) and were prevented with 1.6% dietary methionine.

"Spraddled hindleg" is a problem occasionally seen in newborn pigs, and some evidence suggests that incidence has a strong genetic component. However, the condition is often attributed to choline deficiency and is prevented by supplementation of the vitamin. Whether folacin and B_{12} are involved in the condition is unknown, but under conditions of deficiencies of these vitamins choline requirements are increased. Spraddled legs can be described as a congenital disorder in which the newborn pig cannot stand or walk because of the leg condition (Fig. 14.4). The problem seems to be worse on slippery floors. Nursing is also hindered, thereby affecting weaning weights.

Spraddled leg condition started to appear as swine producers began to decrease feed allowances given sows during gestation from 2.7–3.2 kg daily to 1.4–2.0 kg (Cunha, 1977), which resulted in reduced intakes of both choline and methionine. Studies from Colombia, South America (J. H. Maner, personal communication to Cunha, 1977), revealed death losses due to the spraddled leg condition. Some of these pigs recuperated by the tenth day after birth, which could indicate that the condition can be corrected through the sow's milk. Other reports have indicated that a high proportion of baby pigs affected by spraddled legs were able to recover after a few days, especially if the hindlegs are bound temporarily to allow them to move and suckle.

Research reports have shown that sows without choline had a significantly lower conception rate and farrowing rate and farrowed significantly fewer total

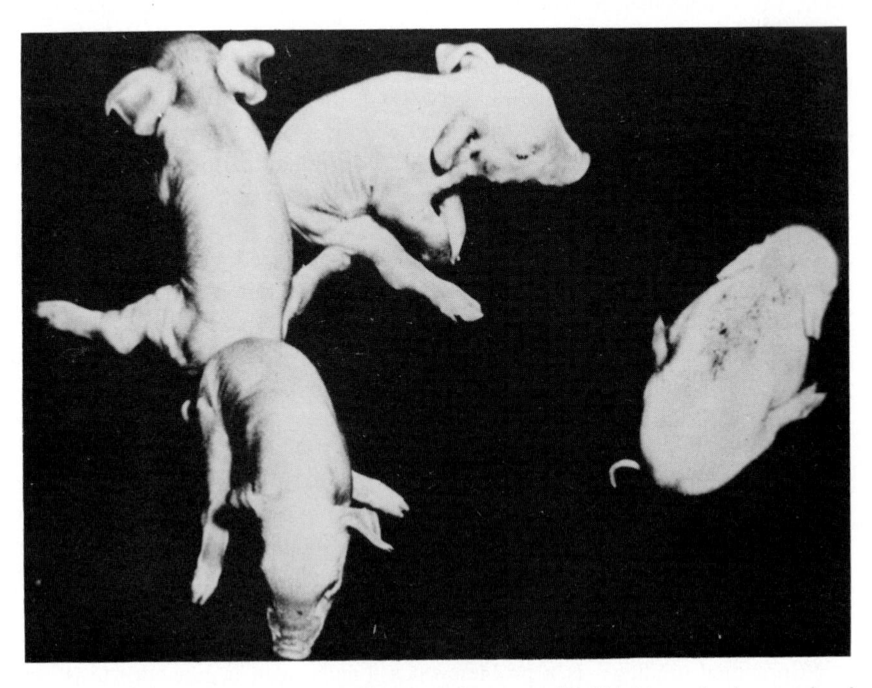

Fig. 14.4 Choline deficiency. The condition shown here—spraddled hind legs—has been produced with a purified ration and is prevented by choline supplementation. Other factors may be involved in this condition. (Courtesy of T.J. Cunha and Washington State University.)

pigs and fewer live pigs per litter. No difference was found in the average birth weight, but sows with choline supplementation weaned significantly more pigs per litter and sows without choline farrowed a slightly higher percentage of pigs with spraddled legs. Pigs from choline-deficient sows were unthrifty in appearance and became increasingly so with age (Ensminger *et al.*, 1947).

The NRC-42 committee on swine nutrition in 1976 evaluated the effects of supplemental choline (770 mg/kg of diet) during gestation and lactation on litter size at birth and at weaning. Nine stations participated with 22 trials and 551 sows. The ration was a 15.0% protein corn–soybean meal type during gestation and 7.5% beet pulp was substituted for an equal amount of corn during the lactation period. Results indicated that sows fed supplemental choline farrowed more total pigs per litter (10.54 vs. 9.89) and more live pigs per litter (9.33 vs. 8.64) and weaned more pigs per litter (7.72 vs. 7.29).

3. Poultry

Growth retardation and perosis result from choline deficiency in young poultry. Perosis is first characterized by pinpoint hemorrhages about the hock joint, followed by an apparent flattening of the tibiometatarsal joint (Scott *et al.*, 1982). Progressively, the Achilles tendon slips from its condyles, thus rendering the bird relatively immobile. Some studies indicated that in prevention of perosis, choline is required for the phospholipids needed for normal maturation of the cartilage matrix of bone (Maynard *et al.*, 1979).

Adult chickens probably synthesize sufficient choline to meet requirements for egg production. However, supplementary choline may be necessary for maintenance of egg size in quail (NRC, 1984b). Contrary to some reports, 500 ppm supplemental choline to leghorn hens increased egg weight while reducing specific gravity (Tapia Romero *et al.*, 1985). Also, the choline growth requirement for quail is apparently higher than that for chicks or poults.

Choline requirement of growing chicks decreases with age and it is generally not possible to produce a deficiency at an age over 8 weeks. It was observed that methylation of aminoethanol to methylaminoethanol seems to be the rate-limiting step in choline biosynthesis for young birds. High levels of dietary methionine or other methyl donors, therefore, cannot completely spare the chick's requirement for dietary choline, which is in contrast to the situation with growing mammals such as the pig or the rat (SYNTEX, 1979).

Apparently, choline requirement of laying hens can be influenced by choline level in the diet of the growing pullet (Scott *et al.*, 1982). Hens that received choline-free diets after 8 weeks of age were able to synthesize all the choline required for good egg production. Those that received choline supplements in the growing diet required supplemental choline in the laying diet for maximum egg production. The deficiency signs noted in these hens were a reduction in egg production and an increase in fat content of liver. Even with choline deficiency, however, choline content of the egg was not affected by low dietary choline.

Despite lack of evidence that laying chickens require a dietary source of choline for maximum egg production, addition of choline to practical diets markedly reduces the amount of fat in the liver (NRC, 1984b). However, a number of reports with chicks and turkey poults did not find fatty livers in chicks deficient in choline (Ruiz *et al.*, 1983). A choline response in laying chickens is likely to occur only if inadequate daily sulfur amino acid is provided.

4. Fish

Dietary choline is required for optimum growth, feed utilization, and prevention of fatty livers, anemia, hemorrhages (in liver, kidney, and intestine), and other clinical signs for salmon, trout, carp, catfish, and bream. Severity and nature of choline deficiency signs, as well as the time required for them to appear, seem to be highly variable and may depend on species or diet or both (Ketola, 1976).

Common carp were found to grow normally on a choline-deficient diet but had a 10% higher liver lipid content (Ogino *et al.*, 1970). With a choline deficiency, Japanese eels became anorexic, had poor growth, and exhibited a white-gray intestine. After 4 weeks, shrimp fed a deficient diet were significantly smaller than controls (NRC, 1983).

5. Other Animal Species

a. Laboratory Animals. Guinea pigs, like young poultry, easily develop choline deficiency even with adequate dietary methionine. Clinical signs of choline deficiency in young guinea pigs include poor growth, anemia, muscular weakness, and some adrenal subcutaneous hemorrhages. Both rats and mice exhibit fatty livers but rats, unlike mice, also develop cirrhosis or fibrosis of the liver. In young rats the kidneys become hemorrhagic, owing presumably to a deficiency of choline for the phospholipid required to build cell structure at this critical growth period. Mice, on the other hand, develop hemorrhagic kidney only with more difficulty and diet manipulation.

b. Rabbits. Choline deficiency in rabbits has been described as retarded growth, fatty and cirrotic liver, and a necrosis of the kidney tubules. A progressive muscular dystrophy has been reported in rabbits fed a low-choline diet for more than 70 days (NRC, 1977).

c. Dogs and Cats. In young puppies, choline deficiency results in fatty metamorphosis of the liver and atrophic changes of the thymus. Choline-deficient dogs with fatty livers show an increased rate of hepatic phospholipid synthesis following choline supplementation (NRC, 1985a). For growing cats, choline deficiency results in perilobular fatty infiltration of the liver, hypoalbuminemia, and decreased growth (NRC, 1986).

d. Monkeys. Cebus monkeys, rhesus monkeys, and baboons fed choline-deficient diets for 1–2 years developed fatty changes in the liver and varying degrees of hepatic fibrosis (NRC, 1978d).

6. HUMANS

In humans, as well as in experimental animals, dietary choline can arrest cirrhosis of the liver and reverse the fatty infiltration (SYNTEX, 1979). However, these conclusions are inconclusive, with some reports showing no benefit of choline in the treatment of cirrhosis (Kuksis and Mookerjea, 1984). It remains uncertain whether alcoholic cirrhosis in humans and animals is caused by one or more nutritional deficiences, but accepted dietary regimens for treatment involve a high-calorie, high-protein, low-fat diet supplemented with choline and perhaps other lipotropic factors, such as methionine and inositol.

The relationship of choline to atherosclerosis has been under study for a number of years. Choline-deficient diets can result in arterial damage, with a high percentage of young rats maintained on low-choline diets developing pathological changes in the major arterial trunks and in the coronary arteries (SYNTEX, 1979). These are changes similar to those observed in early stages of atherosclerosis of the aorta and large arteries in humans. In a 3-year study of coronary thrombosis patients, there was a significant reduction in death rates and lowered blood cholesterol in atheromatous patients administered choline.

Reports indicate that lecithin and/or choline supplements may be useful in preventing age-related memory deficits and certain neurological diseases (Kuksis and Mookerjea, 1984). Evidence suggest that dietary choline and lecithin produce clinical improvement by supplying precursors for formation of the neurotransmitter acetylcholine.

B. Assessment of Status

Research information is very limited on detection methods to determine choline status of animals. Often the best indicator of status or need for choline is observation of clinical signs attributable to choline deficiency (e.g., fatty livers or perosis) for particular species as well as beneficial performance responses when diets are supplemented with the vitamin.

Tissue levels of choline or its functional metabolities can be determined to evaluate choline status. There is evidence of a reduction of acetylcholine in brains, kidneys, and intestines of rats deprived of choline for 6 days after weaning. Choline administered to rats either by injection or by diet causes a dose-related increase in brain acetylcholine (Kuksis and Mookerjea, 1984). Studies on mechanism of liver fat accumulation have suggested that this is related to a lack of phosphatidylcholine synthesis. With a choline deficiency, the hepatic phosphatidylcholine : phosphatidylethanolamine ratio is reduced, and is thus a means of evaluating choline status.

Plasma choline levels in humans reflect dietary consumption (Hirsch *et al.*, 1978). There is the suggestion that fall of free plasma choline below the normal range may be related to a choline deficiency state. After consumption of 3 g choline chloride in a meal, serum choline rose by 86% over the mean fasting level of 11.7 nmol/ml.

X. SUPPLEMENTATION

Choline is made synthetically and in the majority of cases diet supplementation is from synthesized choline salts. Choline is synthesized from natural gas via methanol and ammonia, which are reacted to produce trimethylamine (Griffin and Nyc, 1971). This trimethylamine is subsequently reacted with ethylene oxide to produce choline. For feed supplementation purposes a chloride salt is produced by reacting the alkaline base with hydrochloric acid. Choline is available as chloride (86.8%) and bitartrate (48%) salts. Choline chloride is available for feed use as the 70% liquid or 25–60% dry powder. The 70% liquid is very corrosive and requires special storage and handling equipment. It also is not suitable for inclusion in concentrated vitamin premixes, but rather is most economical to add directly to concentrates.

Response to dietary supplementation of choline will be most dependent on species, age of animals, protein and sulfur amino acid intake, dietary choline, and other choline-sparing nutrients. Unlike most vitamins, choline can be synthesized by various animals, although often in insufficient amounts. It appears that choline deficiency in the young of some species (i.e., chick) may not be due to lack of ability to synthesize choline but more likely to lack of ability to synthesize it at a rate sufficient for animal needs. Age is an important consideration, for example, it is difficult to produce a choline deficiency in growing chicks over 8 weeks of age.

The young of many species (i.e., pig and rat) do not require supplementary choline if dietary methionine level is sufficiently high. On a synthetic milk diet containing 1.6% methionine, young pigs were found not to require supplemental choline (Firth *et al.*, 1953). Neumann *et al.* (1949) reported that the young pig requires 0.1% dietary choline when methionine is present at 0.8–1.0% of the diet.

Methionine can furnish methyl groups for choline synthesis for most species. Choline, however, is effective only in sparing methionine that otherwise would be used to make up for a choline shortage. Methionine is not used up for choline synthesis if there is an adequate level of dietary choline. In formulating typical poultry and swine diets, methionine is frequently one of the most limiting amino acids. Therefore, it would be impractical for marginal quantities of methionine to be wasted for synthesis of the vitamin when supplemental choline can be provided more economically.

Interrelationships of choline and methionine are discussed in Sections VI and VII. In providing supplements of methionine and/or choline, a third nutrient, sulfur, must be considered. Significance of a three-way interrelationship among methionine, choline, and sulfate has been reviewed by Ruiz *et al.* (1983) and Miles *et al.* (1986). Sulfur is present in a number of body metabolites (i.e., mucopolysaccharides) and if not adequately supplied in the diet, sulfur amino acids would likely be degraded. In feeding broilers, supplemental sulfate accompanied by choline or methionine achieved a greater growth response than when either was fed alone (Miles *et al.*, 1983). Data suggest that sulfate must be present for choline to spare a maximum amount of methionine. The practical implication is that sulfate and choline need to be adequately provided in diets so that the more expensive and often marginally deficient nutrient methionine is not used to provide either of these nutrients.

Most choline supplementation studies emphasize production benefits from providing the vitamin to young animals. However, research and observations with adult swine demonstrate improved litter size at weaning and supplementation may keep sows in the producing herd longer (Cunha, 1977). The exact level of choline needed for sow rations is unknown. Until more research data are available, Cunha (1972) suggested the following levels in situations where spraddled hindlegs are likely to occur: (1) during the first part of gestation, a daily level of 3000 mg per sow and (2) for the last month of gestation, daily choline should be increased to 4200–4500 mg.

For humans, in view of the widespread occurrence of choline and methionine in plant and animal foodstuffs, choline supplementation is likely unwarranted except for special dietary regimens, including infants and young children fed diets deficient in protein and high in highly refined products and patients on total parenteral nutrition. Because 5–14 mg total choline per deciliter is present in human milk and is tolerated with no adverse reactions, and because deficiency in young animals produces serious effects, it is recommended that at least 7 mg choline per 100 kcal be included in infant formulas not based on milk (Kuksis and Mookerjea, 1984). Commercially prepared infant formulas and milk products have been supplemented with choline at these levels to assure the presence of choline in an amount approximating that naturally occurring in milk.

Choline is added to feeds as a 70% choline chloride solution or as 25–60% dry powder. Choline choloride is stable in multivitamin premixes but decreases the stability of other vitamins in the premix (Frye, 1978). Choline is stable during processing and storage in pressure-pelleted and extruded feeds. Since the material is hygroscopic, containers containing choline should be kept closed when not in use. Loss during water immersion (e.g., feeding fish) of pellets is less than 10% after 60 min. Synthetically produced choline is relatively inexpensive and readily available.

XI. TOXICITY

Experimental animal toxicity data on clinical signs of choline overdosage include salivation, trembling, jerking, cyanosis, convulsion, and respiratory paralysis. Estimates of the oral LD_{50} of choline chloride in rats varied from 3.4 to 6.7 g/kg (Chan, 1984). Choline levels somewhat above the requirement (868–2000 ppm) were shown to reduce rate and efficiency of gain (Neumann *et al.*, 1949; Southern *et al.*, 1986). Derilo and Balnave (1980) reported a reduced gain and efficiency in young broiler chicks fed a level of choline only slightly in excess of the requirement. Studies with chickens suggest that dietary choline double the requirement is safe, with swine having a higher tolerance for choline (NRC, 1987).

In humans, high intakes of lecithin or choline produce acute gastrointestinal distress, sweating, salivation, and anorexia (Wood and Allison, 1982). However, therapeutic doses of choline chloride and of choline dihydrogen citrate administrated in amounts ranging from 3 to 12 g daily have been used in the treatment of alcoholic cirrhosis for up to 4 months with no toxic effects reported. In treatment of tardive dyskinesia (a movement disorder generally associated with the intake of antipsychotic medication), up to 16 g per day of choline has been used; others have utilized lecithin at doses greater than 100 g daily for more than 4 months with no evidence of ill effects (Kuksis and Mookerjea, 1984).

15

Vitamin C

I. INTRODUCTION

Scurvy, a potentially fatal condition resulting from inadequate vitamin C (ascorbic acid), has been known and feared since ancient times. Its prevention and cure were associated with consumption of fresh fruits, especially citrus, but it was not until 1928 that the antiscorbutic factor was identified. Vitamin C is synthesized in almost all species, the exceptions being the primates, including humans, guinea pigs, fish, fruit-eating bats, insects, and some birds. Animals that cannot synthesize this vitamin need a dietary source for their normal maintenance.

The concept that the sole function of vitamin C is to prevent scurvy has been revised in recent years. Small quantities of vitamin C are sufficient to prevent and cure scurvy, however, larger quantities may be required to maintain good health during adverse environment, physiological stress, and certain disease conditions.

II. HISTORY

Several historical reviews of scurvy and vitamin C are available (Marks, 1975; Vilter, 1978; Jaffe, 1984). A historical record is presented in Table 15.1 (Vilter, 1978).

Scurvy was one of the earliest diseases known, with historical evidence of its existence in Egypt, Greece, and Rome. In the Middle Ages scurvy was endemic in Northern Europe during the late winter months and early spring, because at that period the foodstuffs that provide the chief source of vitamin C (green vegetables) had not been introduced. For hundreds of years it was known or suspected that a dietary factor would protect against scurvy. Real progress was made toward identification of the dietary factor involved when in 1907 Holst and Frolich discovered that guinea pigs could develop scurvy, thus providing an experimental animal.

World exploration and military operations were severely hampered during the era prior to the seventeenth century by ravages of scurvy. On the long sea voyages

TABLE 15.1

Historical Record of Scurvy and Its Relationship to Vitamin C[a]

1550 B.C.	Scurvy described in Ebers Papyrus (Thebes)
600 B.C.	Hippocrates described soldiers afflicted with scurvy
A.D. 1200	Crusades weakened by scurvy
1492–1600	World exploration threatened by scurvy —Magellan lost four-fifths of his crew —Vasco de Gama lost 100 of his 160 men
1536	Jacques Cartier's expedition immobilized by scurvy, learned from Indians the curative value of pine needles and bark
1570	Captain James Lancaster prevented scurvy by giving crew members two jiggers of lemon juice daily
1593	Sir Richard Hawkins used oranges and lemons to treat scurvy in the British Navy Ponsseus referred to the therapeutic use of scurvy grass, watercress, and oranges
1650	Infantile scurvy described by Glisson but was confused with rickets
17th Century	Lime juice used experimentally on ships of the East India Company
1734	Backstrom related scurvy to a deficiency of fresh fruits and vegetables
1740–1744	Lord Anson lost three-fifths of his crew of 1950 men to scurvy
1747	James Lind performed controlled shipboard experiment on the preventive effect of oranges and lemons (published 1753)
1768–1771	Captain James Cook demonstrated that prolonged sea voyages were possible without ravages of scurvy
1789	William Stark induced scurvy in himself by a diet of bread and water for 60 days
1795	Lemon juice made a regular ration in British Navy
1854	Lemon juice made a regular ration in British Merchant Marines
1863	Scurvy epidemic in the opposing armies during the American Civil War
1883	Sir Thomas Barlow differentiated infantile scurvy from rickets
1895	Antiscorbutic ration become official in U.S. Army
1900	Boiling and pasteurization of infant formula increased incidence of scurvy
1906	Hopkins suggested that infantile scurvy was a deficiency disease
1907	Holst and Frolich produced experimental scurvy in guinea pigs by feeding a deficient diet, with pathological changes resembling those in humans
1912	Explorer Captain Scott and his team died of scurvy during their expedition to the South Pole
1928	Szent-Györgyi isolated hexuronic acid from orange juice, cabbage juice, and cattle adrenal glands
1932	Waugh and King isolated hexuronic acid from lemons and identified it as vitamin C
1933	Haworth determined structure of vitamin C
1933	Reichstein synthesized vitamin C
1971	Linus Pauling published book on relationship of vitamin C to the common cold, which stimulated research on the therapeutic and prophylactic uses of megadoses of vitamin C on the common cold, resistance, and various diseases

[a]Adapted and modified from Vilter (1978).

in explorations of the late 1400s and 1500s scurvy frequently occurred. Knowledge of the curative value of certain foods had been known for a considerable time, and American Indians had used an infusion of spruce or pine needles to prevent scurvy for at least four centuries. As early as 1536, Jacques Cartier learned from these Indians that scurvy could be cured and prevented by consuming a drink made from pine needles and bark. Previously, 107 cases of scurvy out of 110 men resulted from Cartier's expedition up the St. Lawrence River. Before the relationship of scurvy to diet was found, there was a tendency to associate scurvy with venereal disease. Mercury was used as a treatment with disastrous results.

In 1747 James Lind, a British fleet physician, carried out one of the first examples of a controlled clinical experiment in which he showed that patients on lemon juice recovered from scurvy while others failed to do so. Lind indeed was responsible for the relief of both scurvy and typhus in the fleet and probably helped as much as Lord Nelson to break the power of Napoleon. By the eighteenth century, it was realized that fresh fruit and vegetables and lime or lemon juice would protect sailors on long voyages. The term "limey," colloquially applied to the English population, stems from lime juice given to sailors in the British navy. Lemon juice had become a routine part of British Navy diets by 1795. However, scurvy was an epidemic problem during the American Civil War and it was not until 1895 that an antiscorbutic diet became official in the U.S. Army.

Szent-Györgyi (1928) isolated a substance that he called hexuronic acid from orange juice, cabbage jucie, and ox adrenal glands. The isolated substance was acidic in nature, with the formula $C_6H_{12}O_6$, and was strongly reducing. The same year, King isolated an antiscorbutic substance in crystalline form from orange juice and demonstrated that it cured scurvy; thus Szent-Györgyi tried his hexuronic acid, found it also cured scurvy, and concluded it must be vitamin C. It was not readily accepted that it represented a vitamin because amounts required were large relative to amounts required of other vitamins (up to 100 mg). But finally in 1933, when hexuronic acid was synthesized by Richstein and shown to have the true activity of the natural product, ascorbic acid was recognized as a vitamin.

III. CHEMICAL STRUCTURE, PROPERTIES, AND ANTAGONISTS

Vitamin C occurs in two forms (Fig. 15.1), namely the reduced ascorbic acid and the oxidized dehydroascorbic acid. Only the L isomer of ascorbic acid has activity. Although the majority of the vitamin exists as ascorbic acid, both forms are biologically active. In foods the reduced form of vitamin C may reversibly oxidize to the dehydro form with dehydroascorbic acid further oxidized to the inactive and irreversible compound of diketogulonic acid. This change takes place readily, and thus vitamin C is very susceptible to destruction through oxidation,

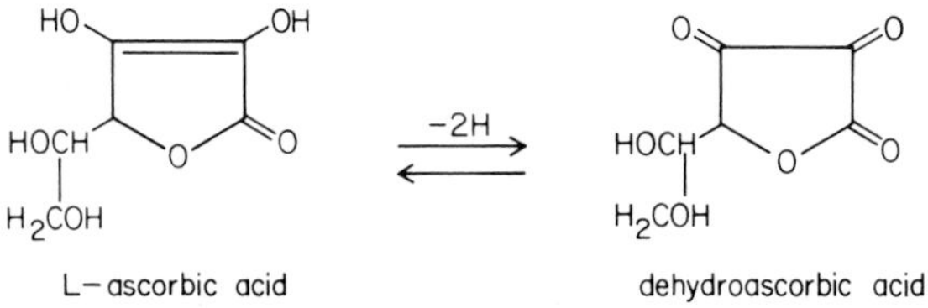

Fig. 15.1 Structures of vitamin C: L-ascorbic acid (reduced form) and dehydroascorbic acid (oxidized form).

a change that is accelerated by heat and light. Diketogulonic acid can be further oxidized to oxalic acid and L-threonic acid. Ascorbic acid is so readily oxidized to dehydroascorbic acid that other compounds may be protected against oxidation. This antioxidant property is used in the addition of the vitamin in canning of certain fruits to prevent oxidation changes that cause darkening. Reversible oxidation–reduction of ascorbic acid with dehydroascorbic acid is the most important chemical property of vitamin C and the basis for its known physiological activities and stabilities (Jaffe, 1984). Vitamin C is the least stable, and therefore most easily destroyed, of all vitamins.

Ascorbic acid is a white to yellow-tinged crystalline powder. It crystallizes out of water solution as square or oblong crystals and is slightly soluble in acetone and lower alcohols. A 0.5% solution of ascorbic acid in water is strongly acid with a pH of 3. The vitamin is more stable in an acid than an alkaline medium. It is not found in dry foods and is markedly destroyed by cooking, particularly when the pH is alkaline. Cooking losses also result because of its solubility. Crystalline ascorbic acid is relatively stable in air without moisture, and small concentrations of metal ions will accelerate destruction of ascorbic acid. Various derivatives and analogs of vitamin C have been prepared that have little if any antiscorbutic activity. Glycoascorbic acid acts as an antimetabolite for vitamin C, which is an ascorbic acid homolog, and contains an added CHOH group that has undergone optical inversion.

IV. ANALYTICAL PROCEDURES

Analysis of vitamin C includes biological, chemical, and physical methods. The biological method for vitamin C determination is specific for antiscorbutic activity and, as such, can be accepted as the final standard of reference in cases where it is suspected that accuracy of chemical or physical procedures may be affected by presence of interfering substances. The biological test measures total amount of vitamin C present, that is, in both the reduced form of ascorbic acid itself and the reversibly oxidized form of dehydroascorbic acid. Applicability of the biological method may be limited only if potency of the test material is too low for it to be measured accurately. Rats cannot be used as test animals for vitamin C assay owing to their ability to synthesize the vitamin, but guinea pigs have proved satisfactory. Biological methods are based on prevention or cure of scurvy in guinea pigs, in addition to dental histology, curative growth, and serum concentrations of alkaline phosphate associated with the deficiency. Procedures with guinea pigs are time-consuming and require about 10 weeks for completion. Bioassays based on microorganisms have not been developed because no organism has been found that has an absolute requirement for L-ascorbic acid (Jaffe, 1984).

Biological analytical procedures have largely been replaced by chemical and physical methods, which provide precise, faster and less expensive assays. Chemical and physical methods require precautions to prevent oxidation (i.e., homogenize under N_2 and avoid copper and other metallic ions). Dye methods are widely used with the reagents 2,6-dichlorophenolindophenol for reduced ascorbic acid and 2,4-dinitrophenylhydrazine for the oxidized form and for total ascorbic acid after oxidation. L-Ascorbic acid absorbs strongly in the ultraviolet, which is the basis of spectrophotometric methods (Tono and Fujita, 1982). Both gas–liquid chromatographic (GLC) and high-performance liquid chromatographic (HPLC) methods have been developed for L-ascorbic acid determination (Rose and Nahrwold, 1982).

V. METABOLISM

Vitamin C is absorbed in a manner similar to carbohydrates (monosaccharides). Intestinal absorption in vitamin C-dependent animals appears to require a Na^+-dependent active transport system. It is assumed that those species that are not scurvy-prone do have an absorption mechanism by diffusion (Spencer *et al.*, 1963). Ascorbic acid is readily absorbed when quantities ingested are small, but limited intestinal absorption occurs when excess amounts of ascorbic acid are ingested. Bioavailability of vitamin C in foods is limited, but apparently 80–90% appears to be absorbed (Kallner *et al.*, 1977). Site of absorption in the guinea pig is located in the duodenal and proximal small intestine, whereas the rat showed highest absorption in the ileum (Hornig *et al.*, 1984).

In its metabolism, ascorbic acid is first converted to dehydroascorbate by a number of enzyme or nonenzymatic processes and is then reduced in cells (Rose *et al*, 1986). Absorbed vitamin C readily equilibrates with the body pool of the vitamin. No specific binding proteins for ascorbic acid have been reported and it is suggested that the vitamin is retained by binding to subcellular structures.

Ascorbic acid is widely distributed throughout the tissues, both in animals capable of synthesizing ascorbic acid as well as in those dependent on an adequate dietary amount of vitamin C. In experimental animals, highest concentrations of vitamin C are found in pituitary and adrenals, with high levels also found in liver, spleen, brain, and pancreas. It tends to localize around healing wounds.

Humans receiving adequate intakes of vitamin C have a body pool of approximately 1.5 g of the vitamin, with 3–4% of the existing body pool utilized daily. The half-life of ascorbic acid is inversely related to daily intake and is 13–40 days in humans and 3 days in guinea pigs, which correlates with the longer time needed by a human to develop scurvy: 3 months for a human on a vitamin C-free diet compared with 3 weeks for the guinea pig (Jaffe, 1984).

Absorbed ascorbic acid is excreted in urine, sweat, and feces. Fecal loss is minimal and even with large intakes in humans only 6–10 mg daily is excreted by this route (Marks, 1975). Loss in sweat is also probably low. In guinea pigs, rats, and rabbits, CO_2 is the major excretory mechanism for vitamin C. Primates do not normally utilize the CO_2 catabolic pathway, with the main loss occurring in the urine. Urinary excretion of vitamin C depends on the body stores, intake, and renal function. Mechanism and mode of elimination are a function of glomerular filtration rate of ascorbic acid and are dependent on plasma ascorbate concentration. Substantial quantities of L-ascorbic acid are excreted in urine after concentration in blood plasma exceeds its usualy threshold of approximately 1.4 mg per 100 ml.

Urine contains numerous metabolites of ascorbic acid, including dehydroascorbic acid, diketogulonic acid, ascorbate-2-sulfate, oxalate, methyl ascorbate, and 2-ketoascorbitol (Sauberlich, 1984). In humans, with physiological doses of 60–100 mg/day, urinary oxalate is the major metabolite, with 30–50 mg/day being formed. But when given in large doses up to 10 g/day, urinary oxalate is increased by only 10–30 mg/day, and the vitamin is excreted largely unmetabolized in urine and feces (Jaffe, 1984). Excretion of ascorbic acid in urine declines to undetectable levels with inadequate intakes of the vitamin or in case of scurvy.

VI. FUNCTIONS

Ascorbic acid has been found to be involved in a number of biochemical processes. Function of vitamin C is related to its reversible oxidation and reduction characteristics, however, the exact role of this vitamin in the living system is not clearly known since a coenzyme form has not yet been reported. Nevertheless, vitamin C plays important roles in many biochemical reactions, such as mixed-function oxidation involving incorporation of oxygen into the substrate.

The most clearly established functional role for vitamin C involves collagen biosynthesis. Collagens are the tough, fibrous, intercellular materials (proteins) that are principal components of skin and connective tissue, the organic substances of bones and teeth, and the ground substances between cells. Impairment of collagen synthesis in vitamin C deficiency appears to be due to lowered ability to hydroxylate lysine and proline. Syntheses of collagens involve enzymatic hydroxylations of proline to form a stable extracellular matrix and of lysine for glycosylation and formation of cross-links in the fibers (Barnes and Kodicek, 1972). The requirement for ascorbic acid is specific, probably protecting the hydroxylase enzymes by oxidation of both the ferrous ions and thiol groups present. Hydroxyproline is found only in collagen (14%) and arises from hydroxylation of proline. In its absence, a nonfibrous collagen precursor is formed instead of fibrous collagen.

Beneficial effects result from ascorbic acid in the synthesis of ''repair'' collagen. Alteration of basement membrane collagen synthesis and its integrity in mucosal epithelial during vitamin C restriction might explain the mechanism by which the capillary fragility is induced in scurvy and also increased incidences of periodontal disease under vitamin C deprivation (Chatterjee, 1978). Failure of wounds to heal and gum and bone changes resulting from vitamin C undernutrition are direct consequences of reduction of insoluble collagen fibers.

Biochemical and physiological functions of vitamin C have been reviewed (Sebrell and Harris, 1967; Marks, 1975; Chatterjee, 1978; Sauberlich, 1984; Jaffe, 1984; Horing *et al.*, 1984). The functional importance of vitamin C, other than the previously mentioned role in collagen synthesis, includes the following:

1. Because of the ease with which ascorbic acid can be oxidized and reversibly reduced, it is probable that it plays an important role in reactions involving electron transfer in the cell. Almost all terminal oxidases in plant and animal tissues are capable of directly or indirectly catalyzing the oxidation of L-ascorbic acid. Such enzymes include ascorbic acid oxidase, cytochrome oxidase, phenolase, and peroxidase. In addition, its oxidation is readily induced under aerobic conditions by many metal ions, hemochromogens, and quinones.
2. Metabolic oxidation of certain amino acids including tyrosine. Tyrosyluria is observed when high levels of tyrosine are being metabolized. Vitamin C appears to prevent the inhibition of the enzyme *p*-hydroxyphenylpyruvic acid oxidase by its substrate, *p*-hydroxyphenylpyruvic acid, in the tyrosine metabolism sequence.
3. Ascorbic acid has a role in metal ion metabolism because of its reducing and chelating properties. It can result in enhanced absorption of minerals from the diet and their mobilization and distribution throughout the body. Ascorbic acid promotes nonheme iron absorption from food and acts by reducing the ferric iron at the acid pH in the stomach and by forming complexes with iron ions that stay in solution at alkaline conditions in the duodenum. It appears to also function in the reduction and release of ferric iron from its tight linkage with plasma protein and its incorporation into ferritin. Ascorbic acid also tends to alleviate toxic effects of transition metals in the body.
4. Carnitine is synthesized from lysine and methionine and is dependent on two hydroxylases, both containing ferrous iron and L-ascorbic acid. Vitamin C deficiency can reduce the formation of carnitine, which can result in accumulation

of triglycerides in blood and of the physical fatigue and lassitude associated with scurvy.

5. Interrelationships of vitamin C to B vitamins are known as tissue levels and urinary excretion of vitamin C are affected in animals with deficiencies of thiamin, riboflavin, pantothenic acid, folacin, and biotin. Vitamin C is active in changing the form of folacin to the tetrahydro derivative, a reduced form. It may also affect the ability of the body to store folacin. When vitamin C is deficient, utilization of folacin and vitamin B_{12} is impaired, resulting in anemia.

6. Ascorbic acid is reported to have a stimulating effect on phagocytic activity of leukocytes, on function of the reticuloendothelial system, and on formation of antibodies. Some of the most controversial topics regarding the interactions of vitamin C with immune functions have been the reduction of common cold symptoms and favorable responses to cancer treatment.

7. Vitamin C has been demonstrated to be a natural inhibitor of nitrosamines, which are potent carcinogens. Action of ascorbate in preventing formation of nitrosamines is reported to result from its direct reaction with nitrate.

8. Synthesis of corticosteroids in adrenal glands may involve ascorbic acid-related hydroxylation steps. On stimulation of the adrenal gland by ACTH (adrenocorticotrophic hormone), a fall in ascorbate concentration was observed and suggestion was made that vitamin C is required for steroidogenesis.

9. Ascorbic acid is found in up to a 10-fold concentration in seminal fluid (as compared to serum levels). Decreasing levels have caused nonspecific sperm agglutination. It has been hypothesized that the vitamin may help to protect sperm from effects of harmful oxidation.

VII. REQUIREMENTS

A wide variety of plant and animal species can synthesize vitamin C from carbohydrate precursors including glucose and galactose. The missing step in the pathway of ascorbic acid biosynthesis in all vitamin C-dependent species has been traced to inability to convert L-gulonolactone to 2-keto-L-gulonate, which is transformed by spontaneous isomerization into its tautomeric form, L-ascorbic acid. Vitamin C dietary-dependent species, therefore, lack the enzyme L-gulonolactone oxidase.

Metabolic need for ascorbic acid is a general one among species, but a dietary need is limited to humans, subhuman primates, guinea pigs, fruit-eating bats, some birds, including the red-vented bulbul and related Passeriformes species, insects, fish, such as coho salmon, rainbow trout, and carp, and perhaps certain reptiles. Even for species that do synthesize vitamin C, it has been shown that the synthesizing capacity of liver microsomal preparations varies strongly from animal to animal (Chatterjee, 1978), suggesting possible dietary need of the vitamin for individuals within a species. Circumstances in which vitamin C deficiencies may occur in domestic animals are discussed in Sections IX and X.

In general, domestic animals such as poultry, ruminants, swine, horses, dogs, and cats have the ability to biosynthesize ascorbic acid within their body and hence there is no recommended requirement established by the National Research Council. However, Marks (1975) proposed the following vitamin C requirements (mg/kg diet) for poultry and swine: poultry, 50–60 mg; starting pigs, 300 mg; and finishing pigs, 150 mg. Itze (1984) suggested 250 mg daily of ascorbic acid for young calves. Vitamin C requirements for cold-water fishes are estimated at 100 mg/kg of diet (NRC, 1981a), while the catfish is reported to require 60 mg/kg (NRC, 1983).

For humans, the RDA (1980) currently recommends 60–100 mg of vitamin C depending on body weight and physiological function (Table 15.2). The human requirement for vitamin C has been the subject of considerable debate with diversity of opinion seeming to be irreconcilable. This controversy is reflected in the wide range of recommended daily allowances established by different countries (ranging from 20 to 200 mg) (Sauberlich, 1984). These values are based on whether the allowance should prevent scurvy and permit a margin of safety or whether more complete tissue saturation is preferable. Human scurvy can be prevented and cured with daily intake of 10 mg of vitamin C. This level of intake, however, permits little or no reserves. Approximately 18–25 mg/day keeps tissues half saturated.

Requirements are increased by pregnancy, lactation, thyrotoxicosis, increased metabolism, decreased absorption, or increased loss. Individuals may require increased vitamin C as a result of stress or unfavorable environmental situations. Cigarette smokers with vitamin C intakes comparable to those of nonsmokers have serum vitamin C levels lower than those of nonsmokers. Smoker requirements for ascorbic acid were estimated to be increased by as much as 50%.

TABLE 15.2

Recommended Vitamin C Daily Dietary Allowances for Humans[a]

Group	Amount (mg)
Under 1 year	35
1–10 years	45
11–14 years	50
15–51 + years	60
Pregnancy, second half	80
Lactation	100

[a]RDA (1980).

Likewise, increased vitamin C intakes are required in subjects exposed to cold or elevated temperatures and other acute stresses, including surgery and trauma (Sauberlich, 1984). Other circumstances in which dietary vitamin C is increased include the elderly, alcoholics, and users of oral contraceptives.

Research during the past decade that established that the sole function of ascorbic acid was to prevent scurvy must be revised. There is evidence suggesting that ascorbic acid participates in extra ''antiscorbutic'' functions for which a higher requirement than necessary to protect against overt scurvy is needed. Publication of a popular book by two-time Nobel Laureate Dr. Linus C. Pauling (1971) stimulated widespread interest in self-medication with ''megavitamin'' doses of ascorbic acid. There are reports in the literature concerning the beneficial effects of megadoses of vitamin C on the common cold, resistance, and various diseases. Pauling (1971) suggests that for optimum health a daily intake of ascorbic acid for an adult man should be 2.3 g, which could be increased to 9–10 g in presence of some ailments. There is great individual variation in vitamin C requirements; some people in excellent health need 250 mg/day, most people need 4–5 g, and for many 10 g is best. Ten grams is over 100 times the daily requirement suggested by the RDA (1980).

Intakes of vitamin C far in excess of physiological requirements have been reported to have beneficial effects that include the following (Pauling, 1971, 1974; Jaffe, 1984):

1. Prevention and reduction of the severity of the common cold.
2. Prevention of cancer and prolonged life of cancer patients.
3. Lowering of serum cholesterol and severity of atherosclerosis.
4. Increased wound repair and normal healing processes.
5. Increased immune response for prevention and treatment of infections.
6. Control of schizophrenia.
7. Inactivation of disease viruses.
8. Prevention of megaloblastic anemia of formula-fed premature infants.

The efficacy of pharmacological levels of the vitamin remains controversial because many of the claims have not been substantiated. Many controlled studies will be required to establish which claims for megadoses of vitamin C are valid. In view of the controversy, some authorities suggest not taking large amounts of vitamin C without medical counsel.

VIII. NATURAL SOURCES

The main sources of vitamin C are fruit and vegetables, but some foods of animal origin contain more than traces of the vitamin (Table 15.3). Vitamin C is present in relatively large amounts in fresh, canned, and frozen citrus fruits

TABLE 15.3

Vitamin C Concentrations in Various Foods (mg/100 g, As Fed Basis)[a]

Food	mg/100 g	Food	mg/100 g
Vegetables		Guavas	300
Asparagus, canned	15	Lemons	80
Beans, runner	5	Limes	250
Brussels sprouts	90	Melons (cantaloupe)	25
Cabbage, red	55	Olives	0
Carrots	2–6	Oranges	40–60
Cauliflower, raw	50–90	Peaches	7–14
Celery, raw	7	Pears	3
Corn	12	Pineapple	25
Oats, whole	0	Rose hips	1000
Onions, raw	10	Strawberries	40–90
Peas, frozen	13	Tangerines	30
Peppers, raw	100	Animal products	
Potatoes, new	18	Fish	5–30
Potatoes, old	5–16	Kidney, lamb	9
Radishes	25	Kidney, pig	11
Rice	0	Liver, calf	13
Rye, whole	0	Liver, pig	15
Spinach	10–60	Milk, cow	1–7
Wheat, whole	0	Milk, human	3–6
Fruit			
Apples, unpeeled	10–30		
Bananas	6–12		
Blackberries	20		
Cherries	5		
Cranberries	12		
Grapefruit	35–45		

[a]Adapted from Nobile and Woodhill (1981) and Jaffe (1984).

and in smaller but important amounts in other fruits, tomatoes, potatoes, and leafy vegetables (Nobile and Woodhill, 1981). It occurs in significant quantities in animal organs such as liver and kidney, but in only small quantities in meat. For untreated fruits and vegetables there is an extremely variable content among skin, pulp, and even between two leaves of the same vegetable or adjacent plants of the same variety. Other factors affecting vitamin C content include variety, maturity, fertility, and season (Snehalatha Reddy and Lakshmi Kumari, 1988). Postharvest storage values vary with time, temperature, damage, and enzyme content.

Fresh tea leaves, some berries, guava, and rose hips are accumulators of ascorbic acid and consequently are rich sources. For practical purposes, raw citrus fruits are good daily sources of ascorbic acid since appreciable amounts in other foods can be destroyed during processing. Ascorbic acid in foods is easily destroyed by oxidation and, therefore, undue exposure to oxygen, copper, and iron and prolonged cooking at high temperatures in the presence of oxygen should be avoided. Vitamin C is relatively stable to normal boiling but losses are substantial with greater oxidation of steaming or pressure cooking. There is only a small vitamin C loss for foods during freezing or dehydration.

IX. DEFICIENCY

A. Effects of Deficiency

Under practical feeding situations only humans, nonhuman primates, guinea pigs, and fish will develop vitamin C deficiency if diets are lacking in the vitamin. Farm livestock synthesize ascorbic acid from glucose in either the liver or kidney and vitamin C deficiency usually does not occur in such animals. In case of well-balanced nutrition their tissues receive endogenous ascorbate continuously and the level in blood and tissues can only with difficulty be affected by exogenous vitamin C. However, with nutritionally unbalanced diets, relative vitamin C deficiency may be induced in ascorbate-synthesizing animals as well (Ginter, 1970). Low blood ascorbic acid can be caused by various types of stress, including metabolic disorders, improper nutrition, insufficient vitamin A or β-carotene intake, and various infectious diseases. Under such conditions, exogenous vitamin C can have a positive effect in ascorbic acid-synthesizing species.

1. Ruminants

All known ruminants can synthesize ascorbic acid, however, clinical cases of scurvy in ruminants have been described. Death of cows and calves due to scurvy was characterized by changes in the oral cavity mucosa, muzzle, and skin accompanied by weight losses and general unthriftiness (Cole *et al.*, 1944; Duncan, 1944). In calves there was an extensive dermatosis accompanied by hair loss and thickening of skin for animals receiving insufficient milk. Blood ascorbic acid was low and the condition was successfully treated with vitamin C injections. Scurvy and lowered blood ascorbic acid content in weaned calves were reported by Martynjuk (1952). Studies of blood vitamin C concentrations in calves receiving the same diet have revealed great individual differences, with variations related to genetic background (Palludan and Wegger, 1984).

Positive effects of ascorbic acid supplementation on milk yield and milk quality were reported by Kuemyj (1955). Studies with bulls housed differently showed a lowering of vitamin C stores as a result of cold stress (Hidiroglou *et al.*, 1977). Ruminants can actually be considered more prone to a vitamin C deficiency because of impaired synthesis than monogastric animals since they cannot rely on exogenous supplies of this vitamin, which is rapidly destroyed by ruminal microflora (Cappa, 1958; Itze, 1984).

Hypovitaminosis C is most often observed in winter and spring and tends to reduce general resistance of the animals, causing infertility, high incidence of retained placenta, low viability of progeny, and several other pathological conditions that can result in economic losses, particularly with calves. Some data from the literature indicate that ascorbic acid-synthesizing capacity of calves during the first phases of life is insufficient to satisfy their requirements. Some researchers even suggest that calves do not synthesize vitamin C before 2–3 weeks after birth (Itze, 1984). Evidence suggests that serum vitamin C content is dependent on nutrition of the dam but also on age and health of the calves. Studies tend to show that low plasma ascorbic acid levels are likely in animals experiencing gastrointestinal disorders. Calves from herds characterized by poor health status generally also had reduced ascorbic acid status during the critical period from birth to 2 weeks of age. When calves were given doses of 1.25–2.5 g/day, infectious disease resistance appeared to increase and respiratory diseases were almost totally eliminated (Itze, 1984; Palludan and Wegger, 1984).

In conclusion, it is generally accepted that adult, healthy ruminants under normal dietary and environmental conditions are able to meet their vitamin C requirements by body synthesis. However, young ruminants are susceptible to a deficiency during the first few weeks of life, particularly when subjected to stress conditions, including cold, damp environments and disease and/or if limited in colostrum consumption.

2. Swine

Swine nutritionists have generally formulated swine diets without vitamin C because the young pig can synthesize ascorbic acid within a week of birth and both sow colostrum and milk provide a plentiful source of the vitamin to the nursing pig. Recently, though, swine researchers have indicated that under certain situations pigs may need supplemental vitamin C for maximum weight gain and feed use (Mahan *et al.*, 1966; Yen and Pond, 1981), however, there is nearly an equal number of reports that are negative (Brown *et al.*, 1970; Yen and Pond, 1984, 1987). Reasons for this inconsistency may be that environmental, physiological, and psychological stresses imposed on animals may increase requirements for ascorbic acid.

Brown (1984) indicated that level of available dietary energy is a major factor in deterimining the amount of ascorbic acid available to the pig. Serum ascorbic acid concentrations as well as urinary output are directly related to the level of energy in the diet. It was also found that a minor stress such as individual penning will evoke a positive growth response from supplementary ascorbic acid, especially in animals fed a ''low-energy'' diet. Dietary energy is able to cause a shift in ascorbic acid synthesis because of restrictions on amount of free glucose available for synthesis of ascorbic acid.

If a need for dietary vitamin C exists in swine, the newly weaned pig would seem to be the class of swine most likely to be deficient. Sow's milk contains a high concentration of vitamin C at parturition, but the level drops dramatically with time toward weaning. For the baby pig, the general consensus is that ascorbic acid blood level increases with colostrum intake, drops at weaning, and slowly increases after 7 weeks to the mature level (Wegger and Palludan, 1984). Handling practices at weaning (especially early weaning), which are generally considered to be stressful, including transport and mixing with unfamiliar animals, have been shown to deplete ascorbate. In view of decreased plasma vitamin C concentration and dramatic changes in nutritional, social, and other environmental factors associated with weaning, it was suggested that beneficial response from supplemental vitamin C with weanling pigs may be related to suppression of postweaning subclinical disease (Yen and Pond, 1981).

Spontaneous scurvy as a result of a genetic defect was observed in a swine production herd among 2- to 3-week-old piglets. Closer observation revealed that all pigs were from the same boar. Analysis of their blood and tissues revealed only a very small concentration of vitamin C. The 3 : 1 ratio between normal and affected pigs was characteristic of simple autosomal recessive inheritance in matings between nonaffected carriers. Liver microsomes were shown to be incapable of synthesizing ascorbic acid *in vitro* even with L-gulonolactone as substrate (Thodejensen and Basse, 1984).

3. Poultry

Like swine, poultry are able to synthesize vitamin C and thus it is assumed they do not require dietary sources of the vitamin. However, for newly hatched poultry there is a slow rate of ascorbate synthesis and this combined with encountered stress increases probability of vitamin C deficiency. The chick is subject to considerable stress conditions such as rapid growth, exposure to hot or cold temperatures, starvation, vaccination, and disease conditions such as coccidiosis. For both stressed mature and newly hatched poultry, several reports have documented a beneficial effect of supplementing the feed with ascorbic acid on growth rate, egg production, egg shell strength and thickness, and fertility and spermatozoan production, on counteracting unfavorable climate and housing conditions, and in case of intoxication or disease (McDonald *et al.*, 1981; Pardue, 1987). On the contrary, many researchers have found no beneficial effect of vitamin C supplementation under any conditions.

For heat-stressed chickens, supplemental vitamin C provided definite improvements in egg production, egg shell strength, and interior egg quality (El-Boushy and Van Albada, 1970). Peebles and Brake (1984) also reported that supplemental ascorbic acid holds promise for increased production during high environmental temperatures or for nutritionally marginal diets. When ascorbic acid was used at levels of 100 ppm or less for commercial layers, there were improvements in livability, egg production, and egg shell quality. Perek and Kendler (1963) carried out experiments in the Jordan Valley, where hens were subjected to hot temperatures, and reported an increase in egg production of 23 and 11.2% in two experiments with the birds given supplemental ascorbic acid. They also reported increased egg weights, decreased culls and mortality, and no shell quality differences. Other researchers were not able to confirm the positive effects of ascorbic acid supplementation. In a study supplementing 2600 ppm of ascorbic acid, egg production, egg shell thickness, egg weight, and mortality were not affected but interior quality was improved (Nockels, 1984).

Disease conditions have been found to affect vitamin C metabolism in poultry. When chicks were infected with fowl typhoid, their plasma vitamin C concentration was reduced (Hill and Garren, 1958). The vitamin C concentrations in plasma and tissue were also reduced in chicks infected with intestinal coccidiosis (Kechik and Sykes, 1979). Dietary ascorbate was shown to prevent this and contributed to intestinal repair.

4. Horses

Horses, like other farm species, synthesize vitamin C but stress situations such as bacterial and viral infection (i.e., influenza, rhinopneumonia) have been reported to lower vitamin C serum levels (Jaeschke, 1984). This is further associated with blood parameters indicating a delayed and/or disturbed collagen metabolism in young horses. Studies have shown that horses of all ages that suddenly show poor performance often have a reduced ascorbic acid serum level (Jaeschke, 1984). Performance of these horses improved after intravenous administration of ascorbic acid.

5. Fish

Intensively fed, caged channel catfish grew slowly and exhibited scoliosis and lordosis (Fig 15.2), broken-back syndrome, and elevated mortality from bacterial infection while receiving an ascorbic acid-deficient diet (Lovell, 1973). Other signs of ascorbic acid deficiency in channel catfish are internal and external hemorrhage, fin erosion, dark skin color, and reduced formation of bone collagen. Japanese eels fed a vitamin C-deficient diet had reduced growth after 10 weeks and hemorrhage in the head and fins after 14 weeks (NRC, 1983). Vitamin C-deficient shrimp exhibit black death syndrome, a condition characterized by melanized lesions in connective tissue under the exoskeleton as well as on the gills, abdomen, and gut (NRC, 1983). For cold-water fish, vitamin C deficiency results in anorexia, lethargy, structural deformities, scoliosis, lordosis, abnormal support cartilage of the eye, gill, and fins, internal hemorrhaging, and anemia (NRC, 1981a).

6. Other Animal Species

Because of body synthesis, the rat, mouse and hamster do not require dietary vitamin C, which classic studies on scurvy were carried out with the guinea pig (NRC, 1978c). Early signs of vitamin C deficiency in guinea pigs are reduced

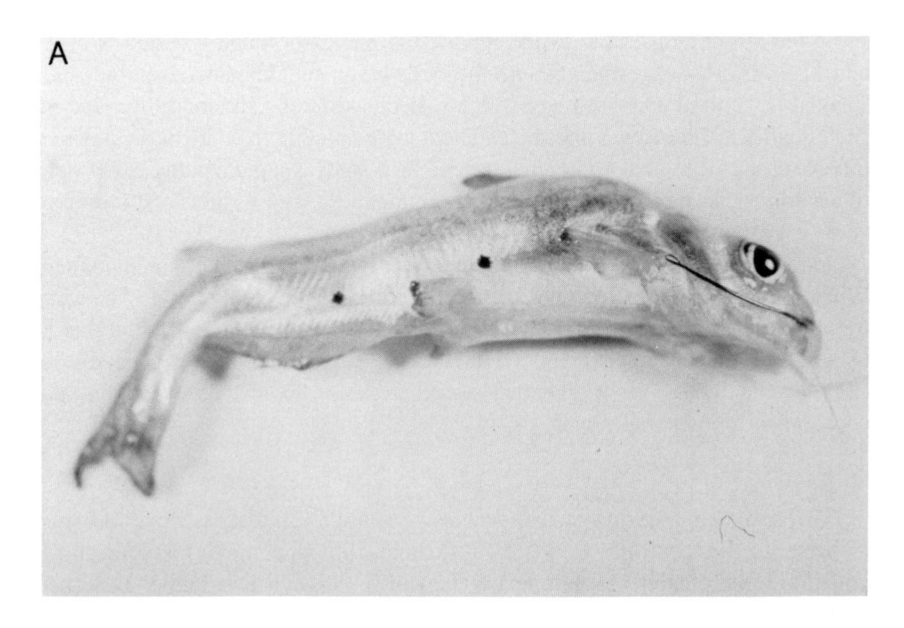

Fig. 15.2 Vitamin C deficiency in catfish. (A) shows fingerling channel catfish fed a vitamin C-free diet for 8 weeks. Note scoliosis and lordosis. (B) shows channel catfish from commercial cage culture where diet was devoid of vitamin C. One fish shows lateral curvature of the spine (scoliosis) and the other shows vertical curvature (lordosis) and a vertical depigmented band at the point of spinal injury, which is characteristic. (Courtesy of R. T. Lovell and Auburn University, Alabama.)

feed intake and weight loss, followed by anemia and widespread hemorrhages. Additional signs include enlarged costochondral junction, disturbed epiphyseal growth centers of long bones, bone demineralization, altered dentine, and gingivitis. On the basis of ascorbic acid supplementation trials, Helgebostad (1984) indicated that the fox and mink are able to synthesize sufficient vitamin C.

7. HUMANS

In humans, gross vitamin C deficiency results in scurvy, a disease characterized by multiple hemorrhages (Fig 15.3). In adults, manifest scurvy is often preceded by lassitude, fatigue, anorexia, muscular pain and greater susceptibility to infection and stress. Scurvy is characterized by anemia and alteration of protein metabolism, weakening of collagenous structures in bone, cartilage, teeth, and connective tissue, swollen, bleeding gums with loss of teeth, fatigue and lethargy, rheumatic pains in the legs, degeneration of muscles, and skin lesions. Bleeding gums gingivitis, and loosening of the teeth are usually the earliest objective signs (Marks, 1975; Vilter, 1978; Jaffe, 1984). Structural defects characteristic of scurvy include:

1. Bones and cartilage—the cartilage cells cease to form matrix.
2. Teeth—odontoblasts, predentine, dentine, and enamel are not formed; interrelated to bone and collagen formation.
3. Muscles—atrophy and necrosis with calcium deposits.
4. Connective tissues—collagen fibers are not formed by fibroblasts.
5. Capillaries and vascular system—fragility of the capillary walls.
6. Blood—hemorrhage causes anemia because red blood cells are rapidly destroyed.
7. Liver and other organs—liver atrophies and is infiltrated with fat; bile secretion is impaired; kidney atrophies; spleen enlarges.
8. Reproductive organs—degeneration of ovaries or germinal epithelium of testes, but may be nonspecific.
9. Other endocrines—thyroid shows hyperemia, hypersecretion, and irregularity of structure; adrenals are abnormal.

The disease may be fatal, particularly in infants and otherwise debilitated adults. Well-defined scurvy is not common in more developed countries, and occurs chiefly in infants fed diets deficient in ascorbic acid. Infantile scurvy, which is usually due to lack of vitamin C in artificial foods, generally occurs between the age of 6 and 18 months. As a rule it is first noticed that the infant cries on being handled, is irritable, and loses appetite and weight. Tenderness of extremities and pain on movement are almost invariably present (Marks, 1975). Manifestations of scurvy appear insidiously, usually after 5–6 months of severe deprivation of vitamin C. In adults, occurrence of scurvy is associated with poverty, alcoholism, famine, and nutritional ignorance. It is commonly found in neglected or alcoholic elderly males.

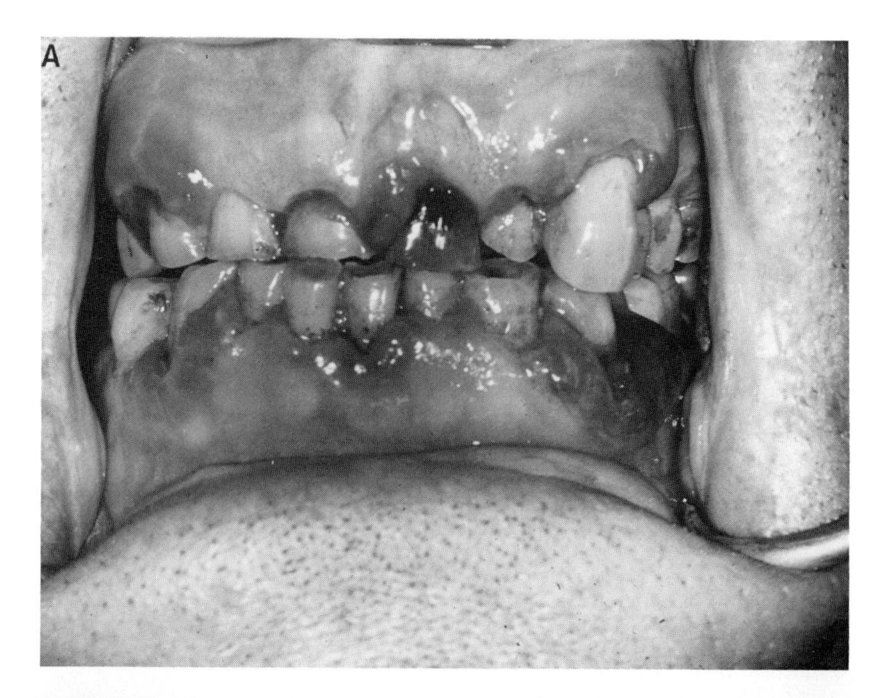

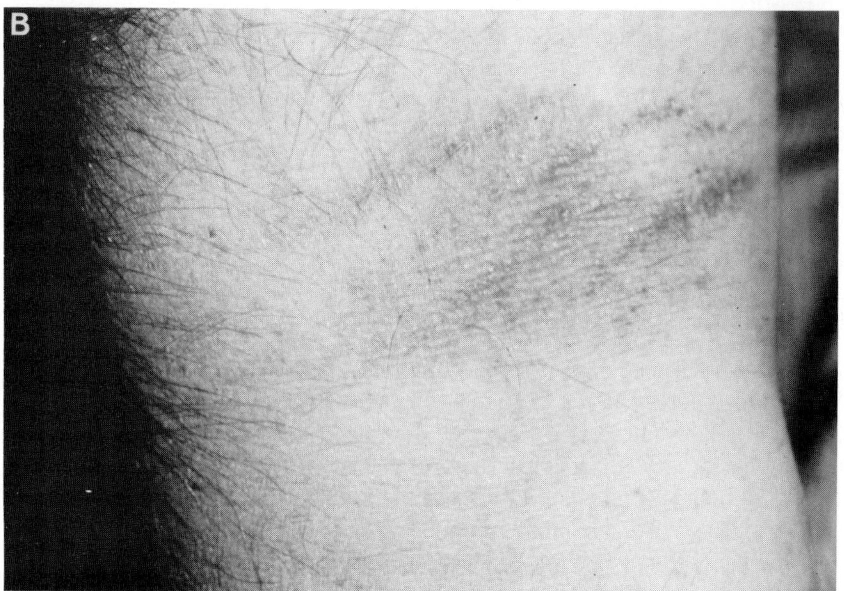

Fig. 15.3 Vitamin C deficiency in humans as represented by scurvy. Severe scurvy is illustrated by the swelling, bleeding, and receding gums (A). (B) represents capillary fragility with small hemorrhagic areas (ecchymoses). (Courtesy of Howerde E. Sauberlich, University of Alabama.)

B. Assessment of Status

Completely satisfactory and reliable procedures to assess vitamin C nutritional status have not been developed because of limited knowledge concerning the vitamin's metabolic functions (Jaffe, 1984). However, information concerning adequacy has been determined by an analysis of vitamin C concentrations is serum (plasma), leukocytes, whole blood, or urine. Precautions need to be taken to protect the vitamin in solution, and to select an assay that measures the vitamin itself and not other substances present (Jaffe, 1984).

Methods of biochemical detection of deficiency include:

1. Tissue content: 28 mg/100 g liver is the saturation level for humans and guinea pigs; lower concentrations will reveal reduced intake.
2. Leukocytes: ascorbate concentration in leukocytes with adequate diets is about 25 mg/100 ml; with less than 20 mg/100 ml in deficiencies.
3. Serum: the normal range of ascorbate (0.5–2.2 mg/100 ml) is often too variable to permit reliable estimation of deficiencies.
4. Urinary excretion: the load test and saturation test generally indicate immediate past intake and not overall nutritional status; nevertheless, amount of a given dose that is excreted is indicative of the tissue stores. Amount of ascorbate excreted during 3 hr following a 100-mg dose will be 50% for a normal, saturated person, 15% for a depleted person, and 5% for a scorbutic patient.
5. Other tests: (a) intradermal test involves rate of decoloration of 2,6-dichlorophenolindophenol injected intradermally, (b) decolorization of dye on the tongue, and (c) serum alkaline phosphatase, excretion of tyrosine metabolism products, or urinary creatine are not specific.

X. SUPPLEMENTATION

Supplementation with vitamin C would not be recommended for common livestock species (ruminants, poultry, swine, and horses) under normal management and feeding regimes. As previously mentioned, stress conditions do affect vitamin C synthesis and supplementation considerations must take this into account. Kolb (1984) summarized various types of stress that apparently increase demands while reducing animals' capability to synthesize vitamin C:

1. Dietary conditions: deficiencies of energy, protein, vitamin E, selenium, iron, etc.
2. Production or performance stress: high production or performance (i.e., rapid growth rates, high milk production, racehorse running).
3. Transportation, animal handling, and new environmental location stress: animals being driven or transported to market, animals placed in new surroundings

(i.e., weaned pigs from different litters placed together), and stressful management practices (castration, vaccination, etc.)

4. Temperature: high ambient temperature or cold trauma.

5. Diseases and parasites: fever and infection reduce blood ascorbic acid, while parasites, particularly of the liver, disturb ascorbic acid synthesis and increase requirements for the vitamin.

During the first weeks of life the calf's requirement for ascorbic acid must be covered by colostrum and milk concentrations and from the inborn storages of the vitamin. One calf study in which reared calves were denied colostrum found that all but one of the experimental animals died of umbilical infections and peritonitis (Palludan and Wegger, 1984). That survivor had a high content of ascorbic acid in its blood at birth. Further investigations showed positive results from supplementation of calves with ascorbic acid.

It is important to note that reserves of ascorbic acid are high at birth but decline rapidly afterward unless exogenous supply is furnished until synthesis can handle the load (Palludan and Wegger, 1984; Itzeova, 1984). Plasma ascorbic acid level in calves fed fresh colostrum twice a day and frozen colostrum of the first milking once a day was high, with indications that the decrease in ascorbic acid content usually seen after birth can be avoided by this practice (Itzeova, 1984). Itze (1984) recommended supplementation with vitamin C to calves reared on milk diets. The initial days of calf rearing are critical, for the calf must adapt itself to a new environment, feeding practice, and housing at an age when its resistance is minimal. Of various programs tested, the author recommends daily oral supplementation with 2.5 g of ascorbic acid in combination with parenteral application of 500 mg ascorbic acid in two doses immediately after moving animals into their new rearing facilities. Lehocky (1981), in his work with dairy calf supplementation, found that only 50% of a calf's requirement for vitamin C is covered when feeding various milk replacers.

Various studies have demonstrated beneficial effects of low doses of 50 to 100 mg/kg ascorbic acid to diets of broilers or laying hens exposed to heat stress (Kolb, 1984). Njoku (1984) reported that 200 mg ascorbic acid/kg fed to broilers helped alleviate heat stress. Egg shell thickness increased for hens (El-Boushy *et al.*, 1968) while livability, weight gain, and immune response improved in broilers (Pardue and Thaxton, 1982) when heat-stressed birds received supplemental vitamin C.

The literature concerning efficacy of supplementation of swine diets with ascorbic acid is conflicting. Young pigs seem to be more likely to respond to supplementation than do adults, much the same as was reported for ruminants. Perhaps the inconsistency of results is due to uncontrolled stress or genetic differences (Brown, 1984).

Early weaning (0–3 weeks) has been shown to decrease ascorbic acid levels in liver and tests seem to indicate that maximal synthesizing capacity is not developed until about 8 weeks, thus indicating a possible advantage with supplementing milk replacer products with vitamin C (Wegger and Palludan, 1984). Sandholm *et al.* (1979) reported that umbilical hemorrhages occurring in piglets immediately after birth can be prevented by supplementing the sows' feed with 1 g of ascorbic acid per day during the last week of gestation (Fig 15.4).

The intensive selection that has taken place for several decades in the swine industry may have altered the enzymatic constitution of animals so that ability to synthesize vitamin C has changed. Furthermore, modern intensive production

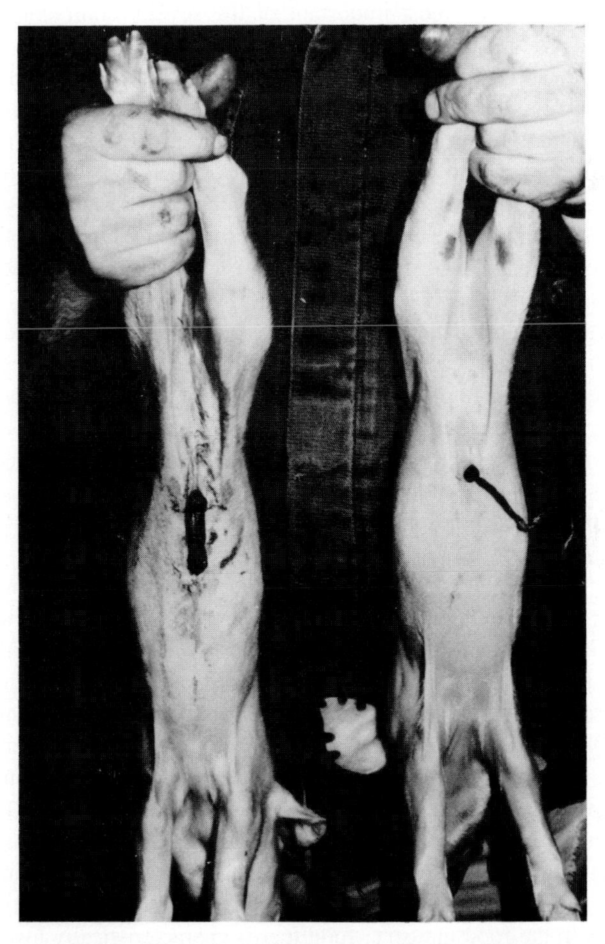

Fig. 15.4 Navel bleeding syndrome. Umbilical cords of a bleeding piglet (left) and of a normal piglet (right), age 10 hr. Prevention of navel bleeding has resulted from preparturient administration of ascorbic acid. (Courtesy of Markus Sandholm, College of Veterinary Medicine, Helsinki, Finland.)

systems and continuous demand for higher productivity may have increased requirement of swine for ascorbic acid. Feeding practice in pig production has also changed, the tendency being to use more processed feedstuffs that practically speaking contain no ascorbic acid.

Therapeutic, as distinguished from nutritional, use of ascorbic acid has been useful in treating infectious diseases of the horse and dog (Kolb, 1984). The importance of ascorbic acid for defense against infections and for phagocytosis has been stressed. Beneficial effects of supplemental vitamin C are reported for puppies with disorders in bone development accompanied by painful swelling of joints when the vitamin is deficient (Kolb, 1984).

L-Ascorbic acid is the most important of the several compounds that have vitamin C activity. Ascorbic acid is commercially available as 100% crystalline, 50% fat-coated, and 97.5% ethylcellulose-coated products and their dilutions. The more soluble sodium salt of ascorbic acid (sodium ascorbate) is also commercially available. When providing supplemental ascorbic acid it is advisable to use a stabilized form, and coating of ascorbic acid crystals with ethylcellulose is a suitable stabilization method. Size of enrobed particles should be 250–700 μm (Kolb, 1984). In storage experiments, ascorbic acid protected in this manner was found to be four times more stable than untreated ascorbic acid crystals (Kolb, 1984).

Adams (1982) reported that coated (ethylcellulose) ascorbic acid showed a higher retention after processing than the crystalline form, 84 versus 48%. Retention of ascorbic acid in mash feed was fairly good, but with time and elevated storage time and temperature stability was poor in crumbled feeds. Although retention of vitamin C activity in feed containing the ethylcellulose-coated product was low, it was 19–32% better than that of the crystalline form. A similar advantage for the coated ascorbic acid product has been shown in extruded feed prepared for catfish.

For humans in developed countries, large quantities of supplemental vitamin C are taken as part of multivitamin pills and separately, sometimes in megavitamin doses. Vitamin C deficiency is likely for individuals who do not consume optimum quantities of fruits (particularly citrus) and vegetables. For certain human foods, enrichment or fortification with vitamin C is practiced, and vitamin C is also used in the food industry for its antioxidant activity. The antioxidant, that is, the reducing property of ascorbic acid, is exploited for preservation of flavor, color, and appearance of certain foods (Nobile and Woodhill, 1981). It is used in commercial preparation of beer, fruit juices, and canned and frozen vegetables and fruits, in meat curing, and in the flour industry to enhance baking qualities and appearance of bread (Jaffe, 1984). As a result of food fortification and antioxidant uses in the food industry, food that is characteristically low in vitamin C will have higher levels than expected, provided processing methods have not destroyed the added vitamin.

XI. TOXICITY

In general, high intakes of vitamin C are considered to be of low toxicity. Safety and tolerance of ascorbic acid in humans at levels as high as 10 g/day have been demonstrated (Koerner and Weber, 1972). Reported toxic effects include possible acidosis, gastrointestinal complaints, glycosuria or sensitivity reactions, mutagenic activity, and adverse effects concerning the metabolism of some minerals and vitamin B_{12} body stores. Barness (1977) reports that toxic effects of megadoses of vitamin C are insignificant, or rare, or troublesome but of little consequence. Effects demonstrated by *in vitro* experiments have often later been found to be nonexistent once definitive *in vivo* studies were conducted. Danger for vitamin C toxicity is minimized and unlikely in humans because of limited intestinal absorption capacity and efficient renal elimination. Nevertheless, prolonged megadose intakes of vitamin C should be avoided as adverse effects may result, particularly for patients with inborn errors of metabolism (i.e., cytinuria, oxalosis, and hyperuricemia).

Chronic toxicity studies generally indicate that ascorbic acid is well tolerated in animals (Jaffe, 1984). Oral ascorbic acid may be administered to most laboratory animals at doses of several grams per kilogram of body weight without appearance of any obvious general effect on health. Rabbits showed only transient subconjunctival hemorrhages without other manifestations after 4 months of daily parenteral injections of 200 mg/kg body weight. Guinea pigs tolerated daily doses of 8.9 g/kg body weight, equivalent to 1800 times the normal requirement of 4–5 mg/kg body weight per day. However, Helgebostad (1984) reports that high doses of 100–200 mg/kg body weight daily were harmful to mink. Male guinea pigs fed 8.7% ascorbic acid for 6 weeks had decreased bone density and decreased urinary hydroxyproline compared to controls (Bray and Briggs, 1984).

16

Vitamin-like Substances

I. INTRODUCTION

In addition to the 14 vitamins discussed in previous chapters, there are other substances that have been classed with the vitamins although their true vitamin character has not been established. For various reasons, the term vitamin has been applied to many substances that do not meet criteria for vitamin status. Vitamins are regarded as essential organic micronutrients that must be supplied in the diet. What is a vitamin for some species may only be an essential metabolite for others as dietary sources are not needed because of tissue synthesis (e.g., vitamin C). *myo*-Inositol and carnitine fit this category, but apparently for only a few species. Carnitine would be a vitamin for certain insects, while vitamin status for *myo*-inositol would apparently be warranted for fish, gerbils, and perhaps other species.

Many compounds described as vitamins in the scientific literature of the 1930s and 1940s have since proven to be identical to other essential nutrients or to be mixtures of various compounds. Many of the substances referred to as "vitamins," "growth factors," or "accessory factors" in early research literature are no longer considered to be vitamins (Cody, 1984). Some compounds, such as *myo*-inositol, carnitine, lipoic acid, coenzyme Q, and polyphenols, exhibit biological activity without being dietary essentials (or vitamins) for most species. Another group of substances are called vitamins by potential promoters for profit; these substances, including pangamate, laetrile, gerovital, and orotic acid, are not dietary essentials and are more properly called pseudovitamins (Cody, 1984). The present chapter will emphasize *myo*-inositol and carnitine since these two substances are vitamins to some species, with only brief mention allotted to other vitamin-like substances.

II. *myo*-INOSITOL (INOSITOL)

A. Introduction

myo-Inositol is a water-soluble growth factor for which no coenzyme function is known. It was first isolated from muscle in 1850 and was identified as a growth factor for yeast and molds, though not for bacteria. A *myo*-inositol deficiency in mice, characterized by inadequate growth, alopecia, and death, was reported. However, these results were challenged as the diets were ill defined and were apparently deficient in some of the B-complex vitamins (Kukis and Mookerjea, 1978). Most evidence would suggest that *myo*-inositol is not a true vitamin for most species. Nevertheless, signs of inositol deficiency have been demonstrated in fish and gerbils (Kroes, 1978). Difficulty of demonstrating a deficiency is related to endogenous synthesis, highly variable turnover rates, and interactions of *myo*-inositol with certain vitamins or other nutrients.

B. Chemical Structure and Properties

Inositol exists in nine forms. It is a cyclohexane compound, with only *myo*-inositol (Fig. 16.1) demonstrating any biological activity. *myo*-Inositol is an alcohol, similar to a hexose sugar. It is a white, crystalline, water-soluble compound having a sweet taste, and is stable to acids and alkalines and heat up to about 250°C. Because of hydroxyl groups it forms various ester, ethers, and acetals. The hexaphosphoric acid ester (combined with six phosphate molecules) of *myo*-inositol is phytic acid, a compound that complexes with phosphorus and other minerals making them less available for absorption (Fig. 16.1). Methods of analysis include the traditional microbiological method, which is being replaced by more rapid methods of gas–liquid, paper, and thin-layer chromatography.

C. Metabolism and Functions

Ingested *myo*-inositol is absorbed by rats and humans at a rate of over 99%, with *myo*-inositol from phytate absorbed at less than 50% in the presence of high dietary calcium (Cody, 1984). *myo*-Inositol is absorbed by active transport from dietary sources or it may be synthesized *de novo* from glucose. From animal studies, *myo*-inositol may also be converted to glucose. *myo*-Inositol appears to

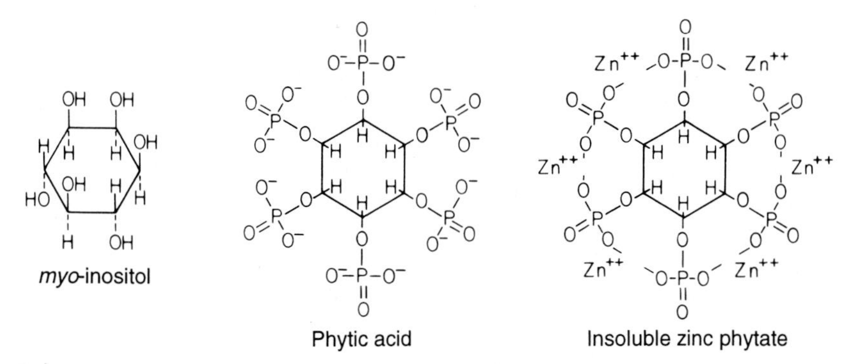

Fig. 16.1 Chemical structures of *myo*-inositol, phytic acid, and phytic acid combined with zinc.

have three metabolic fates: oxidation to CO_2, use in gluconeogenesis, and synthesis of phospholipids.

The function of *myo*-inositol is not completely understood, although its biochemical functions probably relate to its roles as a phospholipid component of membranes and lipoproteins. Assuming that *myo*-inositol is bound as lipositol, it makes up over 25% of total lipid and almost 50% of phospholipid in microsomes of the liver cell. *myo*- Inositol is a structural component of phosphoinositides with many of the possible functions associated with its role as a membrane component. In addition to the general role of maintaining selective permeability of plasma membranes, phosphatidylinositol and its highly charged phosphorylated forms are hypothesized to regulate cell-surface phenomena, such as binding of hormones and transfer of nervous impulses (Michell, 1975). Inositol 1,4,5-triphosphate has been shown to release intracellular Ca^{2+} from the endoplasmic reticulum, providing a link between receptor activation and cellular Ca^{2+} mobilization (Putney *et al.*, 1986). Under certain conditions *myo*-inositol is lipotropic, with the lipotropic activity usually synergistic with that of choline. *myo*-Inositol reduces liver lipids for diets low in protein and fat while choline is lipotropic also for diets containing lipids.

D. Sources, Requirements, and Deficiency

All plants and animals contain measurable amounts of *myo*-inositol (Clements and Darnell, 1980; Cody, 1984). The most concentrated dietary sources of *myo*-inositol are foods that consist of seeds, such as beans, grains, and nuts; cantaloupe and citrus fruits are also good plant sources. The majority of *myo*-inositol from plant sources occurs bound to phosphate as phytate. The best animal sources are the organ meats, which contain *myo*-inositol in free form or as a component of phospholipids. High concentrations are found in heart, kidney, liver, spleen, and thyroid.

There are no known dietary requirements for *myo*-inositol in humans and most species of animals studied. This is most likely due to the availability of *myo*-inositol from dietary sources, from endogenous synthesis, and from bacterial synthesis. Early work suggested that *myo*-inositol was an essential nutrient for several animal species. Effects of *myo*-inositol deficiency included alopecia in rats and mice, fatty liver and "spectacle eye" in rats, and retarded growth in guinea pigs, hamsters, and chickens. Subsequent work using more complete and defined diets with these species failed to show any *myo*-inositol requirement. Early reports were in fact deficiencies of biotin, choline, vitamin E, or other nutrients.

For ruminants, synthesis of *myo*-inositol by microorganisms in the digestive tract, in addition to dietary sources, is presumably sufficient to meet the animals' needs. Since *myo*-inositol deficiency has been associated with a failure of lipid transport in a number of species, it was suggested that *myo*-inositol supplementation might be beneficial in alleviating fatty liver syndrome (hepatic lipidoses) in dairy cattle. However, *myo*-inositol supplementation (17 g daily) to dairy cows was ineffective in altering liver *myo*-inositol concentrations or in improving liver lipoprotein output (Gerloff *et al.*, 1986).

From a review of the *myo*-inositol literature, Kroes (1978) reported that there is little or no conclusive evidence that *myo*-inositol is required by cats, poultry, ruminants, dogs, guinea pigs, hamsters, mice, pigs, and rats. For rats and other species, however, reports show relative requirements for *myo*-inositol when diets are deficient in single other nutrients such as choline or in multiple vitamin and mineral deficiencies (Yagi *et al.*, 1965). The deficiency signs were lipid accumulation for a simultaneous choline deficiency and depressed growth if vitamin and mineral imbalance exists.

Under typical feeding conditions, most evidence indicates that pigs and poultry have no requirement for dietary *myo*-inositol. Lindley and Cunha (1946) demonstrated that addition of 100 mg *myo*-inositol per 100 g of a diet deficient in biotin and supplemented with sulfathalidine could partially relieve signs of biotin deficiency. Presumably, the added *myo*-inositol stimulated intestinal synthesis of biotin. In some studies *myo*-inositol has promoted growth in chicks and has decreased liver fat and increased egg production. These results have been refuted by several other investigators (Kroes, 1978).

Animal species that have shown a dietary need for *myo*-inositol are fish and gerbils. For fish, a deficiency of *myo*-inositol results in anorexia, fin degeneration, edema, anemia, reduced growth, and inefficient feed conversion, as well as a decreased rate of gastric emptying and activity of cholinesterase and certain transaminases (Halver, 1982). A tentative recommended allowance for young salmon and trout is 350–500 mg per kilogram feed (McLaren *et al.*, 1947). *myo*-Inositol-deficient female gerbils were characterized with intestinal lipodystrophy with a resulting hypocholesterolemia, debilitation, and eventual death (Hegsted *et al.*, 1973). Male gerbils were apparently protected by testicular synthesis of *myo*-inositol. Requirement of *myo*-inositol for female gerbils varied from 20 mg/kg of diet when the diet contained predominantly unsaturated fats (20% safflower oil) to 120 mg/kg of diet with saturated fats (20% coconut oil) (Kroes *et al.*, 1973.

A number of factors affect *myo*-inositol status and it is apparent that under certain conditions a need for *myo*-inositol for various species can be shown. The reason for the need under those conditions is not exactly known, however, administration of antibiotics, dietary stress, and physiological stress may influence the need for *myo*-inositol (Cunha, 1971; Cody, 1984). Antibiotics kill *myo*-inositol-producing intestinal flora, thus reducing the exogenous supply of *myo*-inositol to the body. Increasing levels of dietary saturated fatty acids may stress *myo*-inositol-requiring lipid transport systems. For humans, impaired *myo*-inositol metabolism occurs in diabetics, uremics, and premature infants (Cody, 1984).

E. Detrimental Effects of *myo*-Inositol as Phytates

myo-Inositol is of greatest significance to livestock and human nutrition because of its hexaphosphate ester phytic acid (Fig. 16.1). Phytic acid, which is formed from six phosphate molecules combined with *myo*-inositol, hinders intestinal absorption of phosphorus, calcium, and other minerals including iron, manganese, and zinc. In grains and plant protein supplements, about two-thirds of the phosphorus is in the less available phytate form. Utilization of phytate phosphorus is influenced by phytase present in plant materials or synthesized by rumen microflora, and by intake of vitamin D, calcium, and zinc, as well as by such factors as alimentary tract pH and dietary ratio of calcium to phosphorus (NRC, 1979). Calcium exaggerates the inhibition of zinc absorption by phytate while vitamin D needs are higher to counteract phytin when proteins containing high phytates are fed.

Phytin is especially high in bran of cereal grains and oilseed meals. About 20–50% of phytin phosphorus is available to the pig. A good guide is to assume that no more than about 50% of the phosphorus in plant feeds is available to the pig (Cunha, 1977). There is disagreement concerning the ability of poultry to utilize phytin phosphorus (NRC, 1984b). Most data, however, indicate that the utilization of phytin phosphorus, by young or adult poultry, is negligible if dietary calcium concentrations are sufficient to meet the birds' requirements. Some reports suggest, however, that the older bird has ability to use most of the phytin phosphorus. Many cereal grains contain the enzyme phytase, which is capable of splitting phosphorus from *myo*-inositol and leaving the phosphorus, calcium, and other minerals attached to it available for absorption. Rye in particular and also wheat contain enough phytase to lead to considerable destruction of phytic acid. Thus, although 50% or more of the phosphorus in the original whole grain may be in the form of phytic acid, the amount in the final product may be very much less. At the other extreme oats contain little phytase. Phytin phosphorus maybe almost totally unavailable to the pig unless the phytase of grains or other sources is present in the diet. Ruminants utilize phytin phosphorus quite satisfactorily because of consumption of dietary phytases and phytase production by rumen microorganisms.

Many human nutritionists feel that the two minerals most likely to be deficient in typical diets are calcium and iron, both of which are affected by plant phytates. In human nutrition, the amount of calcium absorbed is dependent on bread type. Calcium is much less freely absorbed from diets consisting largely of brown bread than from those consisting largely of white, because whole wheat flour contains much higher levels of phytate. Inhibitory effects of soybean-derived food products and cereal grains on iron absorption as a result of phytates have caused concern among nutritionists and food scientists (Morris, 1983).

The high use of soy protein or soybean meal in human and monogastric diets has led to zinc deficiencies. During extraction of soy protein, phytic acid forms a complex with zinc to form zinc phytate (Fig. 16.1), which is insoluble in the intestinal tract. The zinc requirement of growing pigs receiving semipurified diets containing isolated soybean protein or natural corn–soybean meal diets containing the recommended calcium level is about 50 ppm. However, in the absence of plant phytates, pigs receiving a casein-glucose diet require only 15 ppm zinc (NRC, 1988). Higher levels of calcium are known to further exaggerate the inhibition of zinc absorption by phytate, resulting in the formation of zinc–calcium–phytate complexes (Forbes *et al.*, 1983).

III. CARNITINE

A. Introduction

In 1948, Fraenkel's research on dietary requirements of the mealworm (*Tenebrio molitor*) led to recognition of a new B vitamin, B_T, which in 1932 was identified as carnitine (Friedman and Fraenkel, 1972). Carnitine is of universal

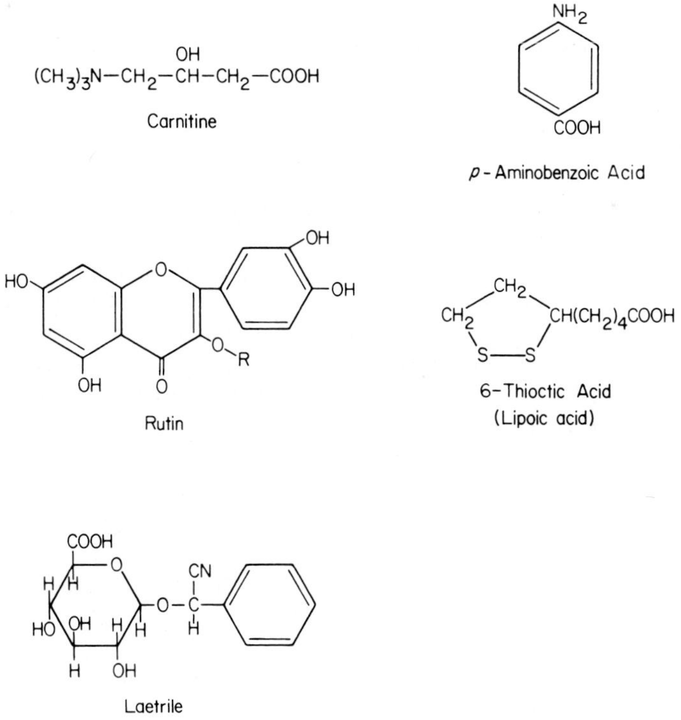

$(CH_3)_3N-CH_2-CH(OH)-CH_2-COOH$

Carnitine

p-Aminobenzoic Acid

Rutin

6-Thioctic Acid (Lipoic acid)

Laetrile

Fig. 16.2 Chemical structures of carnitine, *p*-aminobenzoic acid, rutin, lipoic acid, and laetrile.

occurrence in biological systems and is required as a vitamin for lower forms of life. Clinical studies suggest that endogenous synthesis of carnitine may be inadequate under certain conditions in humans (Borum, 1981).

B. Chemical Structure

Carnitine is a quaternary amine, β-hydroxy-γ-trimethylaminobutyrate (Fig. 16.2). It is a very hygroscopic compound easily soluble in water and has a molecular weight of 161.2. Methods of analysis first utilized the bioassay technique using *T. molitor*. Other methods developed for carnitine determination include chemical, enzymatic, gas chromatography, and radioisotopic procedures (Chan, 1984).

C. Metabolism and Functions

Food carnitine is assumed to be completely absorbed, although neither the mechanism nor site of absorption is known (Mitchell, 1978). Carnitine is not carried in blood in any tightly bound forms, in contrast to many water-soluble vitamins. Cantrell and Borum (1982) reported that carnitine uptake by the heart is facilitated by a cardiac carnitine-binding protein. Free carnitine is excreted in urine, with the principal excretory product being trimethylamine oxide (Mitchell, 1978).

Carnitine is required for transport of long-chain fatty acids into the matrix compartment of mitochondria from cytoplasm for subsequent oxidation by the fatty acid oxidase complex for energy production. Carnitine acyltransferase is the enzyme responsible for this shuttle mechanism. It exists in two forms, carnitine acyltransferase I and carnitine acyltransferase II. After the long-chain fatty acid is activated to acyl-CoA, it is converted to acylcarnitine by the enzyme carnitine acyltransferase I and crosses to the matrix side of the inner mitochondrial membrane. Carnitine acyltransferase II then releases carnitine and the acyl-CoA into the mitochondrial matrix. Acyl-CoA is then catabolized via B-oxidation (Mitchell, 1978; Chan, 1984). Thus, utilization of long-chain fatty acids as a fuel source depends on adequate concentrations of carnitine. Carnitine also has functions in other physiological processes critical to survival such as lipolysis, thermogenesis, ketogenesis, and possibly regulation of certain aspects of nitrogen metabolism (Borum, 1985).

Healthy children and adults can synthesize up to 20 mg of endogenous carnitine daily (Anonymous, 1985a). In mammals, γ-butyrobetaine (YBB), the immediate precursor of carnitine, can be synthesized from the essential amino acids lysine and methionine in most tissues. The ultimate conversion of YBB to carnitine occurs in the liver (Olson and Rebouche, 1987).

D. Sources, Requirements, and Deficiency

In general, foods of plant origin are low in carnitine, whereas animal-derived foods are rich in carnitine (Mitchell, 1978). Carnitine is located principally in skeletal muscle, which has about 40 times the concentration of carnitine in blood. Average muscle content of total carnitine ranges from 10.75 to 19.06 nmol/mg noncollagen protein. Most plant foods that are low in carnitine are also likely to be low in lysine and methionine, the precursors of carnitine. Therefore, a pure vegetarian diet may lack both preformed carnitine and its precursors.

Carnitine is an essential growth factor for some insects, such as the mealworm (*T. molitor*). However, most insects and higher animals, as well as mammals, can synthesize carnitine. Therefore, carnitine deficiency is unlikely, and nutritional requirements for carnitine have not been established. Normal adult humans both synthesize and ingest carnitine in amounts totaling 100 mg per day. Under a variety of circumstances, however, carnitine deficiency may become manifest. Individuals particularly vulnerable are newborn infants, premature infants, persons with inborn metabolic defects, those receiving peritoneal or hemodialysis, alcoholics with liver disease, malnourished infants, and persons receiving total parenteral nutrition (Anonymous, 1985; Rebouche, 1986). Borum (1981) presented evidence that newborns have a critical need for carnitine since they have not attained the full biosynthetic capacity for carnitine and their plasma and tissue concentrations are low. Many infant formulas contain low concentrations of carnitine, with supplemental carnitine recommended to be added to these formulas (Sugiyama *et al.,* 1984).

In carnitine deficiency, fatty acid oxidation is reduced, and fatty acids are diverted into triglyceride synthesis, particularly in the liver. Mitochondrial failure develops in carnitine deficiency when there is insufficient tissue carnitine available to buffer toxic acyl–coenzyme (CoA) metabolites. Toxic amounts of acyl-CoA impair the citrate cycle, gluconeogenesis, the urea cycle, and fatty acid oxidation. Carnitine replacement treatment is safe and induces excretion of toxic acyl groups in the urine (Stumpf *et al.,* 1985).

IV. *p*-AMINOBENZOIC ACID (PABA)

p-Aminobenzoic acid (Fig. 16.2) was originally identified as a growth factor for many species of bacteria and as a required nutrient for lactation in rats and growth for the chick. The only role of PABA in higher animals would appear to be as part of the folacin molecule (see Chapter 12). When sufficient dietary folacin is available, there is little evidence that PABA plays a direct role in the nutrition of higher animals or humans and therefore it cannot be classified as a vitamin. If folacin is lacking, PABA may have its main effect by providing a building block for intestinal synthesis of the vitamin. One of the better recognized properties of PABA is its ability to counteract the bacteriostatic effects of sulfonamides. The chemical structure of PABA is very similar to that of some sulfonamides, which explains why it can counteract inhibition of microbial growth by these drugs.

V. POLYPHENOLS (FLAVONOIDS)

In 1936, Rusznyak and Szent-Györgyi reported that there is a substance in citrus fruits, different from vitamin C, that is essential to prevent fragility of capillaries. The substance was designated as vitamin P. Several reports have shown that catechol, rutin (Fig. 16.2), hesperidin, chalcone, and other nonspecific polyphenols, or flavonoids, which are widely distributed in fruits and vegetables, can provide some protection against capillary fragility under certain conditions. Some of these compounds may have value as a supplement to limited vitamin C intake, particularly under conditions of stress (Maynard *et al.*, 1979).

Flavonoids are colored phenolic substances found in all higher plants; over 800 different flavonoids have been isolated (Cody, 1984). They are the major sources of red, blue, and yellow pigments, except for carotenoids, in the plant kingdom. Although a number of polyphenols exhibit biological activities, including reduction of capillary fragility and protection of biologically important compounds through antioxidant activity, none of the polyphenols has been demonstrated to be essential or to be capable of causing deficiency signs when removed from the diet.

VI. LIPOIC ACID (THIOCTIC ACID)

Lipoic acid (Fig. 16.2) plays an important role in the growth of certain microorganisms. It also is essential in oxidative decarboxylations of α-keto acids, such as pyruvic, in carbohydrate metabolism. There is, however, no clear evidence for an established need in animal nutrition that enables it to be classed as a vitamin, despite several experiments with rats and chicks (Maynard *et al.*, 1979).

VII. COENZYME Q

Coenzyme Q is a collective name for a number of ubiquinones, such as Q_4 and Q_{10}, that play an established role in the respiratory chain in mitochondrial systems. The importance of coenzyme Q as a ubiquitous catalyst for respiration assures its status as an essential metabolite. There is evidence that specific ubiquinones have a sparing effect on vitamin E, resulting in remission of some clinical signs of the vitamin deficiency. Dietary ubiquinone seems on the whole to be unimportant unless it provides the aromatic nucleus for endogenous synthesis. There is no proof that justifies coenzyme Q being classed as a separate vitamin.

VIII. "VITAMIN B_{13}" (OROTIC ACID)

Vitamin B_{13} was isolated from distiller's solubles, with the purified compound orotic acid, which is an intermediate in pyrimidine metabolism. It has been found to stimulate the growth of rats, chicks, and pigs under certain conditions but evidence remains uncertain whether it plays an essential role in an otherwise adequate diet.

IX. "VITAMIN B_{15}" (PANGAMIC ACID)

Vitamin B_{15} is found in rice bran, yeast, blood, and other feeds. It is not a chemically defined substance and there is no evidence that pangamic acid preparations have vitamin activity or offer therapeutic benefit.

X. "VITAMIN B_{17}" (LAETRILE)

Laetrile (amygdalin) is a β-cyanogenic glucoside occurring naturally in the kernels or seeds of most fruits (i.e., apricot). Although many unsupported claims have been made for the therapeutic benefit of laetrile treatment, most publicized claims are for its use in treating cancer. In these claims the two major lines of argument advanced are that the cyanide in laetrile acts specifically to destroy cancer cells and that cancer is a nutritional deficiency disease requiring laetrile treatment for dietary control (Cody, 1984). Extensive animal and clinical trials have not supported these claims, and most nutritionists do not consider laetrile a vitamin.

XI. "VITAMIN H_3" (GEROVITAL)

Vitamin H_3 is a buffered solution of procaine hydrochloride, better known as Novocain, which is used as a pain killer by dentists. It is promoted as a nutritional substance that alleviates symptoms of diseases associated with aging. These claims have not been supported by scientific studies, and it is therefore not recognized as a vitamin.

XII. "VITAMIN U" (CABAGIN)

Vitamin U is claimed to be an antiulcer factor and occurs naturally in cabbage and other green vegetables. The actual active substance is a methylsulfonium salt of methionine. The claims as an antiulcer factor have not been supported by some studies.

XIII. GLUCOSE TOLERANCE FACTOR

Glucose tolerance factor contains the element chromium, which has been reported to be involved in maintaining normal serum cholesterol and in regulating glucose metabolism (Williams and McDowell, 1985). The glucose tolerance factor qualifies as a vitamin since it contains chromium, organic components of nicotinic acid, glycine, glutamic acid, and cysteine and has a much greater biological activity than inorganic sources of chromium alone. This would be comparable to vitamin B_{12} being more metabolically effective than the element cobalt.

XIV. OTHER VITAMIN-LIKE FACTORS

In 1972 Cheldelin and Baich listed "unidentified growth factors" (UGF) and the organisms utilizing them. A total of 255 references were noted. In particular, many references involved unidentified factors present in fish products, alfalfa meal, liver, and whey. Thus, we have such terms as "whey factor," "fish solubles factor," "grass juice factor," and many others.

With the discovery of folacin in the early 1940s, many believed that all of the unidentified factors had been discovered and that there were no more factors needed for optimum nutrition. One reason for this supposition was the fact that completely synthetic diets of known nutrient composition, containing all recognized vitamins and mineral nutrients and adequate in the essential amino acids, would support growth and development in young weanling rats (Scott *et al.*, 1982). However, in 1948 the discovery of vitamin B_{12} as the unknown activity termed the "animal protein factor" responsible for special growth-promoting effects was an excellent example of the fallacy of assuming that no more factors exist simply because animals can survive on synthetic diets.

With the vitamin B_{12} discovery in 1948, the period of active identification and isolation of the major vitamins appeared to be ending. However, even since 1948 many field reports have suggested that practical diets containing sources of UGF are superior to purified or commercial diets. Typically the unidentified factors found in certain feeding ingredients have not been isolated and identified. These factors could be providing vitamins, trace minerals, or a better amino acid balance or counteracting antagonists in the regular diet. For poultry, many of the growth responses obtained from UGF involved relationships between known nutrients, such as natural chelates in corn distiller's dried solubles that improved zinc utilization in a purified diet containing soybean (Scott *et al.*, 1982).

Fish solubles, dried whey, brewer's dried yeast, corn distiller's dried solubles, and other fermentation residues are the major special ingredients often added to poultry diets as potential sources of unrecognized nutritional factors. In swine studies, Cunha (1977) reported that high-quality alfalfa meal and pasture, animal protein concentrates, liver, soil, dried distiller's solubles, fish solubles, grass juice concentrate, dried whey, and other feeds have been shown to contain a factor or factors useful either for the growing pig or for the sow during gestation and lactation.

The use of short, lush, green leafy pastures will minimize vitamin deficiencies in swine (Cunha, 1977). Likewise, most nutritionists recognize the possible benefits of using some UGF supplementation to ensure optimal performance of diets for broilers and breeding hens (Scott *et al.*, 1982). This practice, in addition to pasture use, may have a 2-fold advantage in providing possible unidentified growth factor responses and at the same time supplying additional amounts of some of the known vitamins as UGF supplements usually are good sources of many vitamins. The additional vitamins in the diet may prevent serious losses in cases where there is a loss of potency or omission of an important vitamin from the vitamin premix (Scott *et al.*, 1982).

17

Essential Fatty Acids

I. INTRODUCTION

This chapter is concerned with essential fatty acids (EFA). Although the EFA are certainly not vitamins by definition, a deficiency disease or condition with dietary insufficiency does result and in some ways a similarity to vitamin deficiencies can be seen. The finding that components of fat, other than the fat-soluble vitamins, are dietary essentials is of nutritional and medical importance. Excellent reviews in the literature of EFA have been prepared by Holman (1978a,b).

Knowledge that carbohydrates can be readily converted into fat and that such essential lipid constituents as phospholipids and cholesterol can be made in the body led to the view that dietary lipids were not required. In 1929 Burr and Burr changed this viewpoint by reporting that the total deprivation of fat in the diet of rats induced a syndrome of deficiency that could be corrected by certain components of fat. The EFA originally included linoleic, linolenic, and arachidonic acids. However, arachidonic was later found to be synthesized from linoleic. Most species have a dietary requirement for linoleic while others (e.g., fish) require linolenic. Recent studies are reevaluating the beneficial effects of linolenic acid in species that previously were considered to need only linoleic acid as a dietary essential.

II. HISTORY

The earliest report that components of fat other than fat-soluble vitamins are dietary essentials for rats was made by Evans and Burr (1926). Burr and Burr (1929, 1930) first demonstrated an essential dietary requirement by the rat for a specific unsaturated fatty acid configuration that could not be synthesized by the animal. The name ''essential fatty acids'' was coined to describe these unsaturated fatty acids of linoleic and linolenic. Hume *et al.* (1940) reported that arachidonic acid was also an EFA.

The majority of the history of EFA is associated with Dr. R. T. Holman of the Hormel Institute, University of Minnesota, who has contributed more to the knowledge of essential fatty acids than any other individual. He is responsible for the delineation of metabolic conversions of polyunsaturated fatty acid, interactions among families of fatty acids, and determining quantitative requirements for linoleic and linolenic acids in animals and humans.

III. CHEMICAL STRUCTURE AND PROPERTIES

Chemical structures of linoleic, linolenic, and arachidonic acid as well as other fatty acids associated with EFA are shown in Fig. 17.1. There are three common families of unsaturated, 18-carbon fatty acids and one family of unsaturated, 16-carbon fatty acids. The exact structure of an unsaturated fatty acid is given by three numbers: (1) the number of carbon atoms in the chain, (2) the number of

Palmitoleic (9,hexadecenoic)(16:1ω7)

$CH_3-CH_2-CH_2-CH_2-CH_2-CH_2-CH=CH-CH_2-CH_2-CH_2-CH_2-CH_2-CH_2-CH_2-COOH$

Oleic acid (9,octadecenoic)(18:1ω9)

$CH_3-CH_2-CH_2-CH_2-CH_2-CH_2-CH_2-CH_2-CH=$
$CH-CH_2-CH_2-CH_2-CH_2-CH_2-CH_2-CH_2-COOH$

5,8,11-Eicosatrienoic (20:3ω9)

$CH_3-CH_2-CH_2-CH_2-CH_2-CH_2-CH_2-CH_2-CH=CH-CH_2-CH=$
$CH-CH_2-CH=CH-CH_2-CH_2-CH_2-COOH$

Linoleic acid (9,12 octadecadienoic acid) (18:2ω6)

$CH_3-CH_2-CH_2-CH_2-CH_2-CH=CH-CH_2-CH=$
$CH-CH_2-CH_2-CH_2-CH_2-CH_2-CH_2-CH_2-COOH$

γ-Linolenic acid (6,9,12 octadecatrienoic acid) (18:3ω6)

$CH_3-CH_2-CH_2-CH_2-CH_2-CH=CH-CH_2-CH=$
$CH-CH_2-CH=CH-CH_2-CH_2-CH_2-CH_2-COOH$

8,11,14-Eicosatrienoic acid (20:3ω6)

$CH_3-CH_2-CH_2-CH_2-CH_2-CH=CH-CH_2-CH=$
$CH-CH_2-CH=CH-CH_2-CH_2-CH_2-CH_2-CH_2-CH_2-COOH$

Arachidonic acid (5,8,11,14 eicosatetraenoic acid) (20:4ω6)

$CH_3-CH_2-CH_2-CH_2-CH_2-CH=CH-CH_2-CH=$
$CH-CH_2-CH=CH-CH_2-CH=CH-CH_2-CH_2-CH_2-COOH$

Linolenic acid (9,12,15 octadecatrienoic acid) (18:3ω3)

$CH_3-CH_2-CH=CH-CH_2-CH=CH-CH_2-CH=$
$CH-CH_2-CH_2-CH_2-CH_2-CH_2-CH_2-CH_2-COOH$

4,7,10,13,16,19-Docosahexanoic (22:6ω3)

$CH_3-CH_2-CH=CH-CH_2-CH=CH-CH_2-CH=CH-CH_2-CH=$
$CH-CH_2-CH=CH-CH_2-CH=CH-CH_2-CH_2-COOH$

Fig. 17.1 Structures of essential fatty acids and other unsaturated fatty acids.

double bonds, and (3) the omega (ω) number, which indicates the number of carbon atoms from the terminal methyl group to the carbon atom of the first double bond. The omega system, which was originated by R. T. Holman, designates those unsaturated fatty acids belonging to each series. The ω-9 and ω-7 series can be derived from endogenously synthesized oleic acid (18 : 1 ω-9) and palmitoleic acid (16 : 1 ω-7), respectively. The ω-6 series is derived from linoleic acid (18 : 2 ω-6) and the ω-3 series from linolenic acid (18 : 3 ω3). These latter two fatty acids are considered essential as they are products of plants and cannot be synthesized by animals. Thus, it appears that linoleic acid (18 : 2 ω-6) is essential for most species because of the inability of animals to synthesize a double bond between carbons 6 and 7 counting from the terminal methyl group.

The polyunsaturated fatty acids are liquids at room temperature. Double bonds of natural fatty acids would normally be found in nature as the cis form, while greater concentrations of trans fatty acids are found in hydrogenated fats and oils. Linoleic acid is a colorless oil that melts at $-12°C$. It is soluble in ether, absolute alcohol, and other fat solvents and oils. It has an iodine value of 181 and a molecular weight of 280.44. Arachidonic acid is an oil that melts at $-49.5°C$, has an iodine value of 333.5, and has a molecular weight of 304.46 (Scott *et al.*, 1982).

IV. ANALYTICAL PROCEDURES

Essential and nonessential fatty acids are now readily determined by gas–liquid chromatographic procedures that are much more precise than alkaline isomerization analyses. Methyl esters of the fatty acids are formed before injecting into the column, which are then distributed between a moving gas phase (nitrogen, helium, or argon) and a stationary liquid phase (Hofstetter *et al.*, 1965). Bioassays are available where growth rates of newly weaned rats are dependent on the linoleic acid concentrations of feeds being tested.

V. METABOLISM AND FUNCTIONS

Fats and fatty acid metabolism in relation to digestion, absorption, and excretion are discussed elsewhere (Swenson, 1970; Maynard *et al.*, 1979). Recently fatty acid-binding proteins (FABPs) have been identified as a family of cytosolic proteins found in heart, liver, and epithelial cells lining the small intestine (Anonymous, 1985b). FABPs are believed to be integrally involved in the cellular uptake as well as intracellular transport and/or compartmentalization of fatty acids. It has been postulated that the intestinal FABPs may actually participate in cellular fatty acid transport across the intestinal mucosa, as well as in selected intracellular events.

After fat absorption in monogastric animals, fatty acid composition of body fat is directly related to fatty acid composition of food. In ruminants, however, polyunsaturated fatty acids are hydrogenated to a large extent by ruminal microorganisms, resulting in more saturated body fat of the animal. For all species, certain fatty acids form structural components and serve indispensible biochemical functions.

The relationship between the three common families of unsaturated, 18-carbon fatty acids is shown in Fig. 17.2. Members of a particular family may be metabolically converted to more proximally unsaturated (toward the carboxyl group) or chain-elongated fatty acids, but no conversion from one ω family to another occurs in mammals. For example, linoleic acid may be converted to arachidonic acid (C20 : 4 ω-6) in animals, and linolenic acid (C18 : 3 ω-3) may be converted to eicosapentaenoic (C20 : 5 ω-3) and docosahexenoic acid (C22 : 6 ω-3). Members of the ω-6 and ω-3 families are considered essential fatty acids for mammals, because they cannot be synthesized *de novo*.

Linoleic acid and linolenic acid are the precursors of the entire ω-6 and ω-3 families of polyunsaturated fatty acids, respectively. All members of the ω-6 and ω-3 families are active as essential fatty acids and many have been shown to be more active than their original precursor. Studies employing graded dose levels of arachidonic acid fed to rats have revealed that the deposition of archidonic acid in liver is greater when arachidonic acid itself is fed in the diet than when linoleic acid is fed. This indicates that the conversion of 18 : 2 ω-6 acid to arachidonic acid for deposit in tissue lipids is a less efficient process than is the deposition of dietary 20 : 4 ω-6 acid directly into tissue lipids and that the potency of 20 : 4 ω-6 is greater than that of 18 : 2 ω-6 (Holman, 1978a).

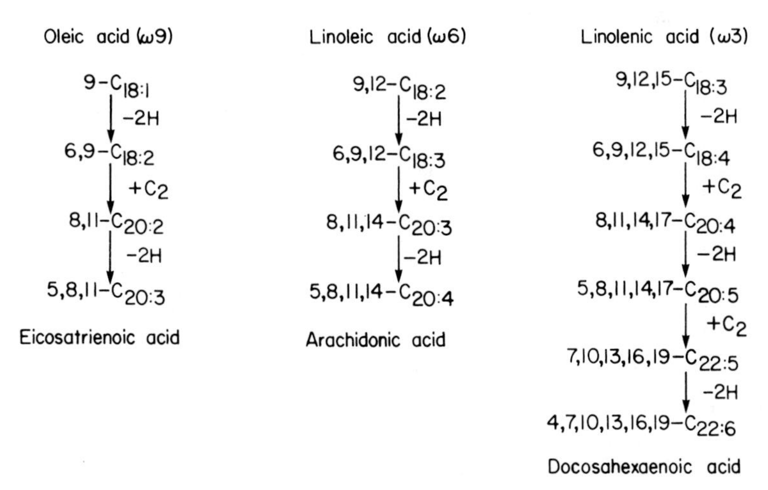

Fig. 17.2 Metabolic transformations of oleic, linoleic, and linolenic fatty acids.

Effect of dietary deficiency of linoleic acid on the fatty acid composition of testicular tissue lipids for swine fed different levels of linoleic acid is illustrated in Fig. 17.3 (Sewell and McDowell, 1966). The same phenomenon has been observed in a large variety of both tissues and different species, the differences being mainly in magnitude (Holman, 1978a, 1986). Changes of greatest magnitude observed thus far occur in heart and liver lipids. At zero intake of linoleic acid, the major differences in fatty acid composition are in the polyunsaturated acids themselves. In EFA deficiency, 18 : 2 ω-6, 20 : 4 ω-6, and 22 : 5 ω-6 are much lower than found in normal animals. Palmitoleic acid, 16 : 1 ω-7, and oleic acid, 18 : 1 ω-9, are higher than normal, but the most striking increase is in 20 : 3 ω-9, which is formed endogenously from oleic acid. This acid, which is a normal component of tissue lipids in trace amounts, increases very dramatically in EFA deficiency. It is found in the phospholipids in the 2 position, the same position in which arachidonic acid, 22 : 5 ω-6, and other polyunsaturated acids are normally found.

Similar studies have been made with graded dose levels of linolenic acid as the sole fatty acid supplement to a fat-free diet (Mohrhauer and Holman, 1963). Supplementation of the fat-free diet with 18 : 3 ω-3 causes dramatic increases in 20 : 5 ω-3, 22 : 5 ω-3, and 22 : 6 ω-3 in comparison with the amounts found in the lipids of fat-deficient animals.

Several studies on dose of a single fatty acid versus response of several fatty acids in tissues have shown that each family of fatty acids suppresses metabolism of other families of polyunsaturated acids (Holman, 1964; Hwang *et al.*, 1988).

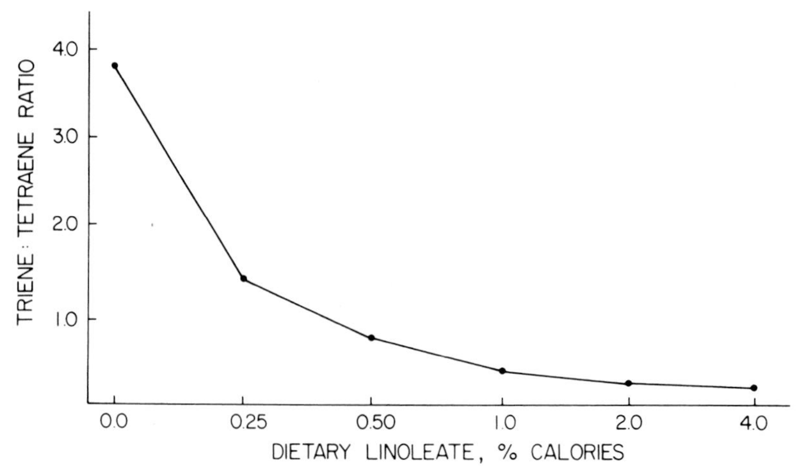

Fig. 17.3 Relationship of triene : tetraene ratio of testes tissue from young swine fed varying levels of dietary linoleic acid. (Modified from Sewell and McDowell, 1966.)

In the absence of the main ω-6 (linoleic) and ω-3 (linolenic) families in the diet, animals are capable of synthesizing some polyunsaturated acids from endogenous precursors. Both oleic (18 : 1 ω-9) and palmitoleic (16 : 1 ω-7) acids themselves and their respective families are enhanced in the tissue lipids. None of these polyunsaturated acids is fully efficacious in meeting the requirement for polyunsaturation, for although they are present in enhanced quantity in EFA deficiency, the animals often die.

The enzymatic systems that perform chain elongation, desaturation, and insertion of fatty acids into various lipid molecules apparently handle all groups of fatty acids, for there is competition between substrates at each step in each of these processes. The ω-3 family effectively suppresses metabolism of the ω-6 family. Likewise, the ω-6 family is able to suppress metabolism of the ω-3 family, but less effectively. The ω-6 family, however, suppresses the formation of polyunsaturated acids from oleic acid as is manifested in EFA deficiency. Ability of the precursor acids to compete for these enzyme systems is in the order linolenic > linoleic > oleic.

Depending on animal species, different EFA are not equal in relationship to requirements or in ability to prevent all signs of EFA deficiency. For most species, linolenic acid does not fully relieve dermal signs of EFA deficiency, even at high levels (Holman, 1978a). However, arachidonic acid (20 : 4 ω-6) is twice as active as its precursor, linoleic acid, in reducing dermal signs attributed to the deficiency. It would appear that the polyunsaturated acids that function as EFA in mammals must be of the linoleic acid family with a terminal structure of ω-6. This requirement may be related to a lipid–enzyme fit or to the binding of polyunsaturated chains to protein. The latter may figure fundamentally in formation of lipoprotein membrane structures that are the sites of many metabolic reactions.

Increased attention has been given in the past few years to the question of possible health benefits of ω-3 (linolenic family) of polyunsaturated fatty acids (Lands, 1986). Whatever the mechanisms involved, epidemiological studies in Greenland Eskimos (who subsist entirely on a marine diet high in ω-3 fatty acids) clearly indicate that their diet does exert potentially antithrombotic effects on platelet function with a low death rate from coronary heart disease (Willis, 1984). From Japan, lower death rate from coronary heart disease has been related to higher fish consumption. Similar results have been obtained in human or animal studies in which fish oils rich in 20 : 5 ω-3 have been administered (Weiner *et al.*, 1986). The most striking result of human studies is marked prolongation of bleeding time. This clearly indicates that marine diets may reduce platelet plug formation in damaged blood vessels and possibly inhibit vessel wall-induced clotting of plasma, as observed in rats (Willis, 1984).

The long-chain ω-3 acids are found in high proportions in reproductive and nervous tissues. The elongated docosahexenoic acid (22 : 6 ω-3) is the most abundant fatty acid in the ethanolamine phospholipids of cerebral gray matter and the retina (Carlson *et al.*, 1986). The need for ω-3 acids in developing visual acuity was presented as evidence for a functional requirement for ω-3 acids in primates (Neuringer *et. al.*, 1984). Differences in physical activity and ability to learn have been related to a low content of 22 : 6 ω-3 in brains of rats produced by feeding a diet low in linolenic acid (Lamptey and Walker, 1976).

Resurgence of interest in action of ω-6 and ω-3 fatty acids on plasma lipoprotein metabolism stems in part from the recent discovery that the prostaglandins derived from eicosapentaenoic acid (20 : 5 ω-3) have biological effects different from those derived from arachidonic acid (20 : 5 ω-6) (Anonymous, 1985b). Recently ω-3 metabolites of linolenic acid were shown to be more effective than ω-6 in lowering cholesterol in rats.

Polyunsaturated fatty acids have a structural function as an integral part of phospholipids, the building unit of biomembranes. This is inferred from the specific composition of the fatty acids in these phospholipids (the β position normally being esterified with the highly unsaturated members of the EFA families) and from the fact that in EFA deficiency these fatty acids are replaced by eicosatrienoic acid (20 : 3 ω-9), biosynthesized from oleic acid (18 : 1 ω-9), with the known concomitant deleterious effects on biomembrane function and integrity.

It has been suggested that EFA deficiency and replacement of the linoleic acid family in membrane structures may cause a disruption in spatial arrangements in mitochondria that results in less efficient oxidative phosphorylation and a derangement of basal metabolism. Such a process may be the partial uncoupling of oxidative phosphorylation in mitochondria. For poultry, presence of linoleic acid may not be absolutely necessary in the body since a deficiency does not result in death. Fatty acids that replace the linoleic acid family in tissue lipids seem to cause a reduction in the metabolic efficiency and functioning of the animal, but life can still be maintained (Scott *et al.*, 1982).

A disturbed water balance is a characteristic defect of EFA deficiency and can include increased water loss through the skin, increased urinary arginine–vasopressin loss, increased water intake, and reduced urine output (Holman, 1968; Hansen and Jensen, 1986; Anonymous, 1986). Increased water loss through skin results from a defect in the permeability barriers of skin, which is an indication that EFA are involved in membrane structure. Histological studies have shown many changes in skin structure as a result of the deficiency.

One of the most important specific metabolic functions of EFA is as precursors for a diverse group of "local hormones" called prostaglandins. These biologically potent compounds seem to play a regulatory role in many cellular processes. Prostaglandins have been shown to be involved in blood clotting, renal free water excretion, renal blood flow, reproduction, bronchoconstriction, gastrointestinal motility and water loss, endocrine function, and neurotransmitter release (Scott *et al.*, 1982). Prostaglandins are intimately involved in regulation of immune function and thus resistance to infection (Boissonneault and Johnston, 1983).

Prostaglandins are formed by elongation and desaturation of linoleic acid to di-homo-γ-linolenic acid (C20 : 3 ω-6) (DHGL) and to arachidonic acid (C20 : 4 ω-6), and from long-chain fatty acids of the linolenic family (20 : 5 ω-3), eicosapentaenoic acid (Anonymous, 1984). These fatty acids are found in membrane phospholipids. Practically all cells are capable either of producing or of being influenced by prostaglandins. A large number of known biologically active prostaglandins have been identified. Prostaglandins formed from DHGL without further desaturation to arachidonate comprise the 1 series of prostaglandins. Prostaglandins formed from arachidonate comprise the 2 series. The 3 series of prostaglandins is formed from eicosapentaenoic acid (C20 : 5 ω-3).

An increase in dietary linoleic acid intake directly influences prostaglandin synthesis (Hwang *et al.*, 1975). This supports the hypothesis that the preventive and curative effects of dietary linoleic acid on the atherosclerotic syndrome can be explained by an increased prostaglandin biosynthesis (Thomasson, 1969). This hypothesis was based on pharmacological data indicating that atherosclerosis-promoting factors such as hypertension and increased thrombotic tendency of blood platelets can be counteracted by the arterial dilatation and the increased water and sodium diuresis induced by certain prostaglandins.

In a review, Vergroesen (1977) summarized the beneficial effects of adequate linoleic acid:

1. decreased blood cholesterol and triglyceride levels;
2. decreased thrombotic tendency of platelets;
3. preventive and curative effects in sodium-induced hypertension;
4. improvement of the physiological function of the heart;
5. normalization of the biochemical abnormalities in obesity and maturity onset diabetes.

Mechanisms of these responses are not clearly established, however, for many of the favorable effects, prostaglandin activities are involved. The role of ω-6 compared to ω-3 fatty acids is under intensive study. It is apparent that populations that consume more ω-3 (marine foods) fatty acid have lower cardiovascular disorders such as thrombosis (Swanson and Kinsella, 1986).

Additional functions of EFA include provision of adequate fluidity to sustain cellular function and for lipid transport (Holman, 1978a). Phospholipids and cholesteryl esters containing an abnormally high proportion of saturated fatty acids would tend to be more rigid or less fluid than would similar compounds with high proportions of polyunsaturated acids. One of the functions of polyunsaturated acids is to provide lipids that are fluid at body temperature. Alloxan diabetes, hyperthyroidism, dietary cholesterol, saturated fat diet, or mineral oil all involve the transport of a "nonessential" lipid in large quantities. These conditions have been found to accelerate EFA deficiency significantly. Studies suggest that one function of polyunsaturated acids is to provide necessary structural components for circulating lipoproteins (Holman, 1978a).

VI. REQUIREMENTS

The dietary essentiality of both linoleic acid (18 : 2 ω-6) and linolenic acid (18 : 3 ω-3) is dependent on species and to a certain extent on the definition of an essential nutrient. Table 17.1 provides EFA requirements for various animals and humans. The EFA requirement of most mammals can be met by linoleic acid and its family of polyunsaturated acids. Determination of linoleic acid requirements have been based on observations of gross dermal lesions as well as variations in tissue polyunsaturated fatty acids. Shifts in fatty acid composition of metabolically active tissues (liver, heart, brain, etc.) that occur during onset of deficiency are very similar in all species. Similarities between EFA deficiencies induced in various species are striking when the biochemical parameters of the deficiency are considered. Quantitative requirements for several species are also strikingly similar when measured with biochemical parameters.

The triene : tetraene ratio is the ratio of abnormally elevated endogenous metabolite of oleic acid, 20 : 3 ω-9, to the metabolic product, 20 : 4 ω-6, derived from linoleic acid (see Section VIII), and it has been used to estimate the minimal linoleic acid requirement (Holman, 1960). The ratio drops from a high value in deficiency to a low and rather constant value in the region between 1 and 2% of calories. The implication from these biochemical parameters is that dietary requirement for linoleic acid lies between 1 and 2% of calories.

Holman (1960) suggested a ratio of 0.4 as the point at which the minimal linoleic acid requirement of the rat, as well as other species, has been met. Figure 17.3 illustrates the plotting of the triene : tetraene ratio versus six dietary linoleic acid levels in 3-week-old swine testes (Sewell and McDowell, 1966). The ratio decreased markedly as dietary level of linoleic acid increased from zero to 1% of dietary calories, with only a slight decrease occurring beyond this level of linoleic acid intake. Requirement for linoleic acid for this age pig is therefore less than 2% of dietary calories. The triene : tetraene ratio at the 1% level of linoleic acid was 0.38, which is comparable to the figure of 0.4 suggested by Holman (1960).

Contrary to other species, evidence has been presented that the cat family (cats and lions) is unable to desaturate linoleic acid and linolenic acids (NRC, 1986). As a result, these species may require a source of preformed longer-chain fatty acids, which means that these animals may exhibit a specific requirement for polyunsaturated lipids of animal origin. Nevertheless, linoleic acid prevents several signs of EFA deficiency, including scaly skin, increased transepidermal water loss, and enlarged fatty livers. Thus linoleic acid has a specific role as an EFA independent of arachidonic acid synthesis.

Most fish, contrary to terrestrial animals, have a definite requirement for linolenic acid (ω-3) while linoleic acid (ω-6) is often of little value. The majority of fish and shellfish species studied require 18 : 3 ω-3, some require a combination

TABLE 17.1

Essential Fatty Acid Requirements for Various Animals and Humans[a]

Animal	Purpose	Requirement	Reference
Dairy cattle	Calf milk replacer	10% fat	NRC (1978a)
Chicken	Leghorn, 0–20 weeks	1% linoleic	NRC (1984b)
	Leghorn, breeding	1% linoleic	NRC (1984b)
	Leghorn, laying[b]	1–1.4% linoleic	Scott *et al.* (1982)
	Leghorn, breeding[b]	1–1.4% linoleic	Scott *et al.* (1982)
Turkey	0–8 weeks	1% linoleic	NRC (1984b)
	8–20 weeks	0.8% linoleic	NRC (1984b)
	Breeding hens	1% linoleic	NRC (1984b)
Japanese quail	Growing–breeding	1% linoleic	NRC (1984b)
Sheep	Growing	<0.32 linoleic as energy	Bruckner *et al.* (1984)
Swine	Growing	1–2% linoleic as energy	Sewell and McDowell (1966)
Cat	Growing[c]	0.5% 18:2 ω-6 and 0.02% 20:4 ω-6	NRC (1986)
Dog	All classes	1% linoleic	NRC (1985a)
Mink	Adult	0.5% linoleic	NRC (1982a)
	Pregnancy-lactation	1.5% linoleic	NRC (1982a)
Fox	All classes	2–3 g linoleic and linolenic daily	NRC (1982a)
Rat	Males	1.3% linoleic as energy	NRC (1978c)
	Females	0.5% linoleic as energy	NRC (1978c)
Guinea pig	Growing	1.9% linoleic as energy	NRC (1978c)
Nonhuman primates	All classes	1–2% linoleic as energy	NRC (1978d)
Fish	Salmon	1% linoleic and 1% linolenic	NRC (1981a)
	Shrimp	1% linolenic	NRC (1983)
	Trout	1% linolenic	NRC (1981a)
Human	All classes	1–2% as energy	RDA (1980)

[a]Often requirements are unknown; for some species, treatments or EFA levels that have been successful are noted. Requirements are expressed either as percentage of diet or as percentage of calories (energy). Values can be either on an as fed (approximately 90% dry matter) or dry basis (see Appendix Table 1).

[b]For laying and breeding hens, 1.4% linoleic acid is required; after maximum egg size is reached, 1% is adequate.

[c]The cat has very limited abilities to desaturate linoleic and linolenic acid to longer-chain fatty acids.

of 18 : 3 ω-3 and 18 : 2 ω-6, and only a few show a preference for 18 : 2 ω-6 (NRC, 1981a, 1983). For some fish species (e.g., rainbow trout), the metabolite of linolenic acid, 22 : 6 ω-3, is more effective in stimulatory growth and other production parameters than is linolenic acid. At least four species of fish require ω-3 fatty acids and 22 : 6 ω-3 best satisfies the requirement. One suggestion to explain the difference in EFA requirements for fish was that ω-3 structure permitted a greater degree of unsaturation, which was necessary in membrane phospholipids to maintain flexibility and permeability characteristics at low temperatures (NRC, 1983).

A number of factors influence development of EFA deficiency and thus requirements for EFA:

1. Age and carryover effects—Animals that had been fed normal diets for a longer time should have larger reserves of polyunsaturated fatty acids and could withstand deficiency for a longer time than weanlings. Inducing dermal signs of EFA deficiency is difficult in adult animals. Linoleic acid requirement of young chicks can be affected markedly by carryover of linoleic acid from the egg to newly hatched chicks. If chicks hatch from eggs low in linoleic acid and are fed purified diets very low in linoleic acid, the dietary requirement may be in excess of 1.4%, compared to the typical requirement of 1.0% (Scott *et al.*, 1982).
2. Dietary fat and hormone imbalance—Animals that practice coprophagy have an additional source of lipids not available to animals that do not. However, diets rich in saturated fatty acids or monounsaturated fatty acids are also known to moderately enhance the development of EFA deficiency. Peifer and Holman (1955) studied the effect of adding 1% cholesterol to the diet of EFA-deficient rats. An EFA deficiency syndrome, judged by growth and dermal signs, occurred within periods of 2 weeks to 1 month. Comparable EFA deficiency signs were observed only after 3 months on fat-free diets without cholesterol. Substances or conditions that induce hypercholesterolemia likewise accelerate EFA deficiency.
3. Growth rate—Any animal that is called upon to grow more rapidly and, therefore, to build more tissue would have a higher requirement for EFA and would consequently exhibit deficiency signs earlier.
4. Humidity and water balance—Low atmospheric humidity hastens onset of dermal signs of EFA deficiency, probably through enhanced loss of water by evaporation, causing additional irritation of skin. Aaes-Jørgensen and Dam (1954) reported an experiment where female rats were raised on diets with various amounts of fat for a period of 16 weeks. They found that the water intake was higher and urine production was lower on diets with hydrogenated peanut oil or hydrogenated whale oil and in the absence of dietary fat than on diets with lard, peanut oil, or coconut oil.
5. Sex of animal—Male animals are known to be more sensitive to EFA deficiency than are females. The requirement for the female rat was found to be

between 10 and 20 mg per day, while the male rat's requirement exceeded 50 mg daily (Greenberg *et al.*, 1950).

6. Pen arrangement—Leat (1962) concluded, from feeding pigs, that whether the animals are penned individually or in groups is important in developing EFA-deficient signs. He found that the skin condition was noticeably better in pigs penned in pairs than in those penned individually. They believed that keeping animals in close proximity with each other may prevent dermatitis from becoming apparent merely through physical contact.

7. Temperature and environment—For fish, EFA requirements may change with temperature and culture conditions. When rainbow trout from the same

TABLE 17.2

Typical Linoleic Acid and Arachidonic Acid in Various Foods and Feedstuffs (as Fed Basis)[a]

Food or feedstuff	Linoleic acid (%)	Arachidonic acid (%)
Alfalfa meal, dehydrated	0.40	—
Barley	0.83	—
Brewer's grains, dehydrated	2.94	—
Coconut oil	1.10	—
Corn gluten meal	3.83	—
Corn oil	55.40	—
Corn, yellow	1.82	—
Cottonseed meal, solvent extracted	0.80	—
Crab meal	0.33	—
Fish meal, anchovy	0.20	—
Fish meal, menhaden	0.15	—
Fish oil, menhaden	2.70	—
Fish solubles, condensed	0.20	—
Lard	18.30	0.3–1.0
Linseed oil	13.90	—
Meat meal	0.34	—
Milk, cow's, dehydrated	0.01	—
Oats	1.49	—
Peanut meal	1.25	—
Poultry by-products meal	1.72	—
Poultry fat (offal)	22.30	0.5–1.0
Rice bran oil	36.50	—
Safflower oil	72.70	—
Sorghum	1.08	—
Soybean meal, solvent extracted	0.35	—
Soybean seed	7.97	—
Tallow	4.30	0–0.2
Wheat	0.58	—
Wheat bran	2.25	

[a]Data adapted from NRC (1982b, 1984b).

source were tested simultaneously in seawater and fresh water, the EFA deficiency was manifested more quickly in seawater (NRC, 1983). Likewise, the fatty acid composition of fish lipids, especially membrane lipids such as phospholipids, are significantly affected by acclimation temperature.

VII. NATURAL SOURCES

The EFA are widely distributed among food fats. For example, vegetable oils of corn, soybean, cottonseed, peanut, and certain others are excellent sources. Safflower oil contains 75% linoleic acid, whereas corn oil, soybean oil, and cottonseed oil all contain approximately 50% linoleic acid. Linolenic acid is particularly high in forage lipids. From the lipids of pasture grasses, 61.0% is reported as linolenic acid (Garton, 1960).

Linoleic acid content of feed ingredients is shown in Table 17.2. Linoleic acid and its dehydrogenation product 18 : 3 ω-6 are found in highest abundance in plants, but more unsaturated and longer-chain members of this family are found principally in animals. Notable exceptions to these generalities are the occurrence of arachidonic acid and other higher members of the group in primitive plants (Schlenk and Gellerman, 1965).

VIII. DEFICIENCY

A. Effects of Deficiency

Induction of EFA deficiency in animals requires rigid exclusion of fat from the diet, and even with supposed low-fat diets for humans, deficiency state in adults is difficult to attain (Holman, 1978b). Clinical signs of EFA deficiency induced by a fat-free diet require almost one-eighth of a rat's normal lifetime to develop, and rarely have humans been subjected to a low-fat diet under observation for a proportionate span of time. Natural diets, even poor ones, usually contain adequate amounts of EFA, and therfore the deficiency is far rarer than deficiencies of protein, vitamins, or minerals. Nevertheless, EFA deficiency does occur when animals or humans receive insufficient dietary fat.

Listing of EFA deficiency signs and other criteria ranges from classic signs such as reduced growth rate, parakeratosis, increased water permeability of skin, increased susceptibility to bacteria, and male and female sterility to recently recognized signs such as decreased prostaglandin biosynthesis, reduced myocardial contractility, abnormal thrombocyte aggregation, and swelling of rat liver mitochondria (Vergroesen, 1977). For all land species studied, the major feature of the deficiency is an impairment of the exterior covering of the animal. Mammals exhibit a dermatitis, chickens exhibit a faulty feathering, and moths are unable to form normal scales on their wings. All the manifestations indicate faulty membrane formation, a feature of deficiency that is common to all tissues and to all species (Holman, 1978a).

1. Ruminants

The EFA deficiency of ruminants has been less extensively researched than that of nonruminants, with the deficiency in adult ruminants not readily demonstrated (Palmquist *et al.,* 1977). The microbial population appears to provide enough EFA to meet the requirements, however, studies with lambs suggest that the required level of EFA may be elevated in the presence of host microflora (Bruckner *et al.,* 1984). Gullickson *et al.* (1942) reported that calves fed a low-fat diet did not develop EFA deficiency signs, but growth was suppressed. Cunningham and Loosli (1954b) reported that calves receiving a fat-free synthetic milk developed leg weakness and muscular twitches within 1–5 weeks and died unless a source of fat was supplied. Lambert *et al.* (1954) also studied the effect of a "lipid-free," semisynthetic milk when fed to dairy calves. They reported the following clinical signs: growth retardation after 3 weeks on trial, scaly dandruff, long dry hair, dull hair coat, excessive loss of hair on the back, shoulders, and tail, and diarrhea.

Cunningham and Loosli (1954a) fed weanling lambs a fat-free diet for 7 months without showing any skin lesions or other clinical signs typical of a fat deficiency. In a second experiment, 2-day-old lambs and kids were given a fat-free synthetic milk. The lambs and kids receiving the fat-free diets became weak and died within 1–7 weeks while controls were raised successfully on the same milk with 2% added lard.

2. Swine

Witz and Beeson (1951) used a diet that contained only 0.06% lipid and produced the following signs: slower growth rate, underdeveloped digestive systems, small gallbladders, enlarged thyroid glands, delayed sexual maturity, scaly dandruff-like dermatitis on the tail, back, and shoulders, loss of hair, with the remaining hair being dull and dry, a brown, gummy exudate on the belly and sides, necrotic areas on the skin around the neck and shoulders, and an unthrifty appearance. Leat (1962) fed pigs from 4.5 to 91 kg liveweight on a diet consisting of 0.07% of the calories as linoleic acid. This diet resulted in a pronounced scaliness of the skin, first noted after about 13 weeks on the diet. Scaliness seemed to be confined to the dorsal surface and was most severe about the shoulders. The hair itself was dry and appeared to stand out from the skin at all angles. When linoleic acid made up 0.5% of the dietary calories, there was little or no flakiness of skin.

Sewell and McDowell (1966) fed 3-week-old male pigs purified diets containing

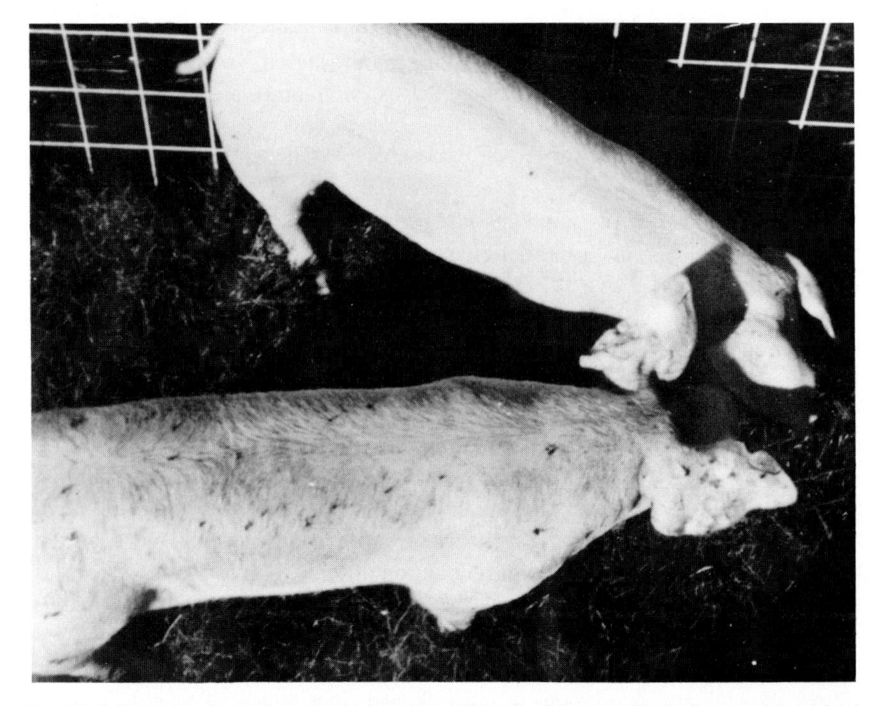

Fig. 17.4 Dermal lesions of pigs receiving diet without linoleic acid. Pig in foreground received no linoleic acid, while the pig in rear received 1.0% of calories as linoleic acid. The black spots on the linoleic-deficient pig are flies that were attracted to the brown, gummy exudate. (Courtesy of L. R. McDowell, R. F. Sewell, and the University of Georgia.)

six levels of linoleic acid. For the 10-week experiment no differences were noted in weight gains, but dermal lesions (Figs. 17.4 and 17.5) were observed after 6–7 weeks on experiment. A scaly, dandruff-like desquamation of the skin over the dorsal surface was the first noticeable sign, and later a brownish, gummy exudate appeared around the ears, axillary spaces, and under the flanks. Skin eruptions were also present about the ears, axillary spaces, and flanks in the severest cases. Lesions were only observed among pigs receiving 0.5% linoleic acid or less; the 1% level and above was free of skin lesions.

3. Poultry

Growing chicks fed a fat-free ration did not survive the fourth week (Reiser, 1950). The most readily observed clinical sign of linoleic acid deficiency in young chicks is a slow growth rate. Machlin and Gordon (1961) found that adding safflower oil or linoleic acid, but not linolenic acid, to purified diets free of unsaturated fatty acids resulted in an immediate (within 7 days) growth response in chickens.

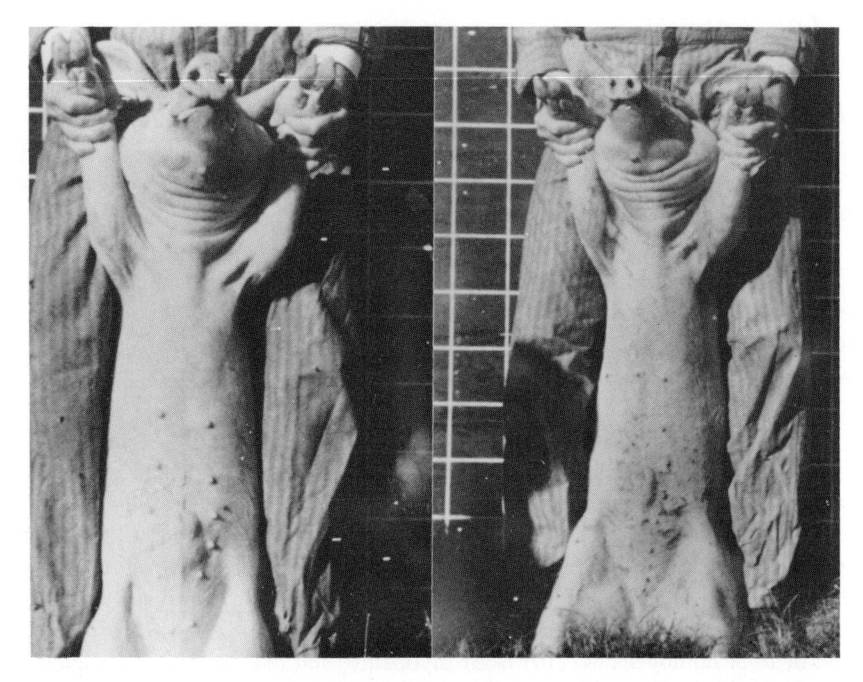

Fig. 17.5 The same two pigs as in Fig. 17.4 with the one on the right illustrating EFA deficiency; dermal lesions were particularly severe at the axillary spaces. (Courtesy of L. R. McDowell, R. F. Sewell, and the University of Georgia.)

Deficiency in chicks has also been reported to result in an enlarged fatty liver, degeneration of testes and subcutaneous edema, and in some cases general edema in the body occurs. Linoleic acid-deficient chicks are more susceptible to respiratory infections (Scott *et al.*, 1982). High mortality resulted from an atypical respiratory infection for chicks fed linoleic acid-deficient diets from hatching to 10–12 weeks of age. Müller *et al.* (1976) showed that some microbial synthesis of ω-6 fatty acids occurs in the gut of quail. On a diet deficient in linoleic acid, germ-free animals had a lower growth rate and had more severe clinical signs than normal ones.

Linoleic acid deficiency in laying hens results in depressed egg production, small egg size, a slight reduction in fertility, and a marked increase in early mortality of the embryo during incubation. Eggs from hens severely deficient in linoleic acid will not hatch. For a severe deficiency in hens it is necessary to feed pullets diets that are very low in linoleic acid from hatching on. Linoleic acid is stored in the body for long periods of time if the animals had been reared on a diet containing adequate linoleic acid.

4. Other Animal Species

a. Cats. Clinical signs attributed to EFA deficiency in cats include listlessness, dry unattractive hair coats, and severe dandruff (NRC, 1986). Growth is poor, and susceptibility to infections is increased. Histological examination of the liver reveals fatty infiltration and parenchymal disorganization. Both males and females lack libido, testes are underdeveloped, and estrous cycles are absent.

b. Dogs. Dogs deficient in EFA have low growth rates, a characteristic dermatitis, a swelling and redness of the paws, and increased susceptibility to infection (NRC, 1985a). Beagle puppies fed a low-fat diet exhibited skin lesions within 2 to 3 months.

c. Fish. Fish nutrition studies have established that ω-3 fatty acids are essential for the maintenance of good health and promotion of rapid growth in rainbow trout and in the red sea bream (NRC, 1981a). Deficiency signs specifically for trout are: (1) poor growth; (2) elevated tissue levels of ω-9 fatty acids (particularly 20 : 3 ω-9); (3) necrosis of the caudal fin; (4) fatty, pale liver; (5) dermal depigmentation; (6) increased muscle water content; (7) syncope accentuated by stress; (8) increased mitochondrial swelling; (9) increased respiration rate of liver homogenates; (10) heart myopathy; and (11) lowered hemoglobin level (NRC, 1981a). Poor growth, low feed efficiency, high mortality, and swollen pale livers were reported in chum salmon (*Oncorhynchus keta*) fed an EFA-deficient diet (Takeuchi *et al.*, 1979).

In certain warm-water fish, common carp, for example, it has been demonstrated that dietary levels of 22 : 6 ω-3 significantly affected egg hatchability (NRC, 1983). Shrimp (*Penaeus setiferus*) would not produce eggs unless the diet contained 20 : 5 ω-3 and 22 : 6 ω-3.

d. Foxes. Foxes on low-EFA diets have clinical signs of hyperkeratosis and dandruff (NRC, 1982a).

e. Monkeys. In monkeys, EFA deficiency resulted in dryness and scaliness of the skin with loss of hair, although many months of the deficient diet were required for appearance of clinical signs (NRC, 1978d).

f. Rabbits. Signs in rabbits of EFA deficiency are reduced growth, hair loss, degenerative changes in seminiferous tubules, and impaired sperm development (Ahluwalia *et al.*, 1967).

g. Rodents. Signs of EFA deficiency in the rat are reduction in growth, which plateaus after about 12–18 weeks, scaly skin, a rough, thin hair coat,

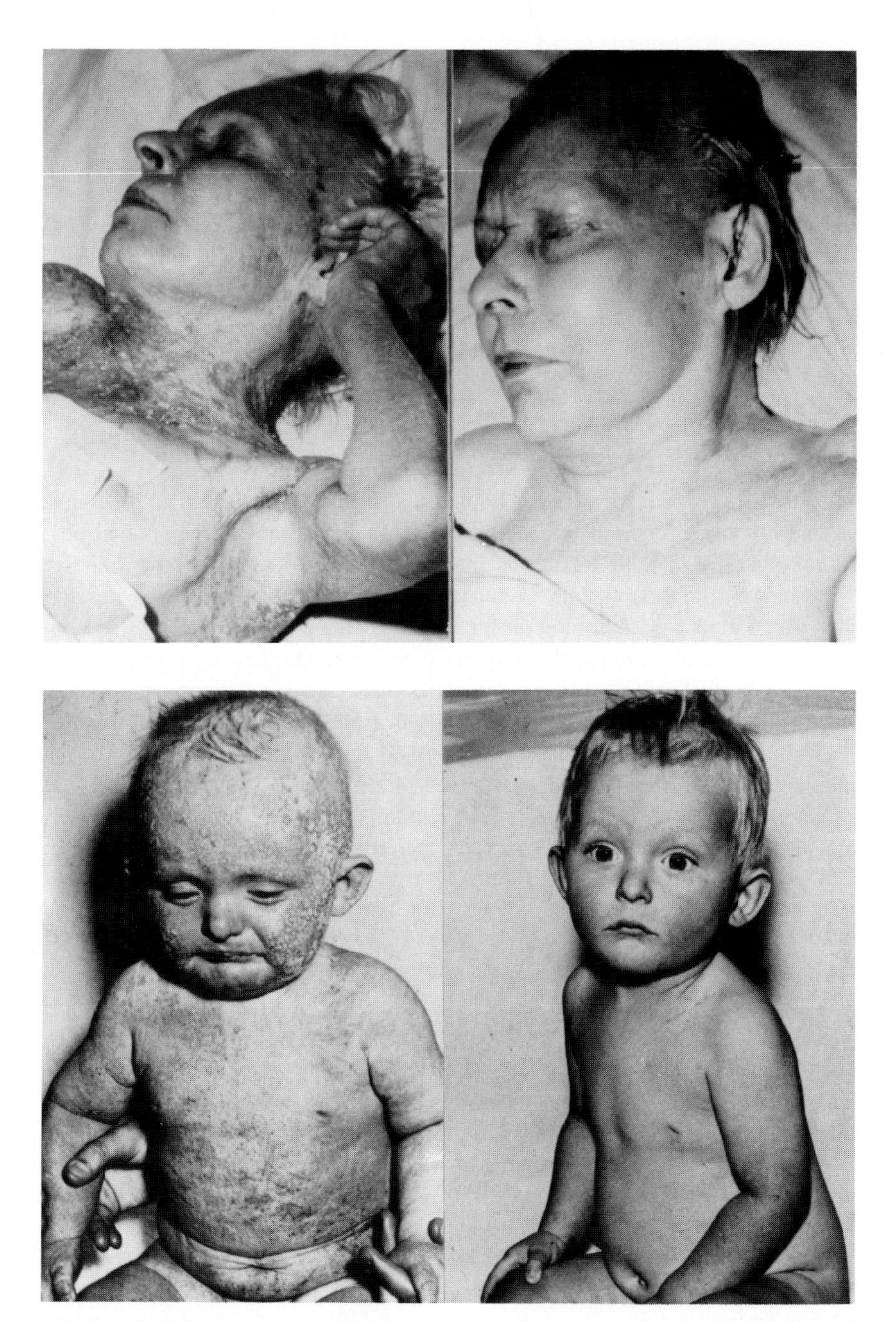

Fig. 17.6 Skin lesions associated with EFA deficiency before and after EFA administration. Top photograph involved fat-free total parenteral nutrition. (Courtesy of R. T. Holman, The Hormel Institute, University of Minnesota, and M. C. Riella, J. W. Broviac, M. Wells, and B. H. Scribner, *Ann. Intern. Med.* **83** (1975), 786–789.)

necrosis of the tail, electrocardiographic abnormalities, fatty liver, renal damage, and death (NRC, 1978c). In young rats fed a fat-free diet, two of the earliest signs of EFA deficiency are increased transepidermal water loss and increased urinary arginine–vasopressin excretion (Hansen and Jensen, 1986).

Mice with EFA deficiency have hair loss, dermatitis with scaling and crusting of skin, and occasional diarrhea. Deficiency in older mice caused infertility without visible skin changes (NRC, 1978c). For the guinea pig with a deficiency, there is weight loss with clinical signs including dermatitis, skin ulcers, fur loss, underdevelopment of spleen, testes, and gallbladder, and enlargement of kidneys, liver, adrenals, and heart.

5. Humans

Essential fatty acid deficiency in humans has primarily been studied in infants. Eczema has been reported in a number of studies for infants maintained on a low-fat diet (Holman, 1978a). Hansen *et al.* (1958) presented data showing that young healthy infants, within a relatively short time, may develop symptoms when given diets extremely low in fat. After several weeks on trial, alterations in the skin were observed in the majority of infants. The first sign detected was dryness, then thickening, and later desquamation with oozing in the intertriginous folds (Fig. 17.6). Addition of linoleic acid, as 2% of the caloric intake, restored the skin to normal, while addition of saturated fatty acid had no effect.

Patients that are maintained by fat-free total parenteral nutrition (TPN) are candidates for EFA deficiency. Premature infants receiving TPN are particularly susceptible to rapid development of EFA deficiency signs (Cooke *et al.*, 1984).

B. Assessment of Status

In addition to various clinical signs and responses to EFA supplementation for a deficiency, the most accurate indicators of EFA status are biochemical changes associated with EFA deficiency. Anatomical signs of EFA deficiency vary from species to species, but biochemical aberrations associated with deficiency in rats are found to be the same in all other species studied thus far (Holman, 1978a). The biochemical signs of EFA deficiency have been known for many years. The chief alterations in various tissues are decreased levels of linoleic (18 : 2 ω-6), arachidonic (20 : 4 ω-6), and docosapentaenoic (22 : 5 ω-6) acids, and increased levels of eicosatrienoic (20 : 3 ω-9) and docosatrienoic (22 : 3 ω-9) acids. With the EFA-deficient diet the tissues attempt to produce unsaturated acids from the oleic acid (18 : 1 ω-9), which results in the accumulation of the trienoic acid, eicosatrienoic acid (20 : 3 ω-9). The triene : tetraene ratio is the ratio of the abnormally elevated endogenous metabolite of oleic acid, 20 : 3 ω-9, to the metabolic product, 20 : 4 ω-6, from the dietary essential linoleic acid (see Section VI). Therefore, it is the ratio of the "abnormal" to "normal"

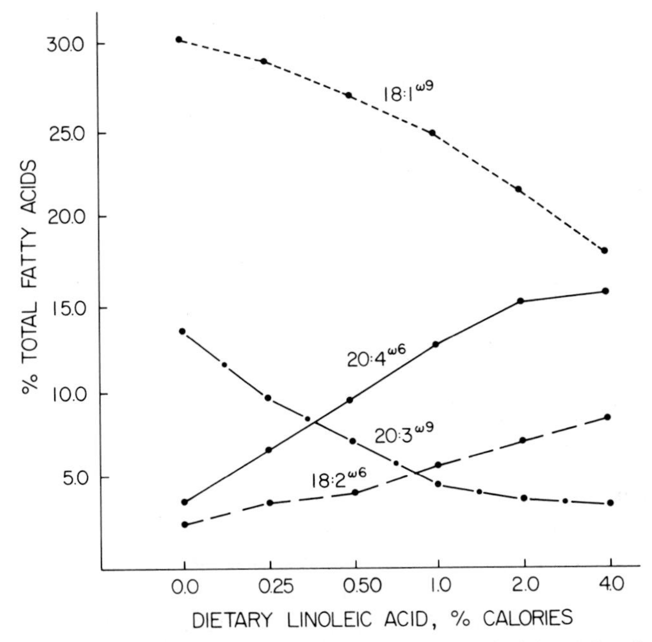

Fig. 17.7 Influence of varying levels of dietary linoleic acid on linoleic (18:2 ω-6), arachidonic (20:4 ω-6), eicosatrienoic (20:3 ω-9), and oleic (18:1 ω-9) acid content of testes tissue from young swine. (Modified from Sewell and McDowell, 1966.)

major polyunsaturated acids in tissue lipids. The curve describing the triene : tetraene ratio versus dietary linoleate had a sharp break at or near 1% of calories, suggesting that 1% of calories of linoleic acid may be a critical amount, the dietary requirement (Holman, 1960). Values of the ratio below 0.4 have been considered normal for animals. Figure 17.7 is an example of a study in which six levels of linoleic acid were fed to young male pigs (Sewell and McDowell, 1966). As dietary linoleic acid is increased, arachidonic acid (20 : 4 ω-6) is elevated and 20 : 3 ω-9 is suppressed.

IX. SUPPLEMENTATION

Fatty acid deficiency is not likely when diets contain appreciable amounts of corn. Yellow corn is the major source of linoleic acid in most feed formulas for swine and poultry. Diets composed of corn and soybean meal with no further supplementation are likely to be just adequate in linoleic acid for chick growth but marginal for maximum egg size. Since linoleic acid has a marked effect on egg size, it is necessary to ensure that a sufficient amount of linoleic acid is included in the diet of laying hens to enable them to lay eggs of maximum size as early as possible during the egg production year (Scott *et al.*, 1982). A typical dietary linoleic acid recommendation for poultry is 1% of total calories. However, for laying and breeding hens the requirement would be 1.4% of calories until egg size is reached (Scott *et al.*, 1982).

Animals that are fed diets containing sorghum, barley, or wheat instead of corn as the major grain may receive suboptimal quantities of linoleic acid. Even more important, when roots and tubers (such as potatoes and cassava) or processed carbohydrates (such as cane sugar) make up a major part of the energy source and when solvent-extracted protein supplements are used exclusively, there is potential danger of skin lesions related to EFA deficiency (McDowell, 1977). For young ruminants, the main supplementation concern is that milk replacers contain adequate concentrations of EFA.

Under typical conditions, EFA deficiency would not be expected for humans. The average content of fat in United States diets is 40% of calories. Likewise, there has been a marked shift in preference from animal fat to vegetable oils, which contain high quantities of linoleic acid. Even if all dietary fat were of one source, it is unlikely that EFA content of the diet would be inadequate. Because polyunsaturated acids of the ω-6 family are nearly ubiquitous in plants and animals and in most natural food sources their content is above the minimum nutrient requirement, even random selection of foods is not likely to induce EFA deficiency (Holman, 1978a).

The main concern for supplementation of EFA for humans would be for infants not receiving mother's milk, but rather a milk product low in EFA. Particular attention is needed for premature infants to ensure that parenteral nutrition includes formulas containing adequate EFA. Research in recent years suggests that marine oil fatty acids (linolenic or ω-3 family) play a role in ameliorating some cardiovascular disorders such as thrombosis (see Section V). Further research is required to determine if supplementing diets with ω-3 fatty acids is warranted.

X. TOXICITY

There are nutritional disadvantages from excessive intakes of EFA. Before the role of antioxidants and the dietary requirement for vitamin E were understood, a large literature accumulated concerning the alleged toxicity of polyunsaturated fats (Holman, 1978a). These readily oxidized acids increase the requirement (see Chapter 4) for vitamin E, which functions as an antioxidant in the body. Several experiments have shown that levels of the vitamin that were normally sufficient to prevent such vitamin E deficiency signs as muscular dystrophy and encephalomalacia proved inadequate as the intakes of EFA were increased. It is difficult to experimentally separate the effects of high levels of polyunsaturated fatty acids from a relative deficiency of tocopherol, but it is obvious that, at least under some circumstances, high levels of polyunsaturated acids may have undesirable effects.

18

Vitamin Supplementation

I. INTRODUCTION

In the early part of this century, pasture, other forages, distiller's solubles or grains, brewer's grains, fermentation products, and meat, milk, and fish by-products were depended on as sources of the vitamins for animal feeding. As animal feeding became more sophisticated, as faster-growing and higher-producing animals were developed, and as the trend toward more intensified operations occurred, it became necessary to add an increasing number of vitamins to properly fortify animal diets.

Vitamins represent only a minute fraction of animal feeds, amounting to less than 0.1% by weight and about 1–2% of the cost for swine and poultry operations depending on the diet used and the level of supplementation required. Yet a balanced vitamin fortification program for meeting requirements of nonruminant animals under a wide range of feeds and different production systems will more than offset the cost of adding vitamins.

Because of proper diet selection and vitamin supplementation, human deficiency diseases attributed to vitamin A (xerophthalmia and night blindness), vitamin D (rickets), vitamin C (scurvy), thiamin (beriberi), niacin (pellagra), and vitamin B_{12} (pernicious anemia) have been eliminated in varying degrees in developing countries but still pose a problem for susceptible groups. Vitamin A deficiency particularly is still a problem in many world areas. In human nutrition, vitamin supplementation should be provided to susceptible population groups, with the precaution of avoiding quantities in extreme excess (megadose) of requirements.

Supplementation guidelines specific for each vitamin have been presented in the respective chapters (Chapters 2–17). The present chapter will discuss general supplementation considerations for both animals and humans.

II. FACTORS RESULTING IN INADEQUATE DIETARY INTAKES OF VITAMINS

Vitamin dietary intake and utilization is influenced by many factors, including particular feed ingredients, bioavailability, harvesting, processing, storage, feed intake, antagonists, least-cost feed formulations, and other factors (NRC, 1973).

A. Agronomic Effects and Harvesting Conditions

Vitamin levels will vary in feed ingredients because of crop location, fertilization, plant genetics, plant disease, and weather. Intensive cropping practices and use of new crop varieties may result in reduced levels of certain vitamins in many feedstuffs. In forage crops, factors that favor production of lush, green plants also favor production of many vitamins, particularly β-carotene, vitamin E, and vitamin K. The vitamin C content of tomatoes depends primarily on intensity of sunlight striking the fruits of the tomato during the immediate preharvest period (Scott, 1973).

Harvesting conditions often play a major role in the vitamin content of many feedstuffs. Vitamin content of corn is drastically reduced when harvest months are not conducive to full ripening. If corn has been subjected to alternate periods of freezing and thawing while it contains a high amount of moisture, fermentation occurs in corn kernels, and there is a loss of vitamin content, particularly of vitamin E and cryptoxanthin. In one study, vitamin E activity in blighted corn was 59% lower than that in sound corn and activity of the vitamin in light-weight corn averaged 21% below that in sound corn (Roche, 1979b). Certain legumes, particularly alfalfa and soybeans, contain the enzyme lipoxidase that, unless quickly inactivated, readily destroys much of the carotenes.

B. Processing and Storage Effects

Many vitamins are delicate substances that can suffer loss of activity due to unfavorable circumstances encountered during processing or storage of premixes and feeds. Stress factors for vitamins include humidity, pressure (pelleting), heat, light, oxidation–reduction, rancidity, trace minerals, pH, and interactions with other vitamins, carriers, enzymes, and feed additives (NRC, 1973).

Humidity is the primary factor that can decrease the stability of vitamins in premixes and feedstuffs. Water softens the matrix, for example, of vitamin A and thus the vitamin becomes more permeable to oxygen. Trace elements, acids, and bases are activated only by water. Humidity augments the negative effects exerted by choline chloride, trace elements, and other chemical reactions that are not found in dry feed. Thus, the water level is responsible for a higher reactivity of vitamins with other feed components. Elevated moisture content or incorrect storage of premixes and feedstuffs are the root of almost all stability problems.

Vitamins mixed and stored with minerals are subject to loss of potency. Hazards to vitamins from minerals are abrasion and direct destruction by certain trace elements, particularly copper. Some abrasion is inevitable in the mixing process but fortunately most minerals contain little moisture with the exception of salt, which is, of course, somewhat hygroscopic if exposed to environmental moisture. Therefore, packaging, careful transport, and storage become important and few companies would willingly use very high levels of salt in a supplement containing fat-soluble vitamins.

Some vitamins are destroyed by light. Riboflavin is stable to most factors involved in processing, however, it is very readily destroyed by either visible or ultraviolet light. Vitamin B_6, vitamin C, and folacin can also be destroyed by light. It is necessary, therefore, to protect premixes of feeds containing these vitamins from light and radiation (Stamberg and Peterson, 1946).

Sun–field curing of cut hay is essential to provide vitamin D activity but results in loss of other vitamin potency. Mangelson *et al.* (1949) showed that mechanical dehydration at 177°C within 1 hr after cutting produced an alfalfa meal that contained 2.5 times as much carotene as did sun-cured alfalfa. There was no loss of riboflavin, pantothenic acid, niacin, or folacin during dehydration. Field-cured alfalfa was lower in riboflavin, and when alfalfa was exposed to rain there was a large loss of pantothenic acid and niacin (Scott, 1973).

Dehydration of alfalfa at 135°C resulted in an average 18% loss of α-tocopherol. When dehydrated alfalfa meal was stored for 12 weeks at 32°C, the α-tocopherol loss averaged 65% (Livingston *et al.*, 1968). Corn is often dried rapidly under high temperatures, resulting in losses of vitamin E activity and other heat-sensitive vitamins. When corn was artificially dried for 40 min at 88°C, losses of α-tocopherol averaged 19%, and when corn was dried for 54 min at 107°C, losses averaged 41% (Adams, 1973).

While pelleting generally improves the value of most energy and protein carriers in a feed, this is not true for some vitamins. During pelleting of feeds, three elements destructive for a number of vitamins are applied in combined action: heat, pressure, and humidity. Gadient (1986) reports that vitamins A, D_3, K_3, thiamin, and C are most likely to show stability problems in pelleted feeds. Pelleting of feed may have a beneficial effect on availability of vitamins such as niacin and biotin, which are often present in bound forms (Scott, 1973).

Processing with the goal of producing better quality fish meals and fish solubles under conditions where putrefaction is prevented has actually resulted in lower levels of vitamin K and vitamin B_{12} in these feedstuffs than were present when the products were allowed to undergo a considerable degree of putrefaction. Early studies on vitamin K and vitamin B_{12} proved that fish meal and rice bran exposed to the action of microorganisms showed increases in content of these vitamins. On the other hand, processing of many raw fishes with heat is required to inactivate a potent thiaminase that destroys thiamin. Also, attempts to preserve fish with nitrates led to the production of carcinogenic nitrosamines.

For human foods, each of the stages of food preparation and storage results in vitamin loss. Fruits and vegetables that are harvested long before use suffer heavy vitamin losses by enzymatic decomposition. Vitamin C is particularly liable to this type of destruction. In apples stored under domestic conditions the vitamin C content may fall to about one-third of the original value after only 2 or 3 months. Blanching vegetables before canning or freezing also results in vitamin loss. Estimates suggest that losses due to blanching fluctuate between 13 and 60% for vitamin C, 2 and 30% for thiamin, and 5 and 40% for riboflavin (Marks, 1975). Vitamin losses during the process of heat sterilization are generally small because oxygen is excluded during this process. Thiamin is the vitamin most likely to suffer because of its labile nature in heat and considerable thiamin losses have been observed from meat. Irradiation results in damage to vitamins, with the most sensitive vitamins being thiamin, riboflavin, vitamin A, and vitamin E, with niacin relatively stable. Storage in cans, freezing, and dehydration are all relatively good methods of preservation as far as vitamin retention is concerned.

C. Reduced Feed Intake

When feed intake is reduced, vitamin allowances should be adjusted to assure adequate vitamin intake for optimum performance. Restricting feed intake practices and/or improved feed conversion will decrease dietary intake of all nutrients, including vitamins. Restricted feeding of broiler breeders, turkey breeder hens, and gestating sows and gilts may result in marginal vitamin intake if diets are not adequately fortified (Roche, 1979b). Reduced feed intake may also result from stress and disease.

Use of high-energy feeds such as fats to provide diets with greater nutrient density for higher animal performance requires a higher vitamin concentration in feeds. Nonruminant species provided diets *ad libitum* consume quantities sufficient to meet energy requirements. Thus, vitamin fortification must be increased for higher energy diets as animals will consume less total feed. Feed consumption comparisons for broilers receiving metabolizable energy ranging from 2800 to 3550 kcal/kg were made (Friesecke, 1975). Feed consumption, and likewise vitamin consumption, was 19.1% lower for the diet with the greater energy density compared to the lowest energy diet.

Ambient temperature also has an important influence on diet consumption, as animals consume greater quantities during cold temperatures and reduced amounts as a result of heat stress. Vitamins, as well as other nutrients, must therefore be adjusted to reflect changing dietary consumptions.

D. Vitamin Variability and Insufficient Analysis

Tables of feed and food composition demonstrate the lack of complete vitamin information, with vitamin levels varying widely within a given feedstuff. Kurnick *et al.* (1972) found that, of the feeds surveyed, information on the niacin and riboflavin content of feedstuffs was more complete than for any other vitamins, whereas values for vitamin B_{12} and vitamin K were most deficient. Thus, 2 to 30% of the ingredients lacked either niacin, riboflavin, or pantothenic acid values, while 89 to 97% did not have values for either carotene, vitamin B_{12}, or vitamin K. Information about the other vitamins was not listed for 36 to 64% of the ingredients.

Variability of vitamin content within ingredients is generally large and difficult to quantify and anticipate. It is well recognized that vitamin levels shown in tables of vitamin composition of feedstuffs represent average values and that actual vitamin content of each feedstuff varies over a fairly wide range. Methods of processing and storage, as previously mentioned, account for variability as well as different analysis techniques. Often it is questionable whether the accuracy of vitamin levels from feedstuffs, calculated using tabular values, can be assured (Kurnick *et al.,* 1972). Proof of this is the statement by the 1982 NRC publication United States–Canadian Tables of Feed Composition that "organic constituents (e.g., crude protein, cell wall constituents, ether extract and amino acids) can vary as much as ±15%, the inorganic constituents as much as ±30% and the energy values as much as ±10%." Therefore, average values in feed composition tables may vary considerably from the nutrient value in a specific group of feeds.

E. Vitamin Bioavailability

Even accurate feedstuff analysis of vitamin concentrations does not provide the bioavailability data needed for certain vitamins. Bound forms of vitamins in natural ingredients often are unavailable to animals. Bioavailability of choline, niacin, and vitamin B_{12} is adequate in some feeds but limited or variable in other ingredients. For example, bioavailability of choline is 100% in corn but varies from 60 to 75% in soybean meal; that of niacin is 100% in soybean meal but zero in wheat and sorghum and varies from 0 to 30% in corn; that of vitamin B_6 is 65% in soybean meal and varies from 45 to 56% in corn (Roche, 1979b). The niacin in cereal grains and their by-products is in a bound form that is virtually unavailable to the pig and chick (Cunha, 1982). For alfalfa meal, corn, cottonseed meal, and soybean meal, bioavailability of biotin is estimated at 100% (Cunha, 1984a). However, biotin availability is variable for other feedstuffs, for example, 20–50% in barley, 62% in corn gluten meal, 30% in fish meal, 20–60% in sorghum, 32% in oats, and 0–62% in wheat. Likewise, ascorbic acid in cooked cabbage is present in the bound form, ascorbinogen, a form that is absorbed very poorly by humans (Marks, 1975).

Some data have been obtained with the pig showing that responses to vitamins may differ depending on whether vitamins are being added to a purified or to a natural diet (Cunha, 1977). Requirements of the pig for niacin, riboflavin, and pantothenic acid were considerably higher on a natural diet than requirements established earlier from experiments using purified diets (McMillen *et al.*, 1949). This shows that results obtained with purified diets must also be verified with natural diets and that bioavailability of vitamins may be greater in purified diets.

F. Computerized, Least-Cost Formulations

Vitamins are not always entered as specifications in computer formulations. Therefore, vitamin-rich feedstuffs, such as alfalfa, distiller's solubles or grains, brewer's grains, fermentation products, and meat, milk, and fish by-products, are often excluded or reduced when least-cost formulations are computed. The resulting least-cost diet consisting of a grain and soybean meal is usually lower in vitamins than a more complex one containing more costly vitamin-rich feeds (Roche, 1979a).

III. FACTORS AFFECTING VITAMIN REQUIREMENTS AND VITAMIN UTILIZATION

A. Physiological Makeup and Production Function

Vitamin needs of animals and humans depend greatly on their physiological makeup, age, health, nutritional status, and function, such as producing meat, milk, eggs, hair, or wool or developing a fetus (Roche, 1979a). For example, dairy cows producing greater volumes of milk have higher vitamin requirements than dry cows or cows producing low quantities. Breeder hens have higher vitamin requirements for optimum hatchability, since vitamin requirements for egg production are generally less than that for egg hatchability. Higher levels of vitamins A, D_3, and E are needed in breeder hen diets than in feeds for rapidly growing broilers. Selection for faster weight gains in swine and increased number of litters per year also demands elevated vitamin requirements (Cunha, 1980a, 1984b).

Different breeds and strains of animals have been shown to vary in their vitamin requirements. Vitamin needs of new strains developed for improved production are higher. Leg problems seen in fast-growing strains of broilers can be corrected in part by higher levels of biotin, folacin, niacin, and choline (Roche, 1979a).

B. Confinement Rearing without Access to Pasture

Moving of swine and poultry operations into complete confinement without access to pasture has had a profound effect on vitamin nutrition (as well as mineral nutrition). Pasture could be depended on to provide significant quantities of most vitamins since, young, lush, green grasses or legumes are good vitamin sources. More available forms of vitamins A and E are present in pastures and green forages containing ample quantities of β-carotene and α-tocopherol versus lower bioavailable forms in grains. Confinement rearing to include poultry in cages and swine on slatted floors has limited animal access to feces (coprophagy), which is rich in many vitamins. Confinement rearing requires producers to pay more attention to higher vitamin requirements needed for this management system (Cunha, 1984b).

C. Stress, Disease, or Adverse Environmental Conditions

Intensified production increases stress and subclinical disease level conditions because of higher densities of animals in confined areas. Stress and disease conditions in animals may increase the basic requirement for certain vitamins. A number of recent studies indicate that the nutrient levels that are adequate for growth, feed efficiency, gestation, and lactation may not be adequate for normal immunity and for maximizing the animal's resistance to disease (Cunha, 1985). Diseases or parasites affecting the gastrointestinal tract will reduce intestinal absorption of vitamins, from both dietary sources and those synthesized by microorganisms. If they cause diarrhea or vomiting this will also decrease intestinal absorption and increase needs. Vitamin A deficiency is often seen in heavily parasitized animals that supposedly were receiving an adequate amount of the vitamin. Mycotoxins are known to cause digestive disturbances such as vomiting and diarrhea as well as internal bleeding, and interfere with absorption of a dietary vitamins A, D, E, and K. In broiler chickens, moldy corn (mycotoxins) has been associated with deficiencies of vitamins D (rickets) and vitamin E (encephalomalacia) in spite of the fact that these vitamins were supplemented at levels regarded as satisfactory.

Mortality from fowl typhoid (*Salmonella gallinarum*) was reduced in chicks fed vitamin levels greater than normal (Hill, 1961). Vitamin E supplementation at a high level decreased chick mortality due to *Escherichia coli* challenge from 40 to 5% (Tengerdy and Nockels, 1975). Scott *et al.* (1982) concluded that coccidiosis produces a triple stress on vitamin K requirements as follows: (1) coccidiosis reduces feed intake, thereby reducing vitamin K intake: (2) coccidiosis injures the intestinal tract and reduces absorption of the vitamin; and (3) treatment with sulfaquinoxaline or other coccidiostats causes an increased requirement for vitamin K.

D. Vitamin Antagonists

Vitamin antagonists (antimetabolites) interfere with the activity of various vitamins, and Oldfield (1987) summarized the action of antagonists. The antagonist could cleave the metabolite molecule and render it inactive, as occurs with thiaminase and thiamin; it could complex with the metabolite, with similar results, as happens between avidin and biotin; or by reason of structural similarity it could occupy reaction sites and thereby deny them to the metabolite, as with dicumarol and vitamin K. The presence of vitamin antagonists in animal and human diets should be considered in adjusting vitamin allowances, as most vitamins have antagonists that reduce their utilization (see Chapters 2–17). Some common antagonists are:

1. Thiaminase, found in raw fish and some feedstuffs, is a thiamin antagonist. Pyrrithiamin is another thiamin antagonist.
2. Dicumarol, found in certain plants, interferes with blood clotting by blocking the action of vitamin K.
3. Avidin, found in raw egg white, and strepavidin, from *Streptomyces* molds, are biotin antimetabolites.
4. Rancid fats inactivate biotin and destroy vitamins A, D, and E and possibly others.
5. Oral contraceptives and drug therapy to control tuberculosis are antagonistic to vitamin B_6.

E. Use of Antimicrobial Drugs

Some antimicrobial drugs will increase vitamin needs of animals by altering intestinal microflora and inhibiting synthesis of certain vitamins. Some sulfonamides may increase requirements of biotin, folacin, vitamin K, and possibly others when intestinal synthesis is reduced. This may be of little significance except when drugs that are antagonistic toward a particular vitamin are added in excess, that is, sulfaquinoxaline versus vitamin K, amprolium versus thiamin, and sulfonamide potentiators versus folacin (Perry, 1978).

F. Levels of Other Nutrients in the Diet

Level of fat in the diet may affect absorption of the fat-soluble vitamins A, D, E, and K, as well as the requirement for vitamin E and possibly other vitamins. Fat-soluble vitamins may fail to be absorbed if digestion of fat is impaired. High cost of fat as an energy source has resulted in minimal fat levels in current, least-cost feed formulations, which may result in reduced absorption of fat-soluble vitamins (Roche, 1979b).

Many interrelationships of vitamins with other nutrients exist and therefore affect requirement. For example, prominent interrelationships exist for vitamin E with selenium, for vitamin D with calcium and phosphorus, for choline with methionine, and for niacin with tryptophan.

G. Body Vitamin Reserves

Body storage of vitamins from previous intake will affect daily requirements of these nutrients. This is more true for fat-soluble vitamins and vitamin B_{12} than for the other water-soluble vitamins. Fat-soluble vitamins are more inclined to remain in the body, which is especially true of vitamin A and/or carotene, which may be stored by an animal in its liver and fatty tissue in sufficient quantities to meet requirements for vitamin A for periods up to 6 months or even longer.

IV. OPTIMUM VITAMIN ALLOWANCES

The National Research Council (NRC) requirements for a vitamin are usually close to minimum levels required to prevent deficiency signs and for conditions of health and adequate performance, provided sufficient amounts of all other nutrients are supplied. Vitamin requirements for various animal species (NRC) and humans (RDA, 1980) are presented in each individual chapter and in Appendix Table 1. Most nutritionists usually consider NRC requirements for vitamins to be close to minimum requirements sufficient to prevent clinical deficiency signs and they may be adjusted upward according to experience within industry in situations where a higher level of vitamins is needed.

Allowances of a vitamin are those total levels from all sources fed to compensate for factors influencing vitamin needs of animals. These ''influencing factors'' include (1) those that may lead to inadequate levels of the vitamin in the diet (see Section II) and (2) those that may affect the animal's ability to utilize the vitamin under commercial production conditions (see Section III). The higher the allowance, the greater is the extent to which it may compensate for the ''influencing factors.'' Thus under commercial production conditions, vitamin allowances higher than NRC requirements may be needed to allow optimum performance (Roche, 1979a).

The concept of optimum vitamin nutrition under commercial production conditions is illustrated in Fig. 18.1 (Roche, 1979a). The marginal zone in Fig. 18.1 represents vitamin levels that are lower than requirements that may predispose animals to deficiency. The requirement zones are minimum vitamin quantities that are needed to prevent deficiency signs, but may lead to suboptimum performance even though animals appear normal. The optimum allowances in Fig. 18.1 permit animals to achieve their full genetic potential for optimum performance. In the excess zone, vitamin levels range from levels still safe, but uneconomical, to concentrations that may produce toxic effects. Usually only vitamins A and D, under practical feeding conditions, pose the possibility of toxicity problems for livestock. Optimum allowances of any vitamin are depicted as a range in Fig. 18.1 because factors influencing vitamin needs are highly variable and optimum allowances to allow maximum response may vary from animal to animal of the same species, type, and age within the same population and from day to day (Roche, 1979a).

It should be emphasized that subacute deficiencies can exist although the actual deficiency signs do not appear. Such borderline deficiencies are both the most costly and the most difficult with which to cope, and often go unnoticed and

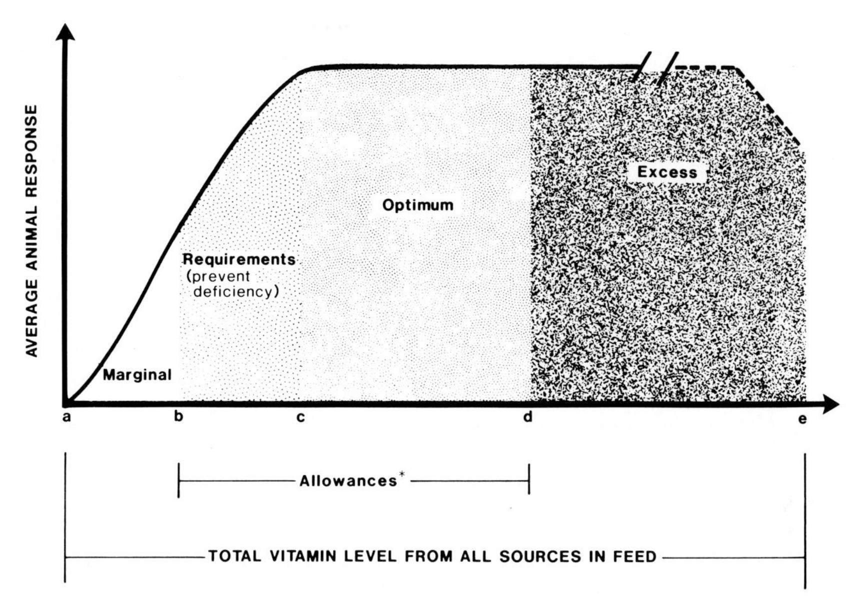

Fig. 18.1 Optimum vitamin nutrition for animals under commercial production conditions. Asterisk indicates total vitamin levels from all sources fed to compensate for factors affecting animals' vitamin needs. (Adapted from Roche, 1979a. Printed with permission of Hoffmann–LaRoche Inc., Nutley, N.J.)

unrectified, yet they may result in poor and expensive gains, impaired reproduction, or depressed production. Also, under farm conditions one will usually not find a single vitamin deficiency. Instead, deficiencies are usually a combination of factors, and often deficiency signs will not be clear-cut. If the NRC minimum requirement for a vitamin is the level that barely prevents clinical deficiency signs, then this level moves in relationship to the level required for optimum production responses. This means that if a greater quantity of a vitamin is required for an optimum response (because of the influencing factors), a greater quantity would also be required to prevent deficiency signs (Fig. 18.2). Similarly, if a lesser quantity is required for an optimum response, less would also be required to prevent deficiencies (Perry, 1978). Optimum animal performance required under modern commercial conditions cannot be obtained by fortifying diets to just meet minimum vitamin requirements. Establishment of adequate margins of safety must provide for those factors that may increase certain dietary vitamin requirements and for variability in active vitamin potencies and availability within individual feed ingredients.

The NRC requirements often do not take into account that certain vitamins have special functions in relation to disease conditions with higher than recommended levels needed for response (Cunha, 1985). In pigs artificially infected with *Treponema hyodysenteriae,* the agent causing diarrhea, high supplementation

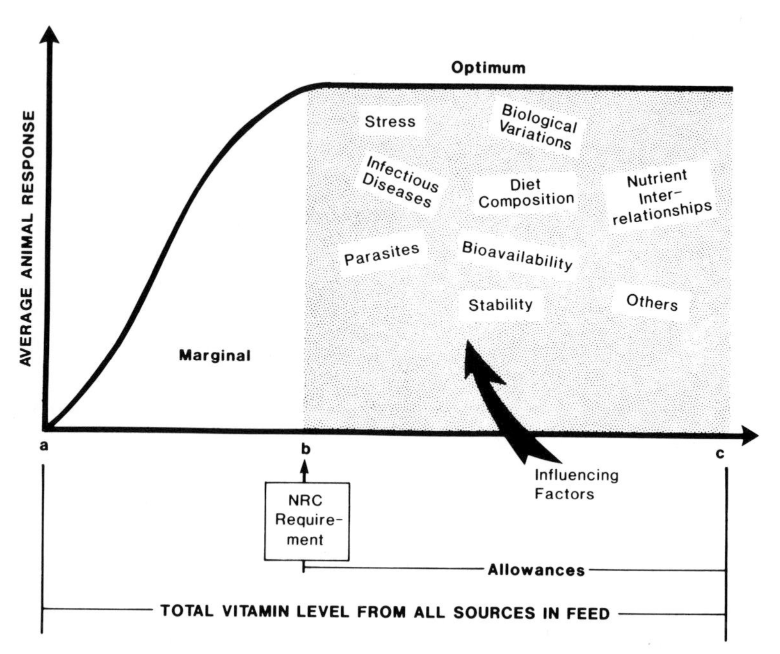

Fig. 18.2 Optimum vitamin nutrition for animals under commercial production conditions and influencing factors. (Adapted from Perry, 1978. Printed with permission of Hoffmann–LaRoche Inc., Nutley, N.J.)

with vitamin E (200 mg/day) in combination with selenium (0.2 mg/day) markedly reduced the number of pigs that became clinically ill (Tiege *et al.,* 1978). Clinical signs and pathological changes were less severe compared with vitamin E-deficient pigs. Thus, high doses of vitamin E increase resistance against disease. In practice, feeds contaminated with mycotoxins increase requirements for fat-soluble and other vitamins (e.g., biotin, folacin, and possibly others) and therefore supplementation should be increased above NRC minimum requirements. Apart from these fat-soluble vitamins, additions of folacin will also improve performance in pigs fed moldy grain (Purser, 1981) and of biotin for pigs fed feeds containing certain molds (Cunha, 1984a). Besides other nutrients, vitamins play a major role in the immune response, the body's defense system against infectious disease. Vitamin supplementation above requirements has been shown to be required for optimum immune responses (Ellis and Vorhies, 1976; Cunha, 1985).

V. VITAMIN SUPPLEMENTATION MOST NEEDED BY LIVESTOCK

Vitamin requirements, as previously noted, are highly variable within the various species and classes of animals. Supplementation allowances need to be set at levels that reflect different management systems and that are high enough to take care of fluctuations in environmental temperatures, energy content of feed, or other factors that might influence feed consumption or the vitamin requirements in other ways (McGinnis, 1986b). This section will briefly discuss vitamins that are normally provided in ruminant, poultry, swine, and horse diets.

A. Ruminants

Grazing ruminants generally only need supplemental vitamin A, if pastures are low in carotene, and possibly vitamin E (influenced by selenium status). Vitamin D is provided by ultraviolet light activity on the skin, while all other vitamins are provided by ruminal or intestinal microbial synthesis.

Ruminants housed under more strict confinement conditions generally require vitamins A and E and may require vitamin D if deprived of sunlight. Under specific conditions, relating to stress and high productivity, ruminants may be benefited by supplemental B vitamins, particularly thiamin (see Chapter 6) and niacin (see Chapter 8). Lee (1984) reports that adding a complete B-vitamin mixture to cattle entering the feedlot during the first month can reduce stress and increase gains. In one study, supplemental B vitamins given feedlot calves tended to reduce morbidity of animals (Zinn *et al.*, 1987). Apparently under stress conditions of feedlots, the microbial population in the rumen is not synthesizing certain B vitamins at adequate levels.

B. Poultry

Poultry under intensive production systems are particularly susceptible to vitamin deficiencies (Scott *et al.*, 1982). Reasons for this susceptibility are (1) poultry derive little or no benefit from microbial synthesis of vitamins in the gastrointestinal tract, (2) poultry have high requirements for vitamins, and (3) the high density concentration of modern poultry operations places many stresses on the birds that may increase their vitamin requirements. Typical grain–oilseed meal (i.e., corn–soybean meal) poultry diets are generally supplemented with vitamins A, D (D_3), E, K, riboflavin, niacin, pantothenic acid, B_{12}, and choline (Scott *et al.*, 1982). Thiamin, vitamin B_6 biotin, and folacin are usually present in adequate quantities in the major ingredients such as corn–soybean meal-based diets.

Vitamins A, D, riboflavin, and B_{12} are usually low in poultry diets. However, adding other vitamins to poultry diets is good insurance. The vitamins D and B_{12} are almost completely absent from diets based on corn and soybean meal. Vitamin K is generally added to poultry diets more than to those for other species because birds have less intestinal synthesis because of a shorter intestinal tract and faster rate of food passage. Birds in cages require more dietary K and B vitamins than those on floor housing because of more limited opportunity for coprophagy.

C. Swine

Vitamin supplementation of swine diets is obviously necessary (Figs. 18.3 and 18.4), with vitamin needs having become more critical in recent years as complete confinement feeding has increased. Swine in confinement, without access to vitamin-rich pasture, and housed on slatted floors, which limits vitamins available from feces consumption, have greater needs for supplemental vitamins. For swine the vitamins most likely to be marginal or deficient in corn–soybean diets are vitamins A, D, E, riboflavin, niacin, pantothenic acid, and B_{12}, and occasionally also vitamin K and choline.

Almost all swine diets in the United States are now fortified with vitamins A, D, B_{12}, riboflavin, niacin, pantothenic acid, and choline. An increasing number of feed manufacturers are adding vitamins E and K, and many are adding biotin and B_6 to diets. Diets are fortified with these vitamins even though not all experiments indicate a need for each of them. Most feed manufacturers add them as a precaution to take care of stress factors, subclinical disease level, and other conditions on the average farm that may increase vitamin needs (Cunha, 1977, 1984b).

Fig. 18.3 A lack of B vitamins causes small, weak pigs at birth. (Courtesy of T. J. Cunha and University of Florida.)

Fig. 18.4 The effect of B vitamin supplementation on deficient pigs obtained from farms in Michigan. The group of pigs in (A) were about 80 days old and averaged 9.1 kg in weight. The (B) group shows these same pigs after 35 days of vitamin supplementation. (Courtesy of R. W. Luecke, Michigan State University.)

D. Horses

There is a lack of experimental information on the level of vitamins required in well-balanced horse diets, as well as on which vitamins need to be added (Cunha, 1980b). The vitamins most likely deficient for all classes of horses are vitamins A and E, with vitamin D also deficient for horses in confinement. Inadequate vitamin D may be provided to racehorses that are exercised only briefly in the early morning, when sunlight provides less antirachitic protection (see Chapter 3). Requirements for vitamins A, D, and E can be met with a high-quality (e.g., green color) sun-cured hay. Deficiencies of vitamin K and the B vitamins appear to be less likely in the mature horse than in other monogastric species as many vitamins occur in the cecum of the horse. It is not known, however, how much of the vitamins synthesized in the cecum are absorbed in the large intestine. Since it is difficult to depend on the intestinal synthesis, many horse owners use B-vitamin supplementation of diets for the young horse and those being developed for racing or performance purposes (Cunha, 1980b).

In recent years, vitamin supplementation has become more critical to the horse as the trend toward total confinement has increased. Presently, few horses receive a high level of vitamin intake from a lush, green pasture or from a high-quality, leafy, green hay. Cunha (1980b) suggested that a vitamin premix for horses contain vitamins A, D, E, K, thiamin, riboflavin, niacin, B_6, pantothenic acid, folacin, B_{12}, and choline. Biotin supplementation is also recommended, as Comben *et al.* (1984) showed a benefit of the vitamin for hoof integrity.

VI. VITAMIN SUPPLEMENTATION FOR HUMANS

People in developed countries consume an average of 3000 calories and almost 40 g of protein daily, while those in developing countries receive two-thirds as many calories and only one-fifth the protein consumption. Thus calorie and protein deficiency are the most important problems in nutrition at the present time. Vitamin deficiencies usually coexist with protein-caloric deficiency. In developed countries, on the other hand, inappropriate eating patterns can lead to poor nutrition and result in vitamin deficiencies (Marks, 1975).

A number of factors influence the vitamin deficiency state in humans. Diseases caused by primary or dietary deficiency result solely from inadequate quantities of the vitamin in the diet. Lack of vitamin intake can result from a number of factors (Marks, 1975), including crop failure, losses during food preparation and storage, poverty, ignorance, loss of appetite, apathy, food taboos or fads, dental problems, and chronic disease. Even social changes can affect dietary intakes. For example, old people, particularly those living alone, tend to abandon ordinary meal preparations and live on small amounts of "easy foods" (coffee, soups, and bread) that can lead to scurvy and other deficiency diseases.

There are vitamin deficiencies, however, even under conditions where diet would normally be considered adequate by most standards. Such deficiencies, usually referred to as secondary or conditioned deficiencies, arise from metabolic stress or organic disease. The increased metabolic demands during pregnancy and lactation are prime examples of metabolic stress states requiring extra vitamin therapy. Use of oral contraceptives and various drugs such as tranquilizers and antibiotics can result in conditioned deficiencies. Particular attention to vitamin supplementation is needed for individuals with inborn errors of metabolism, where requirements are much higher for certain vitamins, and feeding of hospital patients through total parenteral nutrition.

Certain disorders or diseases and some medications may interfere with vitamin intake, digestion, absorption, metabolism, or excretion and thus change requirements. The stress placed on carbohydrate metabolism in the alcoholic is another example of a condition requiring more than normal daily vitamin consumption. A number of B-vitamin deficiencies, including those of thiamin, niacin, B_6, B_{12}, and folacin, have resulted from alcoholism. For example, a major cause of folacin deficiency in the United States is chronic alcoholism. Diets of alcoholics are likely to contain inadequate quantities of a number of nutrients.

Another important cause of conditioned deficiency results from disease. In these cases a defect in the host results in decreased absorption of the vitamin from the diet. Pernicious anemia is a conditioned vitamin B_{12} deficiency because a lack of gastric intrinsic factor severely limits absorption of the vitamin from the gastrointestinal tract. Another example of this type is vitamin K deficiency that results from poor absorption and is caused by inadequate supplies of bile. Vitamin therapy in such cases is usually in the form of parenteral or intravenous injection.

Typical well-balanced diets that include ample quantities of vegetables, fruits, and animal products should provide vitamins in adequate quantities. For most individuals consuming such diets there should be no need for additional supplemental vitamins. Nevertheless, many circumstances, including individual variations, lack of proper diet, older age, and physiological and emotional stress, may warrant vitamin supplementation. Many individuals also consume multiple vitamins (and sometimes minerals) daily, which gives them a sense of well-being and in essence is a type of nutritional insurance against possible deficiencies. In addition to vitamin deficiencies, it is also important to consider possible effects of excessive quantities of vitamins (hypervitaminosis) in human diets, particularly regarding vitamin A and to a lesser degree vitamin D.

The vitamins least likely to be needed for supplementation to human diets are choline, pantothenic acid, and biotin. Conditions in which additional vitamins are warranted for supplementation are:

1. Vitamin A—The deficiency is widespread in the world and is most common in young children who receive inadequate green and yellow vegetables and foods derived from animals.

2. Vitamin D—In developed countries there is vitamin D fortification of human diets, however, most of the world depends on sunlight. Children not receiving sufficient sunlight are at great risk of the deficiency and also older populations need to consider supplementation. Decreased ability of the kidney to convert 25-OHD to 1,25-$(OH)_2D$ is a problem in the aged (Tsai *et al.*, 1984). Thus even with more sunlight and more dietary intake, it is possible that the

aged would not maintain equivalent activity of $1,25(OH)_2D$. Diseases and parasites that impair the liver and kidney's ability to convert vitamin D to $1,25(OH)_2D$ may also be problems especially in developing countries.

3. Vitamin E—This vitamin is most likely deficient in individuals consuming low-fat diets or diets high in unsaturated fats and would be more critical in regions deficient in selenium, as vitamin E supplementation has been claimed to be beneficial for diverse disease conditions.

4. Vitamin K—Newborns are commonly given a single dose of vitamin K to prevent abnormal bleeding. Many persons take aspirin for pain, to minimize coronary problems, and for other reasons. There are indications that this may increase vitamin K needs.

5. Thiamin—Individuals subsisting mostly on refined grains (i.e., polished rice) are most likely in need of thiamin supplementation. Humans in developed countries have for a number of years received food fortified with thiamin (i.e., bread enrichment). Rice enrichment in Far Eastern countries has dramatically reduced the incidence of beriberi.

6. Riboflavin—Often individuals in developing countries do not have the economic resources to consume major dietary sources of riboflavin such as meat, milk, and dairy products.

7. Niacin—Supplementation is important when diets are primarily based on corn and sorghum and for individuals with metabolic defects (i.e., conversion of tryptophan to niacin) or when grain has not been heated or treated sufficiently to make niacin available. Pharmacological effects have included reduction of serum cholesterol and treatment of schizophrenia.

8. Vitamin B_6—Supplementation is most important for young, pregnant, or lactating women. Requirements are increased by oral contraceptives, certain drug therapy, radiation sickness, urinary calculi, and in errors of metabolism. Vitamin B_6 has been used to alleviate nausea during pregnancy and in megadose therapy for a wide variety of conditions, including premenstrual syndrome and behavioral disorders.

9. Folacin—A high percentage of the human population is estimated to be at a moderate to high risk for folacin deficiency (see Chapter 12) during growth, pregnancy, and in old age. Supplementation is often needed for individuals not consuming optimum quantities of fruits and vegetables.

10. Vitamin B_{12}—Organ meats and other animal products are the best sources of vitamin B_{12}. Vegetarians as well as individuals with the malabsorption defect of pernicious anemia need B_{12} supplementation.

11. Vitamin C—Supplementation is likely needed for individuals not consuming optimum quantities of fruits and vegetables. Megavitamin doses of C for most individuals are not harmful.

12. Essential fatty acids—Under typical conditions linoleic acid ($18:\omega 2$) deficiency would not be expected for humans. Recently the need for supplemental

linolenic acid (18 : ω3) has been suggested for prevention of cardiovascular disorders.

VII. PROVIDING VITAMIN SUPPLEMENTS

The physical and chemical forms of various vitamin products currently used to fortify feeds are usually different from the forms found in feedstuffs. Modification of these naturally occurring vitamin forms is required to improve their stability, compatibility, dispersion, and handling characteristics for feed fortification.

Various chemical and physical vitamin forms available for supplementation are presented in Table 18.1 (Adams, 1978). To obtain sufficient stability, compatibility with other feed components, and the properties required for application, vitamin producers try to devise vitamin forms according to the following methods (Schneider and Hoppe, 1986):

Synthesis of stable derivatives;
Addition of stabilizing agents;
Coating;
Absorption of liquid vitamins on suitable carriers;
Transformation of fat-soluble or poorly soluble vitamins into water-soluble or dispersible forms;
Transformation of water-soluble vitamins or derivatives into poorly water-soluble or fat-soluble forms;
Standardization of content;
Providing high bioavailability.

In view of the nutritional importance of vitamins A, D, and E, many commercial vitamin producers have succeeded in enhancing stability of these vitamins in two ways: (1) by mechanical means of enveloping minute droplets of the vitamin or vitamins in a stable fat or gelatin, forming small beads, and thus preventing most of the vitamin from coming into contact with oxygen until it is digested in the animal intestinal tract, and (2) through use of effective antioxidants that markedly prolong the induction period that precedes active vitamin oxidation. The "stabilized" beadlet containing an effective antioxidant will protect these vitamins for storage periods up to 4–8 weeks without much loss of vitamin potency (Scott, 1973). Instability of vitamin D in peroxidizing diets is often overlooked. Studies have shown that very high dietary levels of vitamin D are completely destroyed in diets containing high levels of peroxidizing polyunsaturated fatty acids, and chicks suffer severe rickets by 3 weeks of age (Scott, 1973). Rickets was prevented by a normal level of vitamin D when the diet was supplemented with ethoxyquin at a level of 57 mg/kg of diet.

TABLE 18.1

Product Forms of Vitamins and Application Uses[a]

Vitamin	Product form: Chemical	Product form: Physical	Application uses
A	Alcohol (retinol)	Crystalline or liquid	Liquid and dry oral pharmaceuticals
	Acetate and/or palmitate	Beadlets	Dry feeds; dry oral pharmaceuticals
		Spray- or drum-dried powders	Dry feeds; dry oral pharmaceuticals; water-dispersible vitamin products
		Liquid concentrates	Liquid feed supplements; oral and parenteral pharmaceuticals
		Oil dilutions	Nonpelleted feeds
		Oil absorbates	Nonpelleted feeds
	Propionate	Liquid concentrates	Oral and parenteral pharmaceuticals; liquid feed supplements
D (D_2 and D_3)	Ergocalciferol (D_2) or cholecalciferol (D_3)	Beadlets (with vitamin A)	Dry feeds; dry oral pharmaceuticals
		Spray- or drum-dried powders	Dry feeds; dry oral pharmaceuticals
		Liquid concentrates (with vitamin A)	Liquid feed supplements; oral and parenteral pharmaceuticals
		Oil dilutions	Nonpelleted feeds
		Oil absorbates	Nonpelleted feeds
E	*d*- or *dl*-α-Tocopheryl acetate	Powders or oils	Feeds, foods, and pharmaceuticals
	Mixed tocopherols	Oils	Foods and pharmaceuticals
	d-α-Tocopheryl succinate	Powder	Pharmaceuticals
K	Menadione (K_3), MSB (menadione sodium bisulfite), MSBC (menadione sodium bisulfite complex), or MPB (menadione pyrimidinol bisulfite)	Dry dilutions	Dry feeds or pharmaceuticals
		Water-dispersible powders	Water-dispersible vitamin products
	Phytomenadione (K_1)	Liquid	Parenteral pharmaceuticals

(Continued)

TABLE 18.1 (*Continued*)

Vitamin	Product form		Application uses
	Chemical	Physical	
Thiamin	Thiamin mononitrate	Crystalline	Feeds, foods, and pharmaceuticals
		Dry dilutions	Feeds and foods
	Thiamin hydrochloride	Dry dilutions	Feeds
		Crystalline	Parenteral and oral pharmaceuticals; feeds
Riboflavin	Riboflavin: chemically synthesized crystalline product; fermentation product	High-potency powder; spray-dried powders	Feeds, foods, and dry oral pharmaceuticals
	Riboflavin phosphate	Water-soluble powder	Parenteral and liquid oral pharmaceuticals
Niacin	Niacin (niacinamide; nicotinic acid)	Crystalline	Feeds and pharmaceuticals
		Dry dilutions	Feeds
Vitamin B_6	Pyridoxine hydrochloride	Dry dilution	Feeds
		Crystalline	Oral and parenteral pharmaceuticals
Pantothenic acid	Calcium *d*- or *dl*-pantothenate	Powders	Feeds, foods, and pharmaceuticals
	Calcium *dl*-pantothenate–calcium chloride complex	Powder	Feeds
	d-Panthenol	Liquid	Parenteral and liquid oral pharmaceuticals; cosmetics
Biotin	*d*-Biotin	Crystalline	Feeds, foods, and pharmaceuticals
		Dry dilutions	Feeds
Folacin	Folacin	Crystalline	Feeds, foods, and pharmaceuticals
		Dry dilutions	Feeds
B_{12}	Vitamin B_{12} (cyanocobalamin):		
	(a) Crystalline product from fermentation	Dry dilutions	Feeds
	(b) Chemically synthesized crystalline product	Water-soluble dilutions	Water-dispersible vitamin products; oral and parenteral pharmaceuticals

(*Continued*)

TABLE 18.1 (*Continued*)

Vitamin	Product form		Application uses
	Chemical	Physical	
C	Ascorbic acid	Dry dilutions	Feeds
		Coated products	Feeds and foods
		Crystalline	Parenteral pharmaceuticals; feeds
	Sodium ascorbate	Powder	Antioxidant–preservative for foods
Choline	Choline chloride	70% liquid	Feeds
		25–60% dry powders	Feeds
		Crystalline	Pharmaceuticals
	Choline bitartrate	Water-soluble powders	Water-dispersible vitamin products; pharmaceuticals

[a]Modified from Adams (1978). Printed with permission of Hoffmann–LaRoche Inc., Nutley, N.J.

Stress factors affecting vitamin stability during manufacture and storage of a custom premix may include heat, oxygen, moisture, oxidation, reduction, trace minerals, and pH. Stability of some vitamins is not affected (Frye, 1978). Riboflavin, niacin, *d*-biotin, *d*-pantothenic acid, and choline generally have excellent stability in custom premixes. Other vitamins such as vitamin A acetate, vitamin D_3, and vitamin E acetate are also available in stabilized forms in custom premixes. Vitamin K, unstabilized vitamin A, unstabilized vitamin D, thiamin, folacin, vitamin C, vitamin B_6, and vitamin B_{12} have poor stability in custom premixes under various stress conditions.

Most vitamins are supplemented in dry form, however, liquid vitamin supplements are useful for certain feeding operations. Their appeal is broad, the advantages being lower feed cost per kilogram of gain, conveniences in handling, ready adaptation in formulation, easy control, reduced time and labor, control of feed dust, and good palatability (Perry, 1968). Liquid supplements are not without disadvantages such as physical stability and corrosion potential, most of which are minimal with properly formulated supplements and equipment and adequate feeder use knowledge (Bauernfeind, 1969). Ready availability of the many water-soluble or water-dispersible nutrients such as energy sources (molasses), nonprotein nitrogen (urea), trace minerals, and vitamins (A, D, E) have permitted standard and custom formulation for ruminants in feedlots and on pasture. Special economical liquid emulsions of vitamins A, D, and E, which remain acceptably uniform in physical distribution and with adequate stability patterns, are marketed for this purpose.

Where control of animal feed intake is not complete but where there is good control over water consumption, special vitamin A, D, and E dry beadlet or liquid products are available to add to drinking water. They disperse in cold water into a fine emulsion with desirable physical, taste, and chemical stability characteristics for short time use. This vitamin form is useful in drinking water for poultry and other monogastric animals. Water-soluble vitamins and drugs, in properly formulated products, can also be added to water.

Because of the high variability and unknown bioavailability of vitamins in feeds, feed manufacturers have come to rely to a large extent on synthetic vitamin supplementation, and the vitamins used in human and animal nutrition are also predominantly of synthetic origin. In practice, feed manufacturers usually ignore to a certain degree the contribution of many vitamins in natural feeds and provide most vitamin requirements for a given species with low-cost synthetic vitamins. Bauernfeind (1969) summarized the advantages of synthetic vitamins in animal feeds:

1. Biological and physical characteristics are known; potency is uniform; stability is adequate with few exceptions; supply is usually unlimited.
2. Weight added per ton to the animal diet is small, hence adjustments upward or downward or in different ratios can be made without upsetting the remainder of the diet.
3. Ready-made combinations can be formulated in premix form for quick addition to specific diets for a given species of animal for a defined production objective with assurance of known diet values.
4. Cost is economical, hence vitamin restrictions of individual natural feed ingredients can be removed from computer program, thus increasing the flexibility of programming for least-cost energy and protein needs and shortening computer operation time.
5. Assay costs of determining variability of natural vitamin content of feedstuffs can be decreased by supplementing with the chemically produced nutrients.
6. Assay costs of determining the vitamin content of the final mixed diet can be decreased since assay of one or two components of the premix will give confidence of the mixing adequacy of all the premix vitamins in the final diet.
7. Use of chemically prepared vitamins eliminates unknowns existing at times in natural ingredients or seasonal nutrient variation, physiological nutrient availability, insecticidal residues in products, etc.
8. Chemically prepared vitamins have versatility in form and applications—adaptability to dry feeds, liquid supplements, drinking water solutions, drenches, and parenteral and capsule forms.

VIII. FORMULATING VITAMIN PREMIXES

The custom vitamin premix is generally considered to be a concentrated mixture of vitamins added in the manufacture of complete feeds or feed supplements, and it is customized to meet vitamin specifications of the feed manufacturer.

Use of custom vitamin premixes in feed manufacturing facilitates uniform distribution of vitamins in complete feeds and assures that specifications for vitamin fortification of feeds are met.

Vitamin premixes are mixtures of specific vitamins required in combination with some type of carrier material that is added to feeds at time of mixing to ensure uniform distribution of these ingredients in mixed feeds. Commonly used carriers include (1) soybean meal, (2) ground grain, (3) corn gluten meal, (4) wheat middlings, and (5) several other mill feeds. Premixes are formulated so that they can be packaged for convenient quantity (i.e., 10–20 lb) additions for each ton of complete feed.

Custom vitamin premixes should possess certain physical and chemical properties to ensure optimum quality and dispersion in finished feeds. Desirable properties suggested by Aiello (1978) are:

1. Proper particle size distribution. A good standard is for 100% passing through a #20 U.S. standard sieve.
2. Free of contaminants.
3. Minimum of moisture. Excessive moisture (>12%) can promote destruction of certain vitamins and also promote bacteria and mold growth.
4. Desirable mixing properties: free-flowing, noncaking, nondusty, nonelectrostatic, nonhygroscopic, and nonsegregating.
5. Favorable pH. Most vitamins show greatest stability at approximately pH 5.5.
6. Proper bulk density. Bulk density of 30 ± 5 pounds per cubic foot provides proper particle size and free-flowing properties and facilitates premix handling and packaging.
7. Storage potential. Premixes should also be formulated to assure optimum vitamin levels and stability under practical storage conditions. This includes addition of antioxidants to improve stability of vitamins such as A, D, and E.

Bibliography

Aaes-Jorgensen, E., and Dam, H. (1954). *Br. J. Nutr.* **8,** 290–306.

AAP (American Academy of Pediatrics, Committee on Nutrition) (1977). *Pediatrics* **60,** 19–530.

Abad, A. R., and Gregory, J. F. (1987). *J. Nutr.* **117,** 866–873.

Abe, T. (1969). *J. Vitaminol.* **15,** 339–340.

Abrams, J. T. (1952). *Vet. Rec.* **64,** 151–157.

Abrams, J. T. (1978). *In* "Handbook Series in Nutrition and Food, Section E: Nutritional Disorders" (M. Rechcigl, Jr., ed.), Vol. 2, pp. 179–194. CRC Press, Boca Raton, Florida.

Achuta Murthy, P. N., and Mistry, S. P. (1977). *Prog. Food Nutr. Sci.* **2,** 450–455.

Adams, C. R. (1973). "Effect of Processing on the Nutritional Value of Feeds." pp. 142–150. National Academy of Sciences, Washington, D.C.

Adams, C. R. (1978). *Proc. Roche Vitam. Nutr. Update Meet.,* Hot Springs, Arkansas pp. 54–69.

Adams, C. R. (1979). *Anim. Nutr. Health* June-July, 23–25.

Adams, C. R. (1982). *In* "Vitamins—The Life Essentials." Nutrition Institute, National Feed Ingredient Association, NI-82, 1–9, Des Moines, Iowa.

Adams, C. R., and Zimmerman, C. R. (1984). *Feedstuffs* **56,** 35–41.

Adams, C. R., Richardson, C. E., and Cunha, T. J. (1967). *J. Anim. Sci.* **26,** 903 (Abstr.).

Adamstone, F. B., Krider, J. L., and James, M. F. (1949). *Ann. N.Y. Acad. Sci.* **52,** 260–268.

Ahluwalia, B., Pincus, G., and Holman, T. (1967). *J. Nutr.* **92,** 205–214.

Aiello, R. (1978). *Proc. Roche Vitam. Nutr. Update Meet., Hot Springs, Arkansas* pp. 54–69.

Aitken, F. C., and Hankin, R. G. (1970). "Vitamins in Feeds for Livestock." Commonwealth Bureau of Animal Nutrition, Techn. Comm. No. 25, Central Press, Aberdeen.

Almquist, H. J. (1971). *In* "The Vitamins," (W. H. Sebrell, Jr. and R. S. Harris, eds.), Vol. 3, 2nd Ed., pp. 418–443. Academic Press, New York.

Almquist, H. J. (1978). *In* "Handbook Series in Nutrition and Foods, Section E: Nutritional Disorders" (M. Rechcigl, Jr., ed.), Vol. 2, pp. 195–204. CRC Press, Boca Raton, Florida.

Almquist, H. J., and Stokstad, L. R. (1935). *Nature (London)* **136,** 31.

Anderson, P. H., Berrett, S., and Patterson, D. S. P. (1976). *Vet. Rec.* **99,** 316–318.

Anonymous (1961). "Vitamin Manual," p. 53. Upjohn, Kalamazoo, Michigan.

Anonymous (1969). "Riboflavin," No. 1170. Hoffmann-LaRoche, Basel.

Anonymous. (1972). "Vitamin E," No. 1206. Hoffmann-LaRoche Basel.

Anonymous (1977). *Nutr. Rev.* **35,** 305–309.

Anonymous (1981a). "Rationale for Roche Recommended Vitamin Fortification," RCD 5963/1280. Hoffmann-LaRoche, Nutley, New Jersey.

Anonymous (1981b). "Biotin for Swine," RCD 5737. Roche Chemical Division, Hoffmann-LaRoche, Nutley, New Jersey.

Anonymous (1982). *Nutr. Rev.* **40,** 187–189.

Anonymous (1984a). Vitamin B_{12}. *Roche Tech. Bull.* pp. 1–8.

Anonymous (1984b). *Nutr. Rev.* **42,** 317–318.

Anonymous (1985a). *Nutr. Rev.* **43,** 23–25.

Anonymous (1985b). *Nutr. Rev.* **43,** 268–270.
Anonymous (1985c). *Nutr. Rev.* **43,** 303–305.
Anonymous (1985d). *Nutr. Rev.* **43,** 350–352.
Anonymous (1986). *Nutr. Rev.* **44,** 151–154.
Anonymous (1987). *Nutr. Rev.* **45,** 142–148.
AOAC (Association of Official Agricultural Chemists) (1965). "Official Method of Analysis." AOAC, Washington, D.C.
Arneil, G. C. (1975). *Proc. Nutr. Soc.* **34,** 101–109.
Arnrich, L., Lewis, E. M., and Morgan, A. F. (1952). *Proc. Soc. Exp. Biol. Med.* **80,** 401–404.
Arykroyd, W. R. (1958). *Fed. Proc., Fed. Am. Soc. Exp. Biol.* **17,** 103–143.
Atkins, K. B., Erdman, R. A., and Vandersall, J. H. (1983). *J. Dairy Sci.* **66** (Suppl. 1), 175.
Axelrod, A. E. (1971). *Am. J. Clin. Nutr.* **24,** 265–271.
Babu, S., and Skrikantia, S. G. (1976). *Am. J. Clin. Nutr.* **29,** 376–379.
Badwey, J. A., and Karnovsky, M. L. (1980). *Annu. Rev. Biochem.* **49,** 695–726.
Bailey, L. B., Mahan, C. S., and Dimperio, D. (1980). *Am. J. Clin. Nutr.* **33,** 1997–2001.
Baker, H., Jaslow, S. P., and Frank, O. (1978). *J. Am. Geriatr. Soc.* **26,** 218–221.
Baker, S. J., and DeMaeyer, E. M. (1979). *Am. J. Clin. Nutr.* **32,** 368–417.
Bar, A., Edelstein, S., Eisner, U., Ben-Gal, I., and Hurwitz, S. (1982). *J. Nutr.* **112,** 1779–1786.
Barash, P. G. (1978). *In* "Handbook Series in Nutrition and Food, Section E: Nutritional Disorders" (M. Rechcigl, Jr., ed.), Vol. 1, pp. 97–100. CRC Press, Boca Raton, Florida.
Barboriak, J. J., and Krehl, W. A. (1957). *J. Nutr.* **63,** 601–609.
Barclay, L. L., and Gibson, G. E. (1982). *J. Nutr.* **112,** 1899–1905.
Barnes, M. J., and Kodicek, K. (1972). *Vitam. Horm.* **30,** 1–43.
Barnes, R. H., and Fiala, G. (1958). *J. Nutr.* **65,** 103–114.
Barnes, R. H., and Kwong, E. (1967). *J. Nutr.* **92,** 224–232.
Barnes, R. H., Fiala, G., McGehee, B., and Brown, A. (1957). *J. Nutr.* **63,** 489–498.
Barness, L. A. (1977). *In* "Re-evaluation of Vitamin C" (A. Hanck and G. Ritzel, eds.), pp. 23–29. Huber, Bern.
Barnett, B. J., Young Cho, C., and Slinger, S. J. (1982). *J. Nutr.* **112,** 2011–2019.
Barton, C. R., and Allen, W. M. (1973). *Vet. Rec.* **92,** 288–290.
Batres, R. O., and Olson, J. A. (1987). *J. Nutr.* **117,** 77–82.
Bauernfeind, J. C. (1969). *World Rev. Anim. Prod.* **5,** 20–50.
Bauernfeind, J. C. (1972). *J. Agric. Food Chem.* **20,** 456–473.
Bauernfeind, J. C. (1974). *Feedstuffs* **46**(45), 30–31.
Bauernfeind, J. C. (1980). *In* "Vitamin E. A Comprehensive Treatise" (L. J. Machlin, ed.), p. 99. Dekker, New York.
Bauernfeind, J. C., and DeRitter, E. (1972). *Feedstuffs* **44**(50), 34–41.
Baugh, C. M. and Krumdieck, C. L. (1971). *Ann. N.Y. Acad. Sci.* **186,** 7–28.
Bechgaard, H. and Jespersen, S. (1977). *J. Pharm. Sci.* **66,** 871–872.
Bechtel, H. E. (1964). *Feedstuffs* **36** (45), 18.
Becker, R. B., Henderson, J. R., and Leighty, R. B. (1965). *Fla. Univ. Agric. Exp. Stn. Tech. Bull.* 699.
Beer, A. E., Scott, M. L., and Nesheim, M. C. (1963). *Br. Poult. Sci.* **4,** 243–253.
Beeson, W. M. (1965). *Fed. Proc., Fed. Am. Soc. Exp. Biol.* **24,** 924–926.
Bell, R. G. (1978). *Fed. Proc., Fed. Am. Soc. Exp. Biol.* **37,** 2599–2604.
Berzin, N., and Bauman, V. K. (1987). *Br. J. Nutr.* **57,** 255–268.
Best, C. H., Channon, H. J., and Ridout, J. H. (1934). *J. Physiol. (London)* **78,** 409.
Bhavanishankar, T. N., Shantha, T., and Ramesh, H. P. (1986). *Nutr. Rep. Int.* **33,** 603–612.
Bieri, J. G., and Prival, E. L. (1965). *Proc. Soc. Exp. Med.* **120,** 554–557.
Bindas, E. M., Gwazdauskas, F. C., Aiello, R. J., Herbein, J. H., McGilliard, M. L., and Polan, C. E. (1984). *J. Dairy Sci.* **67,** 1249–1255.

Birdsall, J. J. (1975). "Technology of Fortification of Foods." pp. 19–31. National Academy of Sciences, Washington, D.C.
Blakley, R. L., and Benkovic, S. J. (1984). "Folates and Pterins." Wiley, New York.
Blaxter, K. L. (1962). *Proc. Nutr. Soc.* **21,** 211–216.
Boissonneault, G. A., and Johnston, P. V (1983). *J. Nutr.* **113,** 1187–1194.
Bonjour, J.-P. (1980). *Int. J. Vitam. Nutr. Res.* **50,** 215–230.
Bonjour, J.-P. (1984). *In* "Handbook of Vitamins" (L. J. Machlin, ed.), pp. 403–435. Dekker, New York.
Bordier, P., Rasmussen, H., Marie, P., Miravet, L., Gueris, J., and Ryckwaert, A. (1978). *J. Clin. Endocrinol. Metab.* **46,** 284–294.
Borum, P. R. (1981). *Nutr. Rev.* **39,** 385–390.
Borum, P. R. (1985). *Can. J. Physiol. Pharmacol.* **63,** 571–576.
Bosco, D. (1980). "The People's Guide to Vitamins and Minerals from A to Zinc." Beaverbooks, Ontario, Canada.
Bostedt, H. (1980). *Prakt. Tierarzt (Special Issue Collegium Vet.)* **61,** 45–50.
Bragdon, J. H., and Levine, D. H. (1949). *Am. J. Pathol.* **25,** 265–271.
Braithwaite, G. D. (1974). *Br. J. Nutr.* **31,** 319–331.
Braithwaite, G. D., and Riazuddin, S. H. (1971). *Br. J. Nutr.* **26,** 215–225.
Braithwaite, G. D., Glascock, R. F., and Riazuddin, S. H. (1969). *Br. J. Nutr.* **23,** 827–834.
Braithwaite, G. D., Glascock, R. F., and Riazuddin, S. H. (1972). *Br. J. Nutr.* **27,** 417–424.
Bräunlich, K. (1974). "Vitamin B_6," No. 1451. Hoffmann-LaRoche, Basel.
Bräunlich, K., and Zintzen, H. (1976). "Vitamin B_1," No. 1593 F. Hoffmann-LaRoche, Basel.
Bray, D. L., and Briggs, G. M. (1984). *J. Nutr.* **114,** 920–928.
Brent, B. E. (1985). *Feed Manage.* **36,** 8–24.
Brent, B. E., and Bartley, E. E. (1984). *J. Anim. Sci.* **59,** 813–822.
Brethour, J. R. (1972). *J. Anim. Sci.* **35,** 260 (Abstr.).
Brief, S., and Chew, B. P. (1985). *J. Anim. Sci.* **60,** 998–1004.
Briggs, G. M., Hill, E. G., and Canfield, T. H. (1953). *Poult. Sci.* **32,** 678–680.
Brin, M. (1962). *Ann. N.Y. Acad. Sci.* **98,** 528–541.
Brin, M. (1971). *Am. J. Clin. Nutr.* **24,** 704–708.
Brody, T., Shane, B., and Stokstad, E. L. R. (1984). *In* "Handbook of Vitamins" (L. J. Machlin, ed.), pp. 459–496. Dekker, New York.
Brooks, P. H. (1982). *Pig News Inf.* **3,** 1–4.
Brooks, P. H., Smith, D. A., and Irwin, V. C. R. (1977). *Vet. Rec.* **101,** 46–50.
Brown, E. D., Chen, W., and Smith, J. C. (1976). *J. Nutr.* **106,** 563–568.
Brown, G. M. (1962). *J. Biol. Chem.* **237,** 536–540.
Brown, R. G. (1984). *In* "Proc. Ascorbic Acid in Domestic Animals" (I. Wegger, F. J. Tagwerker, and J. Moustgaard, eds.), pp. 60–67. Danish Agriculture Society, Copenhagen.
Brown, R. G., Sharma, V. D., and Young, L. G. (1970). *Can. J. Anim. Sci.* **50,** 605–609.
Bruckner, G., Gurnewald, K. K., Tucker, R. E., and Mitchell, Jr., G. E. (1984). *J. Anim. Sci.* **58,** 971–977.
Bryant, K. L., Kornegay, E. T., Knight, J. W., Webb, Jr., K. E., and Notter, D. R. (1985). *J. Anim. Sci.* **60,** 145–162.
Buenrostro, J. L., and Kratzer, F. H. (1984). *Poult. Sci* **63,** 1563–1570.
Burr, G. O., and Burr, M. M. (1929). *J. Biol. Chem.* **82,** 345–367.
Burr, G. O., and Burr, M. M. (1930). *J. Biol. Chem.* **86,** 587–621.
Buziassy, C., and Tribe, D. E. (1960). *Aust. J. Agric. Res.* **11,** 989–1001.
Byers, F. M. (1979). *Anim. Nutr. Health* **35,** 20–22.
Campbell, J. E. and Phillips, P. H. (1952). *J. Nutr.* **47,** 621–629.
Cantrell, C. R., and Borum, P. R. (1982). *J. Biol. Chem.* **257,** 10599–10604.
Cappa, C. (1958). *Rev. Zooiatr.* **31,** 299–308.

Carlisle, T. L., Shah, D. V., Schlegel, R., and Suttie, J. W. (1975). *Proc. Soc. Exp. Biol. Med.* **148,** 140–144.
Carlson, L. A., and Rossner, S. (1975). *Atherosclerosis* **22,** 317–323.
Carlson, S. E., Carver, J. D., and House, S. G. (1986). *J. Nutr.* **116,** 718–725.
Carpenter, K. J., Schelstraete, M., Vilicich, V. C., and Wall, J. S. (1988). *J. Nutr.* **118,** 165–169.
Carrillo, B. J. (1973). *Rev. Invest. Agropecu. Ser. 4, Patol. Anim.* **10,** 65–77.
Carroll, F. D., Gross, H., and Howell, C. E. (1949). *J. Anim. Sci.* **8,** 290–299.
Carter, E. G. A., and Carpenter, K. J. (1982). *J. Nutr.* **112,** 2091–2103.
Cartwright, G. E., and Wintrobe, M. M. (1949). *Proc. Soc. Exp. Biol. Med.* **71,** 54–57.
Cartwright, G. E., Tatting, B., Kurth, D., and Wintrobe, M. M. (1952). *Blood* **7,** 992–1024.
Carvalho da Silva, A., deAngelis, R. C., Pontes, M. A., and Mansurguerios, M. F. (1955). *J. Nutr.* **56,** 199–213.
Catron, D. V., Richardson, D., Underkofler, L. A., Maddock, H. M., and Friedland, W. C. (1952). *J. Nutr.* **47,** 461–468.
Chan, M. M. (1984). *In* "Handbook of Vitamins" (L. J. Machlin, ed.), pp. 549–570. Dekker, New York.
Chapman, H. L., Shirley, R. L., Palmer, A. Z., Haines, C. E., Carpenter, J. W., and Cunha, T. J. (1964). *J. Anim. Sci.* **23,** 669–673.
Chatterjee, G. C. (1978). *In* "Handbook Series in Nutrition and Food, Section E: Nutritional Disorders" (M. Rechcigl, Jr., ed.), Vol. 2, pp. 149–176. CRC Press, Boca Raton, Florida.
Cheldelin, V. H., and Baich, A. (1972). *In* "The Vitamins" (W. H. Sebrell, Jr. and R. S. Harris, eds.), Vol. 5, 2nd Ed., pp. 398–412. Academic Press, New York.
Chen, M. F., Hill, J. W., and McIntyre, P. A. (1983). *J. Nutr.* **113,** 2192–2196.
Chen, Jr., P. S., and Bosmann, H. B. (1964). *Science* **150,** 19–22.
Chew, B. P. (1983). *Proc. Pac. Northwest Anim. Nutr. Conf., 18th, Corvallis, Oregon* pp. 77–79.
Chew, B. P., Hollen, L. L., Hillers, J. K., and Herlugson, M. L. (1982). *J. Dairy Sci.* **65,** 2111–2118.
Chew, B. P., Luedecke, L. O., and Holpuch, D. M. (1984). *J. Dairy Sci.* **67,** 2566–2570.
Christensen, S. (1973). *Acta Pharmacol. Toxicol.* **32,** 1–68.
Chung, A. S. M., Pearson, W. N., Darby, W. J., Miller, O. N., and Goldsmith, G. A. (1961). *Am. J. Clin. Nutr.* **9,** 573–582.
Church, D. C. (1979). "Digestive Physiology and Nutrition of Ruminants, Vol. 2, Nutrition" (D. C. Church, ed.), 2nd Ed. O & B Books, Corvallis, Oregon.
Cinnamon, A., and Beaton, J. (1970). *Am. J. Clin. Nutr.* **23,** 696–702.
Clagett, C. O. (1971). *Fed. Proc., Fed. Am. Soc. Exp. Biol.* **30,** 127–129.
Clements, R. S., and Darnell, B. (1980). *Am J. Clin. Nutr.* **33,** 1954–1967.
CNAIN (1981). (Committee on Nomenclature of the American Institute of Nutrition). *J. Nutr.* **111,** 8–15.
Cody, M. M. (1984). *In* "Handbook of Vitamins" (L. J. Machlin, ed.), pp. 571–585. Dekker, New York.
Colby, R. W., Cunha, T. J., Lindley, C. E., Cordy, D. R., and Ensminger, M. E. (1948). *J. Am. Vet. Med. Assoc.* **113,** 589–593.
Cole, C. L., Rasmussen, R. A., and Thorp, F. (1944). *Vet. Med.* **39,** 204–211.
Colman, N. (1982). *Nutr. Rev.* **40,** 225–233.
Comben, N., Clark, R. J., and Sutherland, D. J. B. (1984). *Vet. Rec.* **115,** 642–645.
Contractor, S., and Shane, B. (1970). *Am. J. Obstet. Gynecol.* **107,** 635–640.
Cooke, R. J., Zee, P., and Yeh, Y. Y. (1984). *J. Pediatr. Gastroenterol. Nutr.* **3,** 446–449.
Cooper, J. R., Roth, R. H., and Kini, M. M. (1963). *Nature (London)* **199,** 609–610.
Cooperman, J. M., and Lopez, R. (1984). *In* "Handbook of Vitamins" (L. J. Machlin, ed.), pp. 299–327. Dekker, New York.

Corrigan, J. (1982). *Ann. N.Y. Acad. Sci.* **393,** 361–368.
Craig, J. F., and Davis, G. O. (1943). *J. Comp. Pathol. Ther.* **53,** 196–198.
Crane, S., and Price, J. (1983). *J. Nutr. Sci. Vitaminol.* **29,** 381–387.
Creek, R. D., and Vasaitis, V. (1963). *Poult. Sci.* **42,** 1136–1141.
Cunha, T. J. (1971). *In* "The Vitamins" (W. H. Sebrell, Jr. and R. S. Harris, eds.), Vol. 3, 2nd Ed., pp. 394–395. Academic Press, New York.
Cunha, T. J. (1972). *Feedstuffs* **44**(17), 27–28.
Cunha, T. J. (1977). "Swine Feeding and Nutrition." Academic Press, New York.
Cunha, T. J. (1980a). *J. Anim. Sci.* **51,** 1429–1433.
Cunha, T. J. (1980b). "Horse Feeding and Nutrition" Academic Press, New York.
Cunha, T. J. (1982). "Niacin in Animal Feeding and Nutrition." National Feed Ingredients Association (NFIA), Fairlawn, New Jersey.
Cunha, T. J. (1984a). *Feed Manage.* **35,** 14, 18, 22–23.
Cunha, T. J. (1984b). *Squibb Int. Swine Update* **3,** 1, 4–7.
Cunha, T. J. (1985). *Feedstuffs* **57**(41), 37–42.
Cunha, T. J., Lindley, D. C., and Ensminger, M. E. (1946). *J. Anim. Sci.* **5,** 219–225.
Cunha, T. J., Colby, R. W., Bustad, L. K., and Bone, J. F. (1948). *J. Nutr.* **36,** 215–229.
Cunningham, H. M., and Loosli, J. K. (1954a). *J. Anim. Sci.* **13,** 265–273.
Cunningham, H. M. and Loosli, J. K. (1954b). *J. Dairy Sci.* **37,** 453–461.
Daghir, N. J., and Haddad, K. S. (1981). *Poult. Sci.* **60,** 988–992.
Dale, H. H. and Dudley, H. W. (1929). *J. Physiol. (London)* **68,** 97–98.
Dam, H., Schønheyder, F., and Tage-Hansen, E. (1936). *Biochem. J.* **30,** 1075–1079.
Darby, W. J., McNutt, K. W., and Todhunter, E. N. (1975). *Nutr. Rev.* **33,** 289–297.
Dash, S. K., and Mitchell, D. J. (1976). *Anim. Nutr. Health* **Oct.,** 16–17.
Davie, M., and Lawson, D. E. M. (1980). *Clin. Sci.* **58,** 235–242.
Davies, E. T., Pill, A. H., and Austwick, P. K. A. (1968). *Vet. Rec.* **83,** 681–682.
Dawson, E., Ferguson, T. M., Deyoe, C. W., and Counch, J. R. (1962). *Poult. Sci.* **41,** 1639 (Abstr.).
de Boer-van den Berg, M. A. G., Verstijnen, C. P. H. J., and Vermeer, C. (1986). *J. Invest. Dermatol.* **87,** 377–380.
de Jong, M. F., and Sytsema, J. R. (1983). *Vet. Q.* **5,** 58–67.
De La Huerga, J., and Popper, H. (1951). *J. Clin. Invest.* **30,** 463–470.
De La Huerga, J., and Popper, H. (1952). *J. Clin. Invest.* **31,** 598–603.
DeLuca, H. C. (1988). *FASEB J.* **2,** 224–236.
DeLuca, H. F. (1979). *Nutr. Rev.* **37,** 161–193.
Derilo, Y. L., and Balnave, D. (1980). *Br. Poult. Sci.* **21,** 479–487.
Dobbelstein, H., Korner, W., Mempel, W., Grosse-Wilde, H., and Edel, H. (1974). *Kidney Int.* **5,** 233–239.
Donoghue, S., Donawick, W. J., and Kronfeld, D. S. (1983). *J. Nutr.* **113,** 2197–2204.
Doppelt, S. H., Neer, R. M., Daly, M., Bourvet, L., Schiller, A., and Holick, M. F. (1983). *Orthop. Trans.* **7,** 512–513.
Draper, H. H., and Johnson, B. C. (1952). *J. Nutr.* **46,** 123–131.
Driskell, J. A. (1984). *In* "Handbook of Vitamins" (L. J. Machlin, ed.), pp. 379–401. Dekker, New York.
Duncan, C. W. (1944). *J. Dairy Sci.* **27,** 636–645.
Durgakumari, B., and Adiga, P. R. (1986). *Mol. Cell. Endocrinol.* **44,** 285–292.
Du Vigneaud, V., Chandler, J. P., Moyer, A. W., and Keppel, D. M. (1939). *J. Biol. Chem.* **131,** 57–76.
Eaton, H. D., Lucas, J. J., Neilson, S. W., and Helmboldt, C. F. (1970). *J. Dairy Sci.* **53,** 1775–1779.

Edwards, H. M. (1974). *Proc. Nutr. Conf., Atlanta* pp. 1–13.
Edwin, E. E., and Lewis, G. (1971). *J. Dairy Res.* **38,** 79–90.
Egarova, N. D., and Perepelkin, S. R. (1979). *Gig. Sanit.* **10,** 25.
Ek, J. (1980). *J. Pediatr.* **97,** 288–292.
El-Boushy, A. R., and Van Albada, M. (1970). *Neth. J. Agric. Sci.* **18,** 62–71.
El-Boushy, A. R., Simons, P. C. M., and Wiert, G. (1968). *Poult. Sci.* **47,** 456 (Abstr.).
Ellenbogen, L. (1984). *In* "Handbook of Vitamins" (L. J. Machlin, ed.), pp. 497–546. Dekker, New York.
Ellenbogen, L., and Highley, D. R. (1970). *Fed. Proc., Fed. Am. Soc. Exp. Biol.* **29,** 633 (Abstr.).
Ellis, R. P., and Vorhies, M. W. (1976). *J. Am. Vet. Med. Assoc.* **168,** 231–232.
El Shorafa, W. M., Feaster, J. P., Ott, E. A., and Asquith, R. L. (1979). *J. Anim Sci.* **48,** 882–886.
Elvehjem, C. A., Madden, R. J., Strong, F. M., and Wollen, D. W. (1937). *J. Am. Chem. Soc.* **59** 1767–1768.
Emery, R. S., Burg, N., Braur, L. D., and Blank, G. N. (1964). *J. Dairy Sci.* **47,** 1074–1079.
Ensminger, M. E., and Olentine, C. G. (1978). "Feeds and Nutrition." Ensminger, Clovis, California.
Ensminger, M. E., Bowland, J. P., and Cunha, T. J. (1947). *J. Anim. Sci.* **6,** 409–423.
Ensminger, M. E., Colby, R. W., and Cunha, T. J. (1951). *Wash. Agric. Exp. Stn., Stn. Circ.* 134.
Erdman, R. A., Shaver, R. D., and Vandersall, J. H. (1984). *J. Dairy Sci.* **67,** 410–415.
Evans, H. M. (1925). *Proc. Natl. Acad. Sci. U.S.A.* **11,** 373–377.
Evans, H. M., and Burr, G. O. (1926). *Proc. Soc. Exp. Biol. Med.* **24,** 740–743.
Evans, W. C. (1975). *Vitam. Horm.* **33,** 467–504.
Farid, M. F. A., and Ghanem, Y. S. (1982). *World Rev. Anim. Prod.* **18,** 75–77.
Farquharson, J., and Adams, J. F. (1976). *Br. J. Nutr.* **36,** 127–141.
Fell, H. B., and Thomas, L. (1960). *J. Exp. Med.* **111,** 719–744.
Ferguson, T. M., Whiteside, C. H., Creger, C. R., Jones, M. L., Atkinson, R. L., and Couch, J. R. (1961). *Poult. Sci.* **40,** 1151–1159.
Firth, J., James, M., Chang, S., Mistry, P., and Johnson, B. C. (1953). *J. Anim. Sci.* **12,** 915 (Abstr.).
Fisher, J. H., Willis, R. A., and Haskell, B. E. (1984). *J. Nutr.* **114,** 786–791.
Flavin, M., and Ochoa, S. (1957). *J. Biol. Chem.* **229,** 965–979.
Fleming, A. F., and Dawson, I. (1972). *Br. Med. J.* **4,** 236–237.
Folkers, K. (1982). *In* "B_{12}" (D. Dolphin, ed.), pp. 1–15. Wiley, New York.
Folman, Y., Ascarelli, I., Herz, A., Rosenberg, M., Davidson, M., and Halevi, A. (1979). *Br. J. Nutr.* **41,** 353–359.
Folman, Y., Ascarelli, I., Kraus, D., and Barash, H. (1987). *J. Dairy Sci.* **70,** 357–366.
Foo, S. K., and Shane, B. (1982). *J. Biol. Chem.* **257,** 13587–13592.
Forbes, R. M., Erdman, J. W., Parker, H. M., Kondo, H., and Ketelsen, S. M. (1983). *J. Nutr.* **113,** 205–210.
Fox, H. M. (1984). *In* "Handbook of Vitamins" (L. J. Machlin, ed.), pp. 437–457. Dekker, New York.
Foy, H., and Mbaya, V. (1977). *Prog. Food Nutr. Sci.* **2,** 357–394.
Frank, G. R., Bahr, J. M., and Easter, R. A. (1988). *J. Anim. Sci.* **66,** 47–52.
Franklin, M. C., and Johnstone, I. L. (1948). "Maintenance of Serum Calcium level by Calcium Supplements to the Diet," pp. 63–75. C. S. I. R., Melbourne.
Fraser, D. R. (1984). *In* "Nutrition Reviews, Present Knowledge in Nutrition" (R. E. Olson, ed.), pp. 209–225. Nutrition Foundation, Washington, D.C.
Frick, P. G., Riedler, G., and Brogli, H. (1967). *J. Appl. Physiol.* **23,** 387–389.

Friedman, S., and Fraenkel, G. S. (1972). *In* "The Vitamins" (W. H. Sebrell, Jr. and R. S. Harris, eds.), Vol. 5, 2nd Ed., pp. 329–355. Academic Press, New York.
Friesecke, H. (1975). *In* "Pantothenic Acid," No. 1533. Hoffmann-LaRoche, Basel.
Friesecke, H. (1978). Roche Symp. Anim. Nutr. Events, London No. 1700.
Friesecke, H. (1980). "Vitamin B_{12}." Hoffmann-La Roche, Basel.
Frigg, M. (1976). *Poult. Sci.* **55,** 2310–2318.
Frigg, M. (1984). *Poult. Sci.* **63,** 750–753.
Fritschen, R. D., Peo, Jr., E. R., Lucas, L. E., and Grace, O. D. (1970). *J. Anim. Sci. 31,* 199 (Abstr.).
Fritz, J. C., Archer, W. F., and Barker, D. K. (1942). *Poult. Sci.* **21,** 361–369.
Fritz, J. C., Mislivec, P. B., Pla, G. W., Harrison, B. N., Weeks, C. E., and Dantzman, J. G. (1973). *Poult. Sci.* **52,** 1523–1530.
Fronk, T. J., and Schultz, L. H. (1979). *J. Dairy Sci.* **62,** 1804–1807.
Fry, P. C., Fox, H. M., and Tao, H. G. (1976). *J. Nutr. Sci. Vitaminol.* **22,** 339–346.
Frye, T. M. (1978). *Proc. Roche Vitam. Nutr. Update Meet., Arkansas Nutr. Conf.* pp. 54–79.
Fuller, H. L., and Kifer, P. E. (1959). *Poult. Sci.* **38,** 255–260.
Fuller, H. F., Field, R. C., Roncalli-Amici, A., Dunahoo, W. S., and Edwards, H. M. (1961). *Poult. Sci.* **40,** 249–253.
Gadient, M. (1986). *Proc. Nutr. Conf. Feed Manuf., College Park, Maryland* pp. 73–79.
Gaetani, S., and D'Aquin., M. (1987). *Nutr. Rep. Int.* **35,** 93–101.
Gallop, P. M., Lian, J. B., and Hauschka, P. V. (1980). *N. Engl. J. Med.* **302,** 1460–1466.
Gallo-Torres, H. E. (1980). *In* "Vitamin E. A Comprehensive Treatise" (L. S. Machlin, ed.), pp. 193–267. Dekker, New York.
Garabedian, M., Tanaka, Y., Holick, M. F., and DeLuca, H. F. (1974). *Endocrinology* **94,** 1022–1027.
Garton, G. A. (1960). *Nature (London)* **187,** 511–512.
Gebremichael, A., Levy, E. M., and Corwin, L. M. (1984). *J. Nutr.* **114,** 1297–1305.
Geraci, J. R. (1974). *J. Am. Vet. Med. Assoc.* **165,** 801–803.
Gerloff, B. J., Herdt, T. H., Wells, W. W., Liesman, J. S., and Emery, R. S. (1986). *J. Anim. Sci.* **62,** 1682–1692.
Ghazarian, J. G., Jefcoate, C. R., Knutson, J. C., Orme-Johnson, W. H., and DeLuca, H. F. (1974). *J. Biol. Chem.* **249,** 3026–3033.
Ghishan, F. K., Said, H. M., Wilson, P. C., Murrell, J. E., and Green, H. L. (1986). *Am. J. Clin. Nutr.* **43,** 258–262.
Ghosh, H. P., Sarkar, P. K., and Guha, B. C. (1963). *J. Nutr.* **79,** 451–453.
Ginter, E. (1970). *Acta Med. Acad. Sci. Hung.* **27,** 23–29.
Gitter, M., and Bradley, R. (1978). *Vet. Rec.* **103,** 24–26.
Giugliani, E. R. J., Jorge, S. M., and Goncalves, A. L. (1985). *Am. J. Clin. Nutr.* **41,** 330–335.
Glättli, H. R. (1975). *Zentralbl. Veterinaermed. A* **22,** 102–116.
Glienke, L. R., and Ewan, R. C. (1974). *J. Anim. Sci.* **39,** 975 (Abstr.).
Goff, J. P., Horst, R. L., and Littledike, E. T. (1984). *J. Nutr.* **114,** 163–169.
Goff, J. P., Horst, R. L., Beitz, D. E., and Littledike, E. T. (1988). *J. Dairy Sci.* **71,** 1211–1219.
Goldblatt, H., and Soames, K. M. (1923). *Biochem. J.* **17,** 446–453.
Goldblatt, M. J., Conklin, D. E., and Brown, W. D. (1979). *Proc. World Symp. Finfish Nutr. Fishfeed Technol.* **2,** 117–129.
Gonnerman, W. A., Toverud, S. V., Ramp, W. K., and Mechanic, G. L. (1976). *Proc. Soc. Exp. Biol. Med.* **151,** 453–456.
Goodman, D. S. (1980). *Fed. Proc., Fed. Am. Soc. Exp. Biol.* **39,** 2716–2722.
Goodwin, R. F. W. (1962). *J. Comp. Pathol.* **72,** 214–232.

Gopalan, C. (1946). *Indian Med. Gaz.* **81,** 23–26.
Gopalan, C. (1969). *Lancet* **1,** 197–199.
Gopalan, C., and Rao, K. S. J. (1978). *In* "Handbook Series in Nutrition and Food, Section E: Nutritional Disorders" (M. Rechcigl, Jr., ed.), Vol. 3, pp. 23–31. CRC Press, Boca Raton, Florida.
Gopalan, C., and Srikantia, S. G. (1960). *Lancet* **2**, 954–957.
Goplen, B. P., and Bell, J. M. (1967). *Can. J. Anim. Sci.* **47,** 91–100.
Grace, M. L., and Bernhard, R. A. (1984). *J. Dairy Sci.* **67,** 1646–1654.
Graves-Hoagland, R. L., Hoaglund, T. A., and Woody, C. O. (1988). *J. Dairy Sci.* **71,** 1058–1062.
Green, H. B., Horst, R. L., Beitz, D. C., and Littledike, E. T. (1981). *J. Dairy Sci.* **64,** 217–226.
Green, M. H., Green, J. B., and Lewis, K. C. (1987). *J. Nutr.* **117,** 694–703.
Greenberg, S. M., Colbert, C. E., Savage, E. E., and Deuel, J. (1950). *J. Nutr.* **41,** 473–486.
Gregory, III, J. F., Sartain, D. B., and Day, B. P. F. (1984). *J. Nutr.* **114,** 341–353.
Gries, C. L., and Scott, M. L. (1972). *J. Nutr.* **102,** 1269–1285.
Griffith, W. H., and Nyc, J. F. (1971). *In* "The Vitamins" (W. H. Sebrell, Jr. and R. S. Harris, eds.), Vol. 3, pp. 3–154. Academic Press, New York.
Griminger, P. (1984a). *Feedstuffs* **56**(38), 25–26.
Griminger, P. (1984b). *Feedstuffs* **56**(39), 24–25.
Griminger, P., and Brubacher, G. (1966). *Poult. Sci.* **45,** 512–519.
Griminger, P., and Donis, O. (1960). *J. Nutr.* **70,** 361–368.
Groziak, S. M., and Kirksey, A. (1987). *J. Nutr.* **117,** 1045–1052.
Gubler, C. J. (1984). *In* "Handbook of Vitamins" (L. J. Machlin, ed.), pp. 245–297. Dekker, New York.
Guerin, H. B. (1981). *J. Anim. Sci.* **53,** 758–764.
Guilbert, H. R., Miller, R. F., and Hughes, E. H. (1937). *J. Nutr.* **13,** 543–564.
Gullickson, T. W., Fountaine, F. C., and Fitch, J. B. (1942). *J. Dairy Sci.* **25,** 117–128.
Gullickson, T. W., Palmer, L. S., Boyd, W. L., Nelson, J. W., Olson, F. C., Calverley, C. E., and Boyer, P. D. (1949). *J. Dairy Sci.* **32,** 495–508.
Guss, S. B. (1977). "Management and Diseases of Dairy Goats." Dairy Goat Journal Publ. Corp., Scottsdale, Arizona.
Haddad, J. G., and Walgate, J. (1976). *J. Biol. Chem.* **251,** 4803–4809.
Hadler, M. R., and Shadbolt, R. S. (1975). *Nature (London)* **253,** 275–277.
Haeger, K. (1974). *J. Clin. Nutr.* **27,** 1179–1181.
Haenel, H., Ruttloff, H., and Ackermann, H. (1959). *Biochem. Z.* **331,** 209.
Hall, C. E., and Bieri, J. G. (1953). *Endocrinology* **53,** 661–666.
Halsted, C. H., and Tamura, R. (1979). "Problems in Liver Disease" (C. S. Davidson, ed.), pp. 91–100. Stratton, New York.
Halsted, C. H., Robles, E. A., and Mezey, E. (1973). *Gastroenterology* **64,** 526–532.
Halver, J. E. (1982). *Comp. Biochem. Physiol.* **73,** 43–50.
Hankes, L. V. (1984). *In* "Handbook of Vitamins" (L. J. Machlin, ed.), pp. 329–377. Dekker, New York.
Hansen, A. E., Haggard, M. E., Boelsche, A. N., Adam, D. J., and Wiese, H. F. (1958). *J. Nutr.* **66,** 565–576.
Hansen, H. S., and Jensen, B. (1986). *J. Nutr.* **116,** 198–203.
Hardy, B., and Frape, D. L. (1983). "Micronutrients and Reproduction," No. 1895. Hoffmann-LaRoche, Basel.
Harmon, B. G., Miller, E. R., Hoefer, J. A., Ullrey, D. E., and Leucke, R. W. (1963). *J. Nutr.* **79,** 263–268.

Haroon, Y., Shearer, M. J., Rahim, S., Gunn, W. G., McEnery, G., and Barkhan, P. (1982). *J. Nutr.* **112,** 1105–1117.
Harrington, D. D., and Page, E. H. (1983). *J. Am. Vet. Med. Assoc.* **182,** 1358–1369.
Harris, Jr., B. (1975). *Feedstuffs* **47**(48), 42–43.
Harrison, H. C., and Harrison, H. E. (1963). *Am. J. Physiol.* **205,** 107–111.
Harrison, H. E. (1978). *In* "Handbook Series in Nutrition and Food, Section E: Nutritional Disorders" (M. Rechcigl, Jr., ed.), Vol. 3, pp. 117–122. CRC Press, Boca Raton, Florida.
Harrison, J. H., Hancock, D. D., and Conrad, H. R. (1984). *J. Dairy Sci.* **67,** 123–132.
Hart, L. E., DeLuca, H. F., Yamada, S., and Takayama, H. (1984). *J. Nutr.* **114,** 2059–2065.
Hartroft, W. S. (1955). *Proc. Am. Soc. Exp. Biol.* **14,** 655–660.
Hashimoto, Y., Arai, S., and Nose, T. (1970). *Bull. Jpn. Soc. Sci. Fish.* **36,** 791–797.
Haskell, B. E. (1978). *In* "Handbook Series in Nutrition and Food, Section E: Nutritional Disorders" (M. Rechcigl, Jr., ed.), Vol. 1, pp. 43–45. CRC Press, Boca Raton, Florida.
Hay, A. W. M., and Watson, G. (1976). *Comp. Biochem. Physiol.* **56,** 167–172.
Hazell, K., and Baloch, K. H. (1970). *Gerontol. Clin.* **12,** 10–17.
Heald, F. P. (1975). *Med. Clin. North Am.* **59,** 1329–1336.
Hegazy, E., and Schwenk, M. (1983). *J. Nutr.* **113,** 1702–1707.
Hegazy, E., and Schwenk, M. (1984). *J. Nutr.* **114,** 2217–2220.
Hegsted, D. M., Hayes, K. C., Gallagher, A., and Hanford, H. (1973). *J. Nutr.* **103,** 302–307.
Heinemann, W. W., Ensminger, M. E., Cunha, T. J., and McCulloch, E. C. (1946). *J. Nutr.* **31,** 107–125.
Helgebostad, A. (1984). *In* "Proc. Ascorbic Acid in Domestic Animals" (I. Wegger, F. J. Tagwerker, and J. Moustgaard, eds.), pp. 169–174. Danish Agriculture Society, Copenhagen.
Helgebostad, A., and Ender, F. (1973). *Acta Agric. Scand. Suppl.* **19,** 79–83.
Heller, S., Salkeld, R. M., and Korner, W. F. (1973). *Am. J. Clin. Nutr.* **26,** 1339–1348.
Henderickx, H. K., Teague, H. S., Redman, D. R., and Grifo, A. P. (1964). *J. Anim. Sci.* **23,** 1036–1038.
Henderson, H. E. (1971). *J. Anim. Sci.* **32,** 1233–1238.
Henderson, L. M. (1984). *In* "Nutrition Reviews, Present Knowledge in Nutrition" (R. E. Olson, ed.), pp. 303–317. Nutrition Foundation, Washington, D.C.
Henderson, L. M., and Gross, C. J. (1979). *J. Nutr.* **109,** 654–662.
Henry, H. L., and Norman, A. W. (1978). *Science* **201,** 853–837.
Hentges, J. F., Grummer, R. H., Phillips, P. H., Bohstedt, G., and Sorensen, D. K. (1952). *J. Am. Vet. Med. Assoc.* **120,** 213–216.
Herbert, V. (1962). *Trans. Assoc. Am. Physicians* **75,** 307–320.
Herbert, V. (1967). *Am. J. Clin. Nutr.* **20,** 562–569.
Herbert, V. (1968). *Gastroenterology* **54,** 110–115.
Herbert, V. (1981). *Proc. Fla. Symp. Micronutr. Hum. Nutr., Univ. Fla., Gainesville* pp. 121–138.
Herbert, V. (1984). "Nutrition Reviews: Present Knowledge in Nutrition" (R. E. Olson, H. P. Broquist, C. O. Chichester, W. J. Darby, Albert C. Kolbye, and R. M. Stalvey, eds.), 5th Ed., pp. 347–364. Nutrition Foundation, Washington, D.C.
Herbert, V., and Zalusky, R. (1962). *J. Clin. Invest.* **41,** 1263–1276.
Hercberg, S., Bichon, L., Galan, P., Christides, J. P., Carroget, C., and Potier de Courcy, G. (1987). *Nutr. Rep. Int.* **35,** 915–930.
Herrick, J. B. (1972). *Vet. Med.* **67,** 906, 908–909.
Heuser, G. F., and Norris, L. C. (1929). *Poult. Sci.* **8,** 89–98.
Hibbs, J. W., and Pounden, W. D. (1955). *J. Dairy Sci.* **38,** 65–72.
Hicks, V. A., Gunning, D. B., and Olson, J. A. (1984). *J. Nutr.* **114,** 1327–1333.

Hidiroglou, M., Ivan, M., and Lessard, J. R. (1977). *Can. J. Anim. Sci.* **57,** 519–529.
Hidiroglou, M., Williams, C. J., and Ho, S. K. (1978). *Can. J. Anim. Sci.* **58,** 621–630.
Hidiroglou, M., Proulx, J. G., and Roubos, D. (1979). *J. Dairy Sci.* **62,** 1076–1080.
Hidiroglou, M., McAllister, A. J., and Williams, C. J. (1987). *J. Dairy Sci.* **70,** 1281–1288.
Hidiroglou, N., and McDowell, L. R. (1987). *Int. J. Vitam. Nutr. Res.* **57,** 261–266.
Hidiroglou, N., McDowell, L. R., and Boning, A. (1986). *Int. J. Vitam. Nutr. Res.* **56,** 339–344.
Hidiroglou, N., Laflamme, L. F., and McDowell, L. R. (1988). *J. Anim. Sci.* (in press).
Higginbottom, M., Sweetman, L., and Nyhan, W. (1978). *N. Engl. J. Med.* **299,** 317–323.
Hill, C. H. (1961). *Poult. Sci.* **40,** 762–765.
Hill, C. H., and Garren, H. W. (1958). *Poult. Sci.* **37,** 236–237.
Hill, F. W., Scott, M. L., Norris, L. C., and Heuser, G. F. (1961). *Poult. Sci.* **40,** 1245–1245.
Hill, R. B., Schendel, H. E., Rama Rao, P. B., and Johnson, B. C. (1964). *J. Nutr.* **84,** 259–264.
Hilton, J. W. (1983). *J. Nutr.* **113,** 1737–1745.
Hirsch, A. (1982). "Vitamins—The Life Essentials," pp. 1–20. National Feed Ingredient Association, Des Moines, Iowa.
Hirsch, M. J., Growdon, J. H., and Wurtman, R. J. (1978). *Metabolism* **27,** 953–960.
Hodges, R. E., Ohison, M. A., and Bean, W. B. (1958). *J. Clin. Invest.* **37,** 1642–1657.
Hoffbrand, A. V. (1978). "Handbook Series in Nutrition and Food, Section E: Nutritional Disorders" (M. Rechcigl, Jr., ed.), Vol. 2, pp. 55–68. CRC Press, Boca Raton, Florida.
Hoffmann-LaRoche (1967). "Vitamin A, the Key-stone of Animal Nutrition." Hoffmann-LaRoche, Basel.
Hofstetter, H., Sen, N., and Holman, R. T. (1965). *J. Am. Oil Chem. Soc.* **42,** 537–540.
Holick, M. F. (1986). *Curr. Top. Nutr. Dis.* **5,** 241–248.
Holick, M. F. (1987). *Fed. Proc., Fed. Am. Soc. Exp. Biol.* **46,** 1876–1882.
Holick, M. F., MacLaughlin, J. A., and Doppelt, S. H. (1981). *Science* **211,** 590–593.
Hollander, D. (1973). *Am. J. Physiol.* **225,** 360–364.
Holman, R. T. (1960). *J. Nutr.* **70,** 405–410.
Holman, R. T. (1964). *Fed. Proc., Fed. Am. Soc. Exp. Biol.* **23,** 1062–1067.
Holman, R. T. (1968). *Prog. Chem. Fats Other Lipids* **9,** 279–348.
Holman, R. T. (1978a). *In* "Handbook Series in Nutrition and Food, Section E: Nutritional Disorders" (M. Rechcigl, Jr., ed.), Vol. 3, pp. 491–516. CRC Press, Boca Raton; Florida.
Holman, R. T. (1978b). *In* "Handbook Series in Nutrition and Food, Section E: Nutritional Disorders" (M. Rechcigl, Jr., ed.), Vol. 2, pp. 335–368. CRC Press, Boca Raton, Florida.
Holman, R. T. (1986). *J. Am. Coll. Nutr.* **5,** 183–211.
Hopper, J. H., and Johnson, B. C. (1955). *J. Nutr.* **56,** 303–310.
Horner, J. L., Coppock, C. E., Schelling, G. T., Labore, J. M., and Nave, D. H. (1986). *J. Dairy Sci.* **69,** 3087–3093.
Hornig, D., Glatthaar, B., and Moser, U. (1984). *In* "Proc. Ascorbic Acid in Domestic Animals" (I. Wegger, F. J. Tagwerker, and J. Moustgaard, eds.), pp. 3–24. Danish Agriculture Society, Copenhagen.
Horst, R. L., and Napoli, J. L. (1981). *Fed. Proc., Fed. Am. Soc. Exp. Biol.* **40,** 898 (Abstr.).
Horst, R. L., Jorgensen, N. A., and DeLuca, H. F. (1978). *Am. J. Physiol.* **235,** E634–E637.
Horwitt, M. K. (1980). *Nutr. Rev.* **38,** 105–113.
Hove, E. L., Fry, G. S., and Schwarz, K. (1958). *Proc. Soc. Exp. Biol. Med.* **98,** 27–29.
Howell, C. E., Hart, G. H., and Ittner, N. R. (1941). *Am. J. Vet. Res.* **2,** 60–74.
Hughes, M. R., Baylink, D. J., Jones, P. G., and Haussler, M. R. (1976). *J. Clin. Invest.* **58,** 61–70.
Hughes, S. G. (1984). *J. Nutr.* **114,** 1660–1663.
Hume, E. M., Nunn, L. A., Smedley-MacLean, I., and Smith, H. H. (1940). *Biochem J.* **34,** 879–883.
Hunt, S. M. (1975). *In* "Riboflavin" (R. S. Rivlin, ed.), p. 199. Plenum, New York.

Hutchinson, L. J., Scholz, W., and Drake, T. R. (1982). *J. Am. Vet. Med. Assoc.* **181,** 581–584.
Hwang, D. H., Mathias, M. M., Dupont, J., and Meyer, D. L. (1975). *J. Nutr.* **105,** 995–1002.
Hwang, D. H., Boudreau, M., and Chanmugam, P. (1988). *J. Nutr.* **118,** 427–437.
Imawari, M., Kida, K., and Goodman, D. S. (1976). *J. Clin. Invest.* **58,** 514–523.
Ink, S. L., and Henderson, L. M. (1984). *Annu. Rev. Nutr.* **4,** 455–470.
Itze, L. (1984). *In* "Proc. Ascorbic Acid in Domestic Animals" (I. Wegger, F. J. Tagwerker, and J. Moustgaard, eds.), pp. 120–130. Danish Agriculture Society, Copenhagen.
Itzeova, V. (1984). *In* "Proc. Ascorbic Acid in Domestic Animals" (I. Wegger, F. J. Tagwerker, and J. Moustgaard, eds.), pp. 139–147. Danish Agriculture Society, Copenhagen.
Jaeschke, G. (1984). *In* "Proc. Ascorbic Acid in Domestic Animals" (I. Wegger, F. J. Tagwerker, and J. Moustgaard, eds.), pp. 153–161. Danish Agriculture Society, Copenhagen.
Jaffe, G. M. (1984). *In* "Handbook of Vitamins" (L. J. Machlin, ed.), pp. 199–244. Dekker, New York.
Jaster, E. H., Hartwell, G. F., and Hutjens, M. F. (1983). *J. Dairy Sci.* **66,** 1046–1051.
Jensen, L. S., and McGinnis, J. (1957). *Poult. Sci.* **36,** 212–213.
Johnson, B. C., Mitchell, H. H., Hamilton, T. S., and Nevens, W. B. (1947). *Fed. Proc., Fed. Am. Soc. Exp. Biol.* **6,** 410–411.
Johnson, B. C., Hamilton, T. S., Nevens, W. B., and Boley, L. E. (1948a). *J. Nutr.* **35,** 137–145.
Johnson, B. C., James, M. F., and Krider, J. L. (1948b). *J. Anim. Sci.* **7,** 486–493.
Johnson, B. C., Mitchell, H. H., Pinkos, J. A., and Merrill, C. C. (1951). *J. Nutr.* **43,** 37–48.
Johnston, L., Vaughan, L., and Fox, H. M. (1981). *Am. J. Clin. Nutr.* **34,** 2205–2209.
Jukes, T. H. (1947). *Annu. Rev. Biochem.* **16,** 193–218.
Jukes, T. H. (1980). *Trends Biol. Sci.* **3,** 140–141.
Kabir, H., LeKlem, J. E., and Miller, L. T. (1983). *J. Nutr.* **113,** 2412–2420.
Kallner, A., Hartman, D., and Hornig, D. (1977). *Int. J. Vitam. Nutr. Res.* **47,** 383–388.
Kanazawa, S., Herbert, V., Herzlich, B., and Manusselis, C. (1983). *Lancet* **1,** 707–708.
Karlin, R., and Foisy, C. (1972). *Int. J. Vitam. Nutr. Res.* **42,** 545–554.
Keagy, P. M., and Oace, S. M. (1984). *J. Nutr.* **114,** 1252–1259.
Kechik, I. T., and Sykes, A. H. (1979). *Br. J. Nutr.* **42,** 97–103.
Keesling, P. T., and Morris, J. G. (1975). *J. Anim. Sci.* **41,** 317 (Abstr.).
Keiver, K. M., Draper, H. H., and Ronald, K. (1988). *J. Nutr.* **118,** 332–341.
Kesavan, V., and Noronha, J. M. (1983). *Am. J. Clin. Nutr.* **37,** 262–267.
Ketola, H. G. (1976). *J. Anim. Sci.* **43,** 474–477.
Ketola, H. G., and Nesheim, M. C. (1974). *J. Nutr.* **104,** 1484–1489.
Kiatoko, M., McDowell, L. R., Bertrand, J. E., Chapman, H. L., Pate, F. M., Martin, F. G., and Conrad, J. H. (1982). *J. Anim. Sci.* **55,** 28–37.
Kindberg, C., Suttie, J. W., Uchida, K., Hirauchi, K., and Nakao, H. (1987). *J. Nutr.* **117,** 1032–1035.
King, R. L., Burrows, F. A., Hemken, R. W., and Bashore, D. L. (1967). *J. Dairy Sci.* **50,** 943–944.
Kingsbury, J. M. (1964). "Poisonous Plants of the United States and Canada," pp. 105–113. Prentice-Hall, New York.
Kirksey, A. (1981). *In* "Vitamin B_6 Methodology and Nutrition Assessment" (J. E. LeKlem and R. D. Reynolds, eds.), pp. 269–288. Plenum, New York.
Koch, E. M., and Koch, F. C. (1941). *Poult. Sci.* **20,** 33–35.
Kochanowski, B. A., Smith, A. M., Picciano, M. F., and Sherman, A. R. (1983). *J. Nutr.* **113,** 2471–2478.
Kodicek, E., Ashby, D. R., Muller, M., and Carpenter, K. J. (1974). *Proc. Nutr. Soc.* **33,** 105A–106A.
Koerner, W. F., and Weber, F. (1972). *Int. J. Vitam. Nutr. Res.* **42,** 528–544.

Kohlmeier, R. H., and Burroughs, W. (1970). *J. Anim. Sci.* **30,** 1012–1018.

Kolb, E. (1984). *In* "Proc. Ascorbic Acid in Domestic Animals" (I. Wegger, F. J. Tagwerker, and J. Moustgaard, eds.), pp. 162–168. Danish Agriculture Society, Copenhagen.

Kominato, T. (1971). *Vitamins (Jpn.)* **44,** 76–83.

Konishi, T., Baba, S., and Sone, H. (1973). *Chem. Pharm. Bull.* **21,** 220–224.

Kramer, B., and Gribetz, D. (1971). *In* "The Vitamins" (W. H. Sebrell, Jr. and R. S. Harris, eds.), pp. 259–278. Academic Press, New York.

Kramer, T. R., Briske-Anderson, M., Johnson, S. B., and Holman, R. T. (1984). *J. Nutr.* **114,** 2047–2052.

Kratzer, F. H., Knollman, K., Earl, L., and Buenrostro, J. L. (1988). *J. Nutr.* **118,** 604–608.

Krishnan, S., Bhuyan, U. N., Talwar, G. P., and Ramalingaswami, V. (1974). *Immunology* **27,** 383–392.

Kroes, J. (1978). *In* "Handbook Series in Nutrition and Food, Section E: Nutritional Disorders" (M. Rechcigl, Jr., ed.), Vol. 2, pp. 143–147. CRC Press, Boca Raton, Florida.

Kroes, J. F., Hegsted, D. M., and Hayes, K. C. (1973). *J. Nutr.* **103,** 1448–1453.

Kuemyj, A. (1955). *Vopr. Pitan.* **14,** 3–13.

Kukis, A., and Mookerjea, S. (1978). *Nutr. Rev.* **36,** 233–238.

Kukis, A., and Mookerjea, S. (1984). *In* "Present Knowledge in Nutrition" (R. E. Olson, H. P. Broquist, C. O. Chichester, W. J. Darby, A. C. Kolbye, and R. M. Stalvey, eds.), pp. 383–399. Nutrition Foundation, Washington, D.C.

Kumar, M., and Axelrod, A. E. (1978). *Proc. Soc. Exp. Biol. Med.* **157,** 421–423.

Kurnick, A. A., Hanold, F. J., and Stangeland, V. A. (1972). *Proc. Georgia Nutr. Conf. Atlanta* pp. 107–136.

Lambert, M. R., Jacobson, N. L., Allen, R. S., and Zaletel, J. H. (1954). *J. Nutr.* **52,** 259–272.

Lamptey, M. S., and Walker, B. L. (1976). *J. Nutr.* **106,** 86–93.

Lands, W. E. M. (1986). *Nutr. Rev.* **44** 189–195.

Larkin, P. J., and Yates, R. J. (1964). *E. Afr. Agric. For. J.* **30,** 11–20.

Lassiter, C. A., Ward, G. M., Huffman, C. F., Duncan, C. W., and Webster, H. D. (1953). *J. Dairy Sci.* **36,** 997–1005.

Leat, W. M. (1962). *Br. J. Nutr.* **16,** 559–569.

Lee, H. J. (1963). *In* "Animal Health, Production and Pasture" (A. H. Worden, K. O. Sellers, and D. E. Tribe, eds.), p. 662. Longmans, Green, New York.

Lee, H. M., McCall, N. E., Wright, L. D., and McCormick, D. B. (1973). *Proc. Soc. Exp. Biol. Med.* **142,** 642–644.

Lee, R. (1984). *Feedstuffs* **56** (39), 16.

Lehocky, J. (1981). *Veterinarstvi* **31,** 388–390.

Leichter, J., Switzer, V. P., and Landymore, A. F. (1978). *Nutr. Rep. Int.* **18,** 475–479.

LeKlem, J. E., Miller, L. T., Perera, A. D., and Peffers, D. E. (1980). *J. Nutr.* **110,** 1819–1828.

Leo, M. A., Lowe, N., and Lieber, C. S. (1987). *J. Nutr.* **117,** 70–76.

Lester, G. E. (1986). *Fed. Proc., Fed. Am. Soc. Exp. Biol.* **45,** 2524–2527.

Lewis, N. M., Kies, C., and Fox, H. M. (1986). *Nutr. Rep. Int.* **34,** 495–499.

Lilum, D., Hammond, W., Krauss, D., and Schoonmaker, J. (1981). *Invest. Urol.* **18,** 451–456.

Lindemann, M. D., and Kornegay, E. T. (1986). *J. Anim. Sci.* **63** (Suppl. 1), 35 (Abstr.).

Lindley, C. E., Brugman, H. H., Cunha, T. J., and Warwick, E. J. (1949). *J. Anim. Sci.* **8,** 590–602.

Lindley, D. C., and Cunha, T. J. (1946). *J. Nutr.* **32,** 47–59.

Linerode, P. A. (1966). Studies on the synthesis and absorption of B complex vitamins in the equine. Ph.D. dissertation, Ohio State University, Wooster, Ohio.

Littledike, E. T., and Horst, R. L. (1982). *J. Dairy Sci.* **65,** 749–759.

Livingston, A. L., Nelson, J. W. and Kohler, G. O. (1968). *J. Agric. Food Chem.* **16,** 492–495.

Loew, F. M. (1978). *In* "Handbook Series in Nutrition and Food, Section E: Nutritional Disorders" (M. Rechcigl, Jr., ed.), Vol. 2, pp. 3–25. CRC Press, Boca Raton, Florida.
Lofland, H. B., Goodman, H. O., Clarkson, R. B., and Prichard, R. W. (1963). *J. Nutr.* **79,** 188–194.
Long, J. B., and Shaw, J. N. (1943). *North Am. Vet.* **24,** 234–237.
Loosli, J. K. (1988). *In* "Animal Science Handbook" (P. A. Putnam, ed.). CRC Press, Boca Raton, Florida (in press).
Losada, H., Dixon, F., and Preston, T. R. (1971). *Rev. Cubana Cienc. Agric.* **5,** 369–378.
Losito, R., Owen, C. A., and Flock, E. V. (1967). *Biochemistry* **6,** 62–68.
Lovell, R. T. (1973). *J. Nutr.* **103,** 134–138.
Lovell, R. T. (1987). *Feed Manage.* **38**(1), 28–36.
Lovell, R. T., and Buston, J. C. (1984). *J. Nutr.* **114,** 1092–1096.
Lucas, H. L., Heuser, G. F., and Norris, L. C. (1946). *Poult. Sci.* **25,** 137–142.
Luce, W. G., Peo, E. R., and Hudman, D. B. (1967). *J. Anim. Sci.* **26,** 76–84.
Luecke, R. W., McMillen, W. N., and Thorpe, F. (1950). *J. Anim. Sci.* **9,** 78–82.
Luecke, R. W., Hoefer, J. A., and Thorpe, Jr., F. (1953). *J. Anim. Sci.* **12,** 605–610.
Lui, A., Lumeng, L., and Li, T. (1983). *J. Nutr.* **113,** 893–898.
McCormick, D. B., and Olson, R. E. (1984). *In* "Nutrition Reviews, Present Knowledge in Nutrition" (R. E. Olson, ed.), pp. 365–376. Nutrition Foundation, Washington, D.C.
McDonald, P., Edwards, R. A., and Greenhalgh, J. F. D. (1981). "Animal Nutrition," 3rd Ed., pp. 83–84. Longman, New York.
McDowell, L. R. (1977). "Geographical Distribution on Nutritional Diseases in Animals." Dept. of Animal Science, University of Florida, Gainesville.
McDowell, L. R. (1985a). *In* "Nutrition of Grazing Ruminants in Warm Climates" (L. R. McDowell, ed.), pp. 339–357. Academic Press, Orlando, Florida.
McDowell, L. R. (1985b). "Vitamin Lecture Notes." Dept. of Animal Science, University of Florida, Gainesville.
McDowell, L. R., Froseth, J. H., Piper, R. C., Dyer, I. A., and Kroening, G. H. (1977). *J. Anim. Sci.* **45,** 1326–1333.
McDowell, L. R., Conrad, J. H., Ellis, G. L., and Loosli, J. K. (1983). "Minerals for Grazing Ruminants in Tropical Regions," pp. 89. University of Florida, Gainesville.
McDowell, L. R., Conrad, J. H., and Ellis, G. L. (1984). *In* "Symposium on Herbivore Nutrition in Sub-tropics and Tropics—Problems and Prospects" (F. M. C. Gilchrist and R. I. Mackie, eds.), pp. 67–88. The Science Press, Pretoria.
McDowell, L. R., Conrad, J. H., Ellis, G. L., and Mason, Jr., R. M. (1985). *Proc. Nutr. Conf., Univ. Fla., Gainsville* pp. 79–94.
McElroy, L. W., and Goss, H. (1939). *J. Biol. Chem.* **130,** 437–438.
McElroy, L. W., and Goss, H. (1940). *J. Nutr.* **20,** 527–540.
McFarlane, W. D., Graham, W. R., and Richardson, F. (1931). *Biochem. J.* **25,** 358–366.
McGinnis, C. H. (1986a). "Water Soluble Vitamins." Rhone Poulenc, Atlanta.
McGinnis, C. H. (1986b). "Bioavailability of Nutrients in Feed Ingredients," pp. 1–52. National Feed Ingredient Association (NFIA), Des Moines, Iowa.
McIntosh, G. H., and Tomas, F. M. (1978). *Q. J. Exp. Physiol.* **63,** 119–124.
McKigney, J. I., Wallace, H. D., and Cunha, T. J. (1957). *J. Anim. Sci.* **16,** 35–43.
McLaren, B. A., Keller, E., O. Donnell, D. J., and Elvehjem, C. A. (1947). *Arch. Biochem. Biophys.* **15,** 169–178.
McLaren, D. S. (1982). *Nutr. Rev.* **40,** 303–305.
McLaren, D. S. (1984). *In* "Nutrition Reviews, Present Knowledge in Nutrition" (R. E. Olson, ed.), pp. 192–208. Nutrition Foundation, Washington, D.C.
McLaren, D. S. (1986). *In* "Vitamin A Deficiency and Its Control" (J. C. Baunernfeind, ed.), pp. 1–18. Academic Press, Orlando, Florida.

McManus, E. C., and Judith, F. R. (1972). *Poult. Sci.* **51,** 1835–1836.
McMillen, W. N., Luecke, R. W., and Thorpe, Jr., F. (1949). *J. Anim. Sci.* **8,** 518–523.
McMurray, C. H., and Blanchflower, W. J. (1979). *J. Chromatogr.* **178,** 525–531.
McMurray, C. H. and Rice, D. A. (1984). "Vitamin E and Selenium Deficiency Diseases," No. 1953. Hoffmann-LaRoche, Basel.
McMurray, C. H., Rice, D. A., and Blanchflower, W. J. (1980). *Proc. Nutr. Soc.* **39,** 65 (Abstr.).
Machlin, L. J. (1984). *In* "Handbook of Vitamins" (L. J. Machlin, ed.), pp. 99–145. Dekker, New York.
Machlin, L. J., and Gordon, R. S. (1961). *J. Nutr.* **75,** 157–164.
MacPherson, A. (1982). "Roche Vitamin Symposium: Recent Research on the Vitamin Requirements of Ruminants" pp. 1–18. Hoffmann-LaRoche, Basel, Switzerland.
Madhok, T. C., and DeLuca, H. F. (1979). *Biochem. J.* **184,** 491–499.
Mahan, D. C., Pickett, R. A., Perry, T. W., Curtin, T. M., Featherson, W. R., and Beeson, W. M. (1966). *J. Anim. Sci.* **25,** 1019–1026.
Mangelson, F. L., Draper, C. I., Greenwood, D. A., and Crandall, B. H. (1949). *Poult. Sci.* **28,** 603–609.
Manz, U., and Maurer, R. (1982). *Int. J. Vitam. Nutr. Res.* **52,** 248–252.
Marks, J. (1975). "A Guide to the Vitamins. Their Role in Health and Disease," pp. 73–82. Medical and Technical Publ., Lancaster, England.
Martin, F. J., Ullrey, D. E., Miller, E. R., Kemp, K. E., Geasler, M. R., and Henderson, H. E. (1971). *J. Anim. Sci.* **32,** 1233–1238.
Martius, C., and Alvino, C. (1964). *Biochem. Z.* **340,** 316–319.
Martynjuk, B. F. (1952). *Med. Veter.* **29,** 21–30.
Mason, J. B., Gibson, N., and Kodicek, E. (1973). *Br. J. Nutr.* **30,** 297–311.
Mason, Jr., R. M., Ostrum, P. and McDowell, L. R. (1985). *Fl. Cattlemen* **March,** 93.
Matoth, Y., Pinkas, A., and Sroka, C. (1965). *Am. J. Clin. Nutr.* **16,** 356–359.
Matschiner, J. T., Bell, R. G., Amelotti, J. M., and Knauer, T. E. (1970). *Biochim. Biophys. Acta* **201,** 309–315.
Matte, J. J., Girard, C. L., and Brisson, G. J. (1984a). *J. Anim. Sci.* **59,** 158–163.
Matte, J. J., Girard, C. L., and Brisson, G. J. (1984b). *J. Anim. Sci.* **59,** 1020–1025.
Mayer, G. P., Ramberg, C. F., and Kronfeld, D. S. (1968). *J. Nutr.* **95,** 202–206.
Maynard, L. A., Loosli, J. K., Hintz, H. F., and Warner, R. G. (1979). "Animal Nutrition," 7th Ed., pp. 283–355. McGraw-Hill, New York.
Mehansho, H., and Henderson, L. M. (1980). *J. Biol. Chem.* **255,** 11901–11907.
Mellanby, E. (1947). *J. Physiol. (London)* **105,** 382–399.
Michel, R. L., Whitehair, C. K., and Keahey, K. K. (1969). *J. Am. Vet. Med. Assoc.* **155,** 50–59.
Michell, R. H. (1975). *Biochim. Biophys. Acta* **415,** 81–147.
Mickelsen, O. (1956). *Vitam. Horm.* **14,** 1–95.
Miles, R. D., Ruiz, N., and Harms, H. (1983). *Poult. Sci.* **62,** 495–498.
Miles, R. D., Ruiz, N., and Harms, R. H. (1986). *Professional Anim. Sci.* **2,** 33–36.
Miles, W. H., and McDowell, L. R. (1983). *World Anim. Rev.* **46,** 2–10.
Miller, B. E., and Norman, A. W. (1984). *In* "Handbook of Vitamins" (L. J. Machlin, ed.), pp. 45–97. Dekker, New York.
Miller, B. L., Goodrich, R. D., and Meiske, J. C. (1983). *J. Anim. Sci.* **57**(Suppl.), 454 (Abstr.).
Miller, E. E., and Baumann, C. A. (1945). *J. Biol. Chem.* **157,** 551.
Miller, E. R., Schmidt, D. A., Hoefer, J. A., and Luecke, R. W. (1955). *J. Nutr.* **56,** 423–430.
Miller, L. T. (1986). *J. Nutr.* **116,** 1344–1345.
Miller, R. W., Hemken, R. W., Waldo, D. R., and Moore, L. A. (1969). *J. Dairy Sci.* **52,** 1998–2000.

Miller, W. J. (1979). "Dairy Cattle Feeding and Nutrition." Academic Press, New York.

Mitchell, G. E. (1967). *J. Am. Vet. Med. Assoc.* **151,** 430–436.

Mitchell, M. (1978). *Am. J. Clin. Nutr.* **31,** 293–306.

Mittelholzer, E. (1976). *Int. J. Vitam. Nutr. Res.* **46,** 33–39.

Mobarhan, S., Maiani, G., Zanacchi, E., Ferrini, A. M., Scaccini, C., Sette, S., and Ferro-Luzzi, A. (1982). *Clin. Nutr.* **36c,** 71–79.

Mohrhauer, H., and Holman, R. T. (1963). *J. Lipid Res.* **4,** 151–159.

Molitor, H., and Robinson, H. (1940). *Proc. Soc. Exp. Biol. Med.* **43,** 125–128.

Molitoris, B. A., and Baker, D. H. (1976). *J. Anim. Sci.* **42,** 481–489.

Mookerjea, S. (1971). *Fed. Proc., Fed. Am. Soc. Exp. Biol.* **30,** 143–150.

Moore, T., Sharman, I. M., and Ward, R. J. (1957). *J. Sci. Food Agric.* **8,** 97–104.

Morand-Fehr, P. (1981). *In* "Goat Production" (G. Gall, ed.), pp. 193–232. Academic Press, New York.

Morris, E. R. (1983).*Fed. Proc., Fed. Am. Soc. Exp. Biol.* **42,** 1716–1720.

Morris, K. M. L. (1982). *Vet. Hum. Toxicol.* **24,** 34–48.

Mullenax, C. (1983). *Carta Ganadera* **20**(10), 7–10.

Müller, M., Girard, V., Prabucki, A. L., and Schurch (1976). *Bibl. Nutr. Dieta* **23,** 27–33.

Muralt, A. (1962). *Ann. N.Y. Acad. Sci.* **98,** 499–507.

Murphy, M., Keating, M., Boyle, P., Weir, D. G., and Scott, J. M. (1976). *Biochem. Biophys. Res. Commun.* **71,** 1017–1024.

Naidoo, D. (1956). *Acta Psychiatr. Neurol. Scand.* **31,** 205–216.

Narasinga Rao, B. S., and Gopalan, C. (1984). *In* "Nutrition Reviews, Present Knowledge in Nutrition" (R. E. Olson, ed.), pp. 318–331. Nutrition Foundation, Washington, D.C.

Nelson, R. A. (1978). *In* "Handbook Series in Nutrition and Food, Section E: Nutritional Disorders" (M. Rechcigl, Jr., ed.), Vol. 3, pp. 33–36. CRC Press, Boca Raton, Florida.

Nelson, M. M., and Evans, H. M. (1945). *Proc. Soc. Exp. Biol. Med.* **60,** 319–320.

Neumann, A. L., Krider, J. L., James, M. F., and Johnson, B. C. (1949). *J. Nutr.* **38,** 195–203.

Neuringer, M., Connor, W. E., Van Petten, C., and Barstad, L. (1984). *J. Clin. Invest.* **73,** 272–276.

Nguyen, L. B., and Gregory, J. F. (1983). *J. Nutr.* **113,** 1550–1560.

Njoku, P. C. (1984). *Feedstuffs* **56** (52), 23.

Nobile, S., and Woodhill, J. M. (1981). "Vitamin C." MTP Press (International Medical Publ.), Boston.

Noble, R. E., Moore, J. H., and Harfoot, C. G. (1974). *Br. J. Nutr.* **31,** 99–108.

Nockels, C. F. (1984). In "Proc. Ascorbic Acid in Domestic Animals" (I. Wegger, F. J. Tagwerker, and J. Moustgaard, eds.), pp. 175–184. Danish Agriculture Society, Copenhagen.

Norman, A. W., and DeLuca, H. F. (1963). *Biochemistry* **2,** 1160–1168.

Norvell, M. J., and Nesheim, M. C. (1969). *Poc. Cornell Nutr. Conf.* pp. 31–35.

NRC (1973). "Effect of Processing on the Nutritional Value of Feeds." National Academy of Sciences—National Research Council-Washington, D.C.

NRC (1976). (Committee on Swine Nutrition). *J. Anim. Sci.* **42,** 1211–1216.

NRC (1977). "Nutrient Requirements of Domestic Animals, Nutrient Requirements of Rabbits," 2nd Ed. National Academy of Sciences—National Research Council, Washington, D.C.

NRC (1978a). "Nutrient Requirements of Domestic Animals. No. 3: Nutrient Requirements of Dairy Cattle," 5th Ed. National Academy of Sciences—National Researh Council, Washington, D.C.

NRC (1978b). "Nutrient Requirements of Domestic Animals. No. 6: Nutrient Requirements of Horses," 4th Ed. National Academy of Sciences—National Research Council, Washington, D.C.

NRC (1978c). "Nutrient Requirements of Domestic Animals: Nutrient Requirements of Laboratory Animals." National Academy of Sciences—National Research Council, Washington, D.C.
NRC (1978d). "Nutrient Requirements of Domestic Animals: Nutrient Requirements of Nonhuman Primates." National Academy of Sciences—National Research Council, Washington, D.C.
NRC (1981a). "Nutrient Requirements of Domestic Animals: Nutrient Requirements of Coldwater Fishes." National Academy of Sciences—National Research Council, Washington, D.C.
NRC (1981b). "Nutrient Requirements of Domestic Animals. No. 15: Nutrient Requirements of Goats," 1st Ed. National Academy of Sciences—National Research Council, Washington, D.C.
NRC (1982a). "Nutrient Requirements of Domestic Animals: Nutrient Requirements of Mink and Foxes," 2nd Ed. National Academy of Sciences—National Research Council, Washington, D.C.
NRC (1982b). "United States–Canadian Tables of Feed Composition," 3rd Ed. National Academy of Sciences—National Research Council, Washington, D.C.
NRC (1983). "Nutrient Requirements of Domestic Animals: Nutrient Requirements of Warmwater Fishes and Shellfishes," 2nd Ed. National Academy of Sciences—National Research Council, Washington, D.C.
NRC (1984a). "Nutrient Requirements of Domestic Animals: Nutrient Requirements of Beef Cattle," 6th Ed. National Academy of Sciences—National Research Council, Washington, D.C.
NRC (1984b). "Nutrient Requirements of Domestic Animals: Nutrient Requirements of Poultry," 8th Ed. National Academy of Sciences—National Research Council, Washington, D.C.
NRC (1985a). "Nutrient Requirements of Domestic Animals: Nutrient Requirements of Dogs," 2nd Ed. National Academy of Sciences—National Research Council, Washington, D.C.
NRC (1985b). "Nutrient Requirements of Domestic Animals: Nutrient Requirements of Sheep," 5th Ed. National Academy of Sciences—National Research Council, Washington, D.C.
NRC (1986). "Nutrient Requirements of Domestic Animals: Nutrient Requirements of Cats," 3rd Ed. National Academy of Sciences—National Research Council, Washington, D.C.
NRC (1987). "Vitamin Tolerance of Animals." National Academy of Sciences—National Research Council, Washington, D.C.
NRC (1988). "Nutrient Requirements of Domestic Animals: Nutrient Requirements of Swine," 9th Ed. National Academy of Sciences—National Research Council, Washington, D.C.
Obel, A. L. (1953). *Acta Pathol. Microbiol. Scand. Suppl.* **94,** 5–87.
O'Dell, B. L., and Hogan, A. G. (1943). *J. Biol. Chem.* **149,** 323–337.
Ogino, C., Uki, N., Watanabe, T., Iida, Z., and Ando, K. (1970). *Bull. Jpn. Soc. Sci. Fish.* **36,** 1140–1146.
Ohnuma, N., Bannai, K., Yamaguchi, H., Hashimoto, Y., and Norman, A. W. (1980). *Arch. Biochem. Biophys.* **204,** 387–391.
Okada, K. A., Carrillo, B. J., and Tilley, M. (1977). *Econ. Bot.* **31,** 225–236.
Olcese, O., Couch, J. R., Quisenberry, J. H., and Pearson, P. B. (1950). *J. Nutr.* **41,** 423–431.
Oldfield, J. E. (1987). *J. Nutr.* **117,** 1322–1323.
Olentine, C. (1984). *Feed Manage.* **35**(4), 18–24.
Olson, A. L., and Rebouche, C. J. (1987). *J. Nutr.* **117,** 1024–1031.
Olson, J. A. (1986). *Nutr. Today* **Sept./Oct.,** 26–30.
Olson, R. E. (1959). *Annu. Rev. Biochem.* **28,** 467–498.
Olson, R. E. (1984). *In* "Nutrition Reviews, Present Knowledge in Nutrition" (R. E. Olson, ed.), pp. 377–382. Nutrition Foundation, Washington, D.C.
Olson, R. E., and Suttie, J. W. (1978). *Vitam. Horm.* **35,** 59–108.
Oltjen, R. R., Sirny, R. J., and Tillman, A. D. (1962). *J. Nutr.* **77,** 269–277.
Onwudike, O. C., and Adegbola, A. A. (1984). *Trop. Agric. (Trinidad)* **61,** 205–207.
Orstadius, K., Nordstrom, G., and Lannek, N. (1963). *Cornell Vet.* **53,** 60–71.
Oski, F. A., and Barness, L. A. (1967). *J. Pediatr.* **70,** 211–220.

Palludan, B., and Wegger, I. (1984). *In* "Proc. Ascorbic Acid in Domestic Animals" (I. Wegger, F. J. Tagwerker, and J. Moustgaard, eds.), pp. 131–138. Agriculture Society, Copenhagen.
Palmquist, D. L., Mattos, W., and Stone, R. L. (1977). *Lipids* **12,** 235–238.
Panganamala, R. V. and Cornwell, D. C. (982). *Ann. N.Y. Acad. Sci.* **393,** 376–391.
Pappenheimer, A. M., and Goettsch, M. (1931). *J. Exp. Med* **53,** 11–26.
Pappu, A. S., Fatterpaker, P., and Sreenivasan, A. (1978). *Biochem. J.* **172,** 115–121.
Pardue, S. L. (1987). "Proc., the Role of Vitamins on Animal Performance and Immune Response," pp. 18–23. Hoffmann-LaRoche, Nutley, New Jersey.
Pardue, S. L. and Thaxton, J. P. (1982). *Poult. Sci.* **61,** 1522 (Abstr.).
Parsons, J. L., and Klostermann, H. J. (1967). *Feedstuffs* **39** (45), 74.
Paul, P. K., Duttagupta, P. N., and Agarwal, H. C. (1973). *Curr. Sci.* **42,** 613–615.
Pauling, L. (1971). "Vitamin C and the Common Cold." Freeman, San Francisco.
Pauling, L. (1974). *Am. J. Psychiatr.* **131,** 1251–1257.
Pavcek, P. L., and Shull, G. M. (1942). *J. Biol. Chem.* **146,** 351–355.
Payne, J. M., and Manston, R. (1967). *Vet Rec.* **81.** 214–216.
Pearson, P. B., and Schmidt, H. (1948). *J. Anim. Sci.* **7,** 78–83.
Pearson, P. B., Sheybani, M. K., and Schmidt, H. (1944). *J. Anim. Sci.* **3,** 166–174.
Pearson, P. B., Melass, V. H., and Sherwood, R. M. (1945). *Arch. Biochem.* **7,** 353–356.
Peebles, D., and Brake, J. (1984). *Feedstuffs* **56** (21), 24.
Peifer, J. J., and Holman, R. T. (1955). *Arch. Biochem. Biophys.* **57,** 520–521.
Perek, M., and Kendler, J. (1963). *Br. Poult. Sci.* **4,** 191–200.
Perry, S. C. (1978). *Proc. Roche Vitam. Nutr. Update Meet., Arkansas Nutr. Conf.* pp. 29–69.
Perry, T. W. (1962). "Published Reports of the Value of Vitamin A," pp. 3–7. Hoffmann-LaRoche, Nutley, New Jersey.
Perry, T. W. (1968). *Feedstuffs* **40** (25), 48.
Perry T. W. (1980). "Beef Cattle Feeding and Nutrition," pp. 24–32. Academic Press, New York.
Perry T. W., Beeson, W. M., Smith, W. H., and Mohler, M. T. (1967). *J. Anim. Sci.* **26,** 115–118.
Pesti, G. M., Benevenga, N. J., Harper, A. E., and Sunde, M. L. (1981). *Poult. Sci.* **60,** 425–432.
Peterson, P. A., Nilsson, S. F. Ostberg, L., Rask, L., and Vahlquist, A. (1974). *Vitam. Horm.* **32,** 181–214.
Petrelli, F., Moretti, P., and Paparelli, M. (1979). *Mol. Biol. Rep.* **4,** 247–252.
Pineo, G. F., Gallus, A. S., and Hirsh, J. (1973). *Can. Med. Assoc. J.* **109,** 880–883.
Piper, R. C., Froseth, J. A., McDowell, L. R., Kroening, G. H., and Dyer, J. A. (1975). *Am. J. Vet. Res.* **36,** 273–281.
Pit, G. A. J. (1985). *In* "Fat-Soluble Vitamins" (A. T. Diplock, ed.), pp. 1–75. Technomic Publ., Lancaster, Pennsylvania.
Plaut, G. W. E., Smith, C. M., and Alworth, W. L. (1974). *Annu. Rev. Biochem.* **43,** 899–921.
Polak, D. M., Elliot, J. M., and Haluska, M. (1979). *J. Dairy Sci.* **62,** 697–701.
Pollard, W. O., and Creek, R. D. (1964). *Poult. Sci.* **43,** 1415–1420.
Poston, H. A., Riis, R. C., Rumsey, G. L., and Ketola, H. G. (1977). *Cornell Vet.* **67,** 472–509.
Price, J. M., Brown, R. R., and Larson, F. C. (1957). *J. Clin. Invest.* **36,** 1600–1607.
Pugh, E. L., and Wakil, S. J. (1965). *J. Biol. Chem.* **240,** 4727–4733.
Purser, K. (1981). *Anim. Nutr. Health* **April,** 38.
Putney, J. W., Aub, D. L., Taylor, C. W., and Merritt, J. E. (1986). *Fed. Proc., Fed. Am. Soc. Exp. Biol.* **45,** 263–268.
Quackenbush, F. W. (1963). *Cereal Chem.* **40,** 266–269.
Quaterman, J., Dalgarno, A. C., Adam, A., Fell, B. F., and Boyne, R. (1964). *Br. J. Nutr.* **18,** 65–77.
Rabinowitz, J. C., and Snell, E. E. (1949). *Proc. Soc. Exp. Biol. Med.* **70,** 235–240.

Rao, S. B., Raghuram, T. C., and Krishnaswamy, K. (1975). *Nutr. Metab.* **18,** 318–325.
Ray, S. M. (1963). *Proc. World Conf. Anim. Prod., 1st, Rome* **2,** 109 (Abstr.).
RDA (1980). "Recommended Dietary Allowances," 9th Ed. National Academy of Sciences—National Research Council, Washington, D.C.
Rebouche, C. J. (1986). *J. Nutr.* **116,** 704–706.
Reddy, M. U., and Pushpamma, P. (1986). *Nutr. Rep. Int.* **34,** 393–401.
Reddy, P. G., Morrill, J. J., and Frey, R. A. (1987a). *J. Dairy Sci.* **70,** 123–129.
Reddy, P. G., Morrill, J. L., Minocha,, H. C., and Stevenson, J. S. (1987b). *J. Dairy Sci.* **70,** 993–999.
Reid, J. M., Hove, E. L., Braucher, P. F., and Mickelsen, O. (1963). *J. Nutr.* **80,** 381–385.
Reinhardt, T. A., and Hustmyer, F. G. (1987). *J. Dairy Sci.* **70,** 952–962.
Reiser, R. (1950). *J. Nutr.* **42,** 319–323.
Ribovitch, M. L., and DeLuca, H. F. (1975). *Arch. Biochem. Biophys.* **170,** 529–535.
Rice, D. A., Blanchflower, W. J., and McMurray, C. H. (1981). *Vet. Rec.* **109,** 161–162.
Richards, R. K., and Shapiro, S. (1945). *J. Pharmacol. Exp. Ther.* **84,** 93–104.
Ringler, D. H., and Abrams, G. D. (1970). *J. Am. Vet. Med. Assoc.* **157,** 1928–1934.
Ritchie, H. D., Miller, E. R., Ullrey, D. E., Hofer, J. A., and Lucke, R. W. (1960). *J. Nutr.* **70,** 491–496.
Rivlin, R. S. (1978). *In* "Handbook Series in Nutrition and Food, Section E: Nutritional Disorders" (M. Rechcigl, Jr., ed.), Vol. 1, pp. 25–27. CRC Press, Boca Raton, Florida.
Rivlin, R. S. (1984). *In* "Nutrition Reviews, Present Knowledge in Nutrition" (R. E. Olson, ed.), pp. 285–302. Nutrition Foundation, Washington, D.C.
Robinson, E. H., and Lovell, R. T. (1978). *J. Nutr.* **108,** 1600–1605.
Roche (1979a). "Optimum Vitamin Nutrition." Hoffmann-LaRoche, Nutley, New Jersey.
Roche (1979b). "Rationale for Roche Recommended Vitamin Fortification, Swine Rations." Hoffmann-LaRoche, Nutley, New Jersey.
Roe, D. A. (1973). *Drug Ther.* **3,** 23–32.
Roe, D. A., Bogusz, S., Sheu, J., and McCormick, D. B. (1982). *Am. J. Clin. Nutr.* **35,** 495–501.
Roels, O. A., Trout, M., and Dujacquier, R. (1958). *J. Nutr.* **65,** 115–128.
Roland, D. A., and Edwards, H. M. (1971). *J. Nutr.* **101,** 811–818.
Rose, J. A. (1971). *Lancet* **2,** 453–456.
Rose, R. C., and Nahrwold, D. L. (1982). *Anal. Biochem.* **123,** 389–393.
Rose, R. C., McCormick, D. B., Li, T. K., Lumeng, L., Haddad, J. G., and Spector, R. (1986). *Fed. Proc., Fed. Am. Soc. Exp. Biol.* **45,** 30–39.
Rosenberg, I. H., and Newmann, H. (1974). *J. Biol. Chem.* **249,** 5126–5130.
Rothenberg, S. P., and Cotter, R. (1978). *In* "Handbook Series in Nutrition and Food, Section E: Nutritional Disorders" (M. Rechcigl, Jr., ed.), Vol. 3, pp. 69–88. CRC Press, Boca Raton, Florida.
Rothman, D. (1970). *Am. J. Obstet. Gynecol.* **108,** 149–175.
Rothschild, B. (1982). *Arch Intern. Med.* **142,** 840 (Abstr.).
Rowland, R. (1982). "Vitamins—The Life Essentials," pp. 1–12. National Feed Ingredients Association, Des Moines, Iowa.
Roy, R. B., and Buccafuri, A. (1978). *J. Assoc. Off. Anal. Chem.* **61,** 720–726.
Ruiz, N. (1987). Re-evaluation of some of the B-complex vitamin requirements of broiler chickens and turkey poults fed corn–soybean meal diets. Ph.D. thesis, University of Florida, Gainesville.
Ruiz, N., Miles, R. D., and Harms, R. H. (1983). *WPSA J.* **39,** 185–198.
Rumsey, T. S. (1975). *Feedstuffs* **47,** 30–34.
Russell, W. C. (1929). *J. Biol. Chem.* **85,** 289–297.
Rusznyak, S., and Szent-Györgyi, A. (1936). *Nature (London)* **138,** 27 (Abstr.).
Salkeld, R., Knorr, K., and Korner, W. (1973). *Clin. Chim. Acta* **49,** 195–199.

Sandholm, M., Honkanen-Buzalski, R., and Rasi, V. (1979). *Vet. Rec.* **104,** 337–338.
Sanford, P. A., and Smyth, D. H. (1971). *J. Physiol. (London)* **215,** 769–788.
Sauberlich, H. E. (1984). *In* "Nutrition Reviews, Present Knowledge in Nutrition" (R. E. Olson, ed.), pp. 260–272. Nutrition Foundation, Washington, D.C.
Sauberlich, H. E. (1985). *Prog. Food Nutr. Sci.* **9,** 1–33.
Schachter, D., and Kowarski, S. (1982). *Fed. Proc., Fed. Am. Soc. Exp. Biol.* **41,** 84–87.
Schaumburg, H., Kaplan, J., Windebank, A., Vick, N., Rasmug, S., Pleasure, D., and Brown, M. J. (1983). *N. Engl. J. Med.* **309,** 445–448.
Schendel, H. E., and Johnson, B. C. (1962). *J. Nutr.* **76,** 124–130.
Schlenk, H., and Gellerman, J. L. (1965). *J. Am. Oil Chem. Soc.* **42,** 504–511.
Schneider, J., and Hoppe, P. P. (1986). "Bioavailability of Nutrients in Feed Ingredients," pp. 1–6. National Feed Ingredients Association (NFIA), Des Moines, Iowa.
Schultz, L. H. (1983). *Proc. Annu. Pac. Northwest Anim. Nutr. Conf., 18th* pp. 69–76.
Schweigert, B. S., German, H. L., Pearson, P. B., and Sherwood, R. M. (1948). *J. Nutr.* **35,** 89–102.
Scott, D., and McLean, A. F. (1981). *Proc. Nutr. Soc.* **40,** 257–266.
Scott, M. L. (1966). *Proc. Cornell Nutr. Conf.* pp. 35–55.
Scott, M. L. (1973). *In* "Effect of Processing on the Nutritional Value of Feeds," pp. 119–130. National Academy of Sciences, Washington, D.C.
Scott, M. L. (1980). *Fed. Proc., Fed. Am. Soc. Exp. Biol.* **39,** 2736–2739.
Scott, M. L. (1981). *Feedstuffs* **53** (8), 59–60, 62, 64, 66–67.
Scott, M. L., Nesheim, M. C., and Young, R. J. (1982). "Nutrition of the Chicken," pp. 119–276. Scott, Ithaca, New York.
Scriver, C. R. (1973). *Metabolism* **22,** 1319–1344.
Scriver, C. R., and Whelan, D. T. (1969). *Ann. N.Y. Acad. Sci.* **166,** 83–96.
Sebrell, W. H., and Butler, R. E. (1938). *Public Health Rep.* **53,** 2282–2284.
Sebrell, Jr., W. H., and Harris, R. S. (1967). "The Vitamins," pp. 3–305. Academic Press, New York.
Sebrell, Jr., W. H., and Harris, R. S. (1968). "The Vitamins," pp. 119–359. Academic Press, New York.
Sebrell, Jr., W. H., and Harris, R. S. (1973). "The Vitamins: Chemistry, Physiology, Pathology and Methods," Vol. 5, pp. 98–162. Academic Press, New York.
Seckington, I. M., Huntsman, R. G., and Jenkins, G. C. (1967). *Vet. Rec.* **81,** 158–161.
Selke, M. R., Barnhart, C. E., and Chaney, C. H. (1967). *J. Anim. Sci.* **26,** 759–763.
Senti, F. R., and Pilch, S. M. (1985). *J. Nutr.* **115,** 1398–1402.
Sewell, R. F., and McDowell, L. R. (1966). *J. Nutr.* **89,** 64–68.
Sewell, R. F., Cunha, T. J., Shawver, C. B., Ney, W. A., and Wallace, H. D. (1952). *Am. J. Vet. Res.* **13,** 186–187.
Sewell, R. F., Price, D. G., and Thomas, M. C. (1962). *Fed. Proc., Fed. Am. Soc. Exp. Biol.* **21,** 468 (Abstr.).
Seymour, E. W., Speer, V. C., and Hays, V. W. (1968). *J. Anim. Sci.* **27,** 389–393.
Shearer, M. J., and Barkhan, P. (1973). *Biochim. Biphys. Acta* **297,** 300–312.
Shearer, M. J., Barkhan, P., and Webster, G. R. (1970). *Br. J. Haematol.* **18,** 297–308.
Sheffy, B. E., and Williams, A. J. (1981). "Animal Nutrition Events," 1781, 1–14. Hoffmann-LaRoche, Basel.
Sheppard, A. J., and Johnson, B. C. (1957). *J. Nutr.* **61,** 195–205.
Shideler, C. E. (1983). *Am. J. Med. Technol.* **49,** 17–22.
Shields, D. R., and Perry, T. W. (1982). "Effect of Supplemental Niacin on Protein Utilization by Lambs." National Feed Ingredients Association (NFIA), Fairlawn, New Jersey.
Shields, R. G., Campbell, D. R., Hughes, D. M., and Dillingham, D. A. (1982). *Feedstuffs* **22,** 27–28.

Shim, K. F., and Boey, H. L. (1988). *Nutr. Rep. Int.* **37,** 839–900.

Shinton, N. K. (1972). *Br. Med. J.* **1,** 556–559.

Siddons, R. C. (1978). "Handbook Series in Nutrition and Food, Section E: Nutritional Disorders" (M. Rechcigl, Jr., ed.), Vol. 2, pp. 123–132. CRC Press, Boca Raton, Florida.

Sitrin, M. D., Liberman, F., Jensen, W. E., Noronha, A., Milburn, C., and Addington, W. (1987). *Ann. Intern. Med.* **107,** 51–54.

Slinger, S. J., and Pepper, W. F. (1954). *Poult. Sci.* **33,** 633–637.

Slinger, S. J., Razzaque, A., and Cho, C. Y. (1979). *Proc. World Symp. Finfish Nutr. Fishfeed Technol.* **2,** 425–534.

Smith, M. C. (1979). *J. Am. Vet. Med. Assoc.* **174,** 1328–1332.

Smith, P. J., Tappel, A. L., and Chow, C. K. (1974). *Nature (London)* **247,** 392–393.

Smith, S. E. (1970). *In* "Duke's Physiology of Domestic Animals," 8th Ed., pp. 634–659 (M. J. Swenson, ed.), Cornell Univ. Press, Ithaca, New York.

Snehalatha Reddy, N., and Lakshmi Kumari, R. (1988). Nutr. Rep. Int. **37,** 77–81.

Sommer, A., Tarwotso, I., Hussani, G., Susanto, D., and Soegiharto, T. (1981). *Lancet* **1,** 1407–1408.

Sommer, A., Hussani, G., Tarwotijo, I., and Susanto, D. (1983). *Lancet* **2** 585–588.

Sommer, A., Tarwotijo, I., Djunaedi, E., West, K. P., Loeden, A. A., Tilden, R., and Mele, L. (1986). *Lancet* **1,** 1169–1173.

Sommerfeldt, J. L, Horst, R. L., Littledike, E. T., Beitz, D. C., and Napoli, J. L. (1981). *J. Dairy Sci.* **64,** 157(Abstr.).

Sommerfeldt, J. L., Napoli, J. L., Littledike, E. T., Beitz, D. C., and Horst, R. L. (1983). *J. Nutr.* **113,** 2595–2600.

Southern, L. L., Brown, D. R., Werner, D. D., and Fox, M. C. (1986). *J. Anim. Sci.* **62,** 992–996.

Spencer, R. P., Purdy, S., Hoeldtke, R., Bow, T. M., and Markulis, M. A. (1963). *Gastroenterology* **44,** 768–773.

Squibb, R. L. (1964). *Poult. Sci.* **43,** 1443–1445.

Stamberg, O. E., and Peterson, C. F. (1946). *Poult. Sci.* **25,** 394–395.

Steenbock, H., and Black, A. (1924). *J. Biol. Chem,* **61,** 405–422.

Sterner, R. T., and Price, W. R. (1973). *Am. J. Clin. Nutr.* **26,** 150–160.

Stout, F. M., Adair, J., and Oldfield, J. E. (1966). *Nat. Fur. News* **38,** 13, 14, 19.

Stuart, R. L. (1987). "Proc. The Role of Vitamins on Animal Performance and Immune Response," pp. 67–80. Hoffmann-LaRoche, Nutley, New Jersey.

Stumpf, D. A., Parker, W. D., and Angelini, C. (1985). *Neurology* **35,** 1041–1045.

Sturkie, P. D., Singsen, E. P, Matterson, L. D., Kozeff, A., and Jungherr, E. L. (1954). *Am. J. Vet. Res.* **15,** 457–462.

Sugiyama, N., Suzuki, K., and Wada, Y. (1984). *Jpn. Pediatr. Soc.* **88,** 1968–1971.

Sunde, M. L., Cravens, W. W., Elvehjem, C. A., and Halpin, J. G. (1950). *Poult. Sci.* **29,** 10–14.

Sure, B. (1924). *J. Biol. Chem.* **58,** 693–709.

Suttie, J. W. (1980). *Fed. Proc., Fed. Am. Soc. Exp. Biol.* **39,** 2730–2735.

Suttie, J. W. (1984). *In* "Handbook of Vitamins" (L. J. Machlin, ed.), pp. 147–198. Dekker, New York.

Suttie, J. W., and Jackson, C. M. (1977). *Physiol. Rev.* **57,** 1–70.

Suttie, J. W., and Olson, R. E. (1984). *In* "Nutrition Reviews, Present Knowledge in Nutrition" (R. E. Olson, ed.), pp. 241–259. Nutrition Foundation, Washington, D.C.

Sutton, A. L., and Elliot, J. M. (1972). *J. Nutr.* **102,** 1341–1346.

Sutton, R. A. L., and Dirks, J. H. (1978). *Fed. Proc., Fed. Am. Soc. Exp. Biol.* **37,** 2112–2119.

Suzuki, S. (1979). *J. Perinat. Med.* **7,** 229–232.

Swank, R. L. (1940). *J. Exp. Med.* **71,** 683–702.

Swanson, J. E., and Kinsella, J. E. (1986). *J. Nutr.* **116.** 514–523.
Swenson, M. J. (1970). "Duke's Physiology of Domestic Animals," 8th Ed. Cornell Univ. Press, Ithaca, New York.
Swingle, R. S., and Dyer, I. A. (1970). *J. Anim. Sci.* **31,** 404–408.
Sydenstricker, V. P., Singal, S. A., Briggs, A. P., DeVaughn, N. M., and Isbell, H. (1942). *J. Am. Med. Assoc.* **118,** 1199–1200.
SYNTEX (1979). *In* "Choline," pp. 1–60. SYNTEX AgriBusiness, Nutrition and Chemical Division, Springfield, Missouri.
Tagwerker, R. J. (1983). *Feed Int.* **4,** 22.
Takeuchi, R., Watanabe, T., and Nose, T. (1979). *Bull. Jpn. Soc. Sci. Fish.* **45,** 1319–1323.
Tamura, T., Shane, B., Baer, M. T., King, J. C., Margen, S., and Stokstad, E. L. R. (1978). *Am. J. Clin. Nutr.* **31,** 1984–1987.
Tani, M., and Iwai, K. (1984). *J. Nutr.* **114,** 778–785.
Tapia Romero, E., Rojas, R. E., Arias, L. E., and Avila, G. E. (1985). *In* "Resumenes ALPA 85" (C. F. Chicco, ed.), p. 49. Acapulco, Mexico (Abstr.).
Taylor, L. W. (1947). *Poult. Sci.* **26,** 372–376.
Taylor, T., Hawkins, D. R., Hathway, D. E., and Partington, H. (1972). *Br. Vet. J.* **128,** 500–505.
Taylor, T., Cameron, B. D., Hathway, D. E., and Partington, H. (1974). *Res. Vet. Sci.* **16,** 271–275.
Teague, H. S., Grifo, Jr., A. P., and Palmer, W. M. (1971). *J. Anim. Sci.* **33,** 239 (Abstr.).
Tengerdy, R. P., and Brown, J. C. (1977). *Poult. Sci.* **56,** 957–963.
Tengerdy, R. P., and Nockels, C. F. (1975). *Poult. Sci.* **54,** 1292–1296.
Terroine, T. (1960). *Vitam. Horm.* **18,** 1–42.
Thierry, M. J., Hermodson, M. A., and Suttie, J. W. (1970). *Am. J. Physiol.* **219,** 854–859.
Thodejensen, P., and Basse, A. (1984). *In* "Proc. Ascorbic Acid in Domestic Animals" (I. Wegger, F. J. Tagwerker, and J. Moustgaard, eds.), pp. 87–90. Danish Agriculture Society, Copenhagen.
Thomasson, H. J. (1969). *Nutr. Rev.* **27,** 67–69.
Thompson, J. N., and Scott, M. L. (1969). *J. Nutr.* **97,** 335–342.
Thornton, P. A., and Schutze, J. E. (1960). *Poult. Sci.* **39,** 192–199.
Tiege, J., Saxegaard, F., and Froslie, A. (1978). *Acta Vet. Scand.* **19,** 133–146.
Tindall, B. (1985). *Anim. Health Nutr.* **40** (Nov.), 26.
Tokarnia, C. H., Guimaraes, J. A., Canella, C. F. C., and Döbereiner, J. (1971). *Pesqui Agropecu. Bras.* **6,** 61–77.
Tollersrud, S. (1973). *Acta Agric. Scand. Suppl.* **19,** 124–127.
Tollerz, G., and Lannek, N. (1964). *Nature (London),* **201,** 846.
Tomarelli, R. M., Spence, E. R., and Bernhart, F. W. (1955). *J. Agric. Food Chem.* **3,** 338–341.
Tomas, F. M. (1974). *Aust. J. Agric.* **25,** 495–507.
Tomas, F. M., and Somers, M. (1974). *Aust. J. Agric. Res.* **25,** 475–483.
Tono, T., and Fujita, S. (1982). *Agric. Biol. Chem.* **46,** 2953–2959.
Toskes, P. P., Deren, J. J., Fruiterman, J., and Conrad, M. E. (1973). *Gastroenterology* **65,** 199–204.
Trapp, A. L., Keahey, K. K., Whitenack, D. L., and Whitehair, C. K. (1970). *J. Am. Vet. Med. Assoc.* **157,** 289–300.
Tremblay, G. F., Matte, J. J., Lemieux, L., and Brisson, G. J. (1986). *J. Anim. Sci.* **63,** 1173–1178.
Tsai, K. S., Health, H., Kumar, R., and Riggs, B. L. (1984). *Clin. Res.* **32,** 411A.
Tucker, G., Gagnon, R. E., and Haussler, M. R. (1973). *Arch. Biochem. Biophys.* **155,** 47–57.
Turkki, P. R., and Holtzapple, P. G. (1982). *J. Nutr.* **112,** 1940–1952.
Ullrey, D. E. (1972). *J. Anim. Sci.* **35,** 648–657.
Ullrey, D. E. (1981). *J. Anim. Sci.* **53,** 1039–1056.

Ullrey, D. E. (1982). *Proc. Distill. Feed Conf., Distill. Feed Res. Counc. Cincinnati* **37,** 81–98.
Ullrey, D. E., Becker, D. E., Terrill, S. W., and Notzold, R. A. (1955). *J. Nutr.* **57,** 401–414.
Umbreit, W. W. (1960). "Metabolic Maps," Vol. 2, p. 249. Burgess Press, Ames, Iowa.
Underwood, E. J. (1977). "Trace Elements in Human and Animal Nutrition." Academic Press, New York.
Underwood, E. J. (1979). *Proc. Nutr. Conf., Univ. Fla., Gainesville.* pp. 203–230.
Unna, K., and Antopol, W. (1940). *Proc. Soc. Exp. Biol. Med.* **43,** 116–118.
Unna, K., and Greslin, J. C. (1941). *J. Pharmacol. Exp. Ther.* **73,** 85–90.
Unna, K., and Greslin, J. G. (1942). *J. Pharmacol. Exp. Ther.* **76,** 75–80.
Van der Pol, F., Purnomo, S. U., and Van Rosmalen, H. A. (1988). *Nutr. Rep. Int.* **37,** 785–793.
Vanderslice, J. T., Maire, C. E., Doherty, R. F., and Beecher, G. R. (1980). *J. Agric. Food Chem.* **28,** 1145–1149.
VanVleet, J. F., Crawley, R. R., and Amstutz, H. E. (1977). *J. Am. Vet. Med. Assoc.* **171,** 443–445.
Verbeeck, J. (1975). *Feedstuffs* **47** (36), 4, 45.
Vergroesen, A. J. (1977). *Nutr. Rev.* **35,** 1–5.
Vestal, C. M., Beeson, W. M., Andrews, F. N., Hutchings, L. M., and Doyle, L. P. (1950). *Purdue Agric. Exp. Stn. Mimeo* No. 50.
Vilter, R. W. (1978). *In* "Handbook Series in Nutrition and Food, Section E: Nutritional Disorders" (M. Rechcigl, Jr., ed.), Vol. 3, pp. 91–103. CRC Press, Boca Raton Florida.
Virtanen, A. I. (1966). *Science* **153,** 1603–1614.
Wagle, S. R., Mehta, R., and Johnson, B. C. (1958). *J. Biol. Chem.* **230,** 137–147.
Wagner, A. F., and Folkers, K. (1962). "Vitamins and Coenzymes." Wiley (Interscience), New York.
Wagner, C. (1984). "Nutrition Reviews: Present Knowledge in Nutrition" (R. E. Olson, H. P. Broquist, C. O. Chichester, W. J. Darby, A. C. Kolbye, and R. M. Stalvey, eds.), 5th Ed. Nutrition Foundation, Washington, D.C.
Wagner-Jauregg, T. (1972). *In* "The Vitamins" (W. H. Sebrell and R. S. Harris, eds.), pp. 3–43. Academic Press, New York.
Wagonfeld, J. B., Dudzinsky, D., and Rosenberg, I. H. (1975). *Clin. Res.* **23,** 259 (Abstr.).
Wakeman, D. L., Ott, E. A., Crawford, B. H., and Cunha, T. J. (1975). "Horse Production in Florida," pp. 75–90. Bulletin No. 188, Florida Dept. of Agriculture and Consumer Services, Gainesville.
Waisman, H. A., McCall, K. B., and Elvehjem, C. A. (1945). *J. Nutr.* **29,** 1–11.
Wald, G. (1968). *Science* **162,** 230–239.
Wang, J. Y., Owen, F. G., and Larson, L. L. (1988). *J. Dairy Sci.* **71,** 181-186.
Wasserman, R. H. (1975). *Nutr. Rev.* **33,** 1–5.
Wasserman, R. H. (1981). *Fed. Proc., Fed. Am. Soc. Exp. Biol.* **40,** 68–72.
Waterman, R. A. (1978). *In* "Handbook Series in Nutrition and Food, Section E: Nutritional Disorders" (M. Rechcigl, Jr., ed.), Vol. 1, pp. 29–42. CRC Press, Boca Raton, Florida.
Weber, J. C., Pons, V., and Kodicek, E. (1971). *Biochem. J.* **125,** 147–153.
Wegger, I., and Palludan, B. (1984). *In* "Proc. Ascorbic Acid in Domestic Animals" (I. Wegger, F. J. Tagwerker, and J. Moustgaard, eds.), pp. 68–79. Danish Agriculture Society, Copenhagen.
Wehr, N. B., Adair, J., and Oldfield, J. E. (1980). *J. Anim. Sci.* **50,** 877–885.
Weiner, B. H., Ockene, I. S., Levine, P. H., Cuenoud, H. F., Fisher, M., Johnson, B. F., Daoud, A. S., Jarmolych, J., Hosmer, D., Johnson, M. H., Natale, A., Vaudreuil, C., and Hoogasian, J. J. (1986). *N. Engl. J. Med.* **315,** 841–846.
Weiser, H., and Tagwerker, F. (1982). "Optimization of the Vitamin K Bioassay of Vitamin K Forms for Feed Supplementation," No. 1794. Hoffmann-LaRoche, Basel.
Welch, A. D., Heinle, R. W., Sharpe, G., George, W. L., and Epstein, M. (1947). *Proc. Soc. Exp. Biol. Med.* **65,** 364–368.

Welch, B. E., and Couch, J. R. (1955). *Poult. Sci.* **34,** 217–222.
West, R. (1948). *Science* **107,** 398.
Whanger, P. D. (1981). *In* "Selenium in Biology and Medicine" (J. E. Spallholz, J. L. Martin, and H. E. Ganther, eds.), pp. 230–255. AVI, Westport, Connecticut.
White, W. E., Yielding, K. L., and Krumdieck, C. L. (1976). *Biochim. Biophys. Acta* **429,** 689–693.
Whitehead, C. C. (1978). *In* "Handbook Series in Nutrition and Food, Section E: Nutritional Disorders" (M. Rechcigl, Jr., ed.), Vol. 2, pp. 65–93. CRC Press, Boca Raton, Florida.
Whitehead, C. C., and Bannister, D. W. (1978). *Br. J. Nutr.* **39,** 547–556.
Whitehead, C. C., Armstrong, J. A., and Waddington, D. (1982). *Br. J. Nutr.* **48,** 81–88.
Widdowson, E. M. (1986). *Nutr. Rev.* **44,** 221–227.
Wiese, A. C., Johnson, B. C., and Nevens, W. B. (1946). *Proc. Soc. Exp. Biol. Med.* **63,** 521–522.
Wiese, A. C., Johnson, B. C., Mitchell, H. H., and Nevens, W. B. (1947). *J. Dairy Sci.* **30,** 87–94.
Williams, S. N., and McDowell, L. R. (1985). *In* "Nutrition of Grazing Ruminants in Warm Climates" (L. R. McDowell, ed.), pp. 317–338. Academic Press, New York.
Willis, A. L. (1984). *In* "Nutrition Reviews, Present Knowledge in Nutrition" (R. E. Olson, ed.), pp. 90–115. Nutrition Foundation, Washington, D.C.
Wilson, R. B. (1978). *In* "Handbook Series in Nutrition and Food, Section E: Nutritional Disorders" (M. Rechcigl, Jr., ed.), Vol. 2, pp. 95–121. CRC Press, Boca Raton, Florida.
Wing, J. M. (1969). *J. Dairy Sci.* **52,** 479–483.
Wintrobe, M. M., Stein, H. J., Follis, R. H., and Humphreys, S. (1945). *J. Nutr.* **30,** 395–412.
Witz, W. M., and Beeson, W. M. (1951). *J. Anim. Sci.* **10,** 112–128.
Wohl, M., Levy, H., Szutha, A., and Maldia, G. (1960). *Proc. Soc. Exp. Biol. Med.* **105,** 523–527.
Wolker, N. A., and Carrillo, B. J. (1967). *Nature (London)* **215,** 72–74.
Wood, J. L., and Allison, R. G. (1982). *Fed. Proc., Fed. Am. Soc. Exp. Biol.* **41,** 3015–3021.
Wretlind, B. K., Orstadius, K., and Lindberg, P. (1959). *Zentralbl. Vet. Med.* **6,** 963–970.
Wright, L. D., and Welch, A. D. (1943). *Science* **97,** 426–427.
Wyse, B. W., and Hansen, R. G. (1977). *Fed. Proc., Fed. Am. Soc. Exp. Biol.* **36,** 1169 (Abstr.).
Yagi, K., Kotaki, A., and Yamamoto, Y. (1965). *J. Vitaminol.* **11,** 14–19.
Yen, J. T., and Pond, W. G. (1981). *J. Anim. Sci.* **53,** 1291–1299.
Yen, J. T., and Pond, W. G. (1984). *J. Anim. Sci.* **58,** 132–137.
Yen, J. T., and Pond, W. G. (1987). *J. Anim. Sci.* **64,** 1672–1681.
Yen, J. T., Jensen, A. H., and Baker, D. H. (1976). *J. Anim. Sci.* **42,** 866–870.
Young, L. G., Lun, A., Pos, J., Forshaw, R. P., and Edmeades, D. E. (1975). *J. Anim. Sci.* **40,** 495–499.
Young, L. G., Miller, R. B., Edmeades, D. M., Lun, A., Smith, G. C., and King, G. J. (1978). *J. Anim. Sci.* **47,** 639–647.
Young, R. J., Norris, L. C., and Heuser, G. F. (1955). *J. Nutr.* **55,** 353–362.
Zeisel, S. H., Char, D., and Sheard, N. F. (1986). *J. Nutr.* **116,** 50–58.
Zinn, R. A., Owens, F. N., Stuart, R. L., Dunbar, J. R., and Norman, B. B. (1987). *J. Anim. Sci.* **65,** 267–277.
Zintzen, H. (1974). "Vitamin B_1 (Thiamine) in the Nutrition of the Ruminant," No. 1460. Hoffmann-LaRoche, Basel.

APPENDIX

TABLE 1a

Vitamin Requirements for Ruminants and Horses (Values Expressed in Units per Kilogram of Diet, 100% Dry Basis, unless Otherwise Specified)

Animal	A	D	E	K	Thiamin	Riboflavin	Niacin	B_6	Pantothenic acid	Biotin	Folacin	B_{12}	Choline
Beef cattle[a]	IU	IU	IU										
Feedlot	2200	275	—	MS[b]	MS	MS	MS	MS	MS	MS	MS	MS	MS
Pregnant heifers and cows	2800	275	—	MS	MS	MS	MS	MS	MS	MS	MS	MS	MS
Lactating cows and bulls	3900	275	—	MS	MS	MS	MS	MS	MS	MS	MS	MS	MS
Growing	—	—	15–16	—	MS	MS	MS	MS	MS	MS	MS	MS	MS
Dairy cattle[c]	IU	IU	ppm										
Growing	2200	300	25	MS	MS	MS	MS	MS	MS	MS	MS	MS	MS
Lactating cows and bulls	3200	1000	15	MS	MS	MS	MS	MS	MS	MS	MS	MS	MS
Calf milk replacer	3800	600	40	—	—	6.5 ppm	2.6 ppm	6.5 ppm	13 ppm	0.1 ppm	0.5 ppm	0.07 ppm	0.26%
Sheep[d]	IU	IU/100 kg BW	IU										
Replacement ewes, 60 kg	1567	555	15	MS	MS	MS	MS	MS	MS	MS	MS	MS	MS
Pregnant ewes, 70 kg	3306	555	15	MS	MS	MS	MS	MS	MS	MS	MS	MS	MS
Lactating ewes, 70 kg	2380	555	15	MS	MS	MS	MS	MS	MS	MS	MS	MS	MS
Replacement ram, 70 kg	1979	555	15	MS	MS	MS	MS	MS	MS	MS	MS	MS	MS
Ewes, maintenance, 70 kg	2742	555	15	MS	MS	MS	MS	MS	MS	MS	MS	MS	MS
Goat[e]	IU	IU	IU										
All classes	5000	1400	100	MS	MS	MS	MS	MS	MS	MS	MS	MS	MS
Horse[f]	IU	IU	mg		mg	mg			mg		mg		
Growing	2000	275	15	MS	3	2.2	MS	MS	15	MS	20	MS	MS
Maintenance and working	1600	275	15	MS	3	2.2	MS	MS	15	MS	20	MS	MS
Pregnant	3400	—	—	MS	—	—	MS	MS	—	MS	—	MS	MS
Lactating	2800	—	—	MS	—	—	MS	MS	—	MS	—	MS	MS

[a]NRC (1984).
[b]MS = microbial synthesis.
[c]NRC (1988). ''Nutrient Requirements of Domestic Animals, Nutrient Requirements of Dairy Cattle,'' 6th Ed. National Academy of Sciences–National Research Council, Washington, D. C. *Note:* The 5th edition (NRC, 1978a) was used throughout the text as the 1988 edition was available too late for inclusion.
[d]NRC (1985a).
[e]Morand-Fehr (1981).
[f]NRC (1978b).

TABLE 1b

Vitamin Requirements for Monogastric Animals (Values Expressed in Units per Kilogram of Diet unless Otherwise Specified)

Animal	A	D	E	K	Thiamin	Riboflavin	Niacin	B_6	Pantothenic acid	Biotin	Folacin	B_{12}	Choline
	IU	IU	IU	mg	mg	mg	mg	mg	mg	mg	mg	μg	mg
Swine[a]													
Growing–finishing 1–5 kg	2200	220	16	0.5	1.5	4.0	20.0	2.0	12	0.08	0.30	20	600
5–10 kg	2200	220	16	0.5	1.0	3.5	15.0	1.5	10	0.05	0.30	17.5	500
10–20 kg	1750	200	11	0.5	1.0	3.0	12.5	1.5	9	0.05	0.30	15	400
20–50 kg	1300	150	11	0.5	1.0	2.5	10.0	1.0	8	0.05	0.30	10	300
50–110 kg	1300	150	11	0.5	1.0	2.0	7.0	1.0	7	0.05	0.30	5	300
Breeding													
Gilts and sows													
Young and adult boars	4000	200	22	0.5	1	3.75	10	1	12	0.2	0.30	15	1250
Lactating gilts and sows	2000	200	22	0.5	1	3.75	10	1	12	0.2	0.30	15	1000
Poultry[b]													
Leghorn chickens													
Growing 0–6 weeks	1500	200	10	0.5	1.8	3.6	27	3	10	0.15	0.55	9	1300
6–14 weeks	1500	200	5	0.5	1.3	1.8	11	3	10	0.10	0.25	3	900
14–20 weeks	1500	200	5	0.5	1.3	1.8	11	3	10	0.10	0.25	3	500
Laying	4000	500	5	0.5	0.8	2.2	10	3	2.2	0.10	0.25	4	—
Breeding	4000	500	10	0.5	0.8	3.8	10	4.5	10	0.15	0.35	4	—
Broilers 0–3 weeks	1500	200	10	0.5	1.8	3.6	27	3	10	0.15	0.55	9	1300
3–6 weeks	1500	200	10	0.5	1.8	3.6	27	3	10	0.15	0.55	9	850
6–8 weeks	1500	200	10	0.5	1.8	3.6	11	2.5	10	0.10	0.25	3	500
Turkeys 0–4 weeks	4000	900	12	1	2	3.6	70	4.5	11	0.2	1.0	3	1900
4–8 weeks	4000	900	12	1	2	3.6	70	4.5	11	0.2	1.0	3	1600
8–12 weeks	4000	900	10	0.8	2	3.0	50	3.5	9	0.15	0.8	3	1300
12–16 weeks	4000	900	10	0.8	2	3.0	50	3.5	9	0.125	0.8	3	1100
16–20 weeks	4000	900	10	0.8	2	2.5	40	3	9	0.100	0.7	3	950
20–24 weeks	4000	900	10	0.8	2	2.5	40	3	9	0.100	0.7	3	800
Holding	4000	900	10	0.8	2	2.5	40	3	9	0.100	0.7	3	800
Breeding hens	4000	900	25	1	2	4.0	30	4	16	0.15	1.0	3	1000
Geese													
Starting, 0–6 weeks	1500	200	10	0.5	1.8	4	55	3	15	0.15	0.55	9	1300
Growing, 6+ weeks	1500	200	5	0.5	1.3	2.5	35	3	—	0.10	0.25	3	900
Breeding	4000	200	10	0.5	0.8	4	20	4.5	—	0.15	0.35	4	—
Duck 0–2 weeks	4000	220	—	0.4	—	4	55	2.6	11	—	—	—	—
Growing 2–7 weeks	4000	220	—	0.4	—	4	55	2.6	11	—	—	—	—
Breeding	4000	500	—	0.4	—	4	40	3.0	10	—	—	—	—

Animal	A	D	E	K	Ascorbic acid	Thiamin	Riboflavin	Niacin	B_6	Pantothenic acid	Biotin	Folacin	B_{12}	Choline
Pheasant														
Starting	4000	900	12	1	2	3.5	60	4.5	10	0.2	1	3	1500	
Growing	4000	900	11	1	2	3.0	40	3	10	0.1	1	3	1000	
Breeding	4000	900	25	1	2	—	—	4	—	0.15	1	3	—	
Bobwhite quail														
Starting	1500	200	10	0.5	1.8	3.8	30	3	13	0.15	0.55	9	1500	
Growing	1500	200	5	0.5	1.3	—	—	3	—	0.10	0.25	3		
Breeding	4000	500	10	0.5	0.8	4.0	20	4.5	15	0.15	0.35	4	1000	
Japanese quail														
Starting and growing	5000	1200	12	1	2	4	40	3	10	0.3	1	3	2000	
Breeding	5000	1200	25	1	2	4	20	3	15	0.15	1	3	1500	
Dog[c]	IU/kg BW	IU/kg BW	IU/kg BW		μg/kg BW	μg/kg BW	μg/kg BW	μg/kg BW	μg/kg BW		μg/kg BW	μg/kg BW	mg/kg BW	
Growth	202	22	1.2	—	54	100	450	60	400	—	8	1.0	50	
Maintenance	75	8	0.5	—	20	50	225	22	200	—	4	0.5	25	
Cat[d]	IU	IU	IU	IU	mg	mg	mg	mg	mg	μg	μg	μg	g	
Growing	3333	500	30	100	5	4	40	4	5	70	800	20	2.4	
Mink[e]	IU		mg		mg	mg	mg	mg	mg	mg	mg	μg		
Weaning, 13 weeks	5930	—	27	—	1.3	1.6	20	1.6	8	0.12	0.5	32.6	—	
Fox[e]	IU				μg	mg	mg	μg	mg		μg			
Growth	2440	—	—	—	1	3.7	9.6	1.8	7.4	—	0.2	—	—	
Warm-water fish and shellfish[f]	←							mg						→
Catfish	1000–2000	500–1000	30	—	60	1.0	9	14	3	10–20	—	—	—	—
Common carp	10,000	—	200–300	—	—	—	7	28	5–6	30–50	1	—	—	4000
Red sea bream	—	—	—	—	—	—	—	—	5–6	—	—	—	—	—
Shrimp	—	—	—	—	10,000	120	—	—	120	—	—	—	—	600
Cold-water fish[g]	IU	IU	IU	←				mg						→
Trout	2500–5000	1600–2400	—	10	100	1–10	3–15	95	5–15	10–20	0.05–0.25	1–5	0.002–0.003	1000
Salmon	—	—	30	—	50	—	—	190	5–15	10–20	—	1–5	—	—
Rabbit[h]	IU		mg	ppm				mg	μg/g					mg
Growth, female, and breading male	580	—	40	—	—	—	—	180	39	—	—	—	—	1200
Reproducing	1160	—	40	2	—	—	—	—	—	—	—	—	—	

(Continued)

TABLE 1b *(Continued)*

Animal	A	D	E	K	Ascorbic acid	Thiamin	Riboflavin	Niacin	B_6	Pantothenic acid	Biotin	Folacin	B_{12}	Choline
Laboratory animals[i]	IU	IU	IU	μg		mg	mg	mg	mg	mg		mg	μg	mg
Rat														
Growth, gestation, or lactation	4000	1000	30	50	—	4	3	20	6	8	—	1	50	1000
Mouse	IU	IU	IU	mg	←			mg						→
Growth and reproduction	250–500	150	20	3	—	5	7	10	1.0	10	0.2	0.5	0.01	600
	IU	IU	IU	mg	mg	mg	mg	mg	mg	mg	μg	mg	μg	mg
Gerbil	18,000–32,500	2000–3250	9–1200	0.1–45	0–900	4–22.5	3.7–20	22.5–90.0	4–22.5	25–60	200–400	100–1800	0.18	750–3000
Guinea pig	IU	IU	IU	mg	mg	mg	mg	mg	mg	mg	mg	mg	μg	mg
Growing	23,333	1000	50	5	200	2	3	10	3	20	0.3	4	10	1000
Hamster, Golden	IU	IU	IU	mg	mg	mg	mg	mg	mg	mg	mg	mg	μg	mg
Growing	3636	2484	3	4	—	20	15	90	6	40	0.6	2	10	2000
	IU	IU	IU	IU	←			mg						→
Nonhuman primate[j]	10,000–15,000	2000	50	—	100	—	5	50	2.5	15	0.1	0.2	—	—

[a]NRC (1988). (Diets are 90% dry matter).
[b]NRC (1984b). (Diets are 90% dry matter).
[c]NRC (1985). (Requirements on body weight basis).
[d]NRC (1986). (Diets are 100% dry matter).
[e]NRC (1982a). (Diets are 100% dry matter).
[f]NRC (1983). (Diets are 90% dry matter).
[g]NRC (1981a). (Diets are 90% dry matter).
[h]NRC (1977). (Diets are 90% dry matter).
[i]NRC (1978c). (Diets are 90% dry matter).
[j]NRC (1978d). (Diets are 90% dry matter).

TABLE 1c

Vitamin Requirements for Humans[a]

	A	D	E	K	C	Thiamin	Riboflavin	Niacin	B_6	Pantothenic acid	Biotin	Folacin	B_{12}	Choline
	(IU/ day)	(IU/ day)	(IU/ day)	(μg/ day)	(mg/ day)	(mg/ day)	(mg/ day)	(mg/ day)	(mg/ day)	(mg/ day)	(μg/ day)	(μg/ day)	(μg/ day)	
Male	5000	300	10	70–140	60	1	1.2	13–20	2.2	4–7	100–200	400	3	—
Female	4000	300	10	70–140	60	1	1.2	13–20	2.0	4–7	100–200	400	3	—
Pregnant	5000	500	20	70–140	80	1.4	1.5	13.20	2.6	4–7	100–200	800	4	—
Lactating	5200	500	20	70–140	100	1.5	1.5	13–20	2.5	4–7	100–200	500	4	—
Infants 0–6 months	2100	400	2–5	12	35	1	1.2	6–8	0.3	2	35	30	0.5	—
6–12 months	2000	400	2–5	10–20	35	1	1.2	6–8	0.6	3	50	45	1.5	—
Children and adolescents	4000–5000	400	5–12	15–60	60	1	1.2	9–16	1.1–1.25	3–7	65–120	100	3	—

[a]RDA (1980).

TABLE 2

Composition of Important Feeds (Vitamins Expressed at 100% Dry Matter)[a]

Feed name description	Fat-soluble vitamins					Water-soluble vitamins									International Feed No.
	Carotene provitamin A (mg/kg)	A (IU/g)	D (IU/kg)	E (mg/kg)	K (mg/kg)	Biotin (mg/kg)	Choline (mg/kg)	Folic acid (mg/kg)	Niacin (mg/kg)	Pantothenic acid (mg/kg)	Riboflavin (mg/kg)	Thiamin (mg/kg)	B_6 (mg/kg)	B_{12} (μg/kg)	
Alfalfa hay, sun cured	58.0	—	1575	113.0	21.6	0.20	—	3.6	42	28.6	13.4	3.0	—	—	1-00-078
Alfalfa leaves, sun cured	88.0	—	373	—	—	0.28	1062.0	5.8	47	29.0	20.6	4.6	—	—	1-00-146
Bahia grass, fresh	183	—	—	—	—	—	—	—	—	—	—	—	—	—	2-00-464
Bakery waste, dehydrated	5.0	7.0	—	45.0	—	0.07	1005	0.2	28	9.0	1.5	3.2	4.7	—	4-00-466
Barley, grain	2.0	—	—	25.0	0.2	0.17	1177	0.6	94	9.1	1.8	5.0	7.3	—	4-00-549
Bean, navy, seeds	—	—	—	1	—	0.12	1499	1.4	28	2.3	2.0	7.1	0.3	—	5-00-623
Beet, sugar pulp, dehydrated	—	—	637	—	—	—	902	—	18	1.5	0.8	0.4	—	—	4-00-669
Blood, meal	—	—	—	—	—	0.09	854	0.1	34	2.6	2.2	0.4	4.8	49	5-00-380
Bluegrass, Kentucky, fresh	248	—	—	—	—	—	—	—	66	—	11.0	8.8	—	—	2-00-786
Brewer's grains, dehydrated	—	—	—	29	—	0.68	1757	7.7	47	8.9	1.6	0.7	0.8	—	5-02-141
Buckwheat, common, grain	—	—	—	—	—	—	501	—	21	13.1	5.4	4.2	—	—	4-00-994
Buttermilk, dehydrated (cattle)	—	2.4	—	7	—	0.31	1891	0.4	9	40.1	33.1	3.7	2.6	21	5-01-160
Carrot, root, fresh	678	—	—	60	—	0.07	—	1.2	58	30.1	4.9	5.8	12.0	—	4-01-145
Casein, dehydrated (cattle)	—	—	—	—	—	0.05	229	0.5	1	2.9	1.7	0.5	0.5	—	5-01-162
Cattle, liver, fresh	—	439.1	—	25	—	3.51	5093	8.4	269	164.9	92.2	6.3	18.0	1523	5-01-166
Cattle, spleen, fresh	—	3.0	—	56	—	0.16	2036	4.8	25	8.2	15.3	3.1	1.3	247	5-07-942
Chicken, broilers, whole, fresh	—	30.0	—	—	—	—	—	—	230	—	15.6	2.9	—	—	5-07-945
Chicken, hens, whole, fresh	—	—	—	310	—	0.46	6288	0.5	225	20.4	6.4	2.4	4.6	278	5-07-950
Citrus, dried pulp	—	—	—	—	—	—	867	—	24	15.4	2.5	1.6	—	—	4-01-237
Clover, red hay, sun cured	20	—	1914	—	—	0.11	—	—	43	11.2	17.8	2.2	—	—	1-01-415
Coconut, meats, meal															
mechanically extracted	—	—	—	—	—	—	1036	1.5	26	6.8	3.4	0.8	—	—	5-01-572
solvent extracted	—	—	—	—	—	—	1189	0.3	28	6.9	3.7	0.7	4.8	—	5-01-573
Corn, dent yellow															
cobs, ground	1	—	—	—	—	—	—	—	8	4.2	1.1	1.0	—	—	1-28-234
gluten meal	18	—	—	34	—	0.20	391	0.3	55	11.2	1.8	0.2	8.8	—	5-28-241
grain	3	—	—	25	0.2	0.08	567	0.3	28	6.6	1.4	3.8	5.3	—	4-02-935
silage	43	—	439	—	—	—	—	—	47	—	—	—	—	—	3-02-912

Cotton seeds, meal, mechanically extracted, 41% protein	—	—	—	35	—	1.19	2965	2.3	38	11.2	5.7	7.0	5.4	—	5-01-617
Crab, meal	—	—	—	—	—	0.07	2179	0.1	49	7.0	6.7	0.5	7.2	475	5-01-663
Fish, anchovy meal, mechanically extracted	—	—	—	5.0	—	0.21	4036	0.2	89	10.9	8.2	0.5	5.0	233	5-01-985
Fish, menhaden meal, mechancially extracted	—	—	—	13	—	0.20	3398	0.2	60	9.4	5.2	0.6	5.1	133	5-02-009
Fish, sardine meal, mechanically extracted	—	—	—	—	—	0.11	3518	—	81	11.8	5.8	0.3	—	256	5-02-015
Flax (linseed meal)	—	—	—	9	—	0.36	1962	3.1	41	15.8	3.5	4.6	6.1	—	5-02-045
Livers, meal	—	—	—	—	—	0.02	12,281	6.0	221	31.5	39.1	0.2	—	542	5-00-389
Meat, with blood, meal rendered (tankage)	—	—	—	—	—	—	2391	1.7	40	2.8	2.4	0.4	—	147	5-00-386
Milk, dehydrated (cattle)	—	12.3	353	—	—	0.40	—	—	9	23.8	20.6	3.9	4.9	—	5-01-167
Milk, skimmed, dehydrated	—	—	446	10	—	0.35	1480	0.7	12	38.6	20.5	3.9	4.5	54	5-01-175
Millet, grain	—	—	—	—	—	—	489	—	26	12.2	4.2	8.1	—	—	4-03-120
Molasses, sugarcane	—	—	—	7	—	0.92	1012	0.1	49	50.3	3.8	1.2	5.7	—	4-04-696
Oat, grain	—	—	—	15	—	0.31	1116	0.4	16	8.8	1.7	7.1	2.8	—	4-03-309
Pangola grass, fresh	62	—	—	—	—	—	—	—	—	—	—	—	—	—	2-03-493
Pea, seeds	1	—	—	3	—	0.22	662	0.3	36	21	2.0	5.2	1.7	—	5-03-600
Peanut kernels, meal, mechanically extracted	—	—	—	3	—	0.35	2052	0.7	186	49.7	8.8	6.6	8.0	—	5-03-649
Potato tubers, dehydrated	—	—	—	—	—	0.11	2879	0.7	37	22.0	1.1	—	15.5	—	4-07-850
Poultry, by-products, meal, rendered	—	—	—	2	—	0.09	6451	0.5	50	11.8	11.2	0.2	4.7	322	5-03-798
feathers, hydrolyzed	—	—	—	—	—	0.05	962	0.2	23	9.7	2.1	0.1	3.2	90	5-03-795
Rape, seeds, mechanically extracted	—	—	—	20	—	—	7103	—	168	9.8	3.3	1.9	—	—	5-03-870
Rice bran with germ	—	—	—	66	—	0.47	1357	2.4	330	25.2	2.8	24.7	—	—	4-03-928
grain, ground	—	—	—	11	—	0.09	1076	0.4	39	9.1	1.2	3.2	5.0	—	4-03-938
groats, polished	—	—	—	4	—	—	1018	0.2	17	3.9	0.6	0.7	0.4	—	4-03-942
Rye, grain	—	—	—	17	—	0.06	479	0.7	21	9.1	1.9	4.2	2.9	—	4-04-047
Ryegrass, hay, sun cured	120	—	—	211	—	—	—	—	—	—	—	—	—	—	1-04-077
Safflower seeds, meal, mechanically extracted	—	—	—	1	—	1.54	1287	0.5	—	—	—	—	—	—	5-04-109
Sesame seeds, meal, mechanically extracted	—	—	—	—	—	—	1655	—	20	6.4	3.6	3.0	13.4	—	5-04-220
Sorghum grain	1	—	26	10	0.2	0.42	737	0.2	43	12.5	1.4	4.7	5.0	—	4-04-383
silage	15	—	662	—	—	—	—	—	—	—	—	—	—	—	3-04-323
Soybean seeds	1	—	—	37	—	0.37	2939	3.9	24	17.3	3.1	10.6	—	—	5-04-610
meal, solv. extracted	—	—	—	3	—	0.36	2915	0.7	31	18.2	3.2	6.2	6.7	—	5-04-604

(Continued)

TABLE 2 (*Continued*)

Feed name description	Fat-soluble vitamins					Water-soluble vitamins									International Feed No.
	Carotene provitamin A (mg/kg)	A (IU/g)	D (IU/kg)	E (mg/kg)	K (mg/kg)	Biotin (mg/kg)	Choline (mg/kg)	Folic acid (mg/kg)	Niacin (mg/kg)	Pantothenic acid (mg/kg)	Riboflavin (mg/kg)	Thiamin (mg/kg)	B_6 (mg/kg)	B_{12} (μg/kg)	
Sunflower, common seeds, mechanically extracted	—	—	—	—	—	—	4214	—	293	33.3	3.4	3.4	12.4	—	5-09-340
Swine livers, fresh	—	361.7	—	—	—	2.49	—	6.9	544	77.9	90.3	7.7	10.0	935	5-04-792
Timothy fresh	179	—	—	111	—	—	—	—	—	—	11.5	2.9	—	—	2-04-912
hay, sun cured	28	—	2138	38	—	0.07	811	2.3	29	7.9	10.1	1.7	—	—	1-04-893
Tomato pomace, dehydrated	—	—	—	—	—	—	—	—	—	—	6.7	12.3	—	—	5-05-041
Triticale grain	—	—	—	—	—	—	514	—	—	—	0.5	—	—	—	4-20-362
Turnip roots, fresh	—	—	—	—	—	—	—	2.8	72	19.0	6.5	7.1	—	—	4-05-067
Wheat bran	3	—	—	21	—	0.32	1797	1.6	268	33.5	4.6	7.9	9.6	—	4-05-190
grain	—	—	—	17	—	0.11	1085	0.5	64	11.4	1.6	4.8	5.6	1.0	4-05-211
Whey, dehyd. (cattle)	—	0.5	—	—	—	0.38	1921	0.9	11	49.6	29.4	4.3	3.6	20	4-01-182
Yeast, brewer's, dehydrated	—	—	—	2	—	1.08	4227	10.3	482	118.4	38.1	99.2	39.8	1	7-05-527
Yeast, torula, dehydrated	—	—	—	—	—	1.47	3223	26	525	100.6	47.6	6.6	38.9	4	7-05-534

[a]United States–Canadian Tables of Feed Composition (NRC, 1982b).

TABLE 3

Metric Weights and Measures with Customary Equivalents[a,b]

Capacity or Volume		
1 cubic centimeter	=	0.061 cubic inch
1 cubic meter	=	35.315 cubic feet
	=	1.308 cubic yards
1 milliliter	=	0.0338 fluid ounce (U.S.)
1 liter	=	33.81 fluid ounces (U.S.)
	=	2.1134 pints (U.S.)
	=	1.057 quarts (U.S.)
	=	0.2642 gallon (U.S.)
1 kiloliter	=	264.18 gallons (U.S.)
Weight		
1 gram	=	0.03527 ounce (avdp.)
1 kilogram	=	35.274 ounces (advp.)
	=	2.205 pounds (advp.)
1 metric ton (1000 kg)	=	0.984 ton (long)
	=	1.102 tons (short)
	=	2204.6 pounds (avdp.)
Volume per unit area		
1 liter/hectare	=	0.107 gallon (U.S.)/acre
Weight per unit area		
1 kilogram/square centimeter	=	14.22 pounds (avdp.)/square inch
1 kilogram/hectare	=	0.892 pound (avdp.)/acre
Area per unit weight		
1 square centimeter/kilogram	=	0.0703 square inch/pound (avdp.)
Temperature conversion formulas		
Centigrade (Celsius)	=	5/9 (Fahrenheit − 32)
Fahrenheit	=	9/5 Centigrade (Celsius) + 32

[a]Modified from *J. Anim. Sci.* **25** (1966), 270–271.

[b]When conversions are made, the results should be rounded to a meaningful number of digits relative to the accuracy of original measurements. Values for weights and volumes are based on pure water at 4°C under 760 mm of atmospheric pressure.

Index

W

X

Y

Z